Spring Boot

企业级开发实战

（视频教学版）

迟殿委　赵媛媛　郭德先　侯传杰　著

清华大学出版社
北京

内容简介

Spring 框架目前已成为事实上的 Java EE 企业开发标准框架，从 IoC、AOP 两大核心特性逐渐发展成为包括数据访问、WebMVC、消息模块、测试模块等在内的生态帝国。Spring Boot 是一套快速开发框架，采用约定大于配置的原则，与其他框架的集成也非常简单，可以很快创建一个产品级别的 Spring 应用。本书讲解 Spring Boot 应用开发技术，配套源码、课件与教学视频。

本书分为 10 章，内容包括 Spring 核心基础，Spring MVC 开发基础，Spring Boot 入门，Spring Boot 开发 Web 应用，Spring Boot 原理解读，Spring Boot 数据访问与事务，Spring Boot 高并发，Spring Boot 构建企业级应用，Spring Boot 打包、部署与监控，综合项目实战。

本书内容翔实、讲解细致，适合 Spring Boot 初学者，可作为 Web 开发人员常备案头的参考书，也可作为高等院校、中职学校及培训机构计算机相关专业的教材或者课程设计用书。

图书在版编目（CIP）数据

Spring Boot 企业级开发实战：视频教学版 / 迟殿委等著.—北京：清华大学出版社，2021.8（2023. 4重印）
ISBN 978-7-302-58746-0

Ⅰ. ①S…　Ⅱ. ①迟…　Ⅲ. ①JAVA 语言－程序设计－高等学校－教材　Ⅳ. ①TP312.8

中国版本图书馆 CIP 数据核字（2021）第 146404 号

责任编辑：夏毓彦
封面设计：王　翔
责任校对：闫秀华
责任印制：宋　林

出版发行：清华大学出版社
网　　址：http://www.tup.com.cn，http://www.wqbook.com
地　　址：北京清华大学学研大厦 A 座　　邮　　编：100084
社总机：010-83470000　　邮　　购：010-62786544
投稿与读者服务：010-62776969，c-service@tup.tsinghua.edu.cn
质 量 反 馈：010-62772015，zhiliang@tup.tsinghua.edu.cn

印 装 者：三河市铭诚印务有限公司
经　　销：全国新华书店
开　　本：190mm×260mm　　**印　　张**：30　　**字　　数**：768 千字
版　　次：2021 年 9 月第 1 版　　**印　　次**：2023 年 4 月第 2 次印刷
定　　价：119.00 元

产品编号：091193-01

前 言

本书各章节按照 Spring 框架组件的出现时间来设计，先讲解 Spring 核心组件 IoC 和 AOP，然后是 Spring MVC，最后展开讲解 Spring Boot。本书主要针对 Spring Boot，采用由浅入深的方式，将讲解和案例练习相结合，符合读者的学习曲线。从初体验、基本源码分析、Web 应用、数据访问这些基本模块，到高并发处理、消息队列、企业级应用开发、部署和监控等高级模块，通过综合项目贯穿全书的重点知识模块。本书每个章节都由在本章节相关方面有丰富实战经验的企业一线工程师来设计和编写，每章都有实战案例驱动，重点突出，步骤清晰，表达易懂，尤其是在高并发、秒杀场景设计、分布式缓存、Kafka 消息机制以及企业级复杂应用上分享了作者的经验和体会。

此外，为了方便高校师生使用，本书提供了配套 PPT，并为每部分内容配套了教学视频。每个章节视频均为作者精心录制，针对相关章节中的实战案例及涉及的技术点进行讲解，语言表达力求通俗易懂。本书采用了最新稳定的 Spring Boot 版本，并对最新版本特性做了介绍，符合企业目前开发需要。书中每个案例都有清晰的步骤标注和丰富的图片表达，目的就是为了使读者能够以最快的速度将学到的开发技术应用到实际项目中。

本书整体设计上由浅入深，从简单到复杂（Spring 开发基础→Spring Boot 实战→Spring Boot 核心原理剖析→与主流技术整合→企业级开发→综合实战），并且每章都有实战案例驱动。

本书循序渐进的设计思路和丰富的配套资源，非常适合高等院校广大师生作为教材或教学参考书使用，知识层次的全面性也能满足使用 Spring Boot 开发企业应用的 Java 工程师的学习需要。

本书内容

第 1 章主要讲解 Spring 框架的两大核心（IoC 和 AOP），并通过典型案例来帮忙读者巩固 Spring 基础。

第 2 章讲解 Spring MVC 框架的架构特性和工作流程，并通过典型开发案例巩固 Spring MVC 开发基础知识。

第 3 章讲解 Spring Boot 介绍、特性以及新版本变化。实战方面的内容包括配置 Java、Maven 环境，使用集成开发环境进行 Spring Boot 应用的开发，以及通过不同的方式创建 Spring Boot 应用程序。

第 4 章讲解 Spring Boot 是如何开发 Web 应用的，包括内置容器的原理与应用、如何自动配置 Spring MVC、如何集成模板引擎，最后通过前后端分离应用实战加深对 Web 开发的印象。

第 5 章讲解 Spring Boot 自动配置原理、启动流程，starter 和它的配置，以及内嵌 Web 服务器原理。

第 6 章讲解 Spring Boot 数据访问与事务，数据访问方式包括 Spring Data JDBC、Spring Data JPA

和集成 MyBatis 框架。事务包括事务的类型、特性、并发问题等。

第 7 章讲解高并发处理实战，首先引入高并发中常用的缓存技术和消息队列技术，最后通过模拟两个高并发场景来达到 Spring Boot 应用处理高并发实战的目的。

第 8 章讲解怎样构建一个企业级应用。这一章引入了权限认证框架（讲述 Spring Security 和 Shiro 两种不同的权限框架），同时介绍实现单点登录的 3 种方式。然后讲解如何实现第三方登录，以及如何优雅地生成接口文档、集成日志框架等。

第 9 章主要涉及 Spring Boot 应用程序的打包、部署和监控。其中包括 jar 包和 war 包两种不同的打包方式，以及部署到云服务器、Docker 容器，配置热部署来提高开发效率，最后讲解如何对 Spring Boot 应用进行监控。

第 10 章通过一个综合项目案例（图书管理系统），综合运用 Spring Boot 核心知识和相关技术进行实战开发，以加深读者对 Spring Boot 的理解和运用能力。

源码、课件、教学视频下载与技术支持

本书配套的资源，请用微信扫描下边的二维码获取，可按扫描出来的页面提示把链接转到自己的邮箱中下载。如果学习本书过程中发现问题，请联系 booksaga@163.com，邮件主题为“Spring Boot 企业级开发实战”。

本书作者

迟殿委、赵媛媛、郭德先、侯传杰均为企业一线高级软件工程师，并具有丰富的实战经验。

作 者

2021 年 6 月

目 录

第 1 章

Spring 核心基础

本书主要内容是关于 Spring Boot 的开发基础，越来越多的开发者青睐这个框架，其约定大于配置的特性使得构建一个企业级 JavaEE 项目的成本降低，与其他框架的集成也不需要过多的配置，这都是 Spring Boot 框架的魅力所在。要学习 Spring Boot 必须掌握 Spring 技术的核心开发基础。框架发展本身都有一个过程，不是突然出现的。俗话说，知其然也要知其所以然，Spring 框架无论经历了怎样的演变，其最核心的内容依然是 IoC 和 AOP 这两大部分，理解和掌握 Spring 框架核心，将决定你对 Spring Boot 掌握的深度和使用的灵活度。本章将主要介绍 Spring 框架基础内容，重点讲解 IoC 和 AOP 两个组件。

1.1 Spring 概述

1.1.1 Spring 介绍

Spring 框架是一个开源的 Java 平台。它最初由 Rod Johnson 编写，并于 2003 年 6 月首次在 Apache 2.0 许可下发布。Spring 是企业 Java 最流行的应用程序开发框架，实际上已经成为 JavaEE 项目开发标准。开发人员使用 Spring Framework 创建高性能、易于测试和可重用的代码。

Spring 大约有 20 个模块，这些组件被分别整合在核心容器（Core Container）、AOP（Aspect Oriented Programming）和设备支持（Instrumentation）、数据访问及集成（Data Access/Integeration）、Web、报文发送（Messaging）、Test 这 6 个模块集合中。Spring 5 的模块结构如图 1-1 所示。

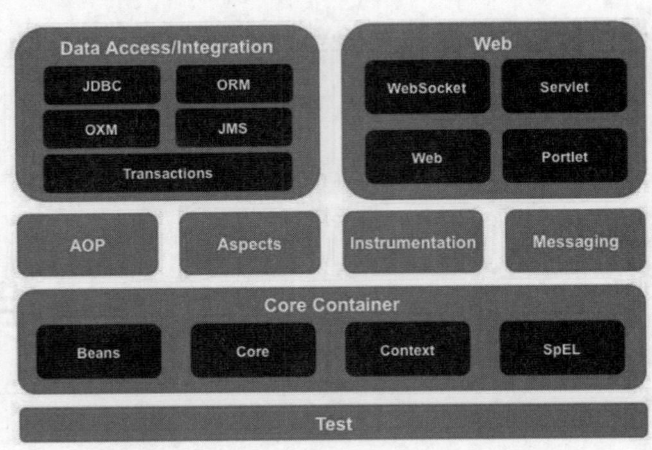

图 1-1

Spring 框架的模块可以独立存在，也可以几个模块进行组合，即可以通过一个或多个模块联合实现。各模块的组成和功能如下。

（1）核心容器：由 spring-beans、spring-core、spring-context 和 spring-expression（Spring Expression Language，SpEL）4 个模块组成。

spring-beans 和 spring-core 模块是 Spring 框架的核心模块，包含了控制反转（Inversion of Control，IoC）和依赖注入（Dependency Injection，DI）。BeanFactory 接口是 Spring 框架中的核心接口，是工厂模式的具体实现。BeanFactory 使用控制反转对应用程序的配置和依赖性规范与实际的应用程序代码进行了分离，但是 BeanFactory 容器实例化后并不会自动实例化 Bean，只有当 Bean 被使用时 BeanFactory 容器才会对该 Bean 进行实例化与依赖关系的装配。

spring-context 模块构架于核心模块之上，扩展了 BeanFactory，为它添加了 Bean 生命周期控制、框架事件体系以及资源加载透明化等功能。此外该模块还提供了许多企业级支持，如邮件访问、远程访问、任务调度等。ApplicationContext 是该模块的核心接口，是 BeanFactory 的超类。与 BeanFactory 不同，ApplicationContext 容器实例化后会自动对所有的单实例 Bean 进行实例化与依赖关系的装配，使之处于待用状态。

spring-expression 模块是统一表达式语言（EL）的扩展模块，可以查询、管理运行中的对象，同时也可以方便地调用对象方法、操作数组、集合等。它的语法类似于传统 EL，但是提供了额外的功能，最出色的要数函数调用和简单字符串的模板函数。这种语言的特性是基于 Spring 产品的需求而设计的，可以非常方便地同 Spring IoC 进行交互。

（2）面向切面编程：我们知道 Java 语言本身最重要的特性就是面向对象的编程，面向对象是实现具体业务时的一种思维方式。在实际项目中，有很多业务流程都包含了一些共性的模块，比如日志模块、数据库事务模块等。这种业务的设计会使用面向切面的编程思维，通俗点说就是将共性的业务拿出来做成对象，供所有业务流程共用。Spring 框架提供了面向切面编程的能力，可以实现一些面向对象编程无法很好实现的操作，例如，将日志、事务与具体的业务逻辑解耦。其主要包含 spring-aop、spring-aspects 组件。spring-aop 是 Spring 的另一个核心模块，是 AOP 主要的实现模块。作为继 OOP 后对程序员影响最大的编程思想之一，AOP 极大地开拓了人们对于编程的思路。Spring 以 JVM 的动态代理技术为基础，设计出一系列 AOP 横切实现，比如前置通知、返回通知、异常通知等，同时利用 Pointcut 接口来匹配切入点（既可以使用现有的切入点来设计横切面，也可以扩展相关方法根据需求进行切入）。

spring-aspects 模块集成自 AspectJ 框架，主要是为 Spring AOP 提供多种 AOP 实现方法。spring-instrument 模块是基于 Java SE 中的“java.lang.instrument”进行设计的，应该算是 AOP 的一个支援模块，主要作用是在 JVM 启用时生成一个代理类。程序员通过代理类在运行时修改类的字节，从而改变一个类的功能，实现 AOP 的功能。

（3）数据访问及集成：Spring 自带了一组数据访问框架，集成了多种数据访问技术。不管直接使用 JDBC，还是 Hibernate、MyBatis 这样的 ORM 框架，Spring 都可以帮助我们消除持久化代码中单调枯燥的数据访问逻辑。Spring 允许我们在持久层选择不同的方案，如 JDBC、MyBatis、Hibernate、Java 持久化 API（Java Persistence API，JPA）以及 NoSQL 数据库。这部分框架由 spring-jdbc、spring-tx、spring-orm、spring-jms 和 spring-oxm 5 个模块组成。

- spring-jdbc 模块：Spring 提供的 JDBC 抽象框架的主要实现模块，用于简化 Spring JDBC，主要提供 JDBC 模板方式、关系数据库对象化方式、SimpleJdbc 方式、事务管理来简化 JDBC 编程，主要实现类是 JdbcTemplate、SimpleJdbcTemplate 以及 NamedParameterJdbcTemplate。
- spring-tx 模块：Spring JDBC 事务控制实现模块，使用 Spring 框架，对事务做了很好的封装，通过 AOP 配置可以灵活地配置在任何一层。在很多的需求和应用中，直接使用 JDBC 事务控制还是有其优势的。事务是以业务逻辑为基础的，一个完整的业务应该对应业务层里的一个方法，如果业务操作失败，则整个事务回滚。所以，事务控制应该放在业务层。持久层的设计应该遵循一个很重要的原则：保证操作的原子性，即持久层里的每个方法都应该是不可分割的。所以，在使用 Spring JDBC 事务控制时，应该注意其特殊性。
- spring-orm 模块：ORM 框架支持模块，主要集成 Hibernate、Java Persistence API（JPA）和 Java Data Objects（JDO）用于资源管理、数据访问对象（DAO）的实现和事务策略。
- spring-jms 模块（Java Messaging Service）：能够发送和接收信息，自 Spring Framework 4.1 以后，还提供了对 spring-messaging 模块的支撑。
- spring-oxm 模块：主要提供一个抽象层以支撑 OXM（Object-to-XML-Mapping，是一个 O/M-mapper，将 Java 对象映射成 XML 数据或者将 XML 数据映射成 Java 对象），例如 JAXB、Castor、XMLBeans、JiBX 和 XStream 等。

（4）Web 模块：由 spring-web、spring-webmvc、spring-websocket 和 spring-webflux 4 个模块组成。

- spring-web 模块：为 Spring 提供了最基础的 Web 支持，主要建立于核心容器之上，通过 Servlet 或者 Listeners 来初始化 IoC 容器，也包含一些与 Web 相关的支持。
- spring-webmvc 模块：是一个 Web-Servlet 模块，实现了 Spring MVC（Model-View-Controller）的 Web 应用。
- spring-websocket 模块：主要是与 Web 前端的全双工通信协议。
- spring-webflux 模块：是一个新的非堵塞函数式 Reactive Web 框架，可以用来建立异步、非阻塞、事件驱动的服务，并且扩展性非常好。

（5）报文发送：spring-messaging 模块。

spring-messaging 是从 Spring 4 开始加入的一个模块，主要职责是为 Spring 框架集成一些基础的报文传送应用。spring-messaging 属于 Spring Framework 项目，其定义了 Enterprise Integration Patterns 典型实现的接口及相关的支持（注解、接口的简单默认实现等）。

消息系统中的 Pipes and Filters 模型如图 1-2 所示。模型主要由两部分组成：用来传递消息的通道（Pipe）和用来对消息进行处理的过滤器（Filter）。这些消息通道将过滤器串联起来，而消息自身则会沿着这些通道流动。

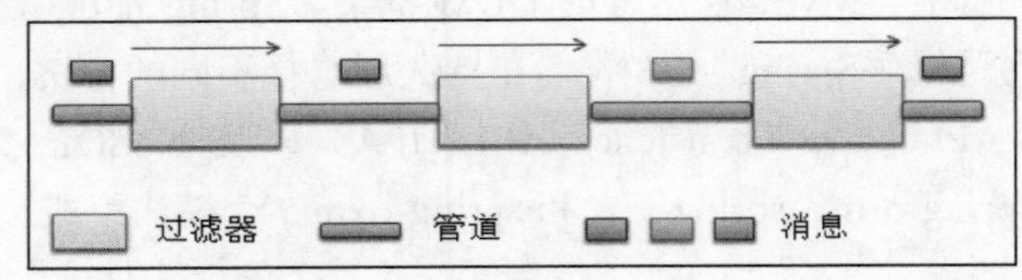

图 1-2

图 1-2 中所显示的是对一个消息进行处理的典型方式。在一个消息被发送到一个管道中后，过滤器将会在具有消息处理能力时从输入管道中接收该消息并开始处理。一旦该消息处理完毕，过滤器就会把处理结果放到输出管道中，并由与该管道相连接的各个过滤器完成消息的后续处理。一个过滤器可能对消息进行处理，也可能仅仅是对消息进行转发。

spring-messaging 定义了以下的接口用来实现 Pipes and Filters 模型：

- Message：消息，包含 MessageHeaders 和 Payload。
- MessageChannel：Pipes and Filters 模型中的 Pipe，其中定义了用来发送消息到此 Pipe 的 send 方法。其根据 MessageHandler 中两种获取消息的方式定义了两个子接口：PollableChannel，用来表示支持拉模型的通道，其中定义了 receive 方法（用来支持 MessageHandler 主动拉取消息）；SubscribableChannel，用来表示支持推模型（发布/订阅模型）的通道，其中定义了 subscribe 和 unsubscribe 方法（用来支持 MessageHandler 的订阅和取消订阅）。
- MessageHandler：Pipes and Filters 模型中的 Filter，定义了 handleMessage 方法用来处理消息。

由于在经典的 Pipes and Filters 模型中，过滤器的输入及输出只能有一个，因此其所包含的各个过滤器并不可以被多条处理逻辑重用。图 1-3 所示的 Enterprise Integration Pattern 模式放松了该约束，使得一个过滤器可以从多个管道接收消息，并向多个管道发送相同或不同的消息。

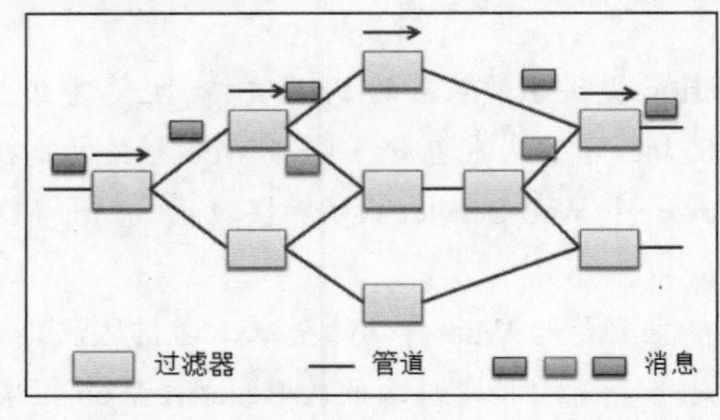

图 1-3

spring-messaging 定义了 DestinationResolver 接口，用来选择消息的目的地。消息的目的地既可以是 Pipe（MessageChannel），也可以是 Filter（MessageHandler）。

（6）测试（Test）：spring-test 模块。

spring-test 模块主要是为测试提供支持的，毕竟在不需要发布（程序）到你的应用服务器或者连接到其他企业设施的情况下能够执行一些集成测试或者其他测试，对于任何企业来说都是非常重要的。

1.1.2 Spring 的优点

Spring 的优点总结起来如下：

- 使用 Spring 的 IoC 容器，将对象之间的依赖关系交给 Spring，降低组件之间的耦合性，让我们更专注于应用逻辑。
- 独立于各种应用服务器，基于 Spring 框架的应用，可以真正实现 "Write Once, Run Anywhere" 的承诺。
- Spring 的 AOP 支持允许将一些通用任务（如安全、事务、日志等）进行集中式管理，从而提供了更好的复用。
- Spring 的 ORM 和 DAO 提供了与第三方持久层框架的良好整合，并简化了底层的数据库访问。
- Spring 的高度开放性，并不强制应用完全依赖于 Spring，开发者可自由选用 Spring 框架的部分或全部。
- Spring 属于轻量级框架，开发不需要涉及大量接口，开发难度降低，而且不需要单独的容器对对象进行管理，资源消耗低。

1.2 Spring 控制反转

1.2.1 IoC 和 DI

IoC（Inversion of Control，控制反转）是 Spring 最核心的要点，并且贯穿始终。IoC 并不是一门技术，而是一种设计思想。在 Spring 框架中，实现控制反转的是 Spring IoC 容器，由容器来控制对象的生命周期和业务对象之间的依赖关系，而不是像传统方式（new 对象）那样由代码来直接控制。程序中所有的对象都会在 Spring IoC 容器中登记，告诉容器该对象是什么类型、需要依赖什么，然后 IoC 容器会在系统运行到适当的时候把它要的对象主动创建好，同时也会把该对象交给其他需要它的对象。也就是说，控制对象生存周期的不再是引用它的对象，而是由 Spring IoC 容器来控制所有对象的创建、销毁。对于某个具体的对象而言，以前是它控制其他对象，现在是所有对象都被 Spring IoC 容器所控制，所以这叫控制反转。

控制反转最直观的表达就是，IoC 容器让对象的创建不用去新建（new）了，而是由 Spring 自动生产，使用 Java 的反射机制，根据配置文件在运行时动态地去创建对象以及管理对象，并调用对象的方法。控制反转的本质是控制权由应用代码转到了外部容器（IoC 容器）。控制权的转移就是反转。控制权的转移带来的好处是降低了业务对象之间的依赖程度，即实现了解耦。

DI（Dependency Injection，依赖注入）是 IoC 的一个别名，其实两者是同一个概念，只是从不同的角度描述罢了（IoC 是一种思想，而 DI 是一种具体的技术实现手段）。

1.2.2 依赖注入实战 XML 方式

本小节涉及的项目参见第 1 章项目源代码文件夹下面的“springioc_xml”。下面讲解在 IDEA 中创建一个基于 XML 配置方式的 Spring 项目的典型步骤。

步骤 01 创建一个 Maven 项目，界面如图 1-4 所示。

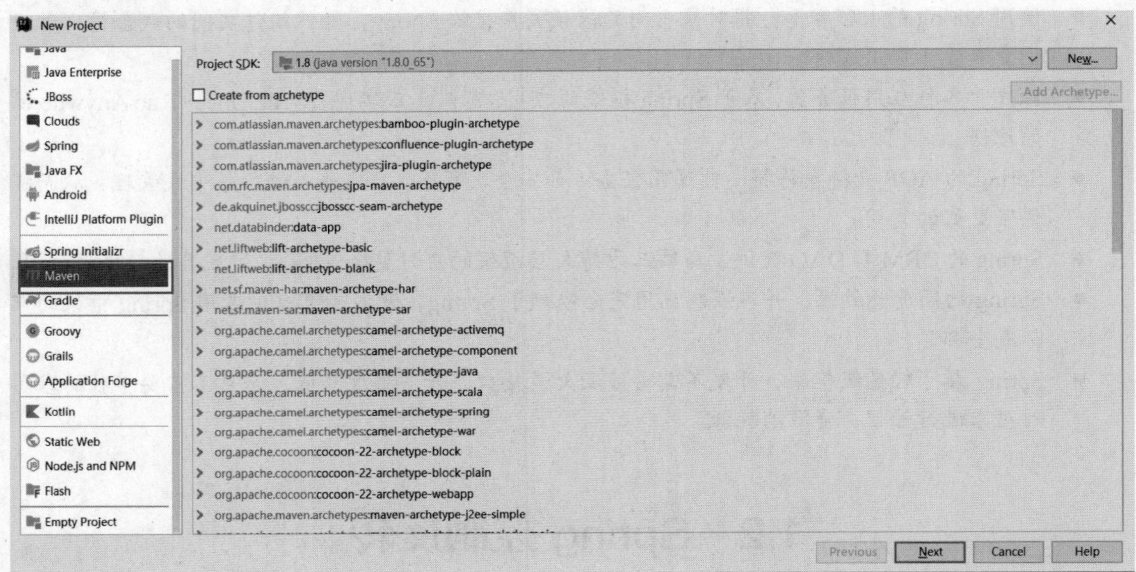

图 1-4

步骤 02 pom.xml 文件中引入 Spring 常用依赖。

示例代码 1-1 pom.xml（部分代码）

```
<dependencies>
    <!-- Spring 常用依赖 -->
    <dependency>
        <groupId>org.springframework</groupId>
        <artifactId>spring-context</artifactId>
        <version>5.2.5.RELEASE</version>
    </dependency>
</dependencies>
```

步骤 03 创建空的 Spring 配置文件。在 resources 文件夹中创建一个名为 applicationContext.xml 的文件。命名并无规定，还有其他的常用命名，比如 spring-context.xml、beans.xml 等。

```
<?xml version="1.0" encoding="UTF-8"?>
<beans xmlns="http://www.springframework.org/schema/beans"
```

```
    xmlns:xsi="http://www.w3.org/2001/XMLSchema-instance"
    xsi:schemaLocation="http://www.springframework.org/schema/beans http://
www.springframework.org/schema/beans/spring-beans.xsd">
</beans>
```

步骤 04 定义 Bean 对象。定义 UserDao 接口，里面包含一个插入用户的方法 insertUser。接口代码参见项目源码。定义一个 Bean 对象，生产该对象并测试该对象内的方法（UserDaoImpl），其中 UserDaoImpl 实现类的定义代码如下。

示例代码 1-2 UserDaoImpl.java

```
import com.mrchi.dao.UserDao;
import com.mrchi.entity.User;
public class UserDaoImpl implements UserDao {
    public int insertUser(User user) {
        System.out.println("------insertUser------");
        return 0;
    }
}
```

步骤 05 完成 UserDao 实现对象的注入。依赖注入在这里可以理解为将要生产的对象注入 Spring 容器中，也就是在 spring-context.xml 文件利用标签注入，这样就可以让 Spring 知道要生产的对象是谁。标签写法：<bean id="唯一标签" class="需要被创建的目标对象全限定名"/>，示例如下。

示例代码 1-3 applicationContext.xml

```
<?xml version="1.0" encoding="UTF-8"?>
<beans xmlns="http://www.springframework.org/schema/beans"
   xmlns:xsi="http://www.w3.org/2001/XMLSchema-instance"
   xsi:schemaLocation="http://www.springframework.org/schema/beans
http://www.springframework.org/schema/beans/spring-beans.xsd">
    <bean id="UserDao" class="com.mrchi.dao.UserDaoImpl"/>
</beans>
```

步骤 06 调用 Spring 工厂创建对象。调用 Spring 工厂 API 接口 ApplicationContext 读取配置 Spring 核心配置文件并创建工厂对象，测试类代码如下。

示例代码 1-4 UserDaoImplTest.java

```
public class UserDaoImplTest {
    /**
     * @MethodName insertUser
     * @Param []
     * @Description 测试使用 Spring 工厂获取对象
     */
    @Test
    public void insertUser() {
        // 读取配置文件所需创建对象中创建的 bean 对象并获取 spring 工厂对象
        ApplicationContext context = new
ClassPathXmlApplicationContext("applicationContext.xml");
        // 通过注入时的唯一标识(id)获取 bean 对象
        UserDao userDao = (UserDao) context.getBean("UserDao");
```

```
        // 使用对象
        userDao.insertUser(null);     // 输出结果为：------insertUser ------
    }
}
```

注意，要使用以上测试类，就必须引入 junit 的依赖，具体依赖代码如下。

```
<dependency>
    <groupId>junit</groupId>
    <artifactId>junit</artifactId>
    <version>4.13</version>
    <scope>test</scope>
</dependency>
```

我们在项目开发中会进行逻辑分层，一般情况下在 Dao 层的上一层会定义业务层来对其进行调用，即 Service 层。Service 层依赖 Dao 对象。

首先，定义 UserService 接口，代码如下。

示例代码 1-5　UserService.java

```
public interface UserService {
    public int insertUser(User user);
}
```

定义接口实现类 UserServiceImpl，代码如下。

示例代码 1-6　UserServiceImpl.java

```
public class UserServiceImpl implements UserService {
    //用 setter 方式实现注入
    private UserDao userDao;
    public UserDao getUserDao() {
        return userDao;
    }
    public void setUserDao(UserDao userDao) {
        this.userDao = userDao;
    }
    public int insertUser(User user) {
        return userDao.insertUser(user);
    }
}
```

我们可以看到，在 Service 中定义了 userDao 属性，并定义了 Setter 方法，这在 Spring 配置文件中通过依赖注入方式实现，配置文件修改如下。

示例代码 1-7　applicationContext.xml

```
<beans xmlns="http://www.springframework.org/schema/beans"
   xmlns:xsi="http://www.w3.org/2001/XMLSchema-instance"
   xsi:schemaLocation="http://www.springframework.org/schema/beans
http://www.springframework.org/schema/beans/spring-beans.xsd">
    <bean id="userDao" class="com.mrchi.dao.impl.UserDaoImpl"/>
    <bean id="userService" class="com.mrchi.service.impl.UserServiceImpl">
    <property name="userDao" ref="userDao"></property>
```

```
    </bean>
</beans>
```

下面定义一个测试方法，获取 UserService 对象，并调用 insertUser 方法。

示例代码 1-8　UserDaoImplTest.java（部分代码）

```
@Test
public void insertUserByUserService() {
    // 读取配置文件所需创建对象中创建的 bean 对象并获取 spring 工厂对象
    ApplicationContext context = new
ClassPathXmlApplicationContext("applicationContext.xml");
    // 通过注入时的唯一标识(id)获取 bean 对象
    UserService userService = (UserService) context.getBean("userService");
    // 使用对象
    userService.insertUser(null);    // 输出结果为：------insertUser ------
}
```

关于注入的方式有多种，比如 Setter 注入、构造函数注入等。上面的案例采用的是基于 Setter 方法实现的依赖注入，这也是推荐使用的注入方式。

1.2.3　依赖注入过程说明

依赖注入是一个过程，在此过程中对象仅通过构造函数参数、工厂方法的参数或从工厂方法构造或返回对象实例后在对象实例上设置的属性来定义其依赖项。然后容器在创建 Bean 时注入这些依赖项。

1. 基于构造函数的依赖注入

基于构造函数的 DI 是通过容器调用具有多个参数的构造函数来实现的，每个参数代表一个依赖项。

2. 基于 Setter 的依赖注入

基于 Setter 的 DI 是在调用无参数构造函数或无参数静态工厂方法来实例化 Bean 之后，容器在 Bean 上调用 Setter 方法来实现的。

在了解了依赖注入方式之后，我们具体看一下依赖解析过程。

（1）创建容器时，Spring 容器验证每个 Bean 的配置。在实际创建 Bean 之前，不会设置 Bean 属性（properties ）本身。创建容器时，将创建单例作用域（在 Bean 作用域中定义）且设置为预实例化（默认）的 Bean，否则只有在请求时才会创建 Bean。

（2）如果主要使用构造函数注入，就可能会创建无法解析的循环依赖场景。例如，类 A 需要一个类 B 通过构造函数注入的实例，而类 B 需要一个类 A 通过构造函数注入的实例。如果将 Bean 配置为类 A 和类 B 相互注入，Spring IoC 容器将在运行时检测这个循环引用，并抛出一个 BeanCurrentlyInCreationException 异常。可以使用 Setter 注入配置循环依赖项。与典型的情况（没有循环依赖关系）不同，Bean A 和 Bean B 之间的循环依赖关系迫使一个 Bean 在完全初始化之前被注入另一个 Bean 中（典型的先有鸡还是先有蛋的场景）。

（3）Spring 在容器加载时检测配置问题，例如对不存在的 Bean 和循环依赖项的引用。在实际创建 Bean 时，Spring 会尽可能晚地设置属性和解析依赖项。

（4）如果不存在循环依赖项，那么当一个或多个协作 Bean 被注入依赖 Bean 时，每个协作 Bean 在被注入依赖 Bean 之前都是完全配置的。这意味着，如果 Bean A 依赖于 Bean B，那么 Spring IoC 容器在调用 Bean A 上的 Setter 方法之前完全配置了 Bean B。换句话说，Bean 是实例化的（如果它不是预先实例化的 Singleton），它的依赖是设置的，并调用相关的生命周期方法（例如配置的 init 方法或 InitializingBean 回调方法）。

3. 懒初始化 Bean（延迟初始化 Bean）

默认情况下，ApplicationContext implementations 将创建和配置所有的单例 Bean 作为初始化过程的一部分。通常，这种预实例化是可取的，因为配置或周围环境中的错误会被立即发现。如果不希望这种行为，则可以通过将 Bean 定义标记为延迟初始化来防止单例 Bean 的预实例化。延迟初始化的 Bean 告诉 IoC 容器在首次请求时（而不是在启动时）创建一个 Bean 实例。比如：

```
    <bean id="lazy" class="com.something.ExpensiveToCreateBean"
lazy-init="true"/>
```

4. 方法注入

在大多数应用程序场景中，容器中的大多数 Bean 都是单例的。当一个单例 Bean 需要与另一个单例 Bean 协作或者一个非单例 Bean 需要与另一个非单例 Bean 协作时，通常通过将一个 Bean 定义为另一个 Bean 的属性来处理依赖关系。当 Bean 生命周期不同时，就会出现一个问题。假设 singleton Bean A 需要使用非 singleton（prototype）Bean B，那么，在程序中容器只会创建一次 singleton Bean A，因此只有一次机会设置属性，容器不能在每次需要 Bean A 时都为它提供一个新的 Bean B 实例。解决的办法是放弃一些控制反转，通过实现 ApplicationContextAware 接口，并在每次 Bean A 需要时调用容器的 getBean("B")来请求 Bean B 实例（通常是一个新的）。

1.2.4 Spring 容器中的 Bean 作用域和对象初始化

开发者主要是使用 Spring 框架做两件事：①开发 Bean；②配置 Bean。对于 Spring 框架来说，它要做的就是根据配置文件来创建 Bean 实例，并调用 Bean 实例的方法完成“依赖注入”，这就是所谓的 IoC 本质。

当通过 Spring 容器创建一个 Bean 实例时，不仅可以完成 Bean 实例的实例化，还可以为 Bean 指定特定的作用域。容器中 Bean 的作用域有很多种，Spring 支持如下五种作用域。

（1）singleton：单例模式，在整个 Spring IoC 容器中，singleton 作用域的 Bean 将只生成一个实例。

只管理一个单例 Bean 的共享实例，所有对 ID 或 ID 与该 Bean 定义匹配的 Bean 的请求，都会导致 Spring 容器返回一个特定的 Bean 实例。换句话说，当你定义一个 Bean 并且它的作用域是一个 singleton 时，Spring IoC 容器正好创建了该 Bean 所定义对象的一个实例。这个单实例存储在这样的单例 Bean 缓存中，该命名 Bean 的所有后续请求和引用都返回缓存的对象。单例作用域是 Spring 中的默认作用域，其工作模式如图 1-5 所示。

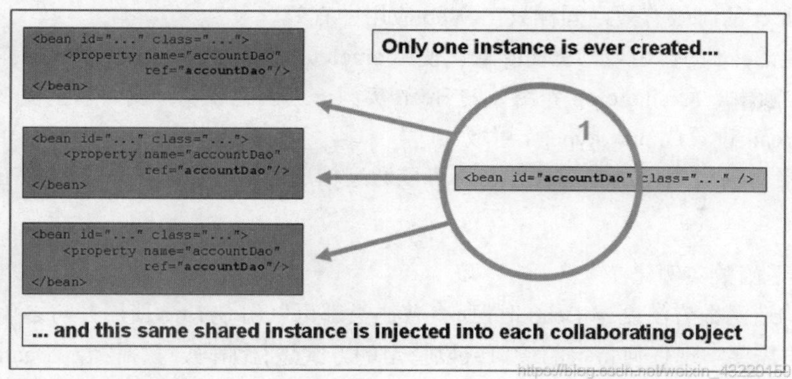

图 1-5

（2）prototype：每次通过容器的 getBean()方法获取 prototype 作用域的 Bean 时都将产生一个新的 Bean 实例。

每次请求特定 Bean 时都创建一个新的 Bean 实例。通常，应将 prototype 作用域用于所有有状态 Bean，将单例作用域用于无状态 Bean。与其他作用域不同，Spring 不管理 prototype Bean 的完整生命周期。容器实例化、配置、以其他方式组装 prototype 对象并将其交给客户端，而不再记录该 prototype 实例。因此，尽管初始化生命周期回调方法在所有对象上都被调用而不管其作用域如何，但对于 prototype 作用域，配置的销毁生命周期回调不会被调用。客户端代码必须清理 prototype 作用域对象并释放 prototype Bean 所拥有的昂贵资源。要让 Spring 容器释放 prototype 作用域 Bean 所拥有的资源，可尝试使用自定义 Bean 后处理器，该处理器保存对需要清理的 Bean 的引用，该作用域工作模式如图 1-6 所示。

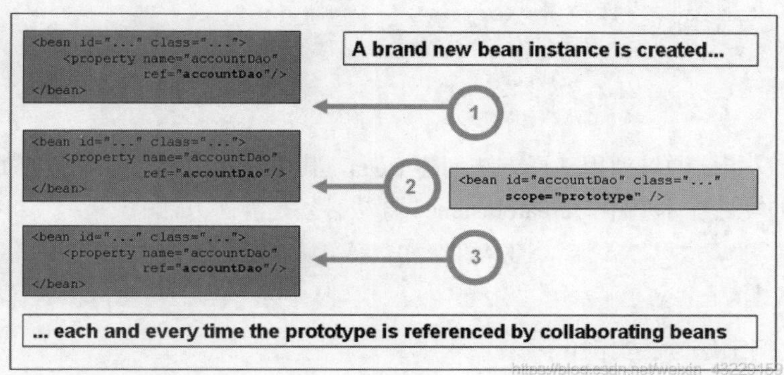

图 1-6

（3）request：对于一次 HTTP 请求，request 作用域的 Bean 将只生成一个实例，这意味着在同一次 HTTP 请求内，程序每次请求该 Bean 得到的都是同一个实例。只有在 Web 应用中使用 Spring 时，该作用域才真正有效。

（4）session：该作用域将 Bean 的定义限制为 HTTP 会话，只在 web-aware Spring ApplicationContext 的上下文中有效。

（5）global session：每个全局的 HTTP Session 对应一个 Bean 实例。在典型的情况下，仅在使用 portlet context 的时候有效，同样只在 Web 应用中有效。

如果不指定 Bean 的作用域，Spring 默认使用 singleton 作用域。prototype 作用域的 Bean 的创建、销毁代价比较大。而 singleton 作用域的 Bean 实例一旦创建成功，就可以重复使用。因此，应该尽量避免将 Bean 设置成 prototype 作用域。

Spring 容器对象实例化也有几种方法，可以通过构造函数、静态工厂、实例工厂方法等来实现。

（1）用构造函数实例化

当你使用构造函数方法创建 Bean 时，所有普通类都可以由 Spring 使用并与之兼容。也就是说，正在开发的类不需要实现任何特定的接口或以特定的方式进行编码，只需指定 Bean 类就足够了。根据你对特定 Bean 使用的 IoC 类型，你可能需要一个默认（空）构造函数。Spring IoC 容器几乎可以管理你希望它管理的任何类，不仅限于管理真正的 JavaBean。使用基于 XML 的配置元数据，你可以如下指定 Bean 类：

```
<bean id="exampleBean" class="examples.ExampleBean"/>
```

（2）静态工厂方法实例化

在定义使用静态工厂方法创建的 Bean 时，使用 class 属性指定包含静态工厂方法的类，使用名为 factory method 的属性指定工厂方法本身的名称。你应该能够调用此方法（如后文所述，使用可选参数）并返回一个活动对象，该对象随后将被视为是通过构造函数创建的。这种 Bean 定义的一个用途是在遗留代码中调用静态工厂。

```
public class ClientService {
    private static ClientService clientService = new ClientService();
    private ClientService() {}
    public static ClientService createInstance() {
        return clientService;
    }
}
```

以下 Bean 定义指定通过调用工厂方法创建 Bean，其中没有指定返回对象的类型（类），只指定包含工厂方法的类。在本例中，createInstance 方法必须是静态方法。

```
    <bean id="clientService" class="examples.ClientService"
factory-method="createInstance"/>
```

（3）使用实例工厂方法实例化

与通过静态工厂方法进行实例化类似，使用实例工厂方法的实例化从容器中调用现有 Bean 的非静态方法来创建新 Bean。一个工厂可以包含一个以上的工厂方法。工厂 Bean 本身可以通过依赖注入进行管理和配置。

```
public class DefaultServiceLocator {
    private static ClientService clientService = new ClientServiceImpl();
    private static AccountService accountService = new AccountServiceImpl();
    public ClientService createClientServiceInstance() {
        return clientService;
    }
```

```
        public AccountService createAccountServiceInstance() {
            return accountService;
        }
    }
    <!-- the factory bean, which contains a method called createInstance() -->
    <bean id="serviceLocator" class="examples.DefaultServiceLocator">
        <!-- inject any dependencies required by this locator bean -->
    </bean>
    <!-- the bean to be created via the factory bean -->
    <bean id="clientService" factory-bean="serviceLocator"
factory-method="createClientServiceInstance"/>
    <bean id="accountService" factory-bean="serviceLocator"
factory-method="createAccountServiceInstance"/>
```

（4）Java 注解方式实例化

Java 配置是 Spring 4.x 之后推荐的配置方式，可以完全替代 XML 配置，主要包含两个注解，分别是@Configuration 和@Bean。Spring 的 Java 配置方式是通过@Configuration 和@Bean 注解实现的：@Configuration 作用于类上，相当于一个 XML 配置文件；@Bean 作用于方法上，相当于 XML 配置中的<bean>。

1.2.5　依赖注入实战 Java 注解配置方式

1.2.2 小节中我们使用 XML 方式进行 Bean 的配置，但 XML 方式配置过于复杂且不好维护，目前大多数企业在开发中采用的都是基于 Java 注解的方式，具体项目参见本章项目源码文件夹下面的"springioc_anno"。

首先需要编写 SpringConfig，用于实例化 Spring 容器，然后打上@Configuration 注解，同时打上@ComponentScan 配置扫描的包。

这里采用的 Bean 对象依然是 1.2.2 小节中的 UserDao 和 UserService 对象，一般做法是在两个类中打上相应的注解，而在 Spring 配置类中对其进行扫描。

一般情况下按照不同业务逻辑含义，不同逻辑层的对象需要对应不同的注解。Dao 层使用@Repository 注解，Service 层使用@Service 注解。这两个注解实际上都是集成@Component 注解。

如果需要在 Spring 配置类中注入 Bean，则可以使用@Bean 注解。@Bean 用于向容器中注入对象，如果在 UserDao 类前面打上@Repository 注解，就不用@Bean 方式。

示例代码 1-9　SpringConfig.java

```
package com.mrchi;
import org.springframework.context.annotation.Bean;
import org.springframework.context.annotation.ComponentScan;
import org.springframework.context.annotation.Configuration;
//通过@Configuration 注解来表明该类是一个 Spring 的配置，相当于一个 XML 文件
@Configuration
@ComponentScan(basePackages = "com.mrchi")
public class SpringConfig {

    // 通过@Bean 注解来表明是一个 Bean 对象，相当于 XML 中的<bean>
```

```
    @Bean
    public UserDao getUserDao() {
        return new UserDaoImpl(); // 直接 new 对象做演示
    }
}
```

注 意

方法名是返回对象的名字，因此一般不带 get，也就是上述放入 Spring 容器的 Bean 的 name 为 getUserDao。

UserService 实现类需要打上@Service 注解，并在属性 userDao 上面打上@Autowire 注解，实现二者的注入关系，将原本在 XML 中的配置集中到 Java 代码中，减少了维护成本。具体 UserServiceImpl.java 代码如下。

示例代码 1-10　UserServiceImpl.java

```
@Service
public class UserServiceImpl implements UserService {
    //用 setter 方式实现注入
    @Autowired
    private UserDao userDao;

    public UserDao getUserDao() {
        return userDao;
    }
    public void setUserDao(UserDao userDao) {
        this.userDao = userDao;
    }
    public int insertUser(User user) {
        return userDao.insertUser(user);
    }
}
```

编写测试方法，用于启动 Spring 容器。

示例代码 1-11　Application.java

```
package com.mrchi;
import java.util.List;
import
org.springframework.context.annotation.AnnotationConfigApplicationContext;
public class Application {
    public static void main(String[] args) {
        // 通过 Java 配置来实例化 Spring 容器
        AnnotationConfigApplicationContext context = new
AnnotationConfigApplicationContext(SpringConfig.class);
        // 在 Spring 容器中获取 Bean 对象
        UserService userService = context.getBean(UserService.class);
        UserService.insertUser();
        // 销毁该容器
        context.destroy();
```

```
    }
}
```

1.3　Spring AOP

1.3.1　AOP 思想

AOP（Aspect-OrientedProgramming，面向切面编程）不是一种新的技术，也不是一种框架，而是一种编程思想，可以理解为 OOP（Object-Oriented Programing，面向对象编程）的补充和完善。OOP 引入封装、继承和多态性等概念来建立一种对象层次结构，用以模拟公共行为的一个集合。当我们需要为分散的对象引入公共行为的时候，OOP 则显得无能为力。也就是说，OOP 允许你定义从上到下的关系，但并不适合定义从左到右的关系。例如日志功能，日志代码往往水平地散布在所有对象层次中，而与它所散布到的对象的核心功能毫无关系。对于其他类型的代码，如安全性、异常处理和透明的持续性也是如此。这种散布在各处的无关代码被称为横切（cross-cutting）代码，在 OOP 设计中，它导致了大量代码的重复，不利于各个模块的重用。

AOP 技术则恰恰相反，它利用一种称为“横切”的技术，剖解开封装的对象内部，并将那些影响了多个类的公共行为封装到一个可重用模块，并将其命名为“Aspect”，即方面。所谓“方面”，简单地说就是将那些与业务无关却为业务模块所共同调用的逻辑或责任封装起来，便于减少系统的重复代码，降低模块间的耦合度，并有利于未来的可操作性和可维护性。AOP 代表的是一个横向的关系，如果说“对象”是一个空心的圆柱体，其中封装的是对象的属性和行为，那么面向方面编程的方法就像一把利刃，将这些空心圆柱体剖开，以获得其内部的消息，而剖开的切面也就是所谓的“方面”了。

1. AOP 涉及的基本概念

（1）Aspect（切面）：通常是一个类，里面可以定义切入点和通知。

（2）JointPoint（连接点）：程序执行过程中明确的点，一般是方法的调用。

（3）Advice（通知）：AOP 在特定的切入点上执行的增强处理，有 before、after、afterReturning、afterThrowing、around。

（4）Pointcut（切入点）：带有通知的连接点，在程序中主要体现为书写切入点表达式。

（5）AOP 代理：AOP 框架创建的对象，代理就是目标对象的加强。Spring 中的 AOP 代理可以是 JDK 动态代理，也可以是 CGLIB 代理，前者基于接口，后者基于子类。

2. 通知方法

（1）前置通知（@Before）：在我们执行目标方法之前运行。

（2）后置通知（@After）：在我们目标方法运行结束之后，不管有没有异常。

（3）返回通知（@AfterReturning）：在我们的目标方法正常返回值后运行。

（4）异常通知（@AfterThrowing）：在我们的目标方法出现异常后运行。

（5）环绕通知（@Around）：动态代理，需要手动执行 joinPoint.procced()（目标方法执行之前相当于前置通知，执行之后就相当于后置通知）。

3. Spring AOP

Spring 中的 AOP 代理还是离不开 Spring 的 IoC 容器。代理的生成、管理及其依赖关系都是由 IoC 容器负责的，Spring 默认使用 JDK 动态代理。在需要代理类而不是代理接口的时候，Spring 会自动切换为使用 CGLIB 代理，不过现在的项目都是面向接口编程的，所以 JDK 动态代理相对来说用得还是多一些。

1.3.2 基于注解的 AOP 实现

以下是一个切面编程的典型应用，即关于日志的横切业务注入具体目标业务方法中，具体代码参见本章源码文件夹下面的“springaop”项目。下面的 Calculator 计算器类将作为目标类，里面有除法运算方法，日志将会在除法运算前后加入相关打印信息。

步骤 01 定义一个日志切面，为其定义切入点 PointCut，并在某些方法上通过@Before、@After 等注解定义通知，具体代码如下。

示例代码 1-12 LogAspects.java

```
import org.aspectj.lang.ProceedingJoinPoint;
import org.aspectj.lang.annotation.After;
import org.aspectj.lang.annotation.AfterReturning;
import org.aspectj.lang.annotation.AfterThrowing;
import org.aspectj.lang.annotation.Before;
import org.aspectj.lang.annotation.Pointcut;
import org.aspectj.lang.annotation.Around;
import org.aspectj.lang.annotation.Aspect;
 //日志切面类
@Aspect
public class LogAspects {
    @Pointcut("execution(public int com.mrchi.aop.Calculator.*(..))")
    public void pointCut(){};
    //@before 代表在目标方法执行前切入，并指定在哪个方法前切入
    @Before("pointCut()")
    public void logStart(){
        System.out.println("除法运行......参数列表是:{}");
    }
    @After("pointCut()")
    public void logEnd(){
        System.out.println("除法结束......");
    }
    @AfterReturning("pointCut()")
    public void logReturn(){
        System.out.println("除法正常返回......运行结果是:{}");
    }
    @AfterThrowing("pointCut()")
    public void logException(){
        System.out.println("运行异常......异常信息是:{}");
    }
    @Around("pointCut()")
```

```
    public Object Around(ProceedingJoinPoint proceedingJoinPoint) throws
Throwable{
        System.out.println("@Arount:执行目标方法之前...");
        Object obj = proceedingJoinPoint.proceed();//相当于开始调用目标方法
        System.out.println("@Arount:执行目标方法之后...");
        return obj;
    }
}
```

SpringAOP 的存在是为了解耦。AOP 可以让一些类共享相同的行为。Spring 支持 AspectJ 的注解式的切面编程，包含@Aspect、@After、@Before、@Around 和@PointCut 注解。所有符合上述代码中 PointCut 表达式的业务方法，都会在合适的时机执行 LogAspects 切面类的各个方法。

以上注解定义来自 aspectjweaver 项目，所以项目的 pom.xml 需要引入依赖，代码如下：

```
<dependency>
<groupId>org.aspectj</groupId>
    <artifactId>aspectjweaver</artifactId>
    <version>1.8.13</version>
</dependency>
```

execution 切点表达式是指通过配置执行类的方法配置不同的切面。例如，“execution(* cn..*Serivice.*(..))”是指对 cn 包下以 Service 约束的类或接口中的所有方法（无论接收多少参数，返回什么类型的值）进行拦截。在注解中使用@Pointcut，在 XML 中声明 AspectJExpressionPointcut 类的 expression 属性来设置切点表达式。

步骤 02 定义目标方法，代码如下。

示例代码 1-13　Calculator.java

```
public class Calculator {
    //业务逻辑方法
    public int div(int i, int j){
        System.out.println("--------");
        return i/j;
    }
}
```

步骤 03 在 Spring IoC 容器中定义 Bean。

示例代码 1-14　AOPBeanConfig.java

```
import org.springframework.context.annotation.Bean;
import org.springframework.context.annotation.Configuration;
import org.springframework.context.annotation.EnableAspectJAutoProxy;
import com.mrchi.aop.Calculator;
import com.mrchi.aop.LogAspects;
@Configuration
@EnableAspectJAutoProxy
public class AOPBeanConfig {
    @Bean
    public Calculator calculator(){
```

```
        return new Calculator();
    }
    @Bean
    public LogAspects logAspects(){
        return new LogAspects();
    }
}
```

@EnableAspectJAutoProxy 注解在@Configuration 类上，用于启用 Spring 的注解 Aspect 功能。

步骤 04 测试类，用于根据注解初始化 Bean 对象，并手动调用 Calculator 的除法运算方法，观察控制台切面代码的输出。

示例代码 1-15 AOPTest.java

```
public class AOPTest {
    @Test
    public void test01(){
        AnnotationConfigApplicationContext app = new
AnnotationConfigApplicationContext(AOPBeanConfig.class);
        Calculator c = app.getBean(Calculator.class);
        int result = c.div(4, 3);
        System.out.println(result);
        app.close();
    }
}
```

运行结果如下：

```
@Arount:执行目标方法之前...
除法运行......参数列表是:{}
---------------
@Aro unt:执行目标方法之后...
除法结束......
除法正常返回......运行结果是:{}
1
```

第 2 章

Spring MVC 开发基础

本章将首先讲解 MVC 架构的基本原理及每层的实际含义，然后详细讲解 Spring MVC 框架的架构特性和工作流程，并通过典型开发案例来贯穿讲述 Spring MVC 开发中的重要基础知识，包括基于配置文件的开发方式、基于 Java 注解的开发方式、Spring MVC 静态资源映射、返回 JSON 字符、拦截器开发、文件上传等内容。

2.1 Spring MVC 概述

2.1.1 MVC 架构简介

MVC（Model-View-Controller，模型-视图-控制器）模式用于应用程序的分层开发，如图 2-1 所示。

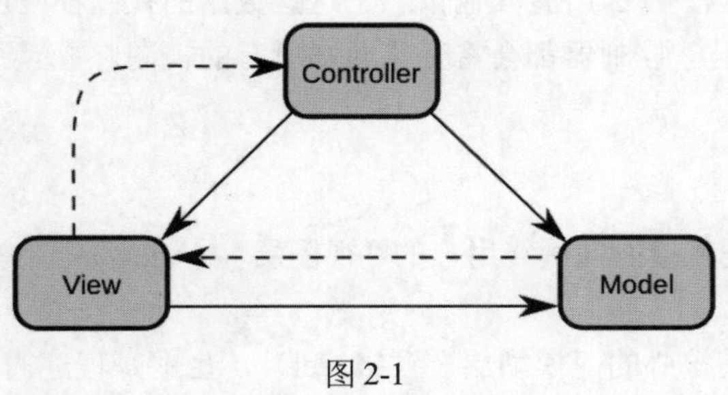

图 2-1

下面详细介绍视图、模型和控制器的具体内容以及它们之间的关系。

1. 视图

视图（View）代表用户交互界面，对于 Web 应用来说，可以概括为 HTML 界面，但有可能为 XHTML、XML 和 Applet。随着应用的复杂性和规模性的增长，界面的处理也变得具有挑战性。一个应用可能有很多不同的视图，MVC 设计模式对于视图的处理仅限于视图上数据的采集和处理，以及用户的请求，而不包括在视图的业务流程的处理。业务流程的处理交予模型（Model）处理。比如一个订单的视图只接受来自模型的数据并显示给用户，以及将用户界面的输入数据和请求传递给控制和模型。

2. 模型

模型（Model）就是业务流程/状态的处理以及业务规则的制定。业务流程的处理过程对其他层来说是黑箱操作，模型接受视图请求的数据并返回最终的处理结果。业务模型的设计可以说是 MVC 的核心。目前流行的 EJB 模型就是一个典型的应用例子，它从应用技术实现的角度对模型做了进一步的划分，以便充分利用现有的组件，但它不能作为应用设计模型的框架。它仅仅告诉你按这种模型设计就可以利用某些技术组件，从而减少了技术上的困难。对一个开发者来说，就可以专注于业务模型的设计。MVC 设计模式告诉我们，把应用的模型按一定的规则抽取出来。抽取的层次很重要，这也是判断开发人员的设计是否优秀的依据。抽象与具体不能隔得太远，也不能太近。MVC 并没有提供模型的设计方法，而只告诉你应该组织管理这些模型，以便于模型的重构和提高重用性。我们可以用对象编程来做比喻，MVC 定义了一个顶级类，告诉它的子类你只能做这些，但无法限制你能做这些。这点对编程的开发人员非常重要。

业务模型还有一个很重要的模型——数据模型，主要是指实体对象的数据保存（持续化）。比如将一张订单保存到数据库，从数据库获取订单。我们可以将这个模型单独列出，所有有关数据库的操作只限制在该模型中。

3. 控制器

控制器（Controller）可以理解为从用户接收请求，将模型与视图匹配在一起，共同完成用户的请求。划分控制层的作用也很明显，它清楚地告诉你，它就是一个分发器，选择什么样的模型，选择什么样的视图，可以完成什么样的用户请求。控制层并不做任何数据处理。例如，用户单击一个链接，控制层接受请求后，并不处理业务信息，只是把用户的信息传递给模型，告诉模型做什么，选择符合要求的视图返回给用户。因此，一个模型可能对应多个视图，一个视图可能对应多个模型。

模型、视图与控制器的分离，使得一个模型可以具有多个显示视图。如果用户通过某个视图的控制器改变了模型的数据，那么所有其他依赖于这些数据的视图都应反映出这些变化。因此，无论何时发生了何种数据变化，控制器都会将变化通知所有的视图，显示更新。这实际上是一种模型的变化——传播机制。

4. 三层之间的关系

（1）最上面的一层是直接面向最终用户的“视图层”（View）。它是提供给用户的操作界面，是程序的外壳。

（2）最下面的一层是核心的“模型层”（Model），也就是程序需要操作的数据或信息。

（3）中间的一层是“控制器层”（Controller），负责根据用户从“视图层”输入的指令选取“模型层”中的数据，然后对其进行相应的操作，产生最终结果。

这三层既是紧密联系在一起的，又是互相独立的，每一层内部的变化不影响其他层。每一层都对外提供接口（Interface），供上面一层调用。这样一来，软件就可以实现模块化，修改外观或者变更数据都不用修改其他层，大大方便了维护和升级。

2.1.2　Spring MVC 框架简介

Spring MVC 是一种基于 Java 的、实现了 MVC 设计模型的请求驱动类型的轻量级 Web 框架，属于 Spring FrameWork 的后续产品，已经融合在 Spring Web Flow 里面。Spring 框架提供了构建 Web 应用程序的全功能 MVC 模块。在使用 Spring 进行 Web 开发时，可以选择使用 Spring MVC 框架或集成其他 MVC 开发框架，如 Struts1（现在一般不用）、Struts2 等。

Spring MVC 已经成为目前主流的 MVC 框架，并且随着 Spring 新版本的发布，全面超越 Struts2，成为优秀的 MVC 框架。

它通过一套注解，让一个简单的 Java 类成为处理请求的控制器，而无须实现任何接口，同时它还支持 RESTful 编程风格的请求。

1. Spring MVC 的优势

（1）清晰的角色划分。框架中的各个角色如下：

- 前端控制器（DispatcherServlet）。
- 请求到处理器映射（HandlerMapping）。
- 处理器适配器（HandlerAdapter）。
- 视图解析器（ViewResolver）。
- 处理器或页面控制器（Controller）。
- 验证器（ Validator）。
- 命令对象（Command 请求参数绑定到的对象）。
- 表单对象（Form Object 提供给表单展示和提交到的对象）。

（2）分工明确，而且扩展点相当灵活，很容易扩展。

（3）命令对象就是一个 POJO，无须继承框架特定 API，可以使用命令对象直接作为业务对象。

（4）和 Spring 其他框架无缝集成，是其他 Web 框架所不具备的。

（5）可适配，通过 HandlerAdapter 可以支持任意的类作为处理器。

（6）可定制性，HandlerMapping、ViewResolver 等能够非常简单地定制。

（7）功能强大的数据验证、格式化、绑定机制。

（8）利用 Spring 提供的 Mock 对象能够非常简单地进行 Web 层单元测试。

（9）本地化、主题的解析的支持，使我们更容易进行国际化和主题的切换。

（10）强大的 JSP 标签库，使 JSP 编写更容易。

另外，还有 RESTful 风格的支持、简单的文件上传、约定大于配置的契约式编程支持、基于注解的零配置支持等。

2. Spring MVC 和 Struts2 的优略分析

（1）它们的共同点如下：

- 它们都是表现层框架，都是基于 MVC 模型编写的。
- 它们的底层都离不开原始 Servlet API。
- 它们处理请求的机制都是一个核心控制器。

（2）它们的区别有以下几点：

- Spring MVC 的入口是 Servlet，而 Struts2 是 Filter。
- Spring MVC 是基于方法设计的，而 Struts2 是基于类的，Struts2 每次执行都会创建一个动作类。所以，Spring MVC 会稍微比 Struts2 快一些。
- Spring MVC 使用更加简洁，同时还支持 JSR303，处理 Ajax 的请求更方便。JSR303 是一套 JavaBean 参数校验的标准，它定义了很多常用的校验注解，我们可以直接将这些注解加在 JavaBean 属性上，在需要校验的时候进行校验。
- struts2 的 OGNL 表达式使页面的开发效率相比 Spring MVC 更高一些，但是执行效率并没有比 JSTL 提升，尤其是 struts2 的表单标签，远没有 HTML 执行效率高。

2.1.3 Spring MVC 工作流程

要把 Spring MVC 流程图搞明白，首先需要知道 6 个组件。

1. DispatcherServlet（中央控制器）

DispatcherServlet 是核心组件。用户在浏览器输入 URL，发起请求，首先会到达 DispatcherServlet，由它来调用其他组件配合工作的完成。DispatcherServlet 的存在大大降低了组件之间的耦合性。

2. HandlerMapping（处理器映射器）

HandlerMapping 记录 URL 与处理器的映射，方式有注解、XML 配置等。

3. HandLer（Controller，处理器）

HandLer 是后端控制器（通俗一点就是 Controller 层所写的业务代码），对用户的请求进行处理。

4. HandlerAdapter（处理器适配器）

通过 HandlerAdapter 对处理器进行处理，这是适配器模式的应用，通过扩展适配器可以对更多类型的处理器进行处理。

5. ViewResolver（视图解析器）

ViewResolver 负责解析 View（视图），并进行渲染（数据填充），将处理结果通过页面展示给用户。

6. View（视图）

View 是一个接口，实现类支持不同的 View 类型（JSP、FreeMarker、Velocity）。

一般情况下，需要通过页面标签或者页面模板技术将模型数据通过页面展示给用户，需要由程序员根据业务需求开发具体的页面。

Spring MVC 典型的工作流程如图 2-2 所示，具体说明如下：

（1）用户发送请求至前端控制器 DispatcherServlet。

（2）前端控制器 DispatcherServlet 收到请求后调用处理器映射器 HandlerMapping。

（3）处理器映射器 HandlerMapping 根据请求的 URL 找到具体的处理器，生成处理器对象 Handler 及处理器拦截器 HandlerIntercepter（如果有则生成），并返回前端控制器 DispatcherServlet。

（4）前端控制器 DispatcherServlet 通过处理器适配器 HandlerAdapter 调用处理器 Controller。

（5）调用处理器（Controller，也叫后端控制器）。

（6）处理器 Controller 执行完后返回 ModelAndView。

（7）处理器映射器 HandlerAdapter 将处理器 Controller 执行返回的结果 ModelAndView 返回给前端控制器 DispatcherServlet。

（8）前端控制器 DispatcherServlet 将 ModelAndView 传给视图解析器 ViewResolver。

（9）视图解析器 ViewResolver 解析后返回具体的视图 View。

（10）前端控制器 DispatcherServlet 对视图 View 进行渲染视图（将模型数据填充至视图中）。

（11）前端控制器 DispatcherServlet 响应用户。

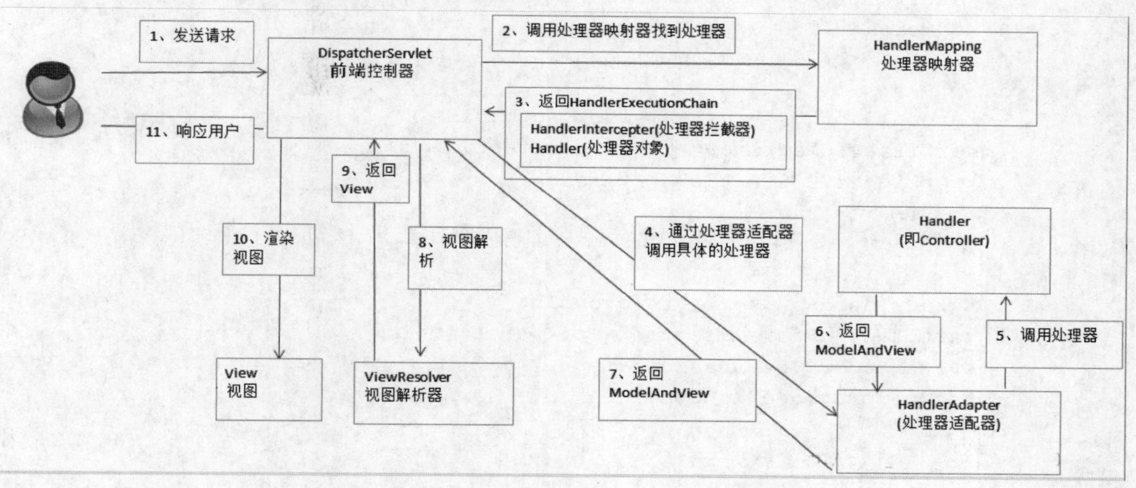

图 2-2

2.2 Spring MVC 开发实战

2.2.1 典型入门程序

Spring MVC 是 MVC 模式的一个应用，本节将看到 Spring MVC 的典型应用案例。

@EnableWebMvc 注解可以让 Spring 启动 MVC 配置。添加了@EnableWebMvc 的类同时要拥有 @Configuration 注解。添加了@EnableWebMvc 注解的类（可选的），可以实现接口 WebMvcConfigurer 或者继承适配器类 WebMvcConfigurerAdapter 来添加更多 Spring MVC 需要配置

的信息，比如添加拦截器、视图解析器等。

非常重要的是，如果不想开发 web.xml 就必须开发 WebApplicationInitializer 接口的子类，以便于注解 Spring 的 DispatcerServlet 类。

接下来开发第一个 Spring MVC 项目，首先是带有 web.xml 文件的，参见项目 SpringMVC_test。

步骤 01 配置 pom.xml。

示例代码 2-1 pom.xml

```
<dependencies>
    <dependency>
        <groupId>junit</groupId>
        <artifactId>junit</artifactId>
        <version>4.11</version>
        <scope>test</scope>
    </dependency>
    <dependency>
        <groupId>org.springframework</groupId>
        <artifactId>spring-web</artifactId>
        <version>5.0.8.RELEASE</version>
    </dependency>
    <dependency>
        <groupId>org.springframework</groupId>
        <artifactId>spring-webmvc</artifactId>
        <version>5.0.8.RELEASE</version>
    </dependency>
    <!--配置 Servlet 依赖  Controller 处理器需要-->
    <dependency>
        <groupId>javax.servlet</groupId>
        <artifactId>javax.servlet-api</artifactId>
        <version>4.0.1</version>
        <scope>provided</scope>
    </dependency>
     <!--配置 jstl 标签需要-->
    <dependency>
        <groupId>javax.servlet</groupId>
        <artifactId>jstl</artifactId>
        <version>1.2</version>
    </dependency>
    <dependency>
        <groupId>taglibs</groupId>
        <artifactId>standard</artifactId>
        <version>1.1.2</version>
    </dependency>
</dependencies>
```

pom.xml 主要加入 Spring 核心依赖和 Spring MVC 依赖，为了页面能够使用 jstl 标签需要添加 jstl 和 standard 依赖。

步骤 02 配置 web.xml，主要是配置 DispatcherServlet 中间类，和 Model、View、Controller 分

别进行联系，降低三者的耦合度，具体配置如下。

示例代码 2-2　web.xml

```
<?xml version="1.0" encoding="UTF-8"?>
<web-app xmlns="http://xmlns.jcp.org/xml/ns/javaee"
         xmlns:xsi="http://www.w3.org/2001/XMLSchema-instance"
         xsi:schemaLocation="http://xmlns.jcp.org/xml/ns/javaee
         http://xmlns.jcp.org/xml/ns/javaee/web-app_4_0.xsd" version="4.0">
    <display-name>Archetype Created Web Application</display-name>
        <servlet>
       <servlet-name>springmvc</servlet-name>
      <servlet-class>org.springframework.web.servlet.DispatcherServlet</ser
      vlet-class>
           <init-param>
               <param-name>contextConfigLocation</param-name>
               <param-value>classpath:springmvc/springmvc.xml</param-value>
           </init-param>
        </servlet>
        <servlet-mapping>
           <servlet-name>springmvc</servlet-name>
           <url-pattern>*.action</url-pattern>
        </servlet-mapping>
      <welcome-file-list>
          <welcome-file>index.jsp</welcome-file>
      </welcome-file-list>
</web-app>
```

web.xml 主要是对 Spring MVC 控制器的配置，控制器是 DispatcherServlet，用来过滤以.action 结尾的请求。

步骤 03 配置实体类。这里省略 set 和 get 方法，具体可以参照源码文件。定义一个实体类 Fruit.java，具体代码如下。

示例代码 2-3　Fruit.java

```
public class Fruit {
    String name;
    Integer price;
    public Fruit(){}
    public Fruit(String name, Integer price) {
        this.name = name;
        this.price = price;
    }
    @Override
    public String toString() {
        return "Fruit{" +
                "name='" + name + '\'' +
                ", price='" + price + '\'' +
                '}';
    }
}
```

步骤04 配置实现 Controller 接口的类。这一步用来定义 Spring MVC 的后端控制器，即 Controller。当请求通过 DispatcherServlet 时，会查找到请求匹配的后端控制器，为了讲解 Spring MVC 的处理流程，本案例没有采用注解的方式，而是采用将 Controller 对应的 Bean 的 name 作为 URL 进行查找，需要配置 Handler 时指定 Bean name，具体代码如下。

示例代码 2-4 FruitController.java

```
import java.util.ArrayList;
import java.util.List;
public class FruitsController implements Controller {
    @Override
    public ModelAndView handleRequest(javax.servlet.http.HttpServletRequest
httpServletRequest, javax.servlet.http.HttpServletResponse httpServletResponse)
throws Exception {
        FruitService fruitService = new FruitService();
        List<Fruit> list = fruitService.queryFruitsList();
        ModelAndView modelAndView = new ModelAndView();
        modelAndView.addObject("fruits",list);
        System.out.println(modelAndView.getModelMap());
        modelAndView.setViewName("/WEB-INF/jsp/fruitsList.jsp");
        return modelAndView;
    }
    class FruitService{
        public List<Fruit> queryFruitsList(){
            List<Fruit> list = new ArrayList<Fruit>();
            list.add(new Fruit("苹果",5));
            list.add(new Fruit("火龙果",7));
            list.add(new Fruit("雪莲果",4));
            return list;
        }
    }
}
```

步骤05 配置要访问的显示界面 JSP。开发 JSP 页面，在页面头部引入 jstl 标签，页面内用标签显示 FruitController 传递过来的水果列表数据，具体代码如下。

示例代码 2-5 fruit_list.jsp

```
<%@ taglib prefix="c" uri="http://java.sun.com/jsp/jstl/core" %>
<%@ page contentType="text/html;charset=UTF-8" language="java" %>
<html>
<head>
    <title>Title</title>
</head>
<body>
<h3>新鲜水果</h3>
<table width="300px" border="1">
    <tr>
        <td>序号</td>
        <td>名称</td>
        <td>价格</td>
```

```
        </tr>
        <c:forEach items="${fruits}" var="fruit" varStatus="varfrt">
            <tr>
                <td>${varfrt.count}</td>
                <td>${fruit.name}</td>
                <td>${fruit.price}</td>
            </tr>
        </c:forEach>
    </table>
    </body>
    </html>
```

步骤 06 配置 springmvc.xml 配置文件。首先，在配置文件中配置处理器映射器，将 Bean 的 name 作为 URL 进行查找，处理映射器名称是 BeanNameUrlHandlerMapping。然后，配置适配器，只要编写实现了 Controller 接口的控制器，适配器就会执行 Controller 的具体方法。最后，配置视图解析器 InternalResourceViewResolver，根据 handler 方法返回 ModelAndView 中的视图的具体位置加载相应的界面，并绑定反馈数据，具体代码如下。

示例代码 2-6 springmvc.xml

```
    <?xml version="1.0" encoding="UTF-8"?>
    <beans xmlns="http://www.springframework.org/schema/beans"
         xmlns:xsi="http://www.w3.org/2001/XMLSchema-instance"
         xsi:schemaLocation="http://www.springframework.org/schema/beans
http://www.springframework.org/schema/beans/spring-beans.xsd">
        <!--配置处理器映射器将 Bean 的 name 作为 URL 进行查找，需要配置 Handler 时指定的
Beanname 就是 URL-->
        <bean class="org.springframework.web.servlet.handler.
BeanNameUrlHandlerMapping"></bean>
        <!--配置适配器，只要编写实现了 Controller 接口的控制器，适配器就会执行 Controller
的具体方法-->
        <bean class="org.springframework.web.servlet.mvc.
SimpleControllerHandlerAdapter"></bean>
        <!--根据 handler 方法返回的 ModelAndView 中的视图的具体位置加载相应的界面，并绑定反
馈数据-->
        <bean class="org.springframework.web.servlet.view.
InternalResourceViewResolver"></bean>
        <bean name="/fruits.action" class="controller.FruitsController"></bean>
    </beans>
```

上面是完整的开发流程，下面说明 Spring MVC 从请求发出到响应的具体流程：

（1）因为 http://localhost:8080/SpringMVC_test/在不存在 Servlet、HTML、JSP 页面时，web.xml 中的<welcome-file-list>标签元素就会指定显示的默认文件 index.jsp 页面。

（2）访问 http://localhost:8080/SpringMVC_test/fruits.action。DispatcherServlet 接收到请求后访问映射器，根据映射器的类型 BeanNameUrlHandlerMapping（将 Bean 的 name 作为 URL 进行查找，需要配置 Handler 时指定 Bean name 为 URL）从 springmvc.xml 的配置文件中找到<bean name="/fruits.action" class="controller.FruitsController"></bean>，返回给 DispatcherServlet。

（3）DispatcherServlet 访问适配器。根据适配器类型 SimpleControllerHandlerAdapter（只要编

写实现了 Controller 接口的控制器，适配器就会执行 Controller 的具体方法），因为 public class FruitsController implements Controller 满足条件就执行它的 handleRequest 类并返回 ModelAndView（包含对应的显示页面和数据）给 DispatcherServlet。

（4）DispatcherServlet 访问视图解析器，根据类型 InternalResourceViewResolver（根据 handler 方法返回的 ModelAndView 中的视图的具体位置，加载相应的界面并绑定反馈数据）访问 fruitsList.jsp 页面，将数据动态加入该页面中，然后返回给 DispatcherServlet，再返回给浏览器。这一步中需要注意以下几点：

- 因为 JSP 中使用了 jstl 标签，所以要进行相关的配置。
- JSP 放在 WEB-INF（项目私有文件夹）中，这样用户就无法通过路径访问该网页，保证了视图的安全性。
- ModelAndView 的视图地址为"/WEB-INF/jsp/fruitsList.jsp"（不需要写 webapp）。

2.2.2 通过注解启动无 web.xml 的 Spring 项目

2.2.1 小节说明了如何通过 web.xml 实现 Spring MVC 控制器的配置，本节将用完全注解代替 XML 方式来实现 Spring MVC 应用。这里的核心关键是需要实现 Spring 提供的接口 org.springframework.web.WebApplicationInitializer，具体实现步骤如下。

步骤 01 开发 Spring 配置类。这里用 Spring 配置类取代 springmvc.xml 中的部分配置，比如视图解析器等。首先为类添加@Configuration 和@EnableWebMvc 注解，具体代码如下。

示例代码 2-7 MvcConfig.java

```
import org.springframework.context.annotation.Bean;
import org.springframework.context.annotation.ComponentScan;
import org.springframework.context.annotation.Configuration;
import org.springframework.web.servlet.config.annotation.EnableWebMvc;
import org.springframework.web.servlet.view.InternalResourceViewResolver;
import org.springframework.web.servlet.view.JstlView;
@Configuration
@EnableWebMvc
@ComponentScan("cn.springmvc")
 /**
 此类用于启动 Spring 注解，并配置 mvc 的一些信息
 @author mrchi
 @version 1.0
 */
 public class MvcConfig {
    @Bean
    public InternalResourceViewResolver viewResolver()
    {
        System.err.println("init....");
        InternalResourceViewResolver viewResolver = new
InternalResourceViewResolver();
        viewResolver.setPrefix("/WEB-INF/classes/views/");
```

```
        viewResolver.setSuffix(".jsp");
        viewResolver.setViewClass(JstlView.class);
        return viewResolver;
    }
}
```

步骤 02 开发 WebApplicationInitializer 的子类注册 DispatcherServlet。为了让项目启动时加载 DispatcherServlet，采用 WebApplicationInitializer 类来代替 web.xml 完成注册，具体代码如下。

示例代码 2-8　WebInitializer.java

```
package cn.springmvc.core;
import javax.servlet.ServletContext;
import javax.servlet.ServletException;
import javax.servlet.ServletRegistration.Dynamic;
import org.springframework.web.WebApplicationInitializer;
import org.springframework.web.context.support.
AnnotationConfigWebApplicationContext;
import org.springframework.web.servlet.DispatcherServlet;
/**
功能：注册 DispatcherServlet<br>
加载 Spring 注解配置类
@author mrchi
@version 1.0
*/
public class WebInitializer implements WebApplicationInitializer {
   @Override
   public void onStartup(ServletContext servletContext) throws
ServletException
   {
      AnnotationConfigWebApplicationContext ctx = new
      AnnotationConfigWebApplicationContext();
      ctx.register(MvcConfig.class);
      ctx.setServletContext(servletContext);
      //添加 DispatcherServlet
      Dynamic servlet = servletContext.addServlet("dispatcher", new
      DispatcherServlet(ctx));
      servlet.addMapping("/");
      servlet.setLoadOnStartup(1);
   }
}
```

步骤 03 开发一个测试控制器。控制器的设计采用注解方式，定义一个普通的 JavaBean，不需要实现任何接口，为类添加@Controller 注解即可。需要为每一个方法定义@RequestMapping 注解来定义与方法匹配的 URL。具体代码如下。

示例代码 2-9　OneController.java

```
package cn.springmvc.controller;
import org.springframework.stereotype.Controller;
import org.springframework.web.bind.annotation.RequestMapping;
```

```
@Controller
public class OneController {
    @RequestMapping("/index")
    public String hello() {
        System.err.println("Hello..." + this);
        return "index";
    }
}
```

这样一个完全使用注解方式开发的 Spring MVC 案例就完成了，效果等同于 2.2.1 小节的开发方式，但明显简化了配置。这种全注解的方式（在后面的 Spring Boot 章节中也能看到），是 Spring Boot 能够做到简化 Spring 开发的基础之一。

2.2.3 Spring MVC 返回 JSON 数据

Spring MVC 返回 JSON 数据可以通过两种方式实现：一是通过在 @EnableMvc 的类上通过 @Bean 返回对应的 HttpMessageConverter；二是在实现 WebMvcConfigurer 接口的类上通过对应的方法 configureMessageConverters 来添加 HttpMessageConverter。

方式 1：使用注解支持 json/string 的返回数据

核心问题是通过@Bean 返回 RequestMappingHandlerAdapter 并添加 MessageConvert，具体代码如下。

示例代码 2-10　MvcConfig.java

```
@Configuration
@EnableWebMvc
@ComponentScan("cn.springmvc")
/**
此类用于启动 Spring 注解，并配置 mvc 的一些信息
@author mrchi
@version 1.0
*/
public class MvcConfig {
    @Bean
    public InternalResourceViewResolver viewResolver()
    {
        System.err.println("init....");
        InternalResourceViewResolver viewResolver = new
        InternalResourceViewResolver();
        viewResolver.setPrefix("/WEB-INF/classes/views/");
        viewResolver.setSuffix(".jsp");
        viewResolver.setViewClass(JstlView.class);
        return viewResolver;
    }
    //注解转换器处理 json/string
    @Bean
    public RequestMappingHandlerAdapter requestMappingHandlerAdapter() {
```

```
        RequestMappingHandlerAdapter r = new RequestMappingHandlerAdapter();
        //注册 Json 转换器，也可以使用其他转换工具
        FastJsonHttpMessageConverter4 json = new
FastJsonHttpMessageConverter4();
        json.setDefaultCharset(Charset.forName("UTF-8"));
        json.setSupportedMediaTypes(Arrays.asList(MediaType.
APPLICATION_JSON_UTF8));
        r.getMessageConverters().add(json);
        //注册 StringMessageConvert
        r.getMessageConverters().add(new
        StringHttpMessageConverter(Charset.forName("UTF-8")));
        System.err.println("创建成功，返回-----" + r);
        return r;
    }
}
```

方式 2：WebMvcConfigurer 配置 MVC

Spring 中有一个 WebMvcConfigurer 类，只要实现此接口或是它的子类 WebMvcConfigurerAdapter 并添加一些配置，即可实现一些 Spring 的设置。另外，还需要在实现类上添加@EnableWebMvc 注解。具体实现代码如下。

示例代码 2-11　MvcConfig.java

```
@Configuration
@EnableWebMvc
@ComponentScan("cn.springmvc")
/**
此类用于启动 Spring 注解，并配置 mvc 的一些信息
@author mrchi
@version 1.0
*/
public class MvcConfig extends WebMvcConfigurerAdapter {
    @Bean
    public InternalResourceViewResolver viewResolver() {
        System.err.println("init....");
        InternalResourceViewResolver viewResolver = new
InternalResourceViewResolver();
        viewResolver.setPrefix("/WEB-INF/classes/views/");
        viewResolver.setSuffix(".jsp");
        viewResolver.setViewClass(JstlView.class); return viewResolver;
    }

    //注解转换器处理 json/string
    @Override
    public void configureMessageConverters(List<HttpMessageConverter<?>>
converters){
        FastJsonHttpMessageConverter4 json = new
FastJsonHttpMessageConverter4();
        json.setDefaultCharset(Charset.forName("UTF-8"));
        json.setSupportedMediaTypes(Arrays.asList(MediaType.
```

```
APPLICATION_JSON_UTF8));
            converters.add(json);
            converters.add(new StringHttpMessageConverter(Charset.
forName("UTF-8")));
        }
    }
```

需要注意的是，这里采用了第三方的 JSON 转换工具，所以需要在项目中引入其依赖，并进行上述两种方式的定义。本节中采用的是阿里巴巴的 fastjson 工具，具体依赖参见本节对应的实例代码。

另外，我们需要在定义 Controller 的方法上面加上@ResponseBody 的注解（Spring MVC @ResponseBody 的作用是把返回值直接写到 HTTP response body 里）。这个 Controller 的方法就是需要关注的地方，它返回 JSON 数据到客户端。另外，这个方法返回的 Object 可以是任何实体类型，在响应到客户端之前都会进行 JSON 格式转换。

2.2.4 静态资源的映射

静态资源需要直接映射，可以在 WebMvcConfigurer 中直接重写 addResourceHandlers 方法来实现。我们以访问 jquery.js 这个静态资源为例来配置静态资源映射，具体步骤如下。

步骤 01 在 main 目录下创建 assets，并将 jquery.js 放到这个目录下。将 jQuery 的 js 文件复制到项目目录的位置，如图 2-3 所示。

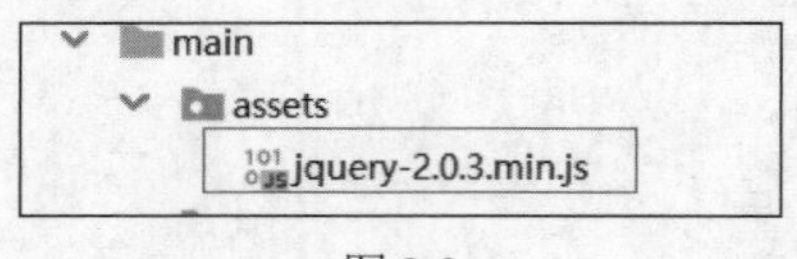

图 2-3

步骤 02 重写 addResourceHandlers 方法，添加静态资源。通过在 MvcConfig 类中重写 addResourceHandlers 方法来定义静态资源映射，具体代码如下。

示例代码 2-12 MvcConfig.java（部分代码）

```
    /**
    注解静态资源解析
    */
    @Override
    public void addResourceHandlers(ResourceHandlerRegistry registry)
{   System.err.println("注册静态资源......");
   //添加一些静态资源解析
   registry.addResourceHandler("*.js").addResourceLocations("classpath:/assets
/"); //所有在这个目录下的资源用于保存某些静态的资源
   registry.addResourceHandler("/assets/**").addResourceLocations("classpath:/
assets/");
    }
```

步骤 03 在页面上引入 js，有以下两种方式：

```
<script type="text/javascript"
src="${pageContext.request.contextPath}/jquery-2.1.4.js"></script>
```

或：

```
<script type="text/javascript"
src="${pageContext.request.contextPath}/assets/jquery-2.1.4.js"></script>
```

说明：

- @EnableWebMvc 开启 Spring MVC 的支持，如果没有此句，那么重写 WebMvcConfigurerAdapter 方法将无效。
- 继承 WebMvcConfigurerAdapter 类，重写其方法，可以对 Spring MVC 进行配置。
- addResourceLocations 是指文件放置的目录，addResourceHandler 是指对外暴露的访问目录。
- 可以通过配置 addMapping(/)将所有请求交由 DispatcherServlet 来处理，此时/会替换容器默认的 Servlet 映射，会导致静态资源无法访问，但 JSP 还可以正常访问，与/*不同。

在配置相同的情况下，DispatcherServlet 配置成/和/*的区别如下：

- /：使用/配置路径，直接访问到 JSP，不经 springDispatcherServlet。
- /*：配置/*路径，不能访问到多视图的 JSP。

例如，对于客户端调用 URL（/user/list）然后返回 user.jsp 视图这种情况，使用/和/*的情况如下：

- /:DispatcherServlet拿到这个请求后返回对应的Controller,然后依据Dispatcher Type为Forward类型转发到 user.jsp 视图，即请求 user.jsp 视图（/user/user.jsp），此时 Dispatcher 没有拦截/user/user.jsp，而是顺利地交给 ModleAndView 去处理显示。
- /*: DispatcherServlet 拿到这个请求后返回对应的 Controller，然后通过 Dispatcher Type 转发到 user.jsp 视图，即请求 user.jsp 视图（/user/user.jsp），此时 Dispatcher 已经拦截/user/user.jsp，会把它当作 Controller 去匹配，没有匹配到就会报 404 错误。

如果让 DispatcherServlet 拦截所有请求，同时不拦截后缀名为.jsp 的请求，就可以通过下面的代码来配置响应所有请求：

```
    Dynamic servlet = servletContext.addServlet("dispatcher", new
DispatcherServlet(ctx));
    servlet.addMapping("/");
    servlet.setLoadOnStartup(1);
```

可以添加一个统一配置来处理所有没有具体配置的请求：

```
    @RequestMapping("/**")
    public void forward(){
        System.err.println("forward....");
    }
```

2.2.5 拦截器的配置

可以让一个普通的 Bean 实现 HandlerInterceptor 接口或是继承 HandlerInterceptorAdapter 来实现自定义拦截器，然后通过重写 WebMvcConfigurerAdapter 的 addInterceptors 方法来注册自定义的拦截器。具体定义和使用拦截器的步骤如下。

步骤 01 在 Spring 中通过实现 HandlerInterceptor 或者继承适配器 HandlerInterceptorAdapter 开发一个拦截器，代码如下。

示例代码 2-13 OneInterceptor.java

```
package cn.springmvc.interceptor;
import javax.servlet.http.HttpServletRequest;
import javax.servlet.http.HttpServletResponse;
import org.springframework.web.servlet.ModelAndView;
import org.springframework.web.servlet.handler.HandlerInterceptorAdapter;
/**
开发一个拦截器
@author mrchi
@version 1.0
*/
public class OneInterceptor extends HandlerInterceptorAdapter {
    @Override
    public boolean preHandle(HttpServletRequest request, HttpServletResponse
response, Object handler)throws Exception {
        System.err.println("uri is:"+request.getRequestURI());
        return true;
    }
    @Override
    public void postHandle(HttpServletRequest request, HttpServletResponse
response, Object handler,ModelAndView modelAndView) throws Exception {
        System.err.println("after ..."+request.getRequestURI()+
",handler:"+handler);
    }
}
```

拦截器 HandlerInterceptAdapter 类中有三个方法 preHandle、postHandle、afterCompletion，具体说明如下：

（1）preHandle

- 调用时间：Controller 方法处理之前。
- 执行顺序：链式 Intercepter 情况下，Intercepter 按照声明的顺序一个接一个执行。
- 注意事项：若返回 false，则中断执行，不会进入 afterCompletion。

（2）postHandle

- 调用前提：preHandle 返回 true。

- 调用时间：Controller 方法处理完之后，DispatcherServlet 进行视图的渲染之前，也就是说在这个方法中你可以对 ModelAndView 进行操作。
- 执行顺序：链式 Intercepter 情况下，Intercepter 按照声明的顺序倒着执行。

（3）afterCompletion

- 调用前提：preHandle 返回 true。
- 调用时间：DispatcherServlet 进行视图的渲染之后。
- 注意事项：多用于清理资源。

步骤 02 在 WebMvcConfigurereAdapter 的 addInterceptors 中注册这个拦截器。addInterceptors 方法用于注册拦截器，将自定义的拦截器通过在 MvcConfig 类中重写 addInterceptors 方法来进行。完整的 MvcConfig 代码如下。

示例代码 2-14 MvcConfig.java

```
    @Configuration
    @EnableWebMvc
    @ComponentScan("cn.springmvc")
    public class MvcConfig extends WebMvcConfigurerAdapter {
        @Bean
        public InternalResourceViewResolver viewResolver() {
            System.err.println("init....");
            InternalResourceViewResolver viewResolver = new
InternalResourceViewResolver();
            viewResolver.setPrefix("/WEB-INF/classes/views/");
            viewResolver.setSuffix(".jsp");
            viewResolver.setViewClass(JstlView.class);
            return viewResolver;
        }
        /*
        注册 messageConvert */
        @Override
        public void configureMessageConverters(List<HttpMessageConverter<?>>
converters)
        {
            FastJsonHttpMessageConverter4 json = new
FastJsonHttpMessageConverter4();
            json.setDefaultCharset(Charset.forName("UTF-8"));
            json.setSupportedMediaTypes(Arrays.asList(MediaType.
APPLICATION_JSON_UTF8));
            converters.add(json);
            converters.add(new StringHttpMessageConverter(Charset.
forName("UTF-8")));
        }
        /**
        注解静态资源解析
        */
        @Override
        public void addResourceHandlers(ResourceHandlerRegistry registry) {
```

```
        System.err.println("注册静态资源......");
        //添加一些静态资源解析
        registry.addResourceHandler("*.js").addResourceLocations
("classpath:/assets/");
        //所有在这个目录下的资源，用于保存某些静态资源
        registry.addResourceHandler("/assets/**").addResourceLocations
("classpath:/assets/");
    }
    //添加拦截器
    @Override
    public void addInterceptors(InterceptorRegistry registry) {
        System.err.println("添加拦截器......");
        registry.addInterceptor(new OneInterceptor())//
        .addPathPatterns("/**")// 配置要求拦截的 URL
        .excludePathPatterns("/**.js");// 配置不需要拦截的 URL
    }
}
```

2.2.6 Spring MVC 文件上传

文件上传是表现层常见的需求。在 Spring MVC 中，底层使用 Apache 的 Commons FileUpload 工具来完成文件上传，对其进行封装，让开发者使用起来更加方便。

步骤 01 导入 common-fileupload 包。

示例代码 2-15 pom.xml（部分代码）

```
<!-- commons-fileUpload -->
<dependency>
   <groupId>commons-fileupload</groupId>
   <artifactId>commons-fileupload</artifactId>
   <version>1.3.1</version>
</dependency>
```

步骤 02 配置文件解析器。在 springmvc.xml 配置文件中加入文件上传解析器来支持文件上传，具体代码如下。

示例代码 2-16 springmvc.xml（部分代码）

```
<!-- 配置文件上传解析器
 注意：必须配置 id，且名称必须为 multipartResolver
-->
<bean id="multipartResolver"
class="org.springframework.web.multipart.commons.CommonsMultipartResolver">
   <!-- 配置限制文件上传大小 (字节为单位)-->
   <property name="maxUploadSize" value="1024000"/>
</bean>
```

注意：

- 必须配置 CommonsMultipartResolver 解析器。

- 该解析器的 id 必须叫 multipartResolver，否则无法成功接收文件。
- 可以通过 maxUploadSize 属性限制文件上传大小。

步骤 03　设计文件上传表单。

示例代码 2-17　fileupload.jsp

```
<%@ page contentType="text/html;charset=UTF-8" language="java" %>
<html>
<head>
    <title>文件上传</title>
</head>
<body>
<h3>SpringMVC 方式文件上传</h3>
<form action="/upload" method="post" enctype="multipart/form-data">
    选择文件：<input type="file" name="imgFile"> <br/>
    文件描述：<input type="text" name="memo"> <br/>
    <input type="submit" value="上传">
</form>
</body>
</html>
```

上传表单需要注意以下几点：

- 表单的 enctype 必须改为 multipart/form-data。
- 表单提交方式必须为 POST，不能是 GET。

步骤 04　编写控制器接收文件及参数。编写 UploadController 接收用户请求，文件会被传递到方法的 MultipartFile 类型参数中，并将参数对应的文件对象进行保存。具体代码如下所示。

示例代码 2-18　UploadController.java

```
/**
 * 演示 Spring MVC 文件上传
 * */
@Controller
public class UploadController {
    /**
     * 接收文件
     */
    @RequestMapping("/upload")
    public String upload(HttpServletRequest request, MultipartFile
imgFile,String memo){
        //1.获取网站 upload 目录的路径：ServletContext 对象
        String upload =
request.getSession().getServletContext().getRealPath("/upload");
        //判断该目录是否存在，不存在就自己创建
        File uploadFile = new File(upload);
        if(!uploadFile.exists()){
            uploadFile.mkdir();
        }
```

```
        //把文件保存到 upload 目录
        //2.生成随机文件名称
        //2.1 原来的文件名
        String oldName = imgFile.getOriginalFilename();
        //2.2 随机生成文件名
        String uuid = UUID.randomUUID().toString();
        //2.3 获取文件后缀
        String extName = oldName.substring(oldName.lastIndexOf(".")); //.jpg
        //2.4 最终的文件名
        String fileName = uuid+extName;
        //3.保存
        try {
            imgFile.transferTo(new File(upload+"/"+fileName));
        } catch (IOException e) {
            e.printStackTrace();
        }
        System.out.println("文件描述: "+memo);
        return "success";
    }
}
```

注 意

这里使用 MultipartFile 对象接收文件，并把文件存放在项目的 upload 目录下，同时还接收了普通参数。

步骤 05 运行测试。文件上传界面如图 2-4 所示。

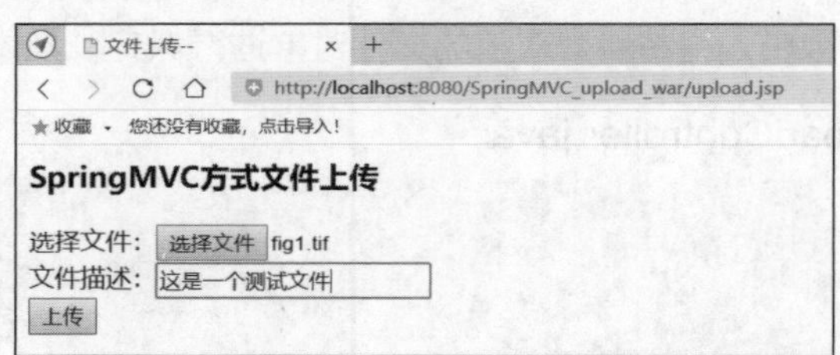

图 2-4

检查项目的 Tomcat 服务器的 upload 目录中是否有文件（文件已成功上传并重命名），结果如图 2-5 所示。

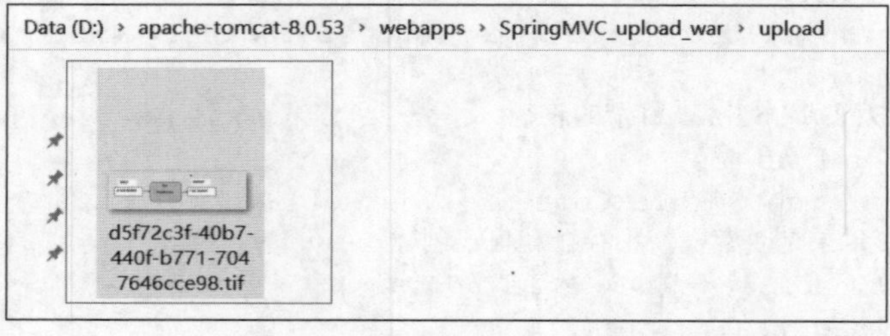

图 2-5

观察控制台输出参数内容，如图 2-6 所示。

```
Output
19-Dec-2020 23:06:48.150 信息 [RMI TCP Connection(3)-127.0.0.1] org.spring
19-Dec-2020 23:06:48.783 信息 [RMI TCP Connection(3)-127.0.0.1] org.spring
19-Dec-2020 23:06:49.103 信息 [RMI TCP Connection(3)-127.0.0.1] org.spring
19-Dec-2020 23:06:49.153 信息 [RMI TCP Connection(3)-127.0.0.1] org.spring
19-Dec-2020 23:06:49.205 信息 [RMI TCP Connection(3)-127.0.0.1] org.spring
19-Dec-2020 23:06:49.273 信息 [RMI TCP Connection(3)-127.0.0.1] org.spring
[2020-12-19 11:06:49,286] Artifact SpringMVC_upload:war: Artifact is deplo
[2020-12-19 11:06:49,286] Artifact SpringMVC_upload:war: Deploy took 2,498
19-Dec-2020 23:06:56.600 信息 [localhost-startStop-1] org.apache.catalina.
19-Dec-2020 23:06:56.636 信息 [localhost-startStop-1] org.apache.catalina.
文件描述：这是一个测试文件
```

图 2-6

第 3 章

Spring Boot 入门

本章将讲解 Spring Boot 的由来、特性和优点，如何配置 Java、Maven 环境，以及使用集成开发环境来进行 Spring Boot 应用的开发，并演示通过不同的方式来创建一个 Spring Boot 应用程序。

3.1 Spring Boot 简介和特性

3.1.1 Spring Boot 简介

Spring 框架自 2003 年发展至今，已经成为事实上的 Java EE 开发标准框架。使用 Spring 可以让简单的 JavaBean 实现之前只有 EJB 才能完成的事情，但是 Spring 不仅仅局限于服务器端开发，任何 Java 应用都能在简单性、可测试性和松耦合性等方面获益。目前的 Spring 框架已经发展成为一个无所不包的“全家桶”。

如果使用 Spring 框架进行过与其他框架的整合，比如较为常见的 SSH 和 SSM 框架，可以回想一下如何创建一个 Spring 应用。以搭建一个 Spring、Spring MVC、MyBatis 框架为例，每一种框架都需要各种配置文件或注解，互相之间的整合也需要配置文件。另外，可能还需要使用 Maven 导入许多依赖、开发测试程序、手动将项目打成 war 包部署到 Servlet 容器上等，相当烦琐。实际上 Spring Boot 就是来简化这些步骤的，它采用约定大于配置、去繁就简的原则，帮助我们快速创建一个产品级别的 Spring 应用，简化 J2EE 开发。针对很多 Spring 应用程序常见的应用功能，Spring Boot 能自动提供相关配置，而且 Spring Boot 本身也整合了许多优秀的框架。可以这样理解，Spring Boot 就像一扇门，打开它，就能看到里面是 Java EE 技术堆栈这座大山。

Spring Boot 是由 Pivotal 团队提供的全新框架，其设计目的是用来简化新 Spring 应用的初始搭建以及开发过程。该框架使用了特定的方式来进行配置，从而使开发人员不再需要定义样板化的配置。采用 Spring Boot 可以大大简化开发模式，所有你想集成的常用框架，它都有对应的组件支持。

Spring Boot 基于 Spring 开发，Spring Boot 本身并不提供 Spring 框架的核心特性以及扩展功能，只是用于快速、敏捷地开发新一代基于 Spring 框架的应用程序。也就是说，它并不是用来替代 Spring 的解决方案，而是和 Spring 框架紧密结合、用于提升 Spring 开发者体验的工具。同时它集成了大量常用的第三方库配置（例如 Redis、MongoDB、JPA、RabbitMQ、Quartz 等），Spring Boot 应用中这些第三方库几乎可以零配置的开箱即用，大部分的 Spring Boot 应用都只需要非常少量的配置代码，开发者能够更加专注于业务逻辑。

Spring Boot 一经推出就受到开源社区的追捧，Spring Boot 官方提供了很多 Starters 方便集成第三方产品，很多主流的框架也纷纷进行了主动的集成，比如 MyBatis。Spring 官方非常重视 Spring Boot 的发展，在 Spring 官网首页进行重点推荐介绍，是目前 Spring 官方重点发展的项目之一。

3.1.2 Spring Boot 的特性和优点

随着 Spring 不断发展，涉及的领域越来越多，项目整合开发需要配合各种各样的文件，慢慢变得不那么易用简单，违背了最初的理念，甚至称为配置地狱。Spring Boot 正是在这样的背景下被抽象出来的，目的是为了让大家更容易使用 Spring 、更容易集成各种常用的中间件、开源软件。另外，Spring Boot 诞生时，微服务概念正在慢慢酝酿中，Spring Boot 的研发融合了微服务架构的理念，实现了在 Java 领域内微服务架构落地的技术支撑。

Spring Boot 作为一套全新的框架，来源于 Spring 大家族，因此 Spring 所具备的功能它都有，而且更容易使用；Spring Boot 以约定大于配置的核心思想帮我们进行了很多设置。多数 Spring Boot 应用只需要很少的 Spring 配置。Spring Boot 开发了很多应用集成包，支持绝大多数开源软件，让我们以很低的成本去集成其他主流开源软件。

Spring Boot 的主要特性如下：

- 使用 Spring 项目引导页面可以在几秒内构建一个项目。
- 方便对外输出各种形式的服务，如 REST API、WebSocket、Web、Streaming、Tasks。
- 非常简洁的安全策略集成。
- 支持关系数据库和非关系数据库。
- 支持运行期内嵌容器，如 Tomcat、Jetty。
- 强大的开发包，支持热启动。
- 自动管理依赖。
- 自带应用监控。
- 支持各种 IDE，如 IntelliJ IDEA、NetBeans。

Spring Boot 的这些特性非常便于快速构建独立的微服务，所以我们使用 Spring Boot 开发项目。如果你使用 Spring Boot 开发过项目，就会被它的简洁高效特性所吸引。

使用 Spring Boot 可以给开发工作带来以下几方面的改进：

- Spring Boot 使编码变简单。Spring Boot 提供了丰富的解决方案，快速集成各种解决方案提升开发效率。
- Spring Boot 使配置变简单。Spring Boot 提供了丰富的 Starters，集成主流开源产品往往只需要简单的配置即可。

- Spring Boot使部署变简单。Spring Boot本身内嵌启动容器，仅仅需要一个命令即可启动项目，结合Jenkins、Docker自动化运维非常容易实现。
- Spring Boot使监控变简单。Spring Boot自带监控组件，使用Actuator轻松监控服务各项状态。

3.2 开发环境配置

Spring Boot 2.x基于Spring Framework 5.x版本进行开发。Spring Framework 5.x对Java最低版本的要求是Java 8。同时，Spring Boot采用了模块化设计，模块化依赖的管理和构建依赖于Apache Maven工具，官方说明要求Apache Maven 3.2及更高版本。

本节将主要介绍Java、Maven运行环境的安装与配置，以及集成开发环境（IDE）的安装和配置。如果你已经熟悉相关的环境安装与配置，可以直接跳过本节内容。

3.2.1 Java环境安装与配置

前面介绍了Spring Boot 2.x要求的最低版本是Java 8，虽然Spring Boot 2.x已经支持到Java 13，但是市面上很多项目依然选择使用Java 8。因为Java 8是一个非常经典的里程碑版本，在进入Java 9以后，Java版本的更新迭代频率加快，版本特性发生了巨大的改变。为了避免引入不必要的麻烦和学习成本，我们依然选择使用Java 8。

如果已经安装JDK，就自行确认版本。如果没有安装，可以按照下面介绍的方法进行安装。

笔者当前使用的操作系统是Windows 10 64位版本，按Win + R快捷键打开运行窗口，输入“cmd”后按确定键打开控制台，输入命令“java -version”确认是否安装Java以及安装的版本。

示例代码 3-1　查看 Java 版本信息

```
C:\Users\Administrator>java -version
java version "1.8.0_201"
Java(TM) SE Runtime Environment (build 1.8.0_201-b09)
Java HotSpot(TM) 64-Bit Server VM (build 25.201-b09, mixed mode)
```

可以看到，笔者机器安装的JDK版本为1.8.0_201，其中Update小版本号为201，如果你安装的是其他小版本号（比如 1.8.0_261），那也是可以使用的，不会造成太大的影响，无须担心小版本差异带来的差别影响。

如果尚未安装 JDK 8，那么可以前往 Oracle 官方网站下载最新的 JDK：https://www.oracle.com/java/technologies/javase-downloads.html。找到Java SE 8，进行下载，单击右侧Oracle JDK下方的JDK Download进行下载，如图3-1所示。

在新打开的页面中选择适合自己操作系统版本的安装包或压缩包进行下载。这里需要注意的是，不同的操作系统的安装方式有所不同，在Windows系统中，选择32位/64位exe安装包下载后直接双击安装即可；在Linux系统中，可选择RPM安装包或者免安装版的压缩包（文件后缀为.tar.gz，注意此版本需要额外配置环境变量）；在Mac OS系统中，选择.dmg文件进行安装。具体安装方式此处不再分别介绍。

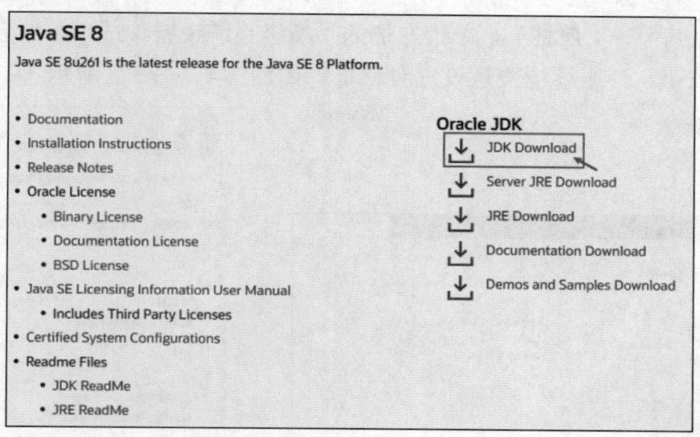

图 3-1

以安装包的方式安装完成之后，不需要再额外单独配置 Java 环境变量，因为在安装过程中默认为你配置好了，直接就可以在控制台验证安装结果了。

上面介绍安装的 JDK 是 Oracle 官方提供的版本，是采用 JRL 协议的，由于 Oracle 的要求，只允许个人研究学习使用，在后续的部分 Update 版本是需要付费获得的，如果想要获得更新支持（比如出现安全漏洞之类的）又没法升级 JDK 版本，就只好选择付费订阅获得更新支持了。有些人有所顾虑或者想要持续获得更新，可以选择 Open JDK。

Open JDK 是 Oracle JDK 的开源版本，两者略微有所不同。Open JDK 采用 GPL 协议，允许商业使用，如果你需要了解 Oracle JDK 和 AdoptOpenJDK 之间的区别以及如何从 Oracle JDK 迁移至 OpenJDK，可以参阅 OpenJDK 官方迁移指南：https://adoptopenjdk.net/migration.html。个人学习研究使用 Open JDK，基本上和 Oracle JDK 没有什么差别。

Open JDK 8 的下载地址为 https://adoptopenjdk.net。选择 Open JDK 8（LTS）+ Hotspot，单击“其他平台”进入下载页面，如图 3-2 所示。

在新打开的页面中找到符合自己本地机器操作系统版本的 JDK 进行下载即可。笔者这里选择的是 Windows x64 压缩版本（.zip 格式，如图 3-3 所示），而不是 MSI 格式的安装包，因为 MSI 格式的安装文件会自动为你配置环境变量，而不需要自己单独配置。作为开发人员，学会自行配置 JDK 环境变量是一个非常必要的基本技能，接下来将会介绍如何进行环境变量的配置。

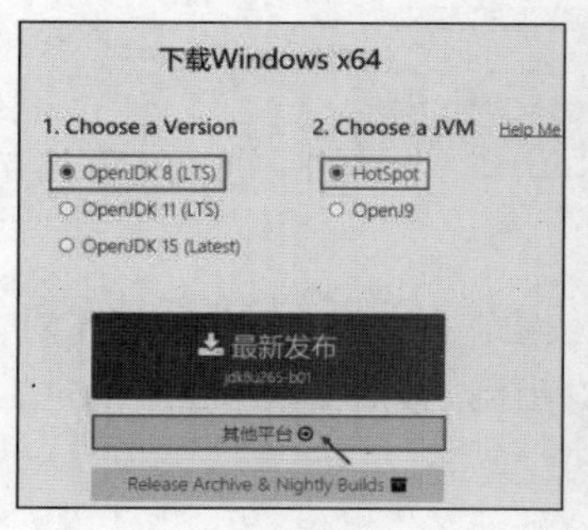

图 3-2

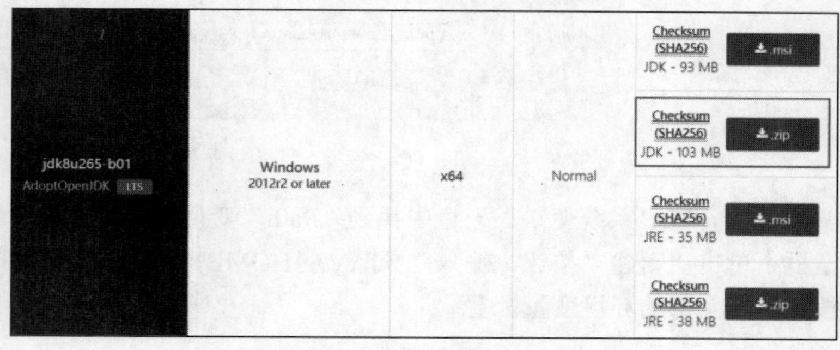

图 3-3

下载完成后，解压到自己的工作目录，这里选择解压到 D:\Develop\Java 目录下。在 Windows

系统中，右击“此电脑”→“属性”，单击左侧的“高级系统设置”选项，然后在弹出的对话框中单击“环境变量”按钮，弹出环境变量设置对话框，如图 3-4、图 3-5 所示。

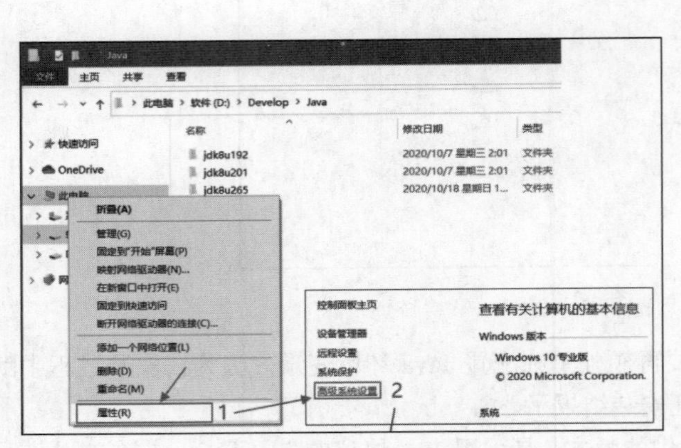

图 3-4

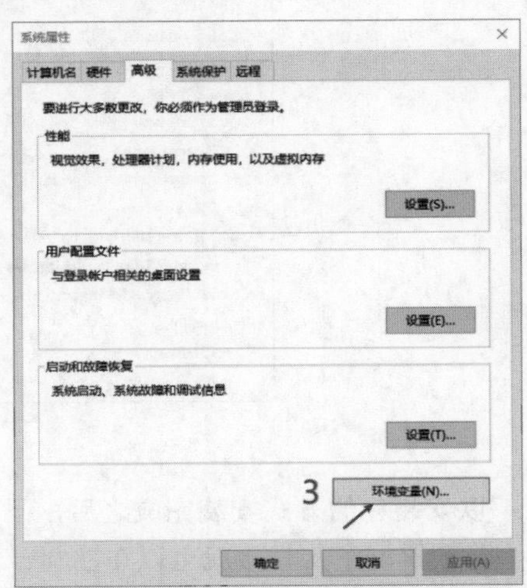

图 3-5

然后在系统变量一栏中单击“新建”按钮，在“变量名”文本框中输入“JAVA_HOME”，在“变量值”文本框中输入“D:\Develop\Java\jdk8u265”（这里输入 JDK 解压后所在的目录即可），单击“确定”按钮，如图 3-6 所示。接着新建变量名为“CLASS_PATH”、变量值为“.;%JAVA_HOME%\lib;%JAVA_HOME%\lib\tools.jar”的变量，单击“确定”按钮，如图 3-7 所示。

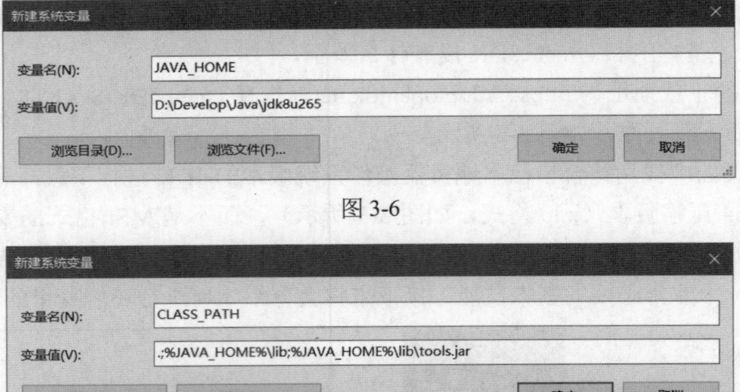

图 3-6

图 3-7

最后，在“系统变量”一栏中找到“Path”变量，单击“编辑”按钮，在“编辑环境变量”对话框中单击“新建“按钮，输入“%JAVA_HOME%\bin”，如图 3-8 所示，最后依次单击“确定”按钮完成全部的环境变量配置。

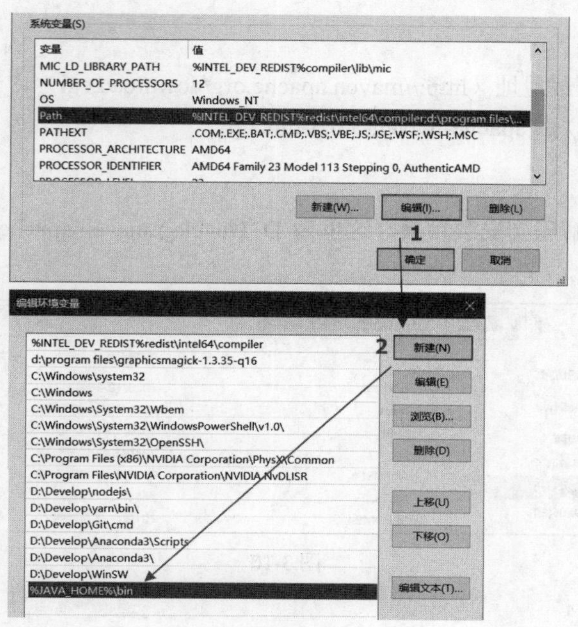

图 3-8

在图 3-8 所示的第 2 步操作中，Windows 某些版本（比如 Windows 7）的系统可能不是这样的展示形式，而是像图 3-9 所示的那样，需要在编辑框最后（或者最前）输入“;%JAVA_HOME%\bin;”。此处一定要注意：Windows 使用英文分号将不同的变量隔开，Linux 使用“:”分隔不同的变量值。

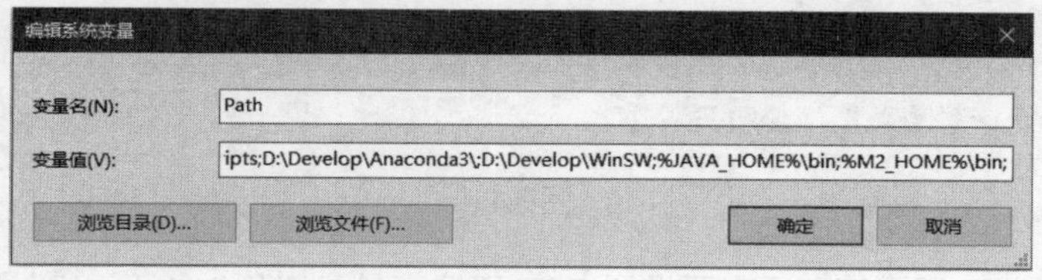

图 3-9

在配置完成环境变量之后，打开控制台，输入“java -version”命令验证是否配置正确。出现下述信息时说明配置成功，否则就需要根据上述步骤重新进行配置。

示例代码 3-2　查看 Java 版本信息

```
C:\Users\Administrator>java -version
openjdk version "1.8.0_265"
OpenJDK Runtime Environment (AdoptOpenJDK)(build 1.8.0_265-b01)
OpenJDK 64-Bit Server VM (AdoptOpenJDK)(build 25.265-b01, mixed mode)
```

3.2.2　Maven 环境安装与配置

Maven 是 Apache 下的一个开源项目，是一个创新的项目管理工具，用于对 Java 项目进行项目构建、依赖管理及项目信息管理。具体关于 Maven 相关的概念和技术知识请读者自行查阅相关图书和资料，这里不再赘述。下面将详细说明安装和配置 Maven 的步骤。

1. 下载 Maven

打开 Maven 官方下载地址（http://maven.apache.org/download.cgi），选择下载二进制归档 zip 文件（Binary zip archive）：apache-maven-3.6.3-bin.zip。

2. 安装 Maven

将下载的 zip 文件解压到安装目录，这里为 D:\Develop\maven\apache-maven-3.6.3，解压后的文件列表如图 3-10 所示。

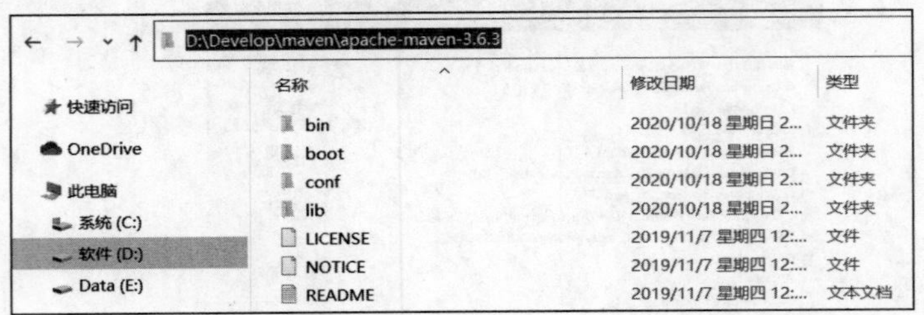

图 3-10

3. 配置环境变量

根据官方推荐的方式设置 M2_HOME 环境变量，类似于 JAVA_HOME 环境变量的设置。“变量值”为第 2 步中解压后的安装目录，即 D:\Develop\maven\apache-maven-3.6.3，如图 3-11 所示。

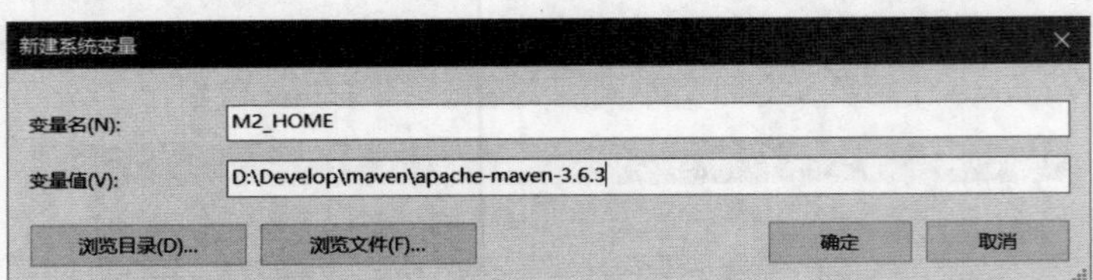

图 3-11

只设置 M2_HOME 还不够，还需要设置 Path 环境变量。打开 Path 环境变量编辑窗口，选择新建一行，输入“%M2_HOME%\bin”，依次单击“确定”按钮完成，如图 3-12 所示。

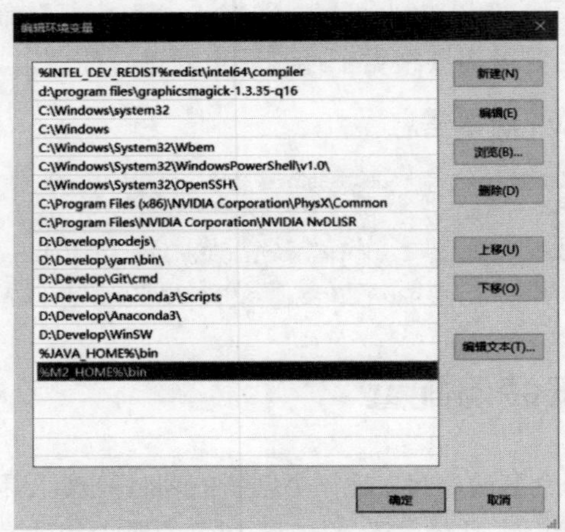

图 3-12

打开控制台，输入“mvn -version”命令检查是否生效。

示例代码 3-3　查看 Maven 版本信息

```
C:\Users\Administrator>mvn -version
Apache Maven 3.6.3 (cecedd343002696d0abb50b32b541b8a6ba2883f)
Maven home: D:\Develop\maven\apache-maven-3.6.3\bin\..
Java version: 1.8.0_265, vendor: AdoptOpenJDK,
runtime: D:\Develop\Java\jdk8u265\jre
Default locale: zh_CN, platform encoding: GBK
OS name: "windows 10", version: "10.0", arch: "amd64", family: "windows"
```

如果控制台输出以上信息，就说明环境变量配置成功了。

4. 修改配置文件

在配置完环境变量之后，还需要修改本地 Maven 仓库的路径和远程 Maven 仓库的地址。Maven 从远程仓库下载依赖包后会存储在本地仓库中。因此需要手动修改配置文件 settings.xml，在 Maven 安装目录下（Windows 中是“%M2_HOME%\conf”，Linux 中是“$M2_HOME$/conf”），这个目录下的 settings.xml 文件是作用于 Maven 全局配置的。如果想让修改的配置仅仅使当前用户生效，可以将此配置文件复制（或修改）到用户家目录下的（Windows 系统为“%USERPROFILE%\.m2”，Linux 系统为“~/.m2”）settings.xml 文件。

Maven 默认提供的官方镜像仓库地址是海外的，在国内下载速度非常慢，推荐使用阿里云提供的 Maven 镜像仓库，这样可以加快依赖包的下载速度和稳定性。

下面提供修改好的配置文件，其中本地 Maven 仓库路径是“D:/Develop/maven/repository”（可以根据自己的情况修改这个路径）。

示例代码 3-4　Maven 配置文件内容

```
<?xml version="1.0" encoding="UTF-8"?>
    <!-- 本地 Maven 仓库 -->
    <localRepository>D:/Develop/maven/repository</localRepository>
    <!-- 远程 Maven 镜像仓库地址 -->
    <mirrors>
        <mirror>
           <id>aliyun-repos</id>
           <name>aliyun repo</name>
           <url>http://maven.aliyun.com/nexus/content/groups/public/</url>
           <mirrorOf>central</mirrorOf>
        </mirror>
    </mirrors>
</settings>
```

上述配置仅仅修改了最常用的项。对于 settings.xml 更多的配置说明，可以参考官方的说明文档：http://maven.apache.org/settings.html。

3.2.3　安装集成开发环境

目前业界最常用的 Java IDE 工具是 Eclipse 和 IntelliJ IDEA。Eclipse 是开源免费的，任何人都可以使用。IntelliJ IDEA 提供了两个版本：社区版（Community）和旗舰版（Ultimate）。两者都可以作为 Spring Boot 应用程序开发的工具，读者可以根据自己的喜好选择。

1. 安装 Eclipse

Eclipse 是著名的跨平台的开源自由集成开发环境（IDE），最初主要用于 Java 语言开发，但是官方也推出了其他版本（或者是插件支持的方式），使其作为其他计算机语言（比如 C++和 Python）的开发工具。

Eclipse 附带了一个标准的插件集，包括 Java 开发工具（Java Development Kit，JDK），配合 Spring 官方提供的 STS 插件（Spring Tools Suite）可以十分便捷地进行 Spring Boot 应用程序的开发。

Eclipse 官方下载地址为 https://www.eclipse.org/downloads/。由于新版 Eclipse 已经全面支持 Java 11，对 JRE 的最低安装要求也是 11，因此建议大家根据自己的操作系统直接下载 Eclipse Installer 进行安装，如图 3-13 所示。在下载页面中，单击“Select Another Mirror”展开列表项，选择任意一个国内的下载源（默认是国外的下载源，速度会很慢）进行下载。具体安装步骤可以参考 https://www.eclipse.org/downloads/packages/installer，安装时选择“Eclipse IDE for Enterprise Java Developers”。

图 3-13

安装完成之后启动 Eclipse，会提示“Select a directory as workspace”，如图 3-14 所示，意思是选择一个自己的代码工程目录作为 Eclipse 的工作空间，之后所有的项目工程默认会保存在这个工作空间中。本示例选择将 D:\code\workspace 作为 Eclipse 的工作空间。

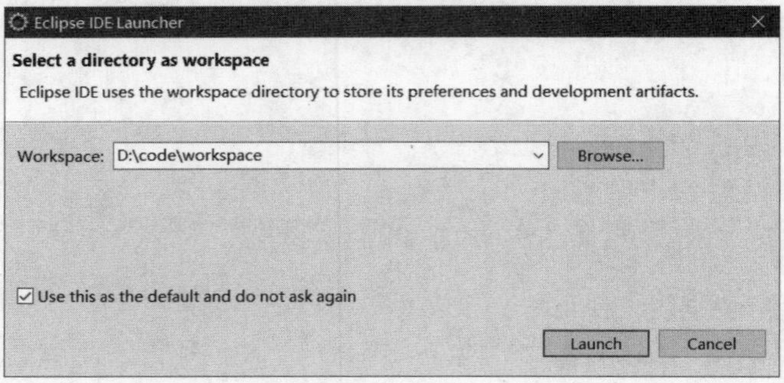

图 3-14

选择好工作空间后单击“Launch”按钮启动，Eclipse 启动后的界面如图 3-15 所示。至此 Eclipse 就算安装完成了。

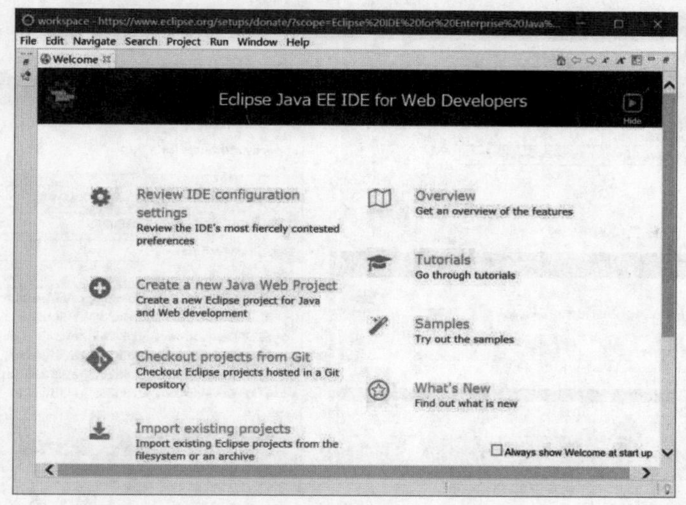

图 3-15

有一个问题要注意，如果下载的是离线安装版本（非 Installer），在运行 eclipse.exe 之后，如果看到如图 3-16 所示的提示，就说明我们安装的 Java 版本太低了，要求 JVM 至少不低于 11（JRE 1.11 版本），而我们安装的版本是 1.8，不符合要求。要解决这个问题，有以下三种方法：

- 下载并安装 JRE 11，然后运行 Eclipse。
- 下载 Eclipse Installer includes a JRE（在线安装版）。
- 选择旧版本的 Eclipse，满足最低版本是 Java 8 的即可。

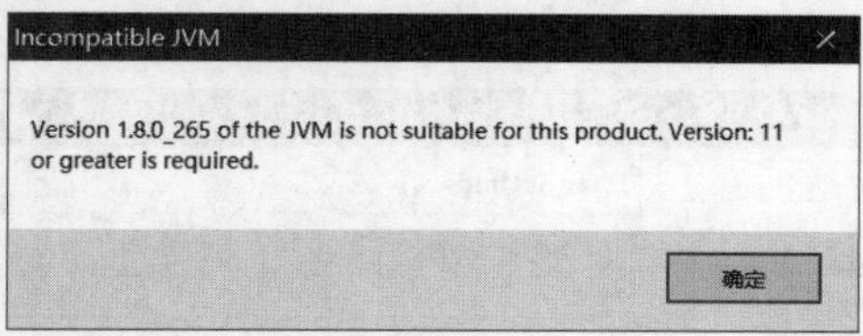

图 3-16

安装完成后还需要做一些额外的配置才可以进入项目的开发，主要是 JDK 环境和 Maven 的配置：

- JDK 配置：依次单击菜单栏“Windows”→“Preferences”，在弹出的对话框左侧列表中选择“Java”→“Installed JREs”，将其中默认的 jre 移除，单击“Add”按钮，在弹出的对话框中选择“Standard VM”，单击“Next”按钮，如图 3-17 所示。然后弹出“JRE Definition”配置框，在“JRE home”配置项右侧单击“Directory...”按钮，如图 3-18 所示。在文件对话框中选择 Java 安装目录（D:\Develop\Java\jdk8u265），然后填写“JRE name”，单击“Finish”按钮，再单击“Apply and Close”按钮，Eclipse 的 Java 环境就配置好了。

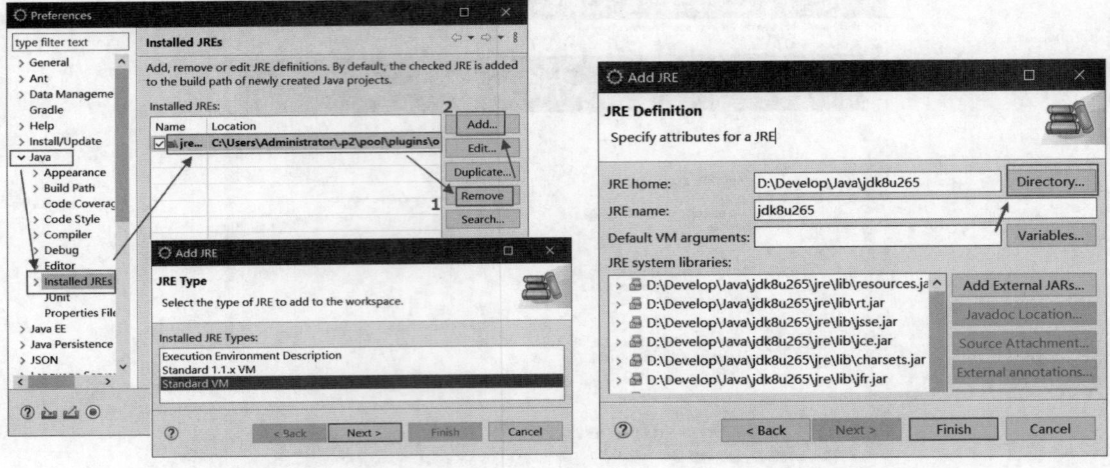

图 3-17　　　　　　　　　　　　　　图 3-18

- Maven 配置：Eclipse 中已经默认集成了 Maven，可以依次单击菜单栏“Windows”→“Preferences”，在弹出的对话框左侧列表中选择“Maven”→“Installations”，就可以看到默认集成的 Maven，其名字为“EMBEDDED”。如果不想使用它，可以添加自己安装的 Maven 版本，方法与添加 JDK 类似。这里主要强调的是修改 Maven 的配置文件，在“Maven”→“User Settings”配置项中（见图 3-19），由于我们在上一小节中已经在用户目录.m2 文件夹下放置了修改后的 settings.xml 文件，因此 Eclipse 会直接从中读取。同时，也可以自行指定配置文件，然后单击“Update Settings”按钮更新配置，最后记得单击“Apply and Close”按钮，否则设置不会生效。

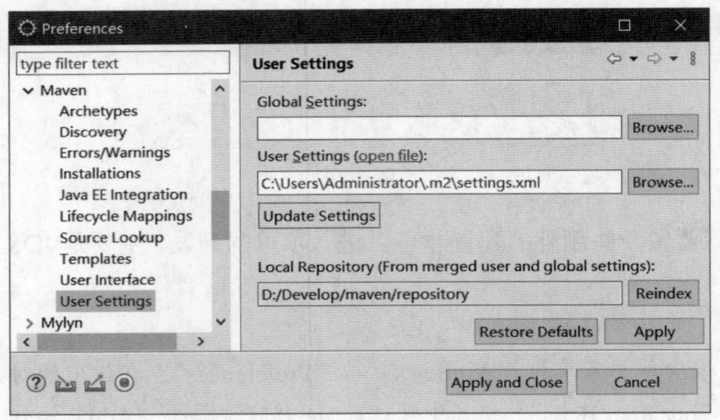

图 3-19

2. 安装 IntelliJ IDEA

在没有 Spring Boot 之前，通常会选择旗舰版做 Java Web 开发，因为旗舰版可以开发 Java Web 项目，比社区版插件更丰富、功能更强大，可以很方便地使用 Java Web 相关的工具进行开发，但是旗舰版是需要商业付费的。

因为 IDEA 授权问题，很多人会选择使用开源免费的 Eclipse 进行开发。不过笔者仍然推荐使用 IntelliJ IDEA，这是因为使用 IDEA 开发更高效。

由于 Spring Boot 内置 Tomcat 等服务器，配合相关插件，IDEA 社区版能够完全胜任 Spring Boot 的开发工作。所以，读者可以选择使用免费社区版来进行学习和开发，如果是学习或者参与开源项目，可以向 JetBrains 公司申请获得 IDEA 旗舰版的 License 授权，不过这仅限于个人学习/开发开源项目使用。

本书后续内容将使用免费社区版 IDEA 作为开发工具，后面不再对此特别说明。IDEA 下载地址为 https://www.jetbrains.com/idea/download，选择“Community”版本下载即可，如图 3-20 所示。

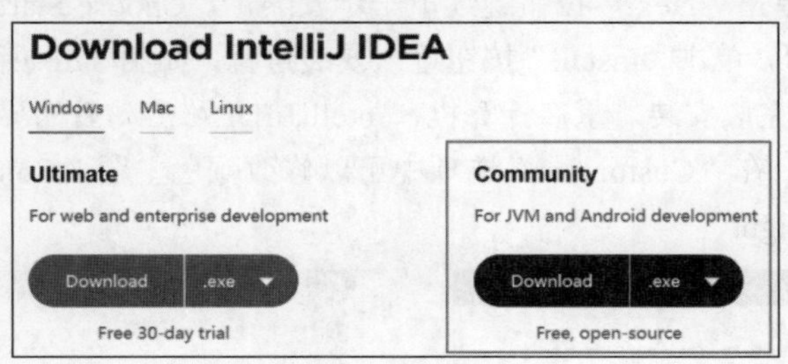

图 3-20

下载完成后单击安装文件进行安装，进入安装引导程序界面，如图 3-21 所示。

单击 Next 按钮，选择安装目录，默认安装在 C 盘，也可以选择安装在其他盘符（本示例的安装路径是“D:\Program Files\JetBrains\IntelliJ IDEA Community Edition 2020.3”），然后单击 Next 按钮，设置安装选项，如图 3-22 所示。

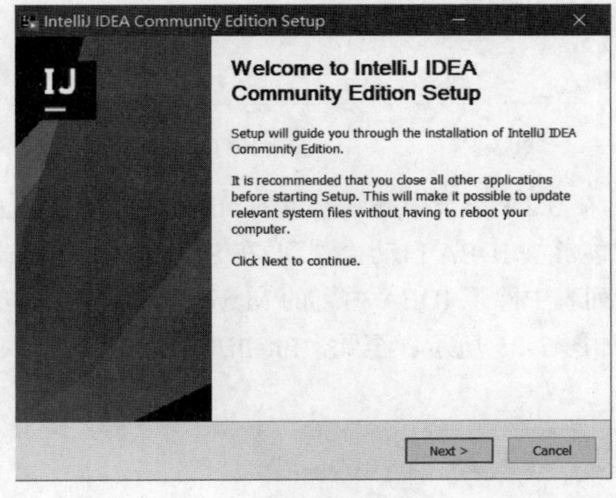

图 3-21

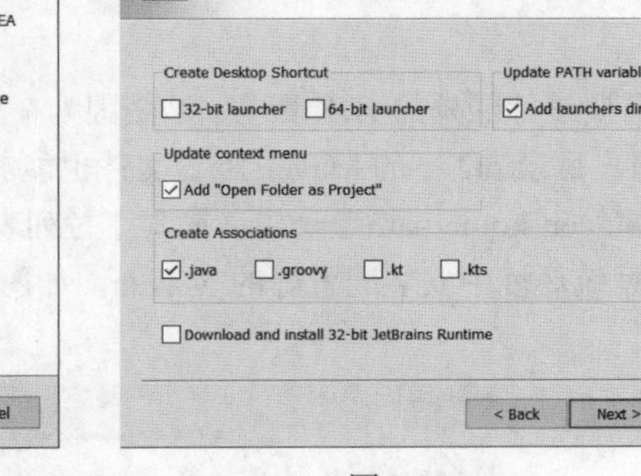

图 3-22

下面对其中的设置选项进行详细说明：

- “Create Desktop Shortcut”选项：可以在桌面创建快捷启动方式，有 32 位和 64 位可选，也可以不选（不在桌面创建快捷方式图标）。
- “Update context menu”选项：勾选“Add ‘Open Folder as Project’”后可以在文件夹中右键快捷菜单中增加“Open Folder as Project”选项，快速从文件夹浏览器所在目录使用 IntelliJ IDEA 打开一个工程。
- “Create Associations”选项：可以建立文件关联，勾选“.java”后，双击点击“.java”文件时会默认选择使用 IDEA 打开。

- “Update PATH variable”选项：勾选“Add launchers dir to the PATH”，会将 IDEA 程序文件所在的目录添加到环境变量，设置此选项后可以实现在控制台（终端）中输入 idea 或者 idea64，直接将终端所在目录作为工程目录打开 IntelliJ IDEA。
- “Download and install 32-bit JetBrains Runtime”选项：勾选后会下载并安装 JetBrains 的 32 位版 JRE，在 3.2.1 小节中我们已经安装配置好了 Java 运行环境，此项不必勾选。

设置好安装选项后单击“Next”按钮继续下一步安装，“Choose Start Menu Folder”是创建开始菜单的目录，默认即可，单击“Install”按钮进入安装进程，如图 3-23 所示。耐心等待安装完成，最后单击“Finish”按钮完成安装。重启计算机，IntelliJ IDEA 启动后的界面如图 3-24 所示，默认是黑色主题“Darclua”，在“Customize”选项中可以修改颜色主题（Color theme），笔者这里选择了亮色主题“IntelliJ Light”。

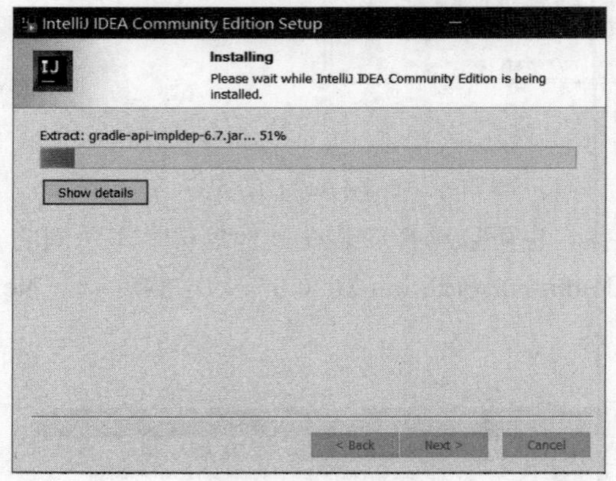

图 3-23

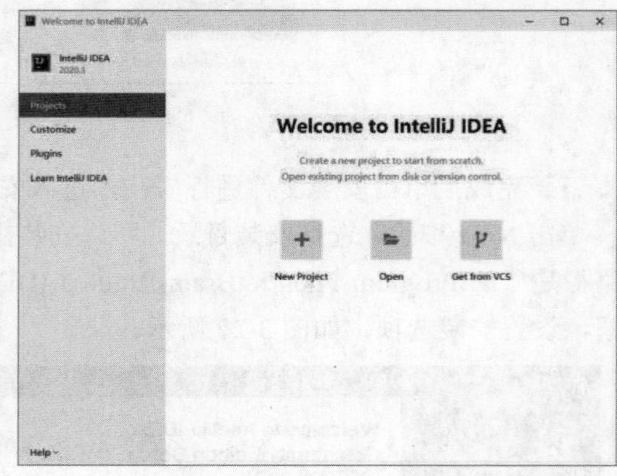

图 3-24

IDEA 不需要进行额外的 JDK 配置，这是因为在 3.2.1 小节中安装配置的 JDK 会被 IDEA 自动检测到，同样用户目录.m2 下的 Maven 配置文件也能够被 IDEA 自动检测到。在 Settings 中的 Maven 设置里面有“Maven home path”一栏，在其下拉列表中除了 IDEA 内置的 Maven 外，它同样自动检测到了我们系统环境中安装配置好的 Maven，如图 3-25 所示。至此，IntelliJ IDEA 开发工具就安装完成了。

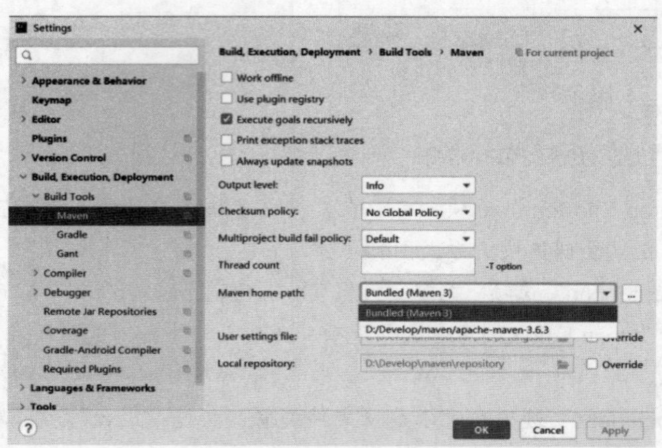

图 3-25

3.3 创建 Spring Boot 应用

本节将介绍4种创建 Spring Boot 应用程序的方法（“使用命令行方式创建”“使用图形化界面创建”“使用 Eclipse STS 插件创建”和“使用 IntelliJ IDEA 创建”），以及如何构建一个可执行 jar 包。

3.3.1 使用命令行方式创建

使用命令行方式创建适用于对传统的命令行方式交互比较熟悉的人群，这种方法比较适合于初学时掌握对 Spring Boot 工程创建步骤的理解，并非完全是为初学者准备的，而是假设读者已经非常熟悉 Java 和 Maven。使用这种方式创建，将不借助于其他第三方工具。

1. 使用 Maven Archetype 插件

我们可以借助 maven-archetype-plugin 插件的相关命令来完成项目的创建。执行之前，确保所在目录具有读写权限。本示例所在目录为 D:\code\spring-boot-example，然后在本目录下打开命令行窗口。在 Windows 环境下，可以按⊞+R（Win+R）快捷键弹出运行窗口，输入 cmd 后按回车键调出命令行窗口；在 Linux 环境下，可以使用 Ctrl+Alt+T 组合键打开终端；在 Mac OS 环境下，按组合键 Control+Space（空格键），输入“term”，按回车键即可调出终端。然后进入 D:\code\spring-boot-example 目录。

示例代码 3-5 Windows 控制台进入 D:\code\spring-boot-example 目录

```
Microsoft Windows [版本 10.0.19042.630]
(c) 2020 Microsoft Corporation. 保留所有权利。
C:\Users\Administrator>D:
D:\>cd D:\code\spring-boot-example
D:\code\spring-boot-example >
```

输入以下命令来创建一个工程目录。

示例代码 3-6 使用 Maven 命令创建工程目录

```
D:\code\spring-boot-example>mvn archetype:generate -DgroupId=com.example
-DartifactId=my-first-spring-boot-app -Dversion=0.0.1-SNAPSHOT
-DinteractiveMode=false -Dpackage=com.example
```

解读构建命令：

- mvn：Maven 命令。
- archetype：指定为使用 maven-archetype-plugin 插件，archetype 是其简称。
- generate：指定 archetype 插件的构建目标（Goal），详细介绍可以参考官方文档（http://maven.apache.org/archetype/maven-archetype-plugin/generate-mojo.html）。
- -D：指定插件参数。和 Java 启动命令一样，Maven 插件也通过-D 来传递所需参数。其中的

-DgroupId 参数是指定工程应用的组名（通常情况下是公司或者组织名的域名倒置，当然也可以不是）；-DartifactId 参数是指定项目的标识符，它和 groupId 一起共同决定了项目命名的唯一性；-Dversion 参数指定版本号；-DinteractiveMode=false 禁用交互模式；-Dpackage 参数指定项目的包名。运行命令后，你将会看到如图 3-26 所示的输出信息，说明工程已经创建成功，工程目录为 D:\code\spring-boot-example\my-first-spring-boot-app。

```
mvnw.cmd
[INFO] --- maven-archetype-plugin:3.2.0:generate (default-cli) @ standalone-pom ---
[INFO] Generating project in Batch mode
[      ] No archetype found in remote catalog. Defaulting to internal catalog
[INFO] No archetype defined. Using maven-archetype-quickstart (org.apache.maven.archetypes:maven-arch
etype-quickstart:1.0)
[INFO] ----------------------------------------------------------------------------
[INFO] Using following parameters for creating project from Old (1.x) Archetype: maven-archetype-quic
kstart:1.0
[INFO] ----------------------------------------------------------------------------
[INFO] Parameter: basedir, Value: D:\code\spring-boot-example
[INFO] Parameter: package, Value: com.example
[INFO] Parameter: groupId, Value: com.example
[INFO] Parameter: artifactId, Value: my-first-spring-boot-app
[INFO] Parameter: packageName, Value: com.example
[INFO] Parameter: version, Value: 0.0.1-SNAPSHOT
[INFO] project created from Old (1.x) Archetype in dir: D:\code\spring-boot-example\my-first-spring-b
oot-app
[INFO] ------------------------------------------------------------------------
[INFO] BUILD SUCCESS
[INFO] ------------------------------------------------------------------------
[INFO] Total time:  1.815 s
[INFO] Finished at: 2020-12-07T21:12:47+08:00
[INFO] ------------------------------------------------------------------------
```

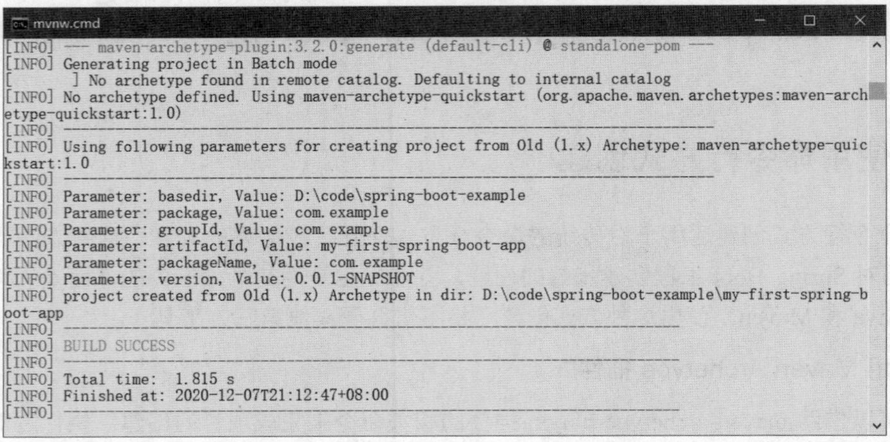

图 3-26

上述命令仅仅是创建了一个标准的 Maven 工程项目，还没有与 Spring Boot 相关的内容，因此还需要加以改造。

首先我们使用 tree 命令来查看该目录结构。

示例代码 3-7　使用 tree 命令查看 Maven 工程目录结构

```
D:\code\spring-boot-example>tree /F my-first-spring-boot-app
│  pom.xml
└─src
    ├─main
    │  └─java
    │      └─com
    │          └─example
    │                  App.java
    └─test
        └─java
            └─com
                └─example
                        AppTest.java
```

观察一下该目录，可以看到根目录下有一个 pom.xml，它是整个工程的主 POM 文件，默认情况下里面仅包含 junit（单元测试套件）的依赖。src 目录是工程的源码目录，一般所有编写的代码都放在此目录下，其中还区分成 main 和 test 两个目录，main 目录下主要存放我们编写的功能代码，而 test 目录存放测试代码，com/example 目录是工程源码目录的包名，存放编写的代码。App.java 和 AppTest.java 分别是程序运行的“主入口类”和“单元测试类”源代码。

接下来需要对 pom.xml 文件进行编辑，在其中增加 Spring Boot 相关的依赖，修改后的 pom.xml

内容如下。

示例代码 3-8 增加 Spring Boot 依赖后的 POM 文件内容

```
    <project xmlns="http://maven.apache.org/POM/4.0.0"
    xmlns:xsi="http://www.w3.org/2001/XMLSchema-instance"
     xsi:schemaLocation="http://maven.apache.org/POM/4.0.0
http://maven.apache.org/maven-v4_0_0.xsd">
        <modelVersion>4.0.0</modelVersion>
        <parent>
            <groupId>org.springframework.boot</groupId>
            <artifactId>spring-boot-starter-parent</artifactId>
            <version>2.3.6.RELEASE</version>
            <relativePath/> <!-- lookup parent from repository -->
        </parent>
        <groupId>com.example</groupId>
        <artifactId>my-first-spring-boot-app</artifactId>
        <packaging>jar</packaging>
        <version>0.0.1-SNAPSHOT</version>
        <name>我的第一个 Spring Boot 程序</name>
        <dependencies>
            <dependency>
                <groupId>org.springframework.boot</groupId>
                <artifactId>spring-boot-starter-web</artifactId>
            </dependency>
            <dependency>
                <groupId>org.springframework.boot</groupId>
                <artifactId>spring-boot-starter-test</artifactId>
                <scope>test</scope>
                <exclusions>
                    <exclusion>
                        <groupId>org.junit.vintage</groupId>
                        <artifactId>junit-vintage-engine</artifactId>
                    </exclusion>
                </exclusions>
            </dependency>
            <dependency>
                <groupId>junit</groupId>
                <artifactId>junit</artifactId>
                <version>3.8.1</version>
                <scope>test</scope>
            </dependency>
        </dependencies>
    </project>
```

其中我们添加了 spring-boot-starter-web、spring-boot-starter-test 两个依赖包，并增加了构建插件 spring-boot-maven-plugin，指定了父 POM 为 spring-boot-starter-parent（<parent>配置节点）。然后进入 my-first-spring-boot-app 目录，执行 mvn dependency:tree 命令来观察项目的依赖树发生了哪些变化。

示例代码 3-9 使用 mvn dependency:tree 命令查看项目依赖树

```
D:\code\spring-boot-example>cd my-first-spring-boot-app
D:\code\spring-boot-example\my-first-spring-boot-app>mvn dependency:tree
-Dinclude=org.springframework*
[INFO] Scanning for projects...
[INFO]
[INFO] ------------< com.example:my-first-spring-boot-app >------------
[INFO] Building 我的第一个 Spring Boot 程序 0.0.1-SNAPSHOT
[INFO] --------------------------------[ jar ]---------------------------------
（省略部分内容）
[INFO] com.example:my-first-spring-boot-app:jar:0.0.1-SNAPSHOT
[INFO] +- org.springframework.boot:spring-boot-starter-web:
jar:2.3.6.RELEASE:compile
[INFO] |  +- org.springframework.boot:spring-boot-starter:jar:
2.3.6.RELEASE:compile（省略部分内容）
[INFO] |  +- org.springframework.boot:spring-boot-starter-json:
jar:2.3.6.RELEASE:compile
（省略部分内容）[INFO] |  +- org.springframework.boot:spring-boot-starter-tomcat:
jar:2.3.6.RELEASE:compile
（省略部分内容）
[INFO] |  +- org.springframework.boot:spring-boot-starter-validation:
jar:2.3.6.RELEASE:compile
[INFO] |  |  +- jakarta.validation:jakarta.validation-api:jar:2.0.2:compile
（省略部分内容）[INFO] |  +-
org.springframework:spring-web:jar:5.2.5.RELEASE:compile
（省略部分内容）[INFO] +- org.springframework.boot:spring-boot-starter-test:jar:
2.3.6.RELEASE:test
[INFO] |  +- org.springframework.boot:spring-boot-test:jar:
2.3.6.RELEASE:test
[INFO] |  +- org.springframework.boot:spring-boot-test-autoconfigure:
jar:2.3.6.RELEASE:test
[INFO] |  +- com.jayway.jsonpath:json-path:jar:2.4.0:test
（省略部分内容）
[INFO] |  +- org.junit.jupiter:junit-jupiter:jar:5.5.2:test
（省略部分内容）
[INFO] |  +- org.mockito:mockito-junit-jupiter:jar:3.1.0:test
[INFO] |  +- org.assertj:assertj-core:jar:3.13.2:test
[INFO] |  +- org.hamcrest:hamcrest:jar:2.1:test
[INFO] |  +- org.mockito:mockito-core:jar:3.1.0:test
（省略部分内容）
[INFO] |  +- org.springframework:spring-core:jar:5.2.5.RELEASE:compile
[INFO] |  |  \- org.springframework:spring-jcl:jar:5.2.5.RELEASE:compile
[INFO] |  +- org.springframework:spring-test:jar:5.2.5.RELEASE:test
[INFO] |  \- org.xmlunit:xmlunit-core:jar:2.6.4:test
[INFO] \- junit:junit:jar:3.8.1:test
[INFO] ------------------------------------------------------------------------
[INFO] BUILD SUCCESS
[INFO] ------------------------------------------------------------------------
[INFO] Total time:  35.791 s
[INFO] Finished at: 2020-12-06T20:49:45+08:00
```

```
[INFO] ------------------------------------------------------------------------
```

通过控制台输出可以发现项目的依赖树发生了变化，增加了 spring-webmvc、tomcat、jackson、logging、mockito 等依赖。上面增加的依赖包将 Spring Boot 基础包都包含了进来，依赖添加完毕，下一步就可以进行 Sprint Boot 应用程序开发了。我们接下来将实现一个输出“Hello Spring Boot!”的简单示例。

找到之前看到的 App.java 文件，打开它可以看到如下内容。

示例代码 3-10　App.java 文件修改之前

```
package com.example;
/**
 * Hello world!
 */
public class App{
    public static void main( String[] args ) {
        System.out.println( "Hello World!" );
    }
}
```

将其内容修改如下。

示例代码 3-11　App.java 文件修改之后

```
package com.example;
import org.springframework.boot.SpringApplication;
import org.springframework.boot.autoconfigure.SpringBootApplication;
import org.springframework.web.bind.annotation.RequestMapping;
import org.springframework.web.bind.annotation.RestController;
/**
 * Hello Spring Boot!
 */
@RestController
@SpringBootApplication
public class App{
    @RequestMapping("/")
    public String index() {
      return "Hello Spring Boot!";
    }
    public static void main(String[] args) {
      SpringApplication.run(App.class, args);
    }
}
```

@RequestMapping 注解用作映射 http(s)请求的路径。在上面的代码中，我们将 index 方法映射到 HTTP 路径的“/”上，当访问“/”时，index 方法就会被调用执行，然后输出“Hello Spring Boot!”。

将工程编译运行的命令为“mvn spring-boot:run”，执行后输出结果如下。

示例代码 3-12　Maven 工程执行编译后的部分输出结果

```
[INFO] Scanning for projects...
（省略部分内容）
```

```
    [INFO] Attaching agents: []
      .   ____          _            __ _ _
     /\\ / ___'_ __ _ _(_)_ __  __ _ \ \ \ \
    ( ( )\___ | '_ | '_| | '_ \/ _` | \ \ \ \
     \\/  ___)| |_)| | | | | || (_| |  ) ) ) )
      '  |____| .__|_| |_|_| |_\__, | / / / /
     =========|_|==============|___/=/_/_/_/
     :: Spring Boot ::        (v2.3.6.RELEASE)
    2020-12-06 21:08:27.681  INFO 32692 --- [           main] com.example.App: 
Starting App on Shawn-Desktop with PID 32692
    （省略部分内容）
    2020-12-06 21:08:28.718  INFO 32692 --- [           main] 
o.a.c.c.C.[Tomcat].[localhost].[/]: Initializing Spring embedded 
WebApplicationContext
    （省略部分内容）
    2020-12-06 21:08:28.921  INFO 32692 --- [           main] 
o.s.b.w.embedded.tomcat.TomcatWebServer: Tomcat started on port(s): 8080 (http) 
with context path ''
    2020-12-06 21:08:28.923  INFO 32692 --- [           main] com.example.App: 
Started App in 1.46 seconds (JVM running for 1.714)
```

上面输出了不少日志信息，其中一部分省略掉了，读者可以自行查看控制台信息。这里只给出了关键的一些信息，我们能够看到程序运行的进程 ID 为 32692，每一行日志的类型（INFO）后面都带有该进程 ID；日志“[Tomcat].[localhost].[/]: Initializing Spring embedded WebApplicationContext”告诉我们 Tomcat 作为内嵌 Web 容器将工程的上下文（context path）映射到了“/”；“Tomcat started on port(s): 8080 (http) with context path”这行信息告诉我们当前的 Spring Boot 应用暴露在 8080 端口提供 HTTP 服务。

然后打开浏览器，访问地址 http://localhost:8080/，验证我们的第一个 Hello Spring Boot 应用是否成功，成功访问结果如图 3-27 所示。

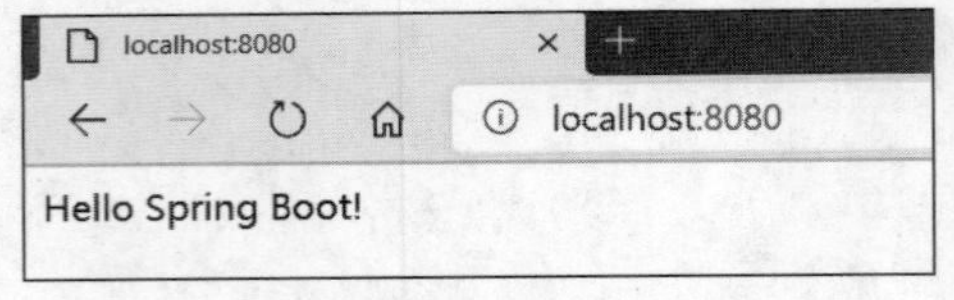

图 3-27

当然，也可以使用 curl 来测试结果，相关命令如下。

示例代码 3-13　使用 curl 命令访问 http://localhost:8080

```
C:\Users\Administrator>curl localhost:8080
Hello Spring Boot!
```

输出内容是“Hello Spring Boot！”，至此我们使用 Maven Achetype 插件命令行方式创建的第一个 Spring Boot 应用程序圆满完成。接下来将使用另外一种命令行工具来创建它。

2. 使用 Spring Boot CLI

Spring Boot CLI 是一个命令行工具，它可以帮助我们快速开发 Spring 应用程序。使用这个工具，我们无须创建太多的样板代码，直接编写 Java 代码文件，然后通过 spring 命令直接运行。首先需要下载 Spring Boot CLI 工具，下载地址为 https://docs.spring.io/spring-boot/docs/current/reference/html/getting-started.html#getting-started-installing-the-cli。

这里提供了 zip 和 tar.gz 两种压缩格式的安装包，选择其一下载即可，然后解压到自己的安装目录，比如 D:\Develop\spring-2.4.0。接下来添加环境变量配置（参照之前 Java 和 Maven 环境变量配置步骤），将“D:\Develop\spring-2.4.0\bin”添加到系统环境变量“path”中，如图 3-28 所示。

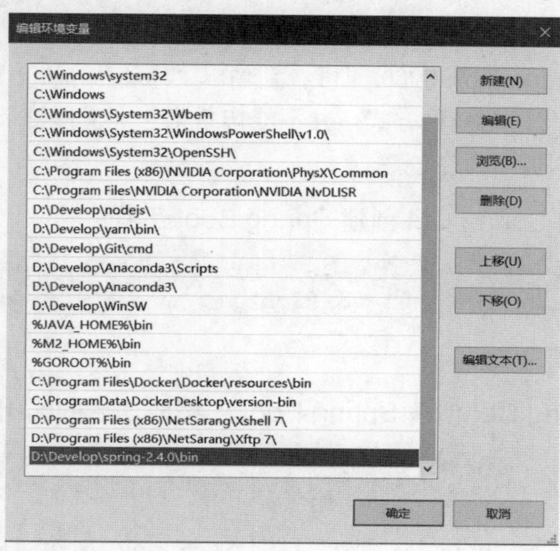

图 3-28

依次单击“确定”按钮关闭环境变量设置窗口。然后重新打开控制台，输入命令“spring –version”来验证是否安装成功，若输出以下内容则说明配置成功。

示例代码 3-14　验证 Spring CLI 是否安装成功

```
C:\Users\Administrator>spring --version
Spring CLI v2.4.0
```

重新进入 D:\code\spring-boot-example 目录，创建一个新的目录 spring-boot-cli，并将前面 my-first-spring-boot-app 工程中的 src/main/com/example/App.java（修改后的）复制到 spring-boot-cli 目录中，执行 spring run App.java 来运行它。

示例代码 3-15　Spring CLI 直接运行 App.java

```
D:\code\spring-boot-example\spring-boot-cli>spring run App.java
Resolving dependencies...............
  .   ____          _            __ _ _
 /\\ / ___'_ __ _ _(_)_ __  __ _ \ \ \ \
( ( )\___ | '_ | '_| | '_ \/ _` | \ \ \ \
 \\/  ___)| |_)| | | | | || (_| |  ) ) ) )
  '  |____| .__|_| |_|_| |_\__, | / / / /
 =========|_|==============|___/=/_/_/_/
 :: Spring Boot ::        (v2.4.0)
2020-12-06 22:10:25.352  INFO 15716 --- [       runner-0]
o.s.boot.SpringApplication           : Starting application using Java 1.8.0_265
```

```
on Shawn-Desktop with PID 15716 (started by Administrator in
D:\code\spring-boot-example\spring-boot-cli)
    （省略部分内容）
    2020-12-06 22:10:26.895  INFO 15716 --- [       runner-0]
o.s.b.w.embedded.tomcat.TomcatWebServer  : Tomcat started on port(s): 8080 (http)
with context path ''
    2020-12-06 22:10:26.904  INFO 15716 --- [       runner-0]
o.s.boot.SpringApplication               : Started application in 1.928 seconds (JVM
running for 20.887)
```

打开浏览器访问 http://localhost:8080/即可看到输出“Hello Spring Boot!”信息。这只是使用 Spring Boot 快速运行 Java 代码的一种方式，并不能初始化一个工程目录。想要初始化一个完整的 Spring Boot 工程，需要使用如下命令创建。

示例代码 3-16　使用 Spring CLI 创建 Spring Boot 工程

```
    spring init -d=web -g=com.example -a=my-cli-spring-boot-app
--package-name=com.example --java-version=1.8 my-cli-spring-boot-app
```

执行完成后输出如下信息。

示例代码 3-17　Spring CLI 创建 Spring Boot 工程成功后的输出

```
    Using service at https://start.spring.io
    Project extracted to 'D:\code\spring-boot-example\my-cli-spring-boot-app'
```

该命令使用 https://start.spring.io 创建了一个 Spring Boot 工程，下载后保存到了 my-cli-spring-boot-app 目录中。使用 tree 命令查看该目录结构，如下所示。

示例代码 3-18　使用 tree 命令查看工程目录结构

```
    D:\code\spring-boot-example>tree /F my-cli-spring-boot-app
    │  .gitignore
    │  HELP.md
    │  mvnw
    │  mvnw.cmd
    │  pom.xml
    ├─.mvn
    │  └─wrapper
    │          maven-wrapper.jar
    │          maven-wrapper.properties
    │          MavenWrapperDownloader.java
    └─src
        ├─main
        │  ├─java
        │  │  └─com
        │  │      └─example
        │  │              DemoApplication.java
        │  └─resources
        │      │  application.properties
        │      ├─static
        │      └─templates
        └─test
```

```
    └─java
        └─com
            └─example
                    DemoApplicationTests.java
```

该目录结构和之前使用 Maven Archetype 创建的相比，多了一些文件，主要是在项目级别（不使用用户自己安装的 Maven）包含了一个可执行的 Maven 程序。这里对其进行简单的介绍：

- .mvn 目录：里面包含了一个 maven-wrapper 的 jar 包，Maven Wrapper 提供了一种简单的 Maven 构建方式，运行构建环境不需要提前安装 Maven 二进制安装包。maven-wrapper.properties 文件定义了 Maven 安装包的下载地址（distributionUrl），默认下载后会将 Maven 安装在 C:\Users\用户名\.m2\wrapper 目录下。
- mvnw：用于 Linux、UNIX 系统，引导 maven-wrapper.jar 下载 Maven 二进制文件，用于项目的构建和编译。
- mvnw.cmd：作用和 mvnw 相同，用于 Windows 系统。
- application.properties：Spring Boot 应用默认的属性配置文件，这是比较重要的一个配置文件，后面章节中也会多次提到，其配置属性可以控制 Spring Boot 的行为。关于其中的属性配置，后面章节会详细介绍。
- HELP.md：一个简单的说明文档。
- .gitignore：git 是一款优秀的开源分布式版本控制软件，.gitignore 用于给出 git 提交时需要忽略的文件或文件夹。关于 git 相关的知识请查看 https://git-scm.com/book/zh/v2。
- DemoApplicationTests.java：这是 Spring Boot 应用的 JUnit 测试代码文件，可以作为一份模板参考，一般测试代码文件与被测试的代码文件保持对应关系，名称上常见的就是被测代码类名称加上 Tests。在本示例中，被测代码类名为 DemoApplication，而这份测试代码类名为 DemoApplicationTests。
- DemoApplication.java：Spring Boot 应用程序的入口代码类文件。

打开 DemoApplication.java 文件，可以看到其中的内容如下。

示例代码 3-19　默认的 DemoApplication.java 文件

```java
package com.example;
import org.springframework.boot.SpringApplication;
import org.springframework.boot.autoconfigure.SpringBootApplication;
@SpringBootApplication
public class DemoApplication {
    public static void main(String[] args) {
        SpringApplication.run(DemoApplication.class, args);
    }
}
```

对其进行修改，增加一个 index 方法来输出“Hello Spring Boot！”。

示例代码 3-20　增加 index 方法后的 DemoApplication.java 文件

```java
package com.example;
import org.springframework.boot.SpringApplication;
import org.springframework.boot.autoconfigure.SpringBootApplication;
```

```
import org.springframework.web.bind.annotation.RequestMapping;
import org.springframework.web.bind.annotation.RestController;
@RestController
@SpringBootApplication
public class DemoApplication {
    @RequestMapping("/")
    public String index() {
        return "Hello Spring Boot!";
    }
    public static void main(String[] args){
        SpringApplication.run(DemoApplication.class, args);
    }
}
```

在 my-cli-spring-boot-app 目录下，执行 mvnw.cmd spring-boot:run 来编译并运行。

示例代码 3-21　mvnw.cmd spring-boot:run 编译运行结果部分输出

```
[INFO] Scanning for projects...
（省略部分内容）
[INFO] Attaching agents: []
  .   ____          _            __ _ _
 /\\ / ___'_ __ _ _(_)_ __  __ _ \ \ \ \
( ( )\___ | '_ | '_| | '_ \/ _` | \ \ \ \
 \\/  ___)| |_)| | | | | || (_| |  ) ) ) )
  '  |____| .__|_| |_|_| |_\__, | / / / /
 =========|_|==============|___/=/_/_/_/
 :: Spring Boot ::                (v2.4.0)
2020-12-07 00:00:59.468  INFO 29140 --- [           main]
com.example.DemoApplication              : Starting DemoApplication using Java
1.8.0_265 on Shawn-Desktop with PID 29140
（省略部分内容）
2020-12-07 00:01:00.636  INFO 29140 --- [           main]
o.s.b.w.embedded.tomcat.TomcatWebServer  : Tomcat started on port(s): 8080 (http)
with context path ''
2020-12-07 00:01:00.643  INFO 29140 --- [           main]
com.example.DemoApplication              : Started DemoApplication in 1.393 seconds
(JVM running for 1.636)
```

使用浏览器访问 http://localhost:8080/即可看到输出内容“Hello Spring Boot!”。这里只简单介绍一下 Spring Boot CLI 是如何创建和运行应用的，详细介绍可以访问官网（https://docs.spring.io/spring-boot/docs/current/reference/html/getting-started.html#getting-started-installing-the-cli），本节不做过多介绍。如果在运行 mvnw 命令时遇到无法执行的问题，就需要为其赋予执行权限，这点需要注意一下。

3.3.2　使用图形化界面创建

Spring 官方提供了一个非常高效和快捷的在线创建 Spring Boot 应用的方式，相对于使用命令行方式，它使用 Web 图形化工具来创建，以一种更加直观的方式来呈现，可以非常方便地选择版本、填写相关配置信息。该在线工具的地址是 https://start.spring.io。

在页面的表单中，Project 选项选择 Maven Project，Language 选择“Java”，Spring Boot 版本选择“2.3.6”，Group 中填写“com.example”，Artifact 中填写“my-initializr-spring-boot-app”，Package Name 中填写“com.example”，Packaging 默认选择“Jar”，Name 和 Description 可以自行填写或者采用默认，在右侧 Dependencies 中单击“ADD DEPENDENCIES”按钮，添加 Spring Web 依赖，如图 3-29 所示。

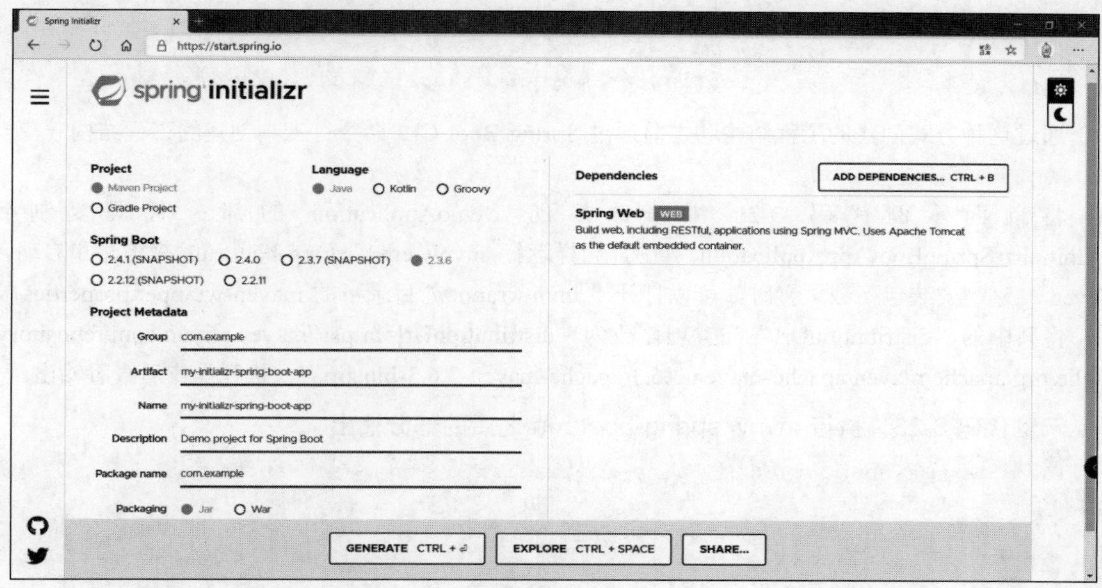

图 3-29

设置完成后，单击“GENERATE”按钮完成创建，提交表单后会生成一份名称为“my-initializr-spring-boot-app.zip”的压缩文件并自动下载。然后将下载的压缩文件解压到示例工程的路径 D:\code\spring-boot-example\，再使用 tree 命令查看一下目录结构。

示例代码 3-22 使用 tree 命令查看目录结构

```
D:\code\spring-boot-example>tree /F my-initializr-spring-boot-app
│  .gitignore
│  HELP.md
│  mvnw
│  mvnw.cmd
│  pom.xml
│
├─.mvn
│  └─wrapper
│          maven-wrapper.jar
│          maven-wrapper.properties
│          MavenWrapperDownloader.java
│
└─src
    ├─main
    │  ├─java
    │  │  └─com
    │  │      └─example
```

```
    │   │               MyInitializrSpringBootAppApplication.java
    │   └─resources
    │       │  application.properties
    │       │
    │       ├─static
    │       └─templates
    └─test
        └─java
            └─com
                └─example
                        MyInitializrSpringBootAppApplicationTests.java
```

通过这种方式创建的工程结构和之前通过 Spring Boot CLI 命令行工具创建的工程结构一模一样。

将前面示例代码 3-20 中修改后的 DemoApplication 的部分代码复制到 MyInitializrSpringBootAppApplication 类中，同样运行 mvnw.cmd spring-boot:run 来构建和启动，Maven 二进制安装包下载太慢时，可以打开“.mvn/wrapper/”目录下的 maven-wrapper.properties 文件，将其中的“distributionUrl”配置项修改为“distributionUrl=https://maven.aliyun.com/repository/public/org/apache/maven/apache-maven/3.6.3/apache-maven-3.6.3-bin.zip”。执行操作后查看输出。

示例代码 3-23　执行 mvnw spring-boot:run 之后的部分输出

```
    D:\code\spring-boot-example\my-initializr-spring-boot-app>mvnw
spring-boot:run
    [INFO] Scanning for projects...
    [INFO] -------------<
com.example:my-initializr-spring-boot-app >--------------
    [INFO] Building my-initializr-spring-boot-app 0.0.1-SNAPSHOT
    [INFO] --------------------------------[ jar ]---------------------------------
    [INFO] >>> spring-boot-maven-plugin:2.3.6.RELEASE:run (default-cli) >
test-compile @ my-initializr-spring-boot-app >>>
    [INFO] --- maven-resources-plugin:3.1.0:resources (default-resources) @
my-initializr-spring-boot-app ---
    [INFO] Using 'UTF-8' encoding to copy filtered resources.
    [INFO] Copying 1 resource
    [INFO] Copying 0 resource
    [INFO] --- maven-compiler-plugin:3.8.1:compile (default-compile) @
my-initializr-spring-boot-app ---
    [INFO] Changes detected - recompiling the module!
    [INFO] Compiling 1 source file to
D:\code\spring-boot-example\my-initializr-spring-boot-app\target\classes
    [INFO] ------------------------------------------------------------------------
    [INFO] BUILD FAILURE
    [INFO] ------------------------------------------------------------------------
    [INFO] Total time:  1.131 s
    [INFO] Finished at: 2020-12-07T22:53:49+08:00
    [INFO] ------------------------------------------------------------------------
    [ERROR] Failed to execute goal
org.apache.maven.plugins:maven-compiler-plugin:3.8.1:compile (default-compile)
on project my-initializr-spring-boot-app: Fatal error compiling: 无效的目标发行版:
```

```
11 -> [Help 1]
    [ERROR]
    [ERROR] To see the full stack trace of the errors, re-run Maven with the -e switch.
    [ERROR] Re-run Maven using the -X switch to enable full debug logging.
    [ERROR]
    [ERROR] For more information about the errors and possible solutions, please
read the following articles:
    [ERROR] [Help 1]
http://cwiki.apache.org/confluence/display/MAVEN/MojoExecutionException
```

查看输出结果发现构建失败了（BUILD FAILURE），并且失败的原因在第一次出现[ERROR]日志的行中已经给出“Failed to execute goal org.apache.maven.plugins:maven-compiler-plugin:3.8.1:compile (default-compile) on project my-initializr-spring-boot-app: Fatal error compiling: 无效的目标发行版: 11 -> [Help 1]”。其中关键错误提示“无效的目标发行版：11”，告诉我们目前安装的 Java 版本不符合要求，要求的目标版本是 Java 11，而我们安装的版本是 Java 8。解决这个问题的办法是打开 pom.xml 文件，找到<properties>节点中的<java.version>配置，将其属性修改为 8。

示例代码 3-24 修改 pom.xml 中 Java 的版本号

```
<properties>
    <java.version>8</java.version>
</properties>
```

再次执行 mvnw spring-boot:run 命令，就可以看到项目启动成功了。打开浏览器，访问 http://localhost:8080/，可以看到“Hello Spring Boot!”的输出结果。

3.3.3 使用 Eclipse STS 插件创建

使用原生的 Eclipse 创建 Spring Boot 工程项目比较麻烦，而 Spring Boot 官方考虑到了广大 Eclipse 用户的这种诉求，专门开发了一个集成开发套件 Spring Tool Suite（简称 STS），用于开发 Spring 应用程序，而且它不仅能支持 Eclipse，还支持 Visual Studio Code 和 Theia IDE。它提供了一个现成的开发环境来实现调试、运行和部署 Spring 应用程序，包括为 Pivotal tc Server（为企业提供了一个安全、支持和扩展的 Java 应用服务器，并与 Apache Tomcat 完全兼容）、Pivotal Cloud Foundry（PCF，关键的云计算平台）、Git、Maven、AspectJ（面向切面编程的框架，是对 AOP 编程思想的一个实现）和最新的 Eclipse 版本提供整合支持。

STS 插件安装可以选择 Eclipse 插件市场在线安装的方式，具体操作步骤如下。

步骤 01 在 Eclipse 菜单栏单击“Help”菜单，在下拉列表中选择“Eclipse Marketplace...”，如图 3-30 所示。

步骤 02 在“Search”选项卡“Find”一栏的输入框里输入“Spring Tools 4”，单击“Go”按钮（或者按回车键）进行搜索，找到“Spring Tools 4 (aka Sprint Tools Suite 4) x.x.x.RELEASE”，单击“Install”按钮进行安装，如图 3-31 所示。

步骤 03 勾选所有组件，单击“Confirm”按钮进行安装，如图 3-32 所示。

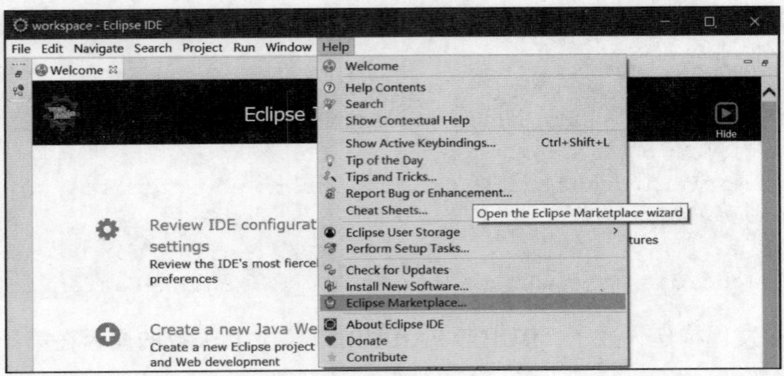

图 3-30

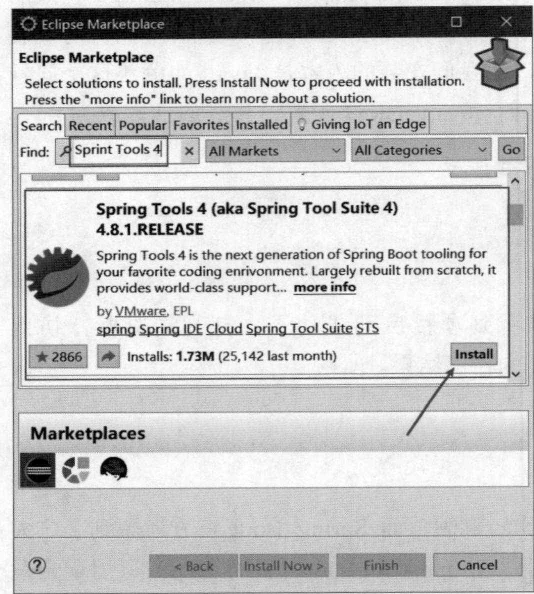

图 3-31

图 3-32

步骤 04 等待上一步完成后，会提示勾选接受协议，选中“I accept the terms of the license agreements”，再单击“Finish”按钮，如图 3-33 所示。然后后续安装步骤会在后台完成，在 Eclipse 右下角状态栏中会显示安装进度，安装完成之后会提示重启 Eclipse。

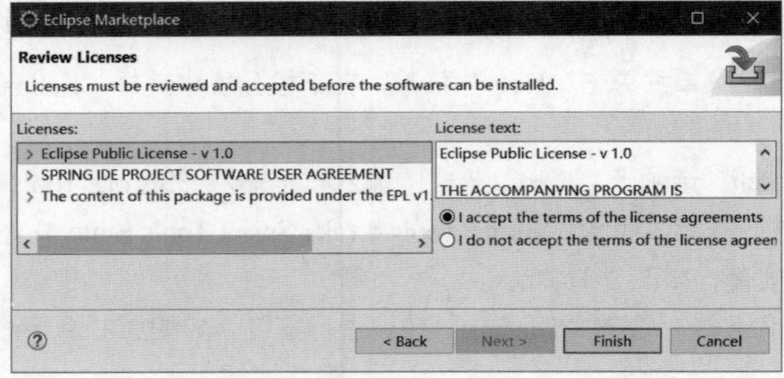

图 3-33

Spring 官方也提供了一个简要的 STS 安装说明文档，供大家安装参考：https://github.com/spring-projects/sts4/wiki/Installation。

安装完成之后，我们使用 Eclipse STS 插件来创建一个 Spring Boot 示例项目。依次单击菜单栏按钮“File”→“New”→“Other…”，在弹出的对话框中找到“Spring Boot”文件夹，展开后选中“Spring Starter Project”，单击“Next”按钮进入下一步，如图 3-34 所示。

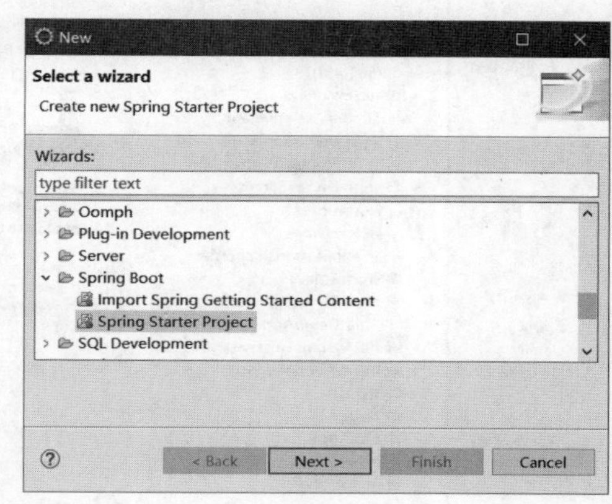

图 3-34

在图 3-35 所示的对话框中，填写相关配置信息，其中包括选择 Java 版本，填写 Name、Group、Artifact、Version、Description、Package、Type 等信息，分别对应表示项目的名称、组织名、工件名、版本号、项目描述、项目包名、工程类型，其中工程类型可以选择 Maven 或者是 Gradle（一个基于 Apache Ant 和 Apache Maven 概念的项目自动化构建开源工具）。这些信息和前面示例中填写的内容基本是一样的。事实上，在对话框中有一项“Service URL”默认填好了“https://start.spring.io”，也就是说 STS 插件创建 Spring Boot 工程项目依然是采用 http://start.spring.io 提供的创建方式，不过是对其做了封装，在 Eclipse 下用 GUI 来展示给我们罢了。

然后选择 Spring Boot 版本 2.3.6，勾选“Spring Web”依赖项，单击“Finish”按钮完成创建，如图 3-36 所示。打开 src/main/java 目录下 com.example.demo 包名下的 DemoApplication 类，参照前面小节的示例工程，修改其中的内容，增加 index 方法，具体项目结构和修改代码如图 3-37 所示。

图 3-35

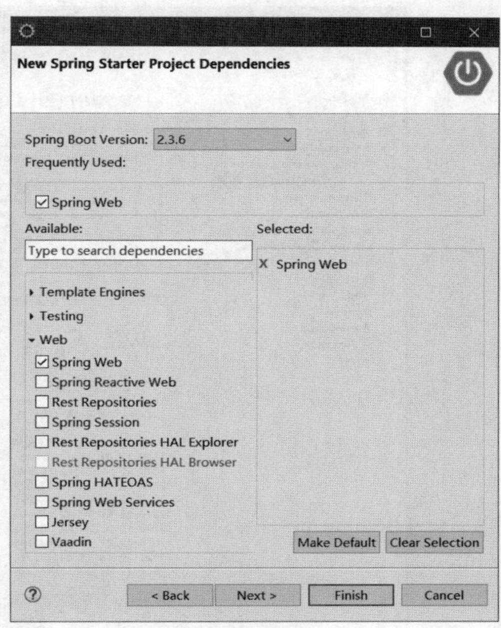

图 3-36

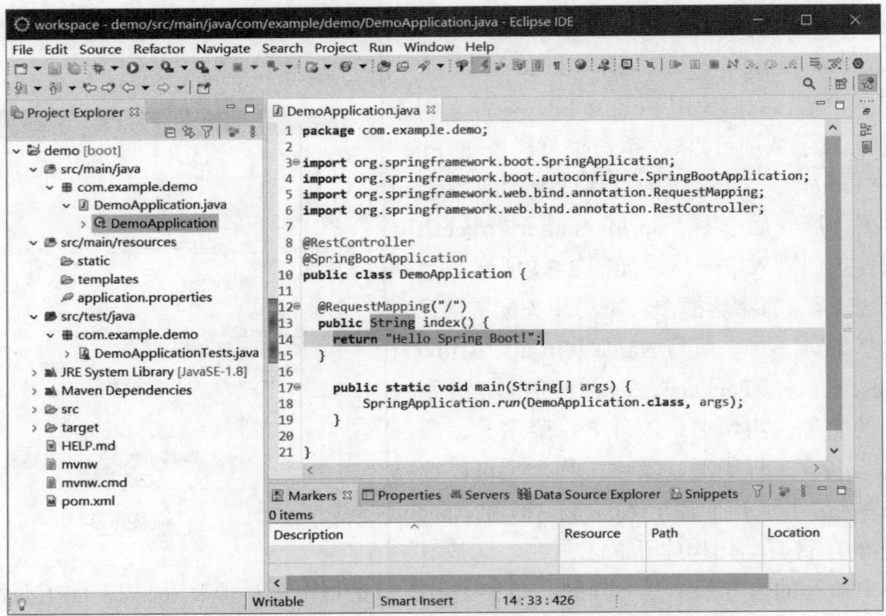

图 3-37

在修改完 DemoApplication 类的代码后，就可以执行构建和运行了。单击 Eclipse 菜单栏中的“Run”→“Run As”→“Spring Boot App”，就会自动构建编译并运行了，在下方“Console”选项卡中会输出相关信息，如图 3-38 所示。打开浏览器访问 https://localhost:8080/，就能看到输出结果“Hello Spring Boot!”。

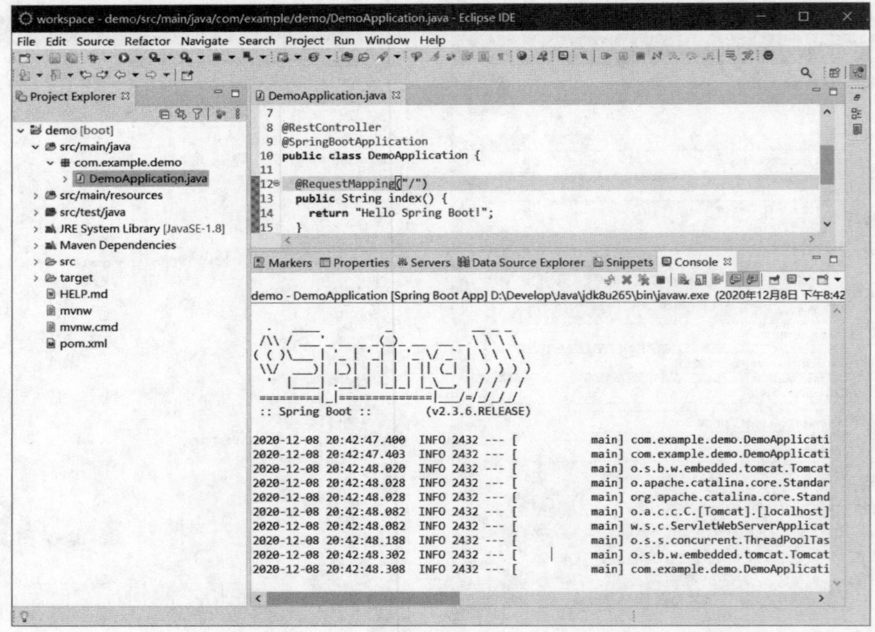

图 3-38

3.3.4　使用 IntelliJ IDEA 创建

旗舰版 IDEA 可以直接使用 Spring Initializr 创建 Spring Boot 应用，而社区版 IDEA 由于功能限制不能直接创建，需要借助一个插件 Spring Assistant，在 IDEA 安装完成之后的界面中，可以在“Plugins”选项中的 Marketplace 中搜索“Spring Assistant”，然后单击“Install”按钮执行安装，如图 3-39 所示。安装完成后重启 IDEA 使插件生效。

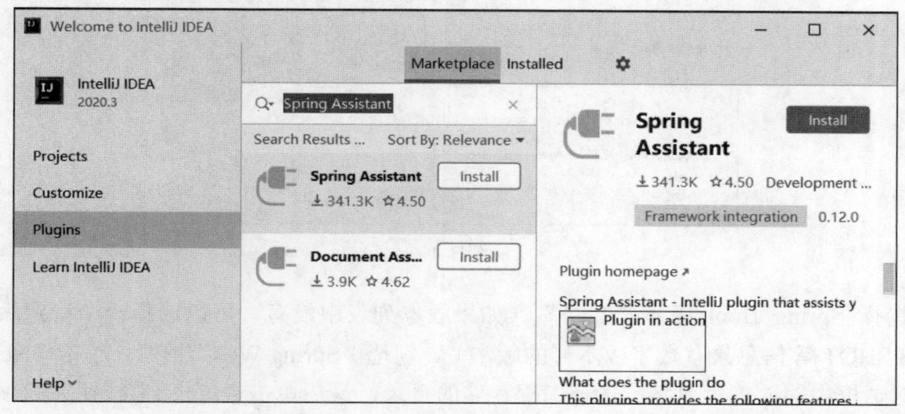

图 3-39

打开 IDEA，单击“New Project”（也可以在菜单栏中单击“File”→“New”→“Project”）进入创建工程界面，在左侧列表中选中“Spring Assistant”（旗舰版是“Spring Initializr”），如图 3-40 所示。IDEA 比 Eclipse 更加智能的是，它会自动探测到本地环境中安装配置好的 JDK。

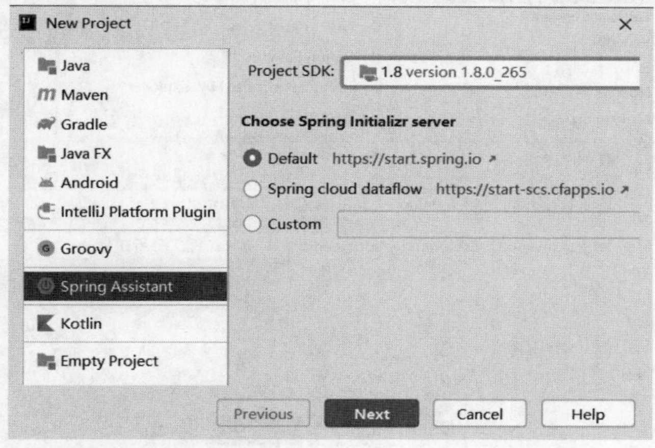

图 3-40

由于我们要创建的是 Spring Boot 应用，因此选择默认的（Default）Spring Initializr server 就可以，单击“Next”按钮进入工程配置界面。在此界面中，同样需要填写（修改）Group Id、Artifact Id、Version、Project Type、Language、Packaging、Java version，如图 3-41 所示。

New Project

Project properties

Group Id: com.example
Artifact Id: idea-demo
Version: 0.0.1-SNAPSHOT
Project type: Maven Project
Language: Java
Packaging: Jar
Java version: 8
Project name: demo
Project description: Demo project for Spring Boot
Package name: com.example.demo

Previous | Next | Cancel | Help

图 3-41

然后选择 Spring Boot 版本“2.3.6”（如果版本列表中没有，可以选择一个稳定版本：不含 M*、SNAPSHOT 等信息只有数字版本号的版本），勾选“Spring Web”依赖，单击“Next”按钮，在弹出的对话框中输入工程名字，选择工程存储的目录，最后单击“Finish”按钮完成项目创建，如图 3-42 所示。

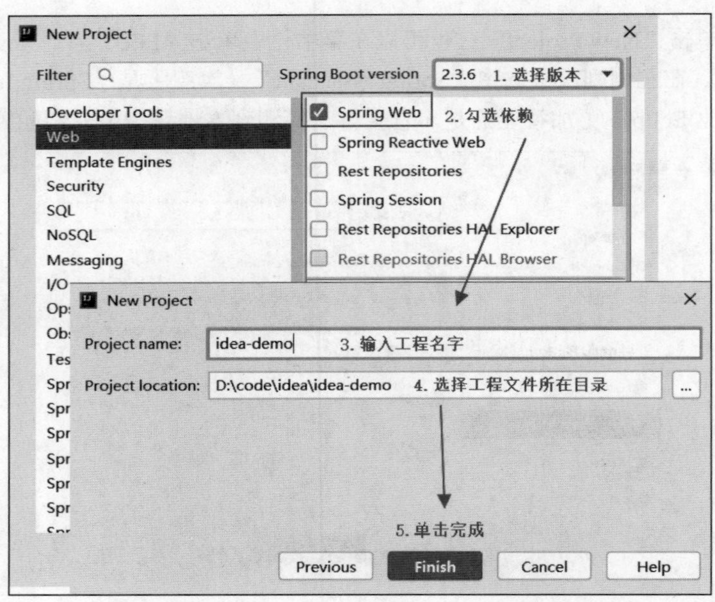

图 3-42

完成创建后会自动打开该工程。将其目录结构展开，查看该目录结构（见图 3-43），发现与之前通过 Spring Boot CLI、图形化界面、Eclipse STS 创建的目录结构和文件是一致的，这是因为这几种方式都是通过 https://start.spring.io 来创建项目的。这里不再做过多介绍，同样对 DemoApplication.java 进行修改，增加 index 方法用于返回“Hello Spring Boot!”。

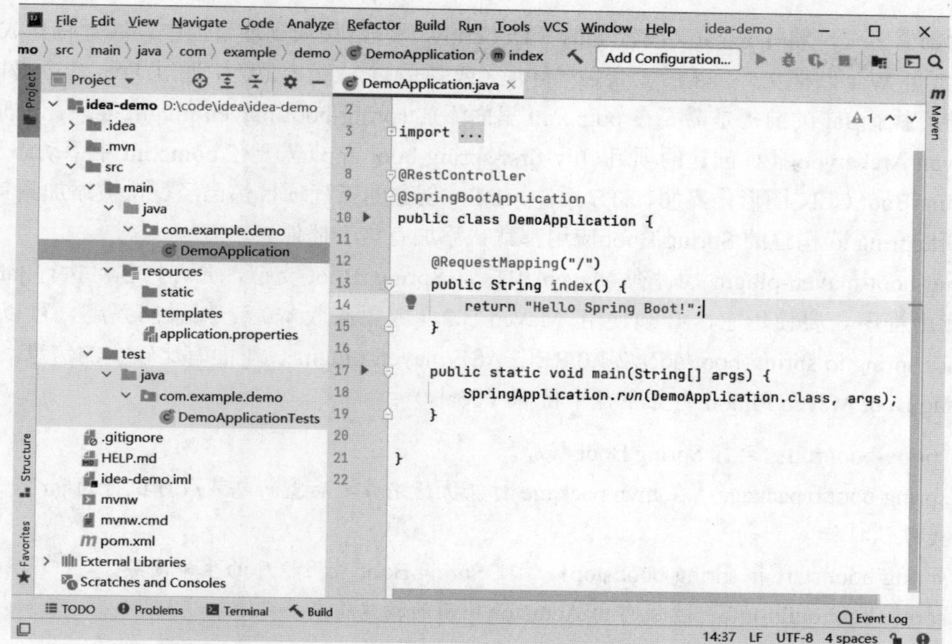

图 3-43

修改完成后，在 DemoApplication.java 文件中单击右键快捷菜单中的“Run 'DemoApplication.main()'”（或者直接按快捷键 Ctrl+Shift+F10），执行构建并运行，在弹出的运行输出窗口中显示运行结果，如图 3-44 所示。通过浏览器访问 http://localhost:8080/，能够看到和前面通过其他方式创建的项目运行所呈现的一样的结果。

```
 :: Spring Boot ::        (v2.3.6.RELEASE)

2020-12-08 21:53:41.823  INFO 36496 --- [           main] com.example.demo.DemoApplication         : Starting DemoApplication on Shawn-Desktop with PID 36496 (D:\code\idea\idea-d
2020-12-08 21:53:41.826  INFO 36496 --- [           main] com.example.demo.DemoApplication         : No active profile set, falling back to default profiles: default
2020-12-08 21:53:43.147  INFO 36496 --- [           main] o.s.b.w.embedded.tomcat.TomcatWebServer  : Tomcat initialized with port(s): 8080 (http)
2020-12-08 21:53:43.157  INFO 36496 --- [           main] o.apache.catalina.core.StandardService   : Starting service [Tomcat]
2020-12-08 21:53:43.157  INFO 36496 --- [           main] org.apache.catalina.core.StandardEngine  : Starting Servlet engine: [Apache Tomcat/9.0.39]
2020-12-08 21:53:43.233  INFO 36496 --- [           main] o.a.c.c.C.[Tomcat].[localhost].[/]       : Initializing Spring embedded WebApplicationContext
2020-12-08 21:53:43.233  INFO 36496 --- [           main] w.s.c.ServletWebServerApplicationContext : Root WebApplicationContext: initialization completed in 1360 ms
2020-12-08 21:53:43.367  INFO 36496 --- [           main] o.s.s.concurrent.ThreadPoolTaskExecutor  : Initializing ExecutorService 'applicationTaskExecutor'
2020-12-08 21:53:43.486  INFO 36496 --- [           main] o.s.b.w.embedded.tomcat.TomcatWebServer  : Tomcat started on port(s): 8080 (http) with context path ''
2020-12-08 21:53:43.494  INFO 36496 --- [           main] com.example.demo.DemoApplication         : Started DemoApplication in 1.96 seconds (JVM running for 2.323)
```

图 3-44

3.3.5 构建可执行 jar 包

在前面我们使用命令 mvn spring-boot:run 或者 mvnw spring-boot:run 来执行项目编译构建和运行。这个命令执行时会先检查项目依赖包，并下载缺失的相关依赖包，然后执行编译、构建和打包，最后才会运行应用程序。这个命令用于在本地构建编译和执行，在生产环境中不会使用，传统的方式是通过构建一个 war 包，部署到生产环境的 Tomcat 等 Web 容器中运行。Spring Boot 改变了这种传统的部署方式，将 Tomcat（或者 Jetty、Undertow）容器集成到项目当中，编译构建成一个独立的 Jar 或者 War 归档文件，然后放到生产环境（或者是 Docker 容器）中通过 Java 命令直接运行。

构建可执行 Jar 的前提是需要在 pom.xml 里面添加 spring-boot-maven-plugin 插件。该插件除了使用 Maven Archetype 插件创建的项目（my-first-spring-boot-app）需要在 pom.xml 中手动添加以外，通过 Spring Boot CLI、图形化界面、STS 插件、IDEA 创建的工程项目都已经默认被添加到其中（由 https://start.spring.io 构建的 Spring Boot 应用都默认添加有这个插件）。

spring-boot-maven-plugin 插件在 Maven 中提供 Spring Boot 支持，允许打包可执行 jar 或 war 和运行应用程序。要使用它，必须使用 Maven 3.2（或更高版本）。如果感兴趣，可以看一下 https://docs.spring.io/spring-boot/docs/2.1.0.RELEASE/maven-plugin/，里面有比较全的信息。

Spring Boot Maven Plugin 有如下几个目标（goals）：

- spring-boot:run：运行 Spring Boot 程序。
- spring-boot:repackage：在 mvn package 打包执行完成之后重新包装成可执行的 jar 或 war，这是默认目标。
- spring-boot:start 和 spring-boot:stop：管理 Spring Boot 应用程序的生命周期。
- spring-boot:build-info：生成可由 Actuator 使用的构建信息。

在页面 https://docs.spring.io/spring-boot/docs/2.1.0.RELEASE/maven-plugin/usage.html 中，可以找到有关如何使用 Spring Boot Maven Plugin 的一般说明。

Spring Boot Maven Plugin 的默认目标是“spring-boot:repackage”。构建可执行 jar 包，就必须在 pom.xml 里面<build>节点配置中添加如下内容：

示例代码 3-25　在 pom.xml 中增加 spring-boot-maven-plugin

```
<build>
    <plugin>
        <groupId>org.springframework.boot</groupId>
        <artifactId>spring-boot-maven-plugin</artifactId>
    </plugin>
    </plugins>
</build>
```

上面的配置没有指定插件的版本号信息，这是因为我们在本章中创建的 Spring Boot 项目的 pom 继承了 spring-boot-starter-parent 的 pom 配置信息，包括使用 Maven Achetype 手工创建的项目 my-first-spring-boot-app。我们可以打开项目的 pom.xml 查看一下，其中包含这项配置。

示例代码 3-26　my-first-spring-boot-app 项目 pom.xml 文件中的<parent>节点配置

```
<parent>
    <groupId>org.springframework.boot</groupId>
    <artifactId>spring-boot-starter-parent</artifactId>
    <version>2.3.6.RELEASE</version>
    <relativePath/> <!-- lookup parent from repository -->
</parent>
```

这个配置指定了项目继承自 spring-boot-starter-parent 项目的配置信息，其中 spring-boot-maven-plugin 插件的版本信息已经在 spring-boot-starter-parent 中声明，因此在我们创建

的项目中就不需要指定版本信息了。如果不是继承自 spring-boot-starter-parent，就需要指定其版本信息，比如 2.3.6.RELEASE。

示例代码 3-27 在 my-first-spring-boot-app 项目的 pom.xml 文件中指定 Spring Boot 版本号

```
<build>
    <plugins>
        <plugin>
            <groupId>org.springframework.boot</groupId>
            <artifactId>spring-boot-maven-plugin</artifactId>
            <version>2.3.6.RELEASE</version>
        </plugin>
    </plugins>
</build>
```

以前面使用 IntelliJ IDEA 创建的示例工程为例，在配置完成后可以执行命令 mvnw clean package 来打包程序。

示例代码 3-28 执行 mvnw clean package 后的部分输出结果

```
D:\code\idea\idea-demo>mvn clean package
[INFO] Scanning for projects...
[INFO]
[INFO] ----------------< com.example:idea-demo >-----------------------
[INFO] Building demo 0.0.1-SNAPSHOT
[INFO] --------------------------------[ jar ]---------------------------
[INFO]
[INFO] --- maven-clean-plugin:3.1.0:clean (default-clean) @ idea-demo ---
[INFO] Deleting D:\code\idea\idea-demo\target
[INFO]
[INFO] --- maven-resources-plugin:3.1.0:resources (default-resources) @
idea-demo ---
[INFO] Using 'UTF-8' encoding to copy filtered resources.
[INFO] Copying 1 resource
[INFO] Copying 0 resource
[INFO]
[INFO] --- maven-compiler-plugin:3.8.1:compile (default-compile) @ idea-demo
---
[INFO] Changes detected - recompiling the module!
[INFO] Compiling 1 source file to D:\code\idea\idea-demo\target\classes
[INFO]
[INFO] --- maven-resources-plugin:3.1.0:testResources (default-testResources)
@ idea-demo ---
[INFO] Using 'UTF-8' encoding to copy filtered resources.
[INFO] skip non existing resourceDirectory
D:\code\idea\idea-demo\src\test\resources
[INFO]
[INFO] --- maven-compiler-plugin:3.8.1:testCompile (default-testCompile) @
idea-demo ---
[INFO] Changes detected - recompiling the module!
[INFO] Compiling 1 source file to D:\code\idea\idea-demo\target\test-classes
```

```
    [INFO]
    [INFO] --- maven-surefire-plugin:2.22.2:test (default-test) @ idea-demo ---
    [INFO]
    [INFO] -------------------------------------------------------
    [INFO]  T E S T S
    [INFO] -------------------------------------------------------
    （省略部分内容）
    [INFO] Results:
    [INFO]
    [INFO] Tests run: 1, Failures: 0, Errors: 0, Skipped: 0
    [INFO] --- maven-jar-plugin:3.2.0:jar (default-jar) @ idea-demo ---
    [INFO] Building jar:
D:\code\idea\idea-demo\target\idea-demo-0.0.1-SNAPSHOT.jar
    [INFO]
    [INFO] --- spring-boot-maven-plugin:2.3.6.RELEASE:repackage (repackage) @
idea-demo ---
    [INFO] Replacing main artifact with repackaged archive
    [INFO] ------------------------------------------------------------------
    [INFO] BUILD SUCCESS
    [INFO] ------------------------------------------------------------------
    （省略部分内容）
```

仔细查看输出的日志信息能够发现，mvn clean package 命令一共执行了 8 个动作：

- maven-clean-plugin:3.1.0:clean (default-clean) ：使用 maven-clean-plugin 插件执行项目清理的处理。
- maven-resources-plugin:3.1.0:resources (default-resources)：负责处理项目执行用资源文件（src/main 目录下的）并复制到输出目录。
- maven-compiler-plugin:3.8.1:compile (default-compile)：编译 src/main 目录下的 Java 源文件。
- maven-resources-plugin:3.1.0:testResources (default-testResources)：负责处理项目测试用资源文件（src/test 目录下的）并复制到输出目录。
- maven-compiler-plugin:3.8.1:testCompile (default-testCompile)：编译 src/test 目录下的测试源文件。
- maven-surefire-plugin:2.22.2:test (default-test)：运行 JUnit 单元测试，创建测试报告。
- maven-jar-plugin:3.2.0:jar (default-jar)：从当前工程中构建 jar 文件。
- spring-boot-maven-plugin:2.3.6.RELEASE:repackage (repackage)：在 maven-jar-plugin 插件构建生成的 jar 文件包中，查找 Manifest 文件，将其中配置的 Main-Class 属性值，设置为程序的主入口类，将其再次打包为可执行的软件包，并将 maven-jar-plugin 生成的软件包重命名为 *.original 结尾的文件。

关于这些 Maven 插件的详细说明，可以参考 Maven 官方文档介绍：http://maven.apache.org/plugins/index.html。

在上面的输出结果中，我们看到了“BUILD SUCCESS”信息，说明 mvn clean package 命令已经成功执行。在 target 目录中，最终生成了 idea-demo-0.0.1-SNAPSHOT.jar 和 idea-demo-0.0.1-SNAPSHOT.jar.original 两个归档文件，其中 idea-demo-0.0.1-SNAPSHOT.jar 可以直

接用来执行，命令如下：

```
java -jar target/idea-demo-0.0.1-SNAPSHOT.jar
```

该命令的执行结果与在IDEA中运行或者使用mvn spring-boot:run命令执行基本无异，并且运行速度不受影响。这两种方式的区别在于，可执行jar包（fat jar）用于放置在生产环境中直接使用java -jar xxx.jar命令来执行，而mvn spring-boot:run方式（或者在IDE中直接单击运行）是开发阶段的运行方式。

第 4 章

Spring Boot 开发 Web 应用

本章将讲解 Spring Boot 是如何开发 Web 应用的，包括内置容器的原理与应用、如何自动配置 Spring MVC、如何集成模板引擎，最后通过前后端分离的应用实战加深对 Web 开发的印象。

4.1 内置容器

4.1.1 内置容器配置

1. Web 容器自动装配原理

使用 Spring Boot 时，首先引人注意的就是配置很少，而且启动方式很简单。我们熟悉的 Web 项目都需要部署到应用服务器上才可以运行。常见的 Web 容器有 Tomcat、WebLogic、Widefly（旧称：JBoss）。Spring Boot 项目可以直接通过 XXApplication.java 中的 main 方法启动的原因是除了高度继承封装了 Spring 一系列框架之外，还内置了 Web 容器。Spring Boot 启动时会根据配置启动相应的上下文环境。下面我们将剖析 Spring Boot 自动转配 Web 容器的原理：

（1）在 spring-boot-autoconfigure 包中，在 META-INF 文件夹下的 spring.factories 文件中有一个 org.springframework.boot.autoconfigure.EnableAutoConfiguration 自动注解的变量，其中 ServletWebServerFactoryAutoConfiguration 类会被 Spring Boot 自动配置类找到并加载。

（2）Spring Boot 启动类上都会有@SpringBootApplication 注解，这是一个组合注解，查看@SpringBootApplication 的源码（见图 4-1、图 4-2）就会发现，自动装配是由其中的@EnableAutoConfiguration 注解实现的。

```
@SpringBootApplication
public class SpringbootWebApplication extends SpringBootServletInitializer {

    public static void main(String[] args) { SpringApplication.run(SpringbootWebApplication.class, args); }

}
```

图 4-1

```
package org.springframework.boot.autoconfigure;

import ...

@Target({ElementType.TYPE})
@Retention(RetentionPolicy.RUNTIME)
@Documented
@Inherited
@SpringBootConfiguration
@EnableAutoConfiguration
@ComponentScan(
    excludeFilters = {@Filter(
    type = FilterType.CUSTOM,
    classes = {TypeExcludeFilter.class}
), @Filter(
    type = FilterType.CUSTOM,
    classes = {AutoConfigurationExcludeFilter.class}
)}
)
public @interface SpringBootApplication {
```

图 4-2

（3）点开@EnableAutoConfiguration 注解，我们会发现其导入了一个实现 DeferredImportSelector 接口的 AutoConfigurationImportSelector 类，如图 4-3、图 4-4 所示。

```
package org.springframework.boot.autoconfigure;

import ...

@Target({ElementType.TYPE})
@Retention(RetentionPolicy.RUNTIME)
@Documented
@Inherited
@AutoConfigurationPackage
@Import({AutoConfigurationImportSelector.class})
public @interface EnableAutoConfiguration {
    String ENABLED_OVERRIDE_PROPERTY = "spring.boot.enableautoconfiguration";

    Class<?>[] exclude() default {};

    String[] excludeName() default {};
}
```

图 4-3

```
public class AutoConfigurationImportSelector implements DeferredImportSelector, BeanClassLoaderAware,
    private static final AutoConfigurationImportSelector.AutoConfigurationEntry EMPTY_ENTRY = new Auto
    private static final String[] NO_IMPORTS = new String[0];
    private static final Log logger = LogFactory.getLog(AutoConfigurationImportSelector.class);
    private static final String PROPERTY_NAME_AUTOCONFIGURE_EXCLUDE = "spring.autoconfigure.exclude";
    private ConfigurableListableBeanFactory beanFactory;
    private Environment environment;
    private ClassLoader beanClassLoader;
    private ResourceLoader resourceLoader;

    public AutoConfigurationImportSelector() {
    }
```

图 4-4

（4）Spring 容器初始化前，会调用 invokeBeanFactoryPostProcessors()→invokeBeanDefinitionRegistryPostProcessors()→BeanDefinitionRegistryPostProcessor.postProcessBeanDefinitionRegistry()，此 BeanDefinitionRegistryPostProcessor 为 Bean 工厂的后置处理器。实现类为 ConfigurationClassPostProcessor，会扫描到用@Import 导入的方法，并把实现 DeferredImportSelector 接口的对象放到

一个 map 中，后面会执行 DeferredImportSelector.getAutoConfigurationEntry()方法。

（5）DeferredImportSelector.getAutoConfigurationEntry()方法会从 META-INF/spring.factories 文件中找到 spring.boot.enableautoconfiguration 变量数组，把定义的变量数组值配置到 Spring 容器中。其中就有一个 ServletWebServerFactoryAutoConfiguration 配置类，配置了 Web 容器。

查看 Spring Boot 自动配置类 ServletWebServerFactoryAutoConfiguration 的源码，其中类名上通过@Import 注解导入了三种容器的配置 Bean，如下源码块所示。

示例代码 4-1 ServletWebServerFactoryAutoConfiguration.class 文件

```
    // Source code recreated from a .class file by IntelliJ IDEA
    // (powered by Fernflower decompiler)
    //
    package org.springframework.boot.autoconfigure.web.servlet;
    @Configuration(
       proxyBeanMethods = false
    )
    @AutoConfigureOrder(-2147483648)
    @ConditionalOnClass({ServletRequest.class})
    @ConditionalOnWebApplication(
       type = Type.SERVLET
    )
    @EnableConfigurationProperties({ServerProperties.class})
    @Import({ServletWebServerFactoryAutoConfiguration.BeanPostProcessorsRegist
rar.class, EmbeddedTomcat.class, EmbeddedJetty.class, EmbeddedUndertow.class})
    public class ServletWebServerFactoryAutoConfiguration {
       public ServletWebServerFactoryAutoConfiguration() {
       }
       @Bean
       public ServletWebServerFactoryCustomizer
servletWebServerFactoryCustomizer(ServerProperties serverProperties) {
          return new ServletWebServerFactoryCustomizer(serverProperties);
       }
       @Bean
       @ConditionalOnClass(
          name = {"org.apache.catalina.startup.Tomcat"}
       )
       public TomcatServletWebServerFactoryCustomizer
tomcatServletWebServerFactoryCustomizer(ServerProperties serverProperties) {
          return new TomcatServletWebServerFactoryCustomizer(serverProperties);
       }
       @Bean
       @ConditionalOnMissingFilterBean({ForwardedHeaderFilter.class})
       @ConditionalOnProperty(
          value = {"server.forward-headers-strategy"},
          havingValue = "framework"
       )
       public FilterRegistrationBean<ForwardedHeaderFilter>
forwardedHeaderFilter() {
          ForwardedHeaderFilter filter = new ForwardedHeaderFilter();
```

```
            FilterRegistrationBean<ForwardedHeaderFilter> registration = new
FilterRegistrationBean(filter, new ServletRegistrationBean[0]);
            registration.setDispatcherTypes(DispatcherType.REQUEST, new
DispatcherType[]{DispatcherType.ASYNC, DispatcherType.ERROR});
            registration.setOrder(-2147483648);
            return registration;
        }
        public static class BeanPostProcessorsRegistrar implements
ImportBeanDefinitionRegistrar, BeanFactoryAware {
            private ConfigurableListableBeanFactory beanFactory;
            public BeanPostProcessorsRegistrar() {
            }
            public void setBeanFactory(BeanFactory beanFactory) throws
BeansException {
                if (beanFactory instanceof ConfigurableListableBeanFactory) {
                    this.beanFactory =
(ConfigurableListableBeanFactory)beanFactory;
                }
            }
            public void registerBeanDefinitions(AnnotationMetadata
importingClassMetadata, BeanDefinitionRegistry registry) {
                if (this.beanFactory != null) {
                    this.registerSyntheticBeanIfMissing(registry,
"webServerFactoryCustomizerBeanPostProcessor",
WebServerFactoryCustomizerBeanPostProcessor.class);
                    this.registerSyntheticBeanIfMissing(registry,
"errorPageRegistrarBeanPostProcessor",
ErrorPageRegistrarBeanPostProcessor.class);
                }
            }
            private void registerSyntheticBeanIfMissing(BeanDefinitionRegistry
registry, String name, Class<?> beanClass) {
                if
(ObjectUtils.isEmpty(this.beanFactory.getBeanNamesForType(beanClass, true,
false))) {
                    RootBeanDefinition beanDefinition = new
RootBeanDefinition(beanClass);
                    beanDefinition.setSynthetic(true);
                    registry.registerBeanDefinition(name, beanDefinition);
                }
            }
        }
    }
```

Spring Boot 支持 Tomcat、Jetty、Undertow 三种容器。另外，Spring Boot 默认使用的容器是 Tomcat，如下代码块所示。

示例代码 4-2　spring-boot-starter-web-2.2.6.RELEASE.pom.xml

```
    <?xml version="1.0" encoding="UTF-8"?>
    <project xmlns="http://maven.apache.org/POM/4.0.0"
```

```
xmlns:xsi="http://www.w3.org/2001/XMLSchema-instance"
        xsi:schemaLocation="http://maven.apache.org/POM/4.0.0
https://maven.apache.org/xsd/maven-4.0.0.xsd">
        <modelVersion>4.0.0</modelVersion>
        <parent>
            <groupId>org.springframework.boot</groupId>
            <artifactId>spring-boot-starter-parent</artifactId>
            <version>2.2.6.RELEASE</version>
            <relativePath/> <!-- lookup parent from repository -->
        </parent>
        <groupId>com.tudou</groupId>
        <artifactId>springboot-web</artifactId>
        <version>0.0.1-SNAPSHOT</version>
        <name>springboot-web</name>
        <description>Demo project for Spring Boot</description>
        <properties>
            <java.version>1.8</java.version>
        </properties>
        <dependencies>
            <dependency>
                <groupId>org.springframework.boot</groupId>
                <artifactId>spring-boot-starter-web</artifactId>
                <exclusions>
                    <!--使用 Jetty 内置 Web 容器，排除默认的 Tomcat 容器-->
                    <exclusion>
                        <groupId>org.apache.tomcat.embed</groupId>
                        <artifactId>tomcat-embed-core</artifactId>
                    </exclusion>
                </exclusions>
            </dependency>
            <dependency>
                <groupId>org.springframework.boot</groupId>
                <artifactId>spring-boot-starter-test</artifactId>
                <scope>test</scope>
                <exclusions>
                    <exclusion>
                        <groupId>org.junit.vintage</groupId>
                        <artifactId>junit-vintage-engine</artifactId>
                    </exclusion>
                </exclusions>
            </dependency>
            <!-- 增加 servlet-api 依赖 -->
            <dependency>
                <groupId>javax.servlet</groupId>
                <artifactId>javax.servlet-api</artifactId>
            </dependency>
        </dependencies>
        <packaging>war</packaging>
        <build>
            <plugins>
```

```
            <plugin>
                <groupId>org.springframework.boot</groupId>
                <artifactId>spring-boot-maven-plugin</artifactId>
            </plugin>
        </plugins>
    </build>
</project>
```

2. Spring Boot 如何配置当前 Web 容器

这里主要讲解一下如何自定义 Tomcat 的一些配置信息。

示例代码 4-3　TomcatWebServerFactoryConfig.java 部分代码

```
@Configuration
public class TomcatWebServerFactoryConfig {
    /**方式一：
     * 配置当前 web 容器 Tomcat
     */
    @Bean
    public TomcatServletWebServerFactory webServerFactory(){
        TomcatServletWebServerFactory factory = new
TomcatServletWebServerFactory();
        factory.setPort(8083);
        return factory;
    }
    /**
     * 方式二：
     * 配置 WebServerFactoryCustomizer，重写 customize()方法，强转相应的
TomcatServletWebServerFactory。
     */
    @Bean
    public WebServerFactoryCustomizer customizer(){
        WebServerFactoryCustomizer customizer = new
WebServerFactoryCustomizer() {
            @Override
            public void customize(WebServerFactory factory) {
                TomcatServletWebServerFactory factory1 =
(TomcatServletWebServerFactory)factory;
                factory1.setPort(8082);
            }
        };
        return customizer;
    }
}
```

代码中，方式一的运行结果如图 4-5 所示。

```
main] c.t.s.SpringbootWebApplication           : Starting SpringbootWebApplication on DESKTOP-246FP3G with PID 12016 (D:\d
main] c.t.s.SpringbootWebApplication           : No active profile set, falling back to default profiles: default
main] o.s.b.w.embedded.tomcat.TomcatWebServer  : Tomcat initialized with port(s): 8083 (http)
main] o.apache.catalina.core.StandardService   : Starting service [Tomcat]
main] org.apache.catalina.core.StandardEngine  : Starting Servlet engine: [Apache Tomcat/9.0.33]
main] o.a.c.c.C.[Tomcat].[localhost].[/]       : Initializing Spring embedded WebApplicationContext
main] o.s.web.context.ContextLoader            : Root WebApplicationContext: initialization completed in 1365 ms
main] o.s.s.concurrent.ThreadPoolTaskExecutor  : Initializing ExecutorService 'applicationTaskExecutor'
main] o.s.b.w.embedded.tomcat.TomcatWebServer  : Tomcat started on port(s): 8083 (http) with context path ''
main] c.t.s.SpringbootWebApplication           : Started SpringbootWebApplication in 2.366 seconds (JVM running for 4.555)
```

图 4-5

代码中，方式二的运行结果如图 4-6 所示。

```
main] c.t.s.SpringbootWebApplication           : Starting SpringbootWebApplication on DESKTOP-246FP3G with PID 19892 (D:\de
main] c.t.s.SpringbootWebApplication           : No active profile set, falling back to default profiles: default
main] o.s.b.w.embedded.tomcat.TomcatWebServer  : Tomcat initialized with port(s): 8082 (http)
main] o.apache.catalina.core.StandardService   : Starting service [Tomcat]
main] org.apache.catalina.core.StandardEngine  : Starting Servlet engine: [Apache Tomcat/9.0.33]
main] o.a.c.c.C.[Tomcat].[localhost].[/]       : Initializing Spring embedded WebApplicationContext
main] o.s.web.context.ContextLoader            : Root WebApplicationContext: initialization completed in 1362 ms
main] o.s.s.concurrent.ThreadPoolTaskExecutor  : Initializing ExecutorService 'applicationTaskExecutor'
main] o.s.b.w.embedded.tomcat.TomcatWebServer  : Tomcat started on port(s): 8082 (http) with context path ''
main] c.t.s.SpringbootWebApplication           : Started SpringbootWebApplication in 2.4 seconds (JVM running for 4.637)
```

图 4-6

4.1.2 替换内置容器

Spring Boot 支持三种 Web 容器，分别是 Tomcat、Jetty、Undertow。虽然 Spring Boot 默认使用 Tomcat，但是我们可以根据自身需求进行切换。下面将示范如何从 Tomcat 容器切换到 Jetty 容器。

1. 配置 JettyWebServerFactory 信息

示例代码 4-4　JettyWebServerFactoryConfig.java

```
/**
* 使用内置的 Jetty Web 容器
 * 1. 配置 Jetty Web 容器
 * 2. 从 POM 中排除默认的 Tomcat 容器
 */
@Configuration
public class JettyWebServerFactoryConfig {
    /**
     * 方式一：
     * 配置当前 Web 容器 Jetty
     */
    @Bean
    public JettyServletWebServerFactory webServerFactory(){
        JettyServletWebServerFactory factory = new
JettyServletWebServerFactory();
        factory.setPort(8084);
        return factory;
    }
```

```
        /**
         * 方式二：
         * 配置 WebServerFactoryCustomizer，重写 customize()方法，强转相应的
JettyServletWebServerFactory。
         */
        @Bean
        public WebServerFactoryCustomizer customizer(){
            WebServerFactoryCustomizer customizer = new
WebServerFactoryCustomizer() {
                @Override
                public void customize(WebServerFactory factory) {
                    JettyServletWebServerFactory factory1 =
(JettyServletWebServerFactory)factory;
                    factory1.setPort(8085);
                }
            };
            return customizer;
        }
    }
```

2. 在 pom.xml 中剔除 Tomcat 内置容器并新增 Jetty 内置容器

示例代码 4-5　pom.xml（核心代码）

```
    <dependency>
        <groupId>org.springframework.boot</groupId>
        <artifactId>spring-boot-starter-web</artifactId>
        <exclusions>
            <!--排除默认的 Tomcat 容器-->
            <exclusion>
                <groupId>org.apache.tomcat.embed</groupId>
                <artifactId>tomcat-embed-core</artifactId>
            </exclusion>
        </exclusions>
    </dependency>
    <!--增加 jetty starter 依赖-->
    <dependency>
        <groupId>org.springframework.boot</groupId>
        <artifactId>spring-boot-starter-jetty</artifactId>
    </dependency>
```

代码中，方式一的运行结果如图 4-7 所示。

```
main] c.t.s.SpringbootWebApplication             : Starting SpringbootWebApplication on DESKTOP-246FP3G with PID 13704 (D:\dev\idea\proje
main] c.t.s.SpringbootWebApplication             : No active profile set, falling back to default profiles: default
main] org.eclipse.jetty.util.log                 : Logging initialized @4684ms to org.eclipse.jetty.util.log.Slf4jLog
main] o.s.b.w.e.j.JettyServletWebServerFactory   : Server initialized with port: 8084
main] org.eclipse.jetty.server.Server            : jetty-9.4.27.v20200227; built: 2020-02-27T18:37:21.340Z; git: a304fd9f351f337e7c0e2a7c
main] o.e.j.s.h.ContextHandler.application       : Initializing Spring embedded WebApplicationContext
main] o.s.web.context.ContextLoader              : Root WebApplicationContext: initialization completed in 1343 ms
main] org.eclipse.jetty.server.session           : DefaultSessionIdManager workerName=node0
main] org.eclipse.jetty.server.session           : No SessionScavenger set, using defaults
main] org.eclipse.jetty.server.session           : node0 Scavenging every 600000ms
main] o.e.jetty.server.handler.ContextHandler    : Started o.s.b.w.e.j.JettyEmbeddedWebAppContext@7f572c37{application,/,[file:///C:/User
main] org.eclipse.jetty.server.Server            : Started @4970ms
main] o.s.s.concurrent.ThreadPoolTaskExecutor    : Initializing ExecutorService 'applicationTaskExecutor'
main] o.e.j.s.h.ContextHandler.application       : Initializing Spring DispatcherServlet 'dispatcherServlet'
main] o.s.web.servlet.DispatcherServlet          : Initializing Servlet 'dispatcherServlet'
main] o.s.web.servlet.DispatcherServlet          : Completed initialization in 5 ms
main] o.e.jetty.server.AbstractConnector         : Started ServerConnector@67dba613{HTTP/1.1, (http/1.1)}{0.0.0.0:8084}
main] o.s.b.web.embedded.jetty.JettyWebServer    : Jetty started on port(s) 8084 (http/1.1) with context path '/'
main] c.t.s.SpringbootWebApplication             : Started SpringbootWebApplication in 2.633 seconds (JVM running for 5.343)
```

图 4-7

代码中，方式二的运行结果如图 4-8 所示。

```
main] c.t.s.SpringbootWebApplication             : Starting SpringbootWebApplication on DESKTOP-246FP3G with PID 5724 (D:\dev\
main] c.t.s.SpringbootWebApplication             : No active profile set, falling back to default profiles: default
main] org.eclipse.jetty.util.log                 : Logging initialized @4219ms to org.eclipse.jetty.util.log.Slf4jLog
main] o.s.b.w.e.j.JettyServletWebServerFactory   : Server initialized with port: 8085
main] org.eclipse.jetty.server.Server            : jetty-9.4.27.v20200227; built: 2020-02-27T18:37:21.340Z; git: a304fd9f351f3
main] o.e.j.s.h.ContextHandler.application       : Initializing Spring embedded WebApplicationContext
main] o.s.web.context.ContextLoader              : Root WebApplicationContext: initialization completed in 1221 ms
main] org.eclipse.jetty.server.session           : DefaultSessionIdManager workerName=node0
main] org.eclipse.jetty.server.session           : No SessionScavenger set, using defaults
main] org.eclipse.jetty.server.session           : node0 Scavenging every 600000ms
main] o.e.jetty.server.handler.ContextHandler    : Started o.s.b.w.e.j.JettyEmbeddedWebAppContext@aaee2a2{application,/,[file:
main] org.eclipse.jetty.server.Server            : Started @4501ms
main] o.s.s.concurrent.ThreadPoolTaskExecutor    : Initializing ExecutorService 'applicationTaskExecutor'
main] o.e.j.s.h.ContextHandler.application       : Initializing Spring DispatcherServlet 'dispatcherServlet'
main] o.s.web.servlet.DispatcherServlet          : Initializing Servlet 'dispatcherServlet'
main] o.s.web.servlet.DispatcherServlet          : Completed initialization in 8 ms
main] o.e.jetty.server.AbstractConnector         : Started ServerConnector@121c54fa{HTTP/1.1, (http/1.1)}{0.0.0.0:8085}
main] o.s.b.web.embedded.jetty.JettyWebServer    : Jetty started on port(s) 8085 (http/1.1) with context path '/'
main] c.t.s.SpringbootWebApplication             : Started SpringbootWebApplication in 2.449 seconds (JVM running for 4.814)
```

图 4-8

4.1.3 采用外部容器

本书中的外部容器以 Tomcat 9.0.37 为例。

（1）设置 pom.xml 依赖。

示例代码 4-6　pom.xml（部分代码）

```
<!--以 war 形式打包 -->
<packaging>war</packaging>
<dependency>
    <groupId>org.springframework.boot</groupId>
    <artifactId>spring-boot-starter-web</artifactId>
    <exclusions>
        <!--排除默认的 Tomcat 容器-->
        <exclusion>
            <groupId>org.apache.tomcat.embed</groupId>
            <artifactId>tomcat-embed-core</artifactId>
        </exclusion>
```

```
        </exclusions>
    </dependency>
    <!-- 增加 servlet-api 依赖 -->
    <dependency>
        <groupId>javax.servlet</groupId>
        <artifactId>javax.servlet-api</artifactId>
    </dependency>
```

（2）入口类继承 SpringBootServletInitializer.java。

示例代码 4-7　SpringbootWebApplication.java

```
@SpringBootApplication
public class SpringbootWebApplication extends SpringBootServletInitializer {
    public static void main(String[] args) {
        SpringApplication.run(SpringbootWebApplication.class, args);
    }
    @Override
    protected SpringApplicationBuilder configure(SpringApplicationBuilder
builder) {
        // 注意这里要指向原先用 main 方法执行的 Application 启动类
        return builder.sources(SpringbootWebApplication.class);
    }
}
```

（3）编写一个 Controller 测试方法。

示例代码 4-8　WebServerController.java

```
@RestController
public class WebServerController {
    @RequestMapping("/hello")
    public String sayHello(){
        return "hello world!";
    }
}
```

（4）在 IDEA 中配置外部 Tomcat 服务、端口、热部署等信息，如图 4-9 所示。

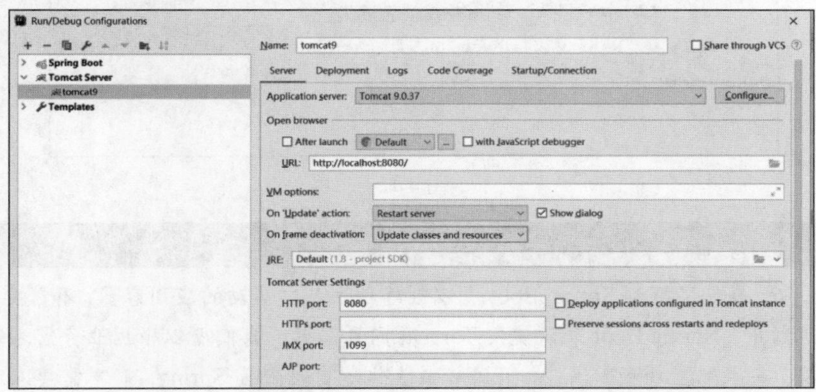

图 4-9

（5）采用 maven install 进行项目打包，如图 4-10 所示。

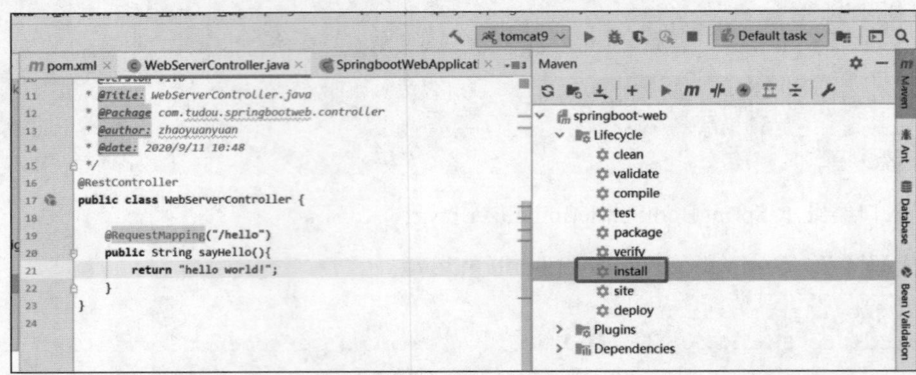

图 4-10

（6）将打包好的项目部署至 Tomcat 9 中，并且可以根据自身需求配置上下文路径，此处上下文路径以“/”为例，如图 4-11 所示。

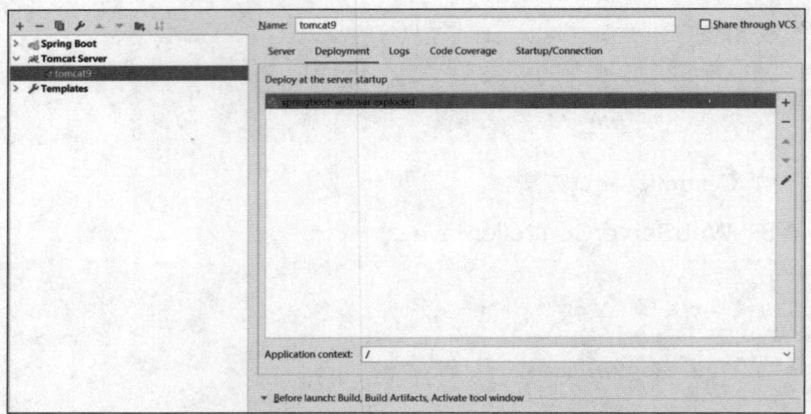

图 4-11

（7）运行项目，访问测试路径 http://localhost:8080/hello，结果如图 4-12 所示。

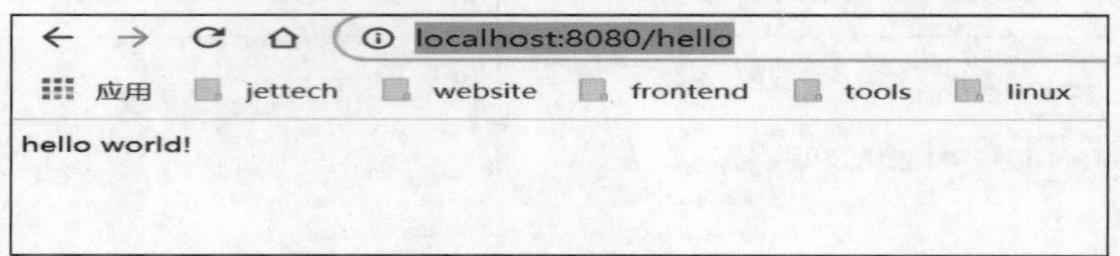

图 4-12

小结

通过以上分析，我们了解到 Spring Boot 可以创建独立且自启动的应用容器，不需要构建 war 包发布到容器中。Spring Boot 默认使用 Tomcat 内置容器，我们可以根据业务需求进行内置容器的切换，甚至配置外置容器，并且可以最大化地自动配置 Spring，而不需要人工配置各种参数等。这些特性为我们的系统开发提供了很大的便利。

4.2　Spring MVC 支持

4.2.1　视图解析器

Spring MVC 的视图解析器组件通过配置好视图的前缀和后缀，根据方法的返回值来得到视图的 View 对象，从而让视图的访问变得方便、安全。视图也就是页面，有传统的 JSP 页面、模板引擎页面和静态 HTML 页面等。前面已经论述过 Spring MVC 的一些知识，这里不再赘述，下面通过一些典型的例子来看看 Spring Boot 如何对 Spring MVC 的视图解析器进行自动配置。

首先，我们创建好 src/main/webapps 文件夹，用于存放 Web 资源。在 webapps 文件夹下新建 html 文件夹，然后在 html 文件夹下新建一个 student.html 用于测试视图解析器，如图 4-13 所示。student.html 的内容无关紧要，显示“学生信息”这几个字即可。

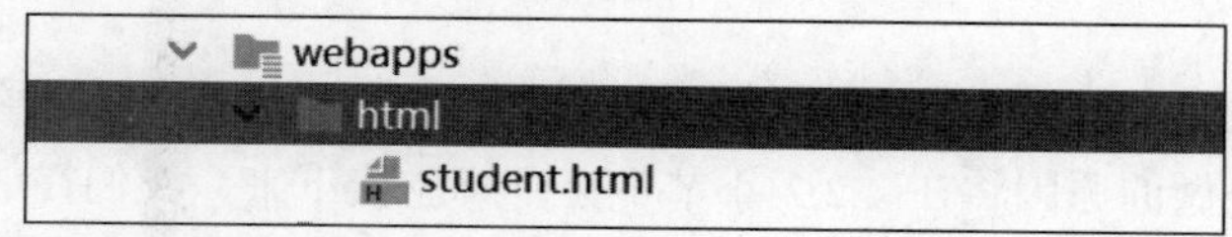

图 4-13

我们新建了 webapps 资源文件夹，如果不做特殊的配置，webapps 文件夹的所有文件是不会编译进 target/classes 目录下的，更不会被打包到 jar 包中。我们需要在 pom.xml 文件的 build 标签中（如果没有就增加一个）新增两个 resource 标签，分别用于打包 src/main/resources 资源文件夹和 src/main/webapps 资源文件夹，如示例代码 4-9 所示。

示例代码 4-9　pom.xml（部分代码）

```
<build>
    <plugins>
        <plugin>
            <groupId>org.springframework.boot</groupId>
            <artifactId>spring-boot-maven-plugin</artifactId>
        </plugin>
    </plugins>
    <resources>
        <resource>
            <directory>src/main/resources</directory>
            <includes>
                <include>**/**</include>
            </includes>
        </resource>
        <resource>
            <directory>src/main/webapps</directory>
            <targetPath>META-INF/resources</targetPath>
            <includes>
                <include>**/**</include>
            </includes>
```

```
        </resource>
    </resources>
</build>
```

通过 Maven 编译后发现 src/main/webapps 下的文件被成功编译进 target/classes 目录下，如图 4-14 所示。

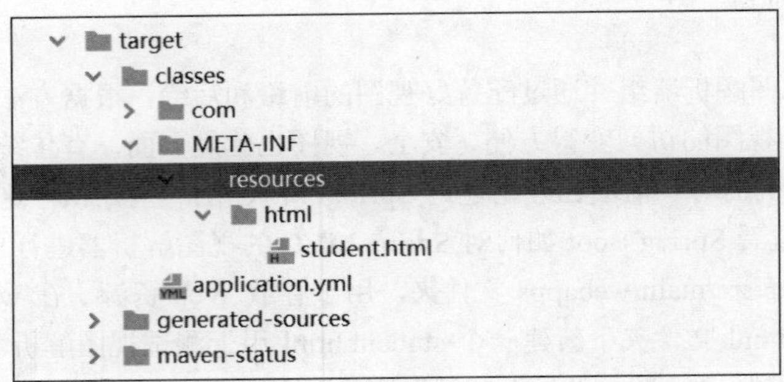

图 4-14

需要注意的是，src/main/webapps 目录下的所有文件要打包到 META-INF/resources 目录下，这个目录不能随意，这样做的原因将在 4.2.2 小节中会讲解。接下来，我们在 application.yml 里配置视图解析器的前缀和后缀，参考示例代码 4-10。

示例代码 4-10　application.yml（部分代码）

```
spring:
  mvc:
    view:
      prefix: /html/
      suffix: .html
```

新建 StudentController 类，并写一个 toStu 方法，注意类上面不能用@RestController 注解替代@Controller 注解，否则视图解析器将不会生效。访问 toStu 方法即可跳转到 student.html 页面。

示例代码 4-11　StudentController.java

```
@Controller
public class StudentController {
    @RequestMapping("/toStu")
    public String toStu(){
        return "student";
    }
}
```

启动项目，在 Chrome 浏览器上访问 http://localhost:8080/toStu，可以看到通过访问 Controller 成功地跳转到了 HTML 页面，如图 4-15 所示。这就是视图解析器作用的体现。

图 4-15

4.2.2　支持静态资源

HTML 是静态资源，用 JS 实现的一些动态效果也是静态资源，css 样式文件、image 图片文件等都是静态资源。静态资源的一个特征就是一成不变（静态文件不会随着服务端数据的变化而变化），即使是用 JS 实现了一些非常炫酷的动态效果。

典型的动态资源就是我们熟知的 JSP 页面。JSP 的本质是 Servlet，我们在访问 JSP 页面的时候，JSP 会被编译成 Servlet，既然 Servlet 是服务端的，那么 JSP 也是服务端的，我们可以在 JSP 中通过四大作用域 Page、Request、Session、Application 来读写服务端的数据。关于 JSP 的知识，如果大家有兴趣可以去阅读其他图书。JSP 访问慢，严重影响服务器性能，逐渐被 HTML 所取代。这里将重点讲解静态资源。

通过 ResourceProperties 类的源码可知，该类中定义了一个 String 数组类型的变量 CLASSPATH_RESOURCE_LOCATIONS，Spring Boot 支持将静态资源文件放在如下目录：

```
classpath:/META-INF/resources/
classpath:/resources/
classpath:/static/
classpath:/public/
```

ResourceProperties 类的源码如图 4-16 所示。

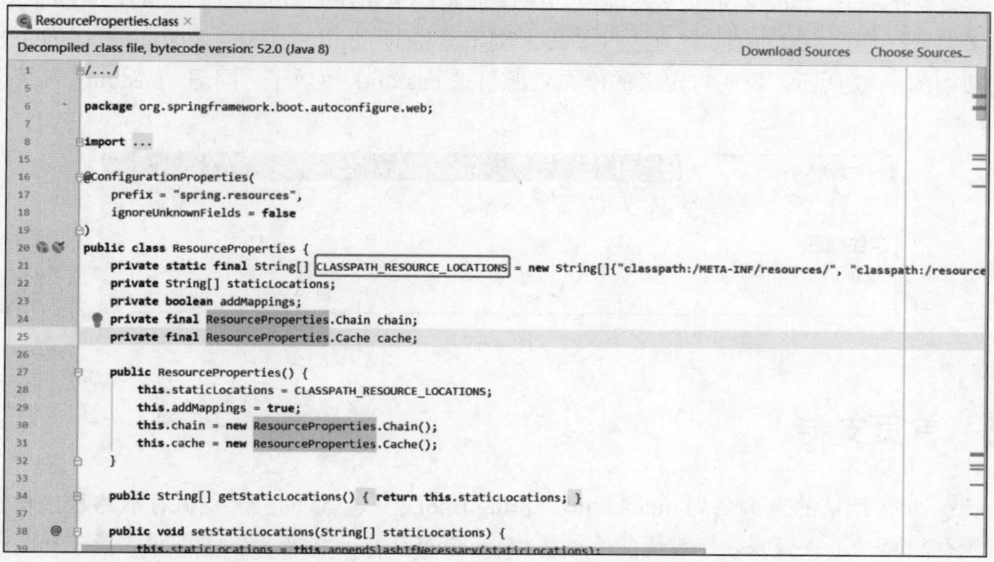

```
ResourceProperties.class ×
Decompiled .class file, bytecode version: 52.0 (Java 8)    Download Sources    Choose Sources...
/.../

package org.springframework.boot.autoconfigure.web;

import ...

@ConfigurationProperties(
    prefix = "spring.resources",
    ignoreUnknownFields = false
)
public class ResourceProperties {
    private static final String[] CLASSPATH_RESOURCE_LOCATIONS = new String[]{"classpath:/META-INF/resources/", "classpath:/resource
    private String[] staticLocations;
    private boolean addMappings;
    private final ResourceProperties.Chain chain;
    private final ResourceProperties.Cache cache;

    public ResourceProperties() {
        this.staticLocations = CLASSPATH_RESOURCE_LOCATIONS;
        this.addMappings = true;
        this.chain = new ResourceProperties.Chain();
        this.cache = new ResourceProperties.Cache();
    }

    public String[] getStaticLocations() { return this.staticLocations; }

    public void setStaticLocations(String[] staticLocations) {
        this.staticLocations = this.appendSlashIfNecessary(staticLocations);
```

图 4-16

当客户端想要访问一个静态资源时，会按先后顺序依次访问上面的四个路径，如果在 classpath:/META-INF/resources 中找到了该静态资源，就不用往下找了。

另外，Spring Boot 提供了 spring.resources.static-locations 配置，如果上面的四个路径不够用，或是有特殊需求，可以用该配置增加静态资源存放的路径，且多个路径之间用逗号分隔，比如增加两个新的资源存放路径 classpath:/META-INF/custom1 和 classpath:/META-INF/custom2，如下代码所示。

示例代码 4-12　application.yml

```
spring:
  resources:
    static-locations: classpath:/META-INF/custom1/,classpath:
/META-INF/custom2/
```

然后，我们修改一下 pom.xml，将 webapps 下的静态文件打包到 custom1 和 custom2 目录下，如下代码所示。

示例代码 4-13　pom.xml 部分代码

```
<resource>
    <directory>src/main/webapps</directory>
    <targetPath>META-INF/custom1</targetPath>
</resource>
<resource>
    <directory>src/main/webapps</directory>
    <targetPath>META-INF/custom2</targetPath>
</resource>
```

用于测试的静态资源文件 student.html 在 4.2.1 小节中已经提到过，接下来我们重新用 Maven 编译一次，目的是让 student.html 文件能成功编译进 META-INF/custom1 和 META-INF/custom2 目录。编译完成之后，启动项目，并通过 Chrome 浏览器访问 http://localhost:8080/html/student.html。访问成功如图 4-17 所示，说明自定义的静态资源目录 custom1 生效了。同理，custom2 目录也会生效。

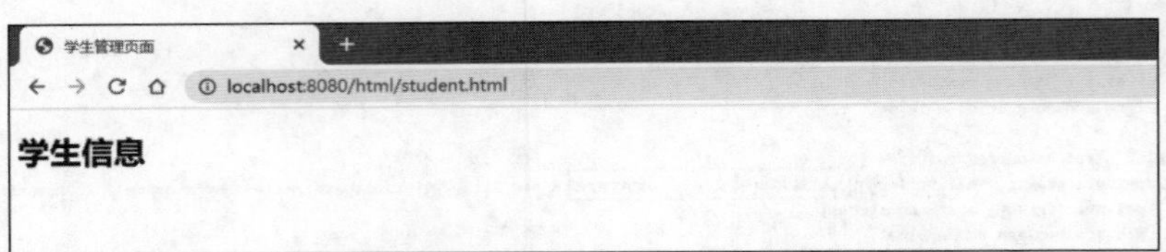

图 4-17

4.2.3　首页支持

Spring Boot 默认的首页名为 index.html。Spring Boot 会从默认的静态资源存放路径或自定义的静态资源存放路径下去查找，一般情况下放在 src/main/webapps 的根目录下即可。新建 index.html，如图 4-18 所示。由于只是测试，因此其内容只显示 welcome 即可。

图 4-18

启动项目，在 Chrome 浏览器上访问 http://localhost:8080，即可看到首页的内容，如图 4-19 所示。

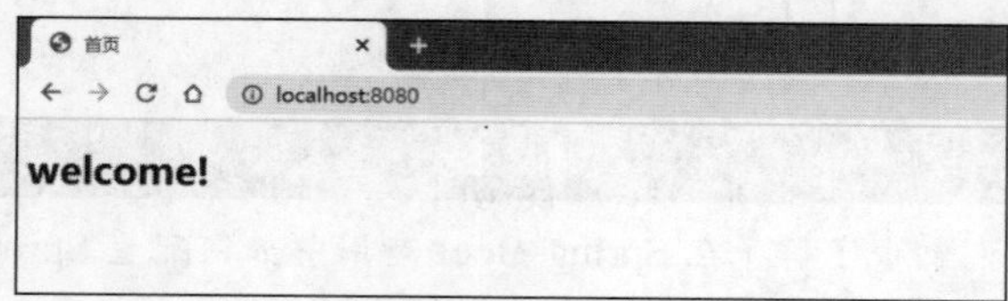

图 4-19

4.2.4 网站 logo 设置

一个成熟的网站离不开 logo 图标，比如访问 Tomcat 资源的时候会在浏览器的标签页上看到一个汤姆猫的 logo 图标。对于 Spring Boot 2.x 版本应用而言，如果我们不设置 logo 图标，那么默认是没有的。

Spring Boot 会从默认的静态资源存放路径或自定义的静态资源路径下去查找 logo 图标文件，默认为 favicon.ico，如果存在这样的文件，就将自动用作应用程序的 favicon。

准备一张图片，可以通过现成的网站（比如 http://www.bitbug.net）去制作 ico 文件，如图 4-20 所示。

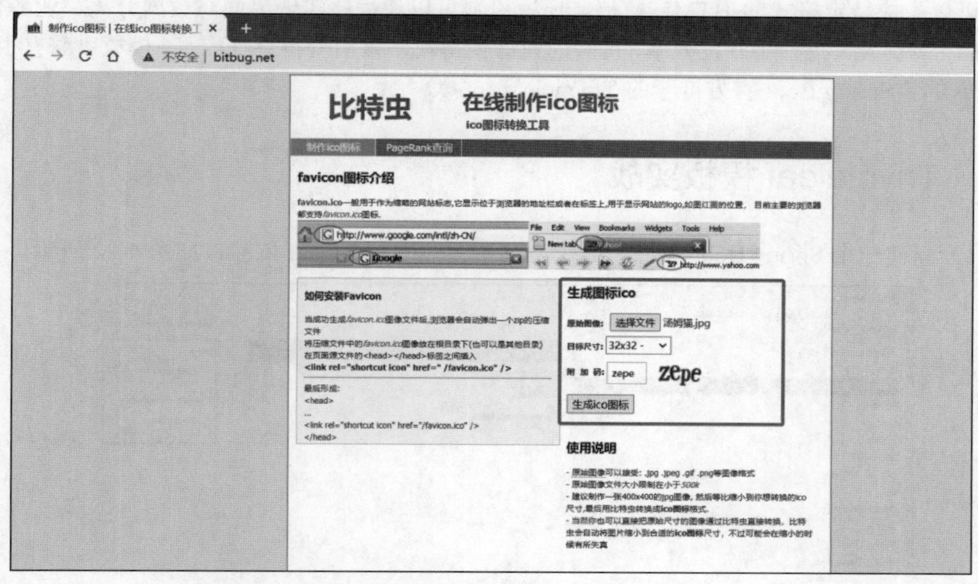

图 4-20

将生成的 ico 文件命名为 favicon.ico，并将其放在 src/main/webapps 下，启动项目，通过浏览器访问首页，可以看到汤姆猫的图标成功展示出来了，如图 4-21 所示。

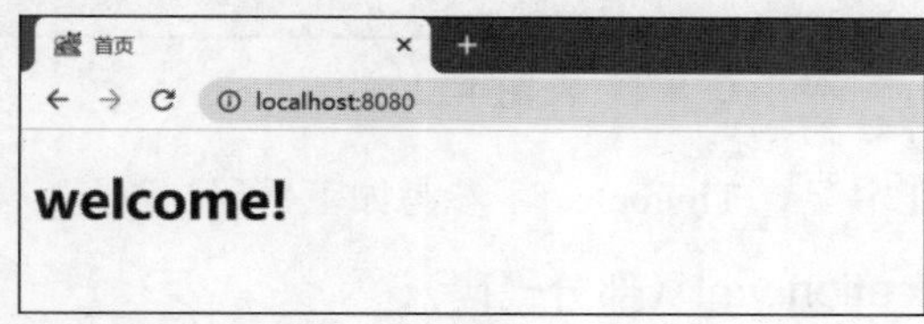

图 4-21

如果看不到图标，可以使用 Ctrl+F5 快捷键刷新网页。

小　结

通过本节的学习，我们初步了解了在 Spring Boot 应用中如何配置 Spring MVC 的视图解析器，以及静态资源、应用首页、网站 logo 的配置，这些都是 Web 开发中常用且比较基础的知识点。当然，仅仅了解这些知识点是远远不够的，更多深层次的知识点将在后面继续为大家讲解。

4.3 模板引擎集成

4.3.1 概述

这一节主要讲的是 Web 开发的模板引擎。模板引擎是为了使用户界面与业务数据分离而产生的，它可以生成特定格式的 HTML 文档。模板引擎可以让程序实现界面与数据分离、业务代码与逻辑代码分离，大大提升了开发效率，良好的设计使得代码重用变得更加容易。模板引擎有很多种，这里主要介绍 Spring Boot 官方推荐的 Thymeleaf。

4.3.2 Thymeleaf 模板实战

（1）新建一个 Spring Boot 项目，选择 Web 依赖和 Thymeleaf 依赖，如图 4-22 所示。

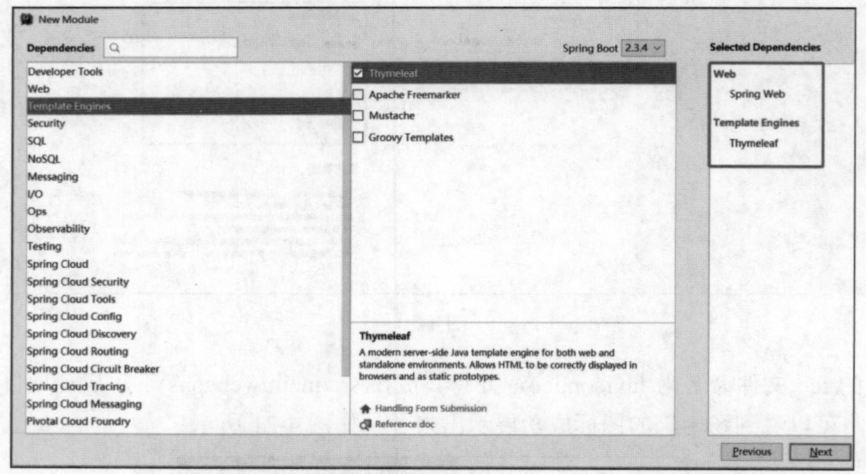

图 4-22

（2）在 application.yml 中配置 Thymeleaf，参考如下代码。

示例代码 4-14　application.yml（部分代码）

```
spring:
```

```
    thymeleaf:
      prefix: classpath:/templates/
      suffix: .html
      encoding: UTF-8
      mode: HTML5
      cache: false
      servlet:
        content-type: text/html
  check-template-location: true
```

其中，prefix 与 suffix 相当于模板页面的视图解析配置，比如 Controller 方法返回了字符串"product"，那么视图解析器就会到找到 classpath/templates 路径下的 product.html，并返回给浏览器。

（3）编写实体类，参考代码如下。

示例代码 4-15　Product.java（部分代码）

```
public class Product {
    private Integer id;
    private String name;
    private BigDecimal price;
    private Integer amount;
    public Product(Integer id, String name, BigDecimal price, Integer amount){
        this.id = id;
        this.name = name;
        this.price = price;
        this.amount = amount;
    }
    public Integer getId() {return id; }
    public void setId(Integer id) {this.id = id; }
    public String getName() {return name; }
    public void setName(String name) {this.name = name; }
    public BigDecimal getPrice() {return price; }
    public void setPrice(BigDecimal price) {this.price = price;}
    public Integer getAmount() {return amount;}
    public void setAmount(Integer amount) {this.amount = amount;}
}
```

（4）编写 Controller 接口，参考代码如下。

示例代码 4-16　ProductController.java（部分代码）

```
@Controller
public class ProductController {
    @RequestMapping("/product/list")
    public String list(Model model) {
        Product p1 = new Product(1,"华为 P10",BigDecimal.valueOf(2599),666);
        Product p2 = new Product(2,"Thinkpad
T480",BigDecimal.valueOf(8999),1900);
        Product p3 = new Product(3,"耳机",BigDecimal.valueOf(30),388);
        Product p4 = new Product(4,"键盘",BigDecimal.valueOf(65),554);
        Product p5 = new Product(5,"鼠标",BigDecimal.valueOf(48),1268);
```

```
        List<Product> data = new ArrayList<>();
        data.add(p1);
        data.add(p2);
        data.add(p3);
        data.add(p4);
        data.add(p5);
        model.addAttribute("data",data);
        return "product";
    }
}
```

当浏览器客户端请求这个接口时，接口返回值“product”会被视图解析器翻译为 classpath:/templates/product.html，而 Thymeleaf 模板解析包会将 Model 中的数据填充到模板中，最终返回给浏览器客户端。

（5）编写模板页面。

示例代码 4-17 product.html（部分代码）

```
<!DOCTYPE html>
<html lang="en" xmlns:th="http://www.thymeleaf.org">
    <head>
        <meta charset="UTF-8">
        <title>商品详情</title>
        <script th:src="@{/jquery-3.4.1.js}"></script>
        <link type="text/css" rel="stylesheet"
th:href="@{/bootstrap/css/bootstrap.min.css}">
    </head>
    <body>
        <table class="table">
            <tr>
                <th>ID</th>
                <th>商品名称</th>
                <th>商品价格</th>
                <th>商品数量</th>
            </tr>
            <tr th:each="p : ${data}">
                <td th:text="${p.id}">默认 ID</td>
                <td th:text="${p.name}">默认商品名称</td>
                <td th:text="${p.price}">默认商品价格</td>
                <td th:text="${p.amount}">默认商品数量</td>
            </tr>
        </table>
    </body>
</html>
```

首先可以看到，在 html 标签内部加入了 xmlns:th=http://www.thymeleaf.org，以此来引入 Thymeleaf 命名空间，此时在 html 模板文件中动态的属性使用“th:”命名空间修饰，这样才可以在其他标签里面使用“th:”语法，这是其他语法的前提。

引入脚本文件的语法是 th:src="@{相对路径或绝对路径}"，而引入样式文件的语法是

th:href="@{相对路径或绝对路径 }"。注意，如上面的代码所示，这里用的是绝对路径，脚本文件和样式文件都是静态资源，因此 Spring Boot 会从 4.2.2 节中讲的静态资源目录中寻找。

为了让前端样式不过于单调，这里引入前端框架 BootStrap 来装饰 table 表格，只需要在 table 标签上加上 class="table"即可简单使用 BootStrap。

由于后端返回的数据是一个集合，因此需要用到 Thymeleaf 的遍历语法。比如 th:each="p : ${data}"，其中${data}用来获取 Model 中的数据，p 是变量名，可以用${p.属性名}来获取属性值，然后用 th:text 语法来给 td 元素赋值。

（6）运行后端项目，访问 http://localhost:8080/product/list，数据以 bootstrap 表格的样式展示出来，如图 4-23 所示。

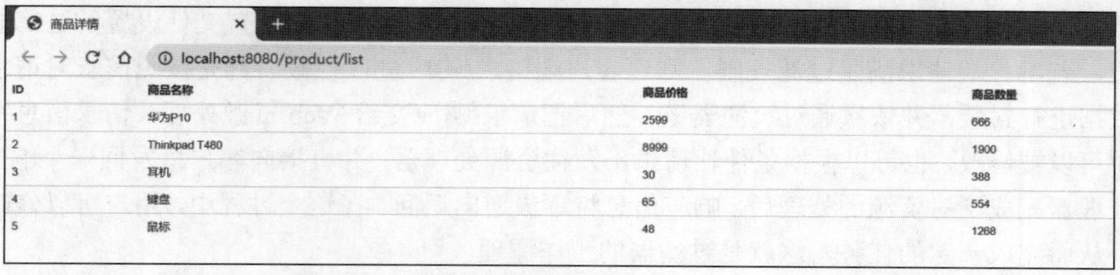

商品详情

localhost:8080/product/list

ID	商品名称	商品价格	商品数量
1	华为P10	2599	666
2	Thinkpad T480	8999	1900
3	耳机	30	388
4	键盘	65	554
5	鼠标	48	1268

图 4-23

小　结

本节的主要内容是集成模板引擎，首先叙述了模板引擎是什么，与传统的 JSP 技术对比有什么优势（为什么要用它）；然后介绍了目前主流的几种模板引擎技术，以及为什么要用 Thymeleaf；最后通过集成 Thymeleaf 模板引擎，以实战的角度来让大家体会模板引擎开发的一般步骤。

4.4　过滤器、拦截器与监听器

4.4.1　过滤器

1. 概述

过滤器是 Web 开发中很实用的一项技术，开发人员可以通过过滤器对 Web 服务管理的资源静态 HTML 文件、静态图片、JSP、Servlet 等进行拦截，从而实现一些特殊的需求，比如设置 URL 的访问权限、过滤敏感词汇、压缩响应信息等。过滤器还适用于对用户请求和响应对象进行检查和修改，但是 Filter 本身并不生成请求和响应对象，只是提供过滤功能。Filter 的完整工作流程如图 4-24 所示。

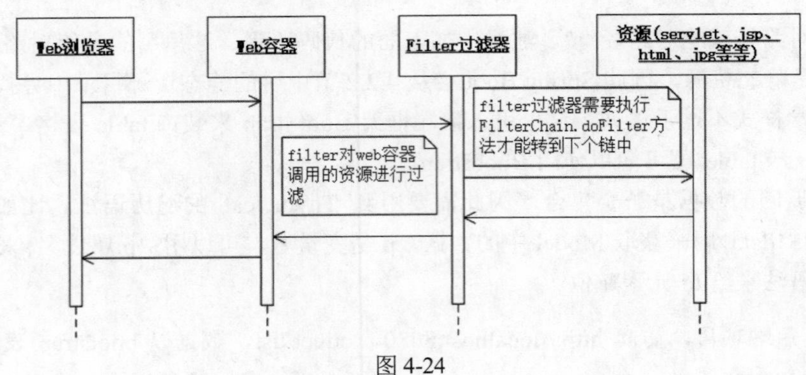

图 4-24

当客户端发出对 Web 资源的请求时，Web 服务器会根据应用程序配置文件设置的过滤规则进行检查，若客户端请求满足过滤规则，则对客户端请求/响应进行拦截。首先按照需求对请求头和请求数据进行封装，并依次通过过滤器链，然后把请求/响应交给 Web 资源处理，请求信息在过滤器链中可以被修改，也可以根据条件让请求不发往资源处理器，并直接向客户机发回一个响应。当资源处理器完成了对资源的处理后，响应信息将逐级逆向返回。在这个过程中，用户可以修改响应信息，从而完成一定的任务。这就是过滤器的工作原理。

另外，过滤器的生命周期也是由 Web 服务器进行负责的，但是相比真正的 Servlet 又有区别。Filter 的生命周期大致分为以下三个阶段：

（1）实例化：Web 容器在部署 Web 应用程序时对所有过滤器进行实例化，此时 Web 容器调用的是它的无参构造方法。

（2）初始化：实例化完成之后，马上进行初始化工作。Web 容器回调 init()方法。请求路径匹配过滤器的 URL 映射时，Web 容器回调过滤器的 doFilter()方法，此方法也是过滤器的核心方法。

（3）销毁：Web 容器在卸载 Web 应用程序前回调 doDestory 方法。

2. @WebFilter 实现自定义过滤器

自定义过滤器常用的有两种方式，这里主要讲解通过@WebFilter 和@ServletComponentScan 实现过滤器。定义的过滤器 WebTestFilter.java 必须实现 javax.servlet.Filter 接口，根据自身需求重写 doFilter()方法。另外，在启动类上加一个注解@ServletComponentScan。其中，urlPatterns 是配置过滤器要过滤的 URL，可支持模糊匹配；filterName 只是定义的过滤器名字；@Order（int）注解主要用于多个过滤器时定义执行顺序，值越小越先执行（后面将会详细讲解配置多个 Filter 的执行顺序）。参考代码如下。

示例代码 4-18　WebTestFilter.java（部分代码）

```
/**
 * 测试单个过滤器
 */
@WebFilter(urlPatterns = "/html/*",filterName = "WebFilter")
@Order(1)
public class WebTestFilter implements Filter {
    @Override
    public void init(FilterConfig filterConfig) throws ServletException {
    }
```

```
    @Override
    public void doFilter(ServletRequest servletRequest, ServletResponse
servletResponse, FilterChain filterChain) throws IOException, ServletException {
        System.out.println("WebTestFilter");
        filterChain.doFilter(servletRequest,servletResponse);
    }
    @Override
    public void destroy() {
    }
}
```

启动项目，访问要过滤的资源（此处以 http://localhost:8080/html/test.html 为例），就可以看到控制台打印的相关内容。如图 4-25 所示。

```
2020-10-02 17:13:13.323  INFO 18852 --- [nio-8080-exec-1] o.a.c.c.C.[Tomcat].[localhost].[/]       : Initializing Spring DispatcherServlet 'dispatcherServlet'
2020-10-02 17:13:13.323  INFO 18852 --- [nio-8080-exec-1] o.s.web.servlet.DispatcherServlet        : Initializing Servlet 'dispatcherServlet'
2020-10-02 17:13:13.327  INFO 18852 --- [nio-8080-exec-1] o.s.web.servlet.DispatcherServlet        : Completed initialization in 4 ms
WebTestFilter
```

图 4-25

3. 自定义配置类配置过滤器

自定义配置类配置过滤器简而言之就是定义一个过滤器，实现 javax.servlet.Filter 接口，然后将此过滤器以@Bean 的形式注册到配置类中。

过滤器类的参考代码如下。

示例代码 4-19　RequestTest1Filter.java

```
/**
 * 过滤器，测试案例 1
 */
public class RequestTest1Filter implements Filter {
    @Override
    public void init(FilterConfig filterConfig) throws ServletException {
        String filterName = filterConfig.getInitParameter("name");
        System.out.println("过滤器测试案例 1,name:"+filterName);
    }
    @Override
    public void doFilter(ServletRequest servletRequest, ServletResponse
servletResponse, FilterChain filterChain) throws IOException, ServletException {
        System.out.println("test1-filter");
        filterChain.doFilter(servletRequest,servletResponse);
    }
    @Override
    public void destroy() {
    }
}
```

过滤器配置类的参考代码如下。

示例代码 4-20　WebFilterConfig.java

```
/**
```

```
 * 配置过滤器
 */
@Configuration
public class WebFilterConfig  {
   @Bean
   public FilterRegistrationBean filterRegistTest1(){
      FilterRegistrationBean<Filter> frBean = new
FilterRegistrationBean<>();
      RequestTest1Filter requestTest1Filter = new RequestTest1Filter();
      frBean.setFilter(requestTest1Filter);
      frBean.addUrlPatterns("/html/*");
      frBean.addInitParameter("name","filter-test1");
      frBean.setName("requestTest1Filter");
      frBean.setOrder(4);
      return frBean;
   }
}
```

此时只要保证 WebFilterConfig.java 被扫描到即可，即配置类上增加@Configuration 注解，然后启动 Spring Boot 项目即可看到过滤器里定义的内容。此处请求 http://localhost:8080/html/test.html 测试，结果如图 4-26 所示：第一行打印的是过滤器名字，第二行是拦截时执行的内容。

```
过滤器测试案例1,name:filter-test1
2020-10-02 18:19:30.222  INFO 14804 --- [           main] o.s.s.concurrent.ThreadPoolTaskExecutor  : Initializing ExecutorService 'applicationTaskExecutor'
2020-10-02 18:19:30.380  INFO 14804 --- [           main] o.s.b.w.embedded.tomcat.TomcatWebServer  : Tomcat started on port(s): 8080 (http) with context path
2020-10-02 18:19:30.383  INFO 14804 --- [           main] c.t.s.SpringbootrestApplication          : Started SpringbootrestApplication in 2.35 seconds (JVM ru
2020-10-02 18:19:42.576  INFO 14804 --- [nio-8080-exec-1] o.a.c.c.C.[Tomcat].[localhost].[/]       : Initializing Spring DispatcherServlet 'dispatcherServlet'
2020-10-02 18:19:42.576  INFO 14804 --- [nio-8080-exec-1] o.s.web.servlet.DispatcherServlet        : Initializing Servlet 'dispatcherServlet'
2020-10-02 18:19:42.581  INFO 14804 --- [nio-8080-exec-1] o.s.web.servlet.DispatcherServlet        : Completed initialization in 5 ms
test1-filter
```

图 4-26

4. 配置多个过滤器

对于通过@WebFilter 定义的多个过滤器，我们真的能通过@Order(int)注解实现自定义执行顺序吗？我们可以通过下面的例子测试一下。

示例代码 4-21　AFilter.java

```
/**
 * 用于测试 order 定义的过滤器顺序
 */
@WebFilter(urlPatterns = "/html/*",filterName = "AFilter")
@Order(2)
public class AFilter implements Filter {
   @Override
   public void init(FilterConfig filterConfig) throws ServletException {
   }
   @Override
   public void doFilter(ServletRequest servletRequest, ServletResponse
servletResponse, FilterChain filterChain) throws IOException, ServletException {
      System.out.println("AFilter");
      filterChain.doFilter(servletRequest,servletResponse);
   }
```

```
    @Override
    public void destroy() {
    }
}
```

运行项目，访问 http://localhost:8080/html/test.html，对比 AFilter.java 和 WebTestFilter.java 两个过滤器执行的顺序，结果如图 4-27 所示。我们发现先执行的是过滤器 AFilter.java，后执行的是 WebTestFilter.java。

```
2020-10-02 18:31:56.655  INFO 16288 --- [nio-8080-exec-1] o.a.c.c.C.[Tomcat].[localhost].[/]       : Initializing Spring DispatcherServlet 'dispatcherServlet
2020-10-02 18:31:56.655  INFO 16288 --- [nio-8080-exec-1] o.s.web.servlet.DispatcherServlet        : Initializing Servlet 'dispatcherServlet'
2020-10-02 18:31:56.660  INFO 16288 --- [nio-8080-exec-1] o.s.web.servlet.DispatcherServlet        : Completed initialization in 5 ms
AFilter
WebTestFilter
```

图 4-27

按理说，通过@Order 注解指定一个 int 值，越小越先执行，但是 AFilter.java 中定义的执行顺序是 order(2)，WebTestFilter.java 中定义的执行顺序是 order(1)，结果显示先执行的是 AFilter.java 过滤器，说明@Order(int)注解和@WebFilter 注解一起使用的时候并未生效。我们可以通过源码发现，@WebFilter 修饰的过滤器在加载时并未使用@Order 注解，而是使用类名来自定义 Filter 的执行顺序，正如当下两个比对的过滤器，执行顺序 AFilter.java > WebTestFilter.java。感兴趣的读者可以学习@ServletComponentScan 和@WebFilter 相关源码，此处不再赘述。

从上面的例子我们会发现，如果想以@WebFilter 方式定义多个过滤器并且限定过滤器执行顺序，就必须限定 Filter 的类名了。以上情况对开发者来说并不友好，所以建议使用配置类注册过滤器的方式配置多个过滤器。

（1）新增一个过滤器，参考代码如下。

示例代码 4-22 RequestTest2Filter.java

```
/**
 * 过滤器，测试案例 2
 */
public class RequestTest2Filter implements Filter {
    @Override
    public void init(FilterConfig filterConfig) throws ServletException {
        String filterName = filterConfig.getInitParameter("name");
        System.out.println("过滤器测试案例 2,name:"+filterName);
    }
    @Override
    public void doFilter(ServletRequest servletRequest, ServletResponse
servletResponse, FilterChain filterChain) throws IOException, ServletException {
        System.out.println("test2-filter");
        filterChain.doFilter(servletRequest,servletResponse);
    }
    @Override
    public void destroy() {
    }
```

（2）对应修改配置类 WebFilterConfig.java，如下代码所示。将两个过滤器

RequestTest1Filter.java 和 RequestTest2Filter.java 注册到配置类 WebFilterConfig.java 中，指定 RequestTest1Filter.java 的执行顺序 order 为 2、RequestTest1Filter.java 的执行顺序 order 为 1。

示例代码 4-23　WebFilterConfig .java

```
/**
 * 配置过滤器，测试过滤器执行顺序
 */
@Configuration
public class WebFilterConfig  {
    @Bean
    public FilterRegistrationBean filterRegistTest1(){
        FilterRegistrationBean<Filter> frBean = new
FilterRegistrationBean<>();
        RequestTest1Filter requestTest1Filter = new RequestTest1Filter();
        frBean.setFilter(requestTest1Filter);
        frBean.addUrlPatterns("/html/*");
        frBean.addInitParameter("name","filter-test1");
        frBean.setName("requestTest1Filter");
        frBean.setOrder(2);
        return frBean;
    }
    @Bean
    public FilterRegistrationBean filterRegistTest2(){
        FilterRegistrationBean<Filter> frBean = new
FilterRegistrationBean<>();
        RequestTest2Filter requestTest2Filter = new RequestTest2Filter();
        frBean.setFilter(requestTest2Filter);
        //frBean.addUrlPatterns("/html/*");
        //配置多个过滤规则
        ArrayList<String> urlList = new ArrayList<>();
        urlList.add("/html/test.html");
        urlList.add("/html/login.html");
        frBean.setUrlPatterns(urlList);
        frBean.addInitParameter("name","filter-test2");
        frBean.setName("requestTest2Filter");
        frBean.setOrder(1);
        return frBean;
    }
}
```

运行项目，访问 http://localhost:8080/html/test.html，测试结果如图 4-28 所示，正如我们配置的顺序，先执行的是过滤器 RequestTest2Filter.java，后执行的是 RequestTest1Filter.java。

```
2020-10-02 18:57:09.949  INFO 20596 --- [           main] o.s.web.context.ContextLoader            : Root WebApplicationContext: initialization completed in 1389 ms
过滤器测试案例1,name:filter-test1
过滤器测试案例2,name:filter-test2
2020-10-02 18:57:10.193  INFO 20596 --- [           main] o.s.s.concurrent.ThreadPoolTaskExecutor  : Initializing ExecutorService 'applicationTaskExecutor'
2020-10-02 18:57:10.375  INFO 20596 --- [           main] o.s.b.w.embedded.tomcat.TomcatWebServer  : Tomcat started on port(s): 8080 (http) with context path ''
2020-10-02 18:57:10.378  INFO 20596 --- [           main] c.t.s.SpringbootrestApplication          : Started SpringbootrestApplication in 2.546 seconds (JVM running for 4.85)
2020-10-02 19:02:54.441  INFO 20596 --- [nio-8080-exec-1] o.a.c.c.C.[Tomcat].[localhost].[/]       : Initializing Spring DispatcherServlet 'dispatcherServlet'
2020-10-02 19:02:54.441  INFO 20596 --- [nio-8080-exec-1] o.s.web.servlet.DispatcherServlet        : Initializing Servlet 'dispatcherServlet'
2020-10-02 19:02:54.446  INFO 20596 --- [nio-8080-exec-1] o.s.web.servlet.DispatcherServlet        : Completed initialization in 5 ms
test2-filter
test1-filter
```

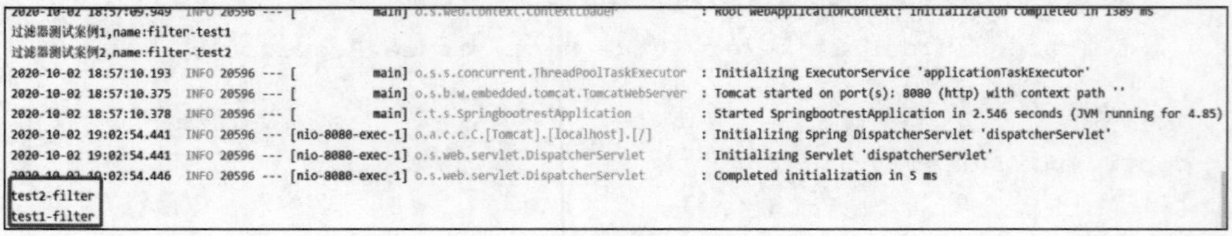

图 4-28

另外，对上面的代码做一些说明：

（1）对于 frBean.addInitParameter("name"," filter-test2")，设置的参数在 Filter 的 init 方法里的 FilterConfig 对象里可以获取，即 filterConfig.getInitParameter("name")，但是初始化时获取的过滤器名字和设定的执行顺序 order 无关，而是根据默认的类加载顺序而来的。

（2）对于 frBean.setUrlPatterns(urls)，可以设置多个 URL 匹配规则，setUrlPatterns 接收一个 List<String>类型的参数。其实现代码在 WebFilterConfig.java 中注册 RequestTest2Filter.java 时已经给出相关示例。

（3）当不设置 setOrder 次序时，过滤器的执行顺序默认是 Bean 的加载顺序。在当前 WebConfig 类中，先加载 RequestTest1Filter.java 过滤器，后加载 RequestTest2Filter.java 过滤器。

4.4.2　拦截器

1. 概述

Spring MVC 拦截器作为一个附加功能，其重要性远不如 struts 中的拦截器。Spring MVC 拦截器主要是帮我们按照一定规则拦截请求，类似于 Servlet 开发过程中使用的过滤器 Filter，拦截器主要是针对某个业务功能进行预处理和后处理。拦截器的执行流程如图 4-29 所示。

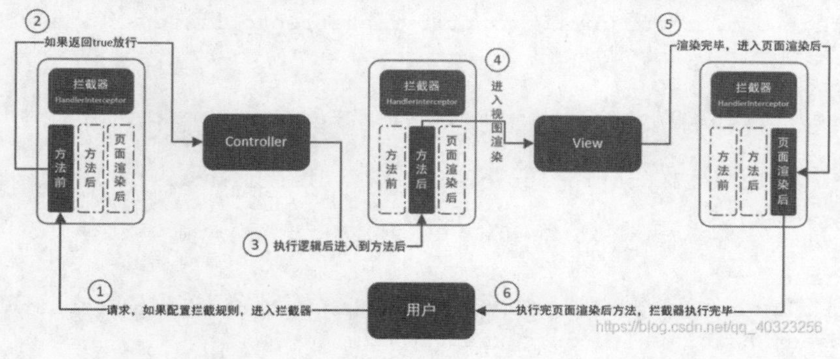

图 4-29

拦截器的本质是 AOP 面向切面编程，也就是说符合横切关注点的功能都可以考虑使用拦截器实现。拦截器有以下几种常见应用场景。

（1）权限检查

例如，用户登录检测，访问项目内部接口时，可以通过拦截器检测用户是否登录，如果未登录，则直接返回用户登录页面。

（2）日志记录

拦截器可以用来连接一些重要的请求，从而记录这些请求的重要信息、仪表与信息的监控、信息统计以及 page view 的计算等。

（3）性能监控

可以记录某个请求在通过拦截器进入处理器的开始时间和在处理器处理完之后记录结束时间，从而得到该请求的耗时，可以用于性能的监控，准确定位到哪些功能卡顿，便于日后进行功能的调优。

（4）通用行为

拦截器还可以读取 cookie、获取用户信息，并将用户信息存放在请求中，方便后续业务流程使用。

2. 自定义拦截器

在 Spring Boot 项目中自定义拦截器有两个关键步骤：一是创建一个拦截器，实现 HandlerInterceptor 接口，并且按照自身需求重写接口中的方法；二是创建一个拦截器的配置类，实现 WebMvcConfigurer.java，并且重写 addResourceHandlers()方法，定义拦截规则。

自定义拦截器，参考代码如下。

示例代码 4-24 LoginInterceptor.java

```
/**
 * 登录拦截器
 * Spring MVC 拦截器作为一个附加功能，重要性远不及 struts 中的拦截器
 * 拦截器由 dispatcherServlet 调用，而 JSP 不穿过 dispatcherServlet，所以拦截器不会拦截 JSP
 * @version V1.0
 * @Title: LoginInterceptor.java
 * @Package com.tudou.springbootfilterinterceptorlistener.interceptor
 * @date: 2020/10/2 20:41
 */
public class LoginInterceptor implements HandlerInterceptor {
    /**
     * 在请求到达具体的 Controller 方法之前执行，且当返回 true 时才执行 Controller 和另外两个方法
     */
    @Override
    public boolean preHandle(HttpServletRequest request, HttpServletResponse response, Object handler) throws Exception {
        System.out.println("LoginInterceptor-执行了 preHandle()方法！");
        //为测试不拦截请求设置的测试数据
        request.getSession().setAttribute("username","bobo");
        String username = (String) request.getSession().getAttribute("username");
        if (username == null) {
            response.sendRedirect("/html/login.html");
            return false;
        }
        return true;
    }
    /**
     * 在执行完 Controller 具体方法之后、返回页面之前执行，而且此时还可以修改 ModelAndView
     */
    @Override
    public void postHandle(HttpServletRequest request, HttpServletResponse
```

```
response, Object handler, ModelAndView modelAndView) throws Exception {
            System.out.println("LoginInterceptor-执行了 postHandle()方法！");
        /**
         * 在即将到达页面之前执行，此时不可以修改 ModelAndView
         */
        @Override
        public void afterCompletion(HttpServletRequest request,
HttpServletResponse response, Object handler, Exception ex) throws Exception{
            System.out.println("LoginInterceptor-执行了 afterCompletion()方法！");
        }
    }
```

HandlerInterceptor 接口中的三个方法说明如下：

（1）preHandle()：Controller 接口方法执行前用于拦截请求的方法。

①可以使用 HttpServletRequest 或 HttpServletResponse 跳转到指定页面。

②该方法返回 true 放行，执行下一个拦截器，如果没有拦截器，就执行请求 Controller 的接口方法。

③该方法返回 false 不放行，不会执行 Controller 中的方法。

（2）postHandle()：在 Controller 中的接口方法执行后、JSP 视图执行前执行。

①可以使用 HttpServletRequest 或 HttpServletResponse 跳转到指定页面。

②如果指定了跳转的页面，那么 Controller 中方法跳转的页面就不生效了。

（3）afterCompletion()：在 JSP 执行后执行，不能通过 HttpServletRequest 或 HttpServletResponse 跳转页面。

拦截器配置类，参考代码如下。

示例代码 4-25　InterceptorConfig.java

```
/**
 * 拦截器配置类
 */
@Configuration
public class InterceptorConfig implements WebMvcConfigurer {
    @Override
    public void addInterceptors(InterceptorRegistry registry) {

        //注册拦截器 LoginInterceptor
        registry.addInterceptor(new LoginInterceptor())
                //拦截所有路径
                .addPathPatterns("/**")
                //添加不被拦截的路径、登录路径和静态资源路径
                .excludePathPatterns("/html/login.html")
                .excludePathPatterns("/css/**");
    }
}
```

Controller 接口测试类，参考代码如下。

示例代码 4-26　InterceptorController.java

```
/**
 * 拦截器接口测试类
 */
@RestController
@RequestMapping("/interceptor")
public class InterceptorController {
    @GetMapping("/test")
    public String testInterceptor(){
        return "hello interceptor";
    }
}
```

启动项目，访问路径 http://localhost:8080/interceptor/test，控制台拦截器的执行效果如图 4-30 所示。

```
2020-10-04 21:52:47.485  INFO 13916 --- [nio-8080-exec-1] o.s.web.servlet.DispatcherServlet
2020-10-04 21:52:47.490  INFO 13916 --- [nio-8080-exec-1] o.s.web.servlet.DispatcherServlet
interceptor1-执行了preHandle()方法!
interceptor1-执行了postHandle()方法!
interceptor1-执行了afterCompletion()方法!
```

图 4-30

3. 多拦截器配置

在实际开发过程中，我们可能会需要配置多个拦截器，此时我们首先会考虑到两个问题：第一，应该怎么注册多个拦截器；第二，怎么自定义拦截器的执行顺序。Spring Boot 为我们提供了非常简便的方式，即在配置类中可以通过注册拦截器的顺序来自定义拦截器的执行顺序，详细实现说明如下。

（1）新增拦截器，参考代码如下。

示例代码 4-27　SecondInterceptor.java

```
/**
 * 拦截器(用于测试拦截器执行顺序)
 */
public class SecondInterceptor implements HandlerInterceptor {
    /**
     * 在请求到达具体的 Controller 方法之前执行，且当返回 true 时才执行 Controller 和另
外两个方法
     */
    @Override
    public boolean preHandle(HttpServletRequest request, HttpServletResponse
response, Object handler) throws Exception {
        System.out.println("SecondInterceptor-执行了 preHandle()方法！");
        return true;
    }

    /**
     * 在执行完 Controller 具体方法之后、返回页面之前执行，而且还可以修改 ModelAndView
     */
```

```
    @Override
    public void postHandle(HttpServletRequest request, HttpServletResponse
response, Object handler, ModelAndView modelAndView) throws Exception {
        System.out.println("SecondInterceptor-执行了postHandle()方法!");
    }

    /**
     * 在即将到达页面之前执行，此时不可以修改ModelAndView
     */
    @Override
    public void afterCompletion(HttpServletRequest request,
HttpServletResponse response, Object handler, Exception ex) throws Exception {
        System.out.println("SecondInterceptor-执行了afterCompletion()方法!");
    }
}
```

（2）修改配置类 InterceptorConfig.java。在配置类中注册拦截器 SecondInterceptor，并且指定先执行拦截器 LoginInterceptor 再执行拦截器 SecondInterceptor。参考代码如下。

示例代码 4-28 InterceptorConfig.java

```
/**
 * 拦截器配置类
 */
@Configuration
public class InterceptorConfig implements WebMvcConfigurer {
    @Override
    public void addInterceptors(InterceptorRegistry registry) {
        /*可以通过registry.addInterceptor()方法控制拦截器注册顺序，
         从而达到自定义拦截器顺序的目的 */
        //先注册拦截器LoginInterceptor
        registry.addInterceptor(new LoginInterceptor())
                //拦截所有路径
                .addPathPatterns("/**")
                //添加不被拦截的路径、登录路径和静态资源路径
                .excludePathPatterns("/html/login.html")
                .excludePathPatterns("/css/**");

        //再注册拦截器SecondInterceptor()
        registry.addInterceptor(new SecondInterceptor())
                //拦截所有路径
                .addPathPatterns("/**")
                //添加不被拦截的路径、登录路径和静态资源路径
                .excludePathPatterns("/html/login.html")
                .excludePathPatterns("/js/**")
                .excludePathPatterns("/img/**");
    }
}
```

启动项目，在浏览器中访问路径 http://localhost:8080/interceptor/test。两个拦截器的执行结果如图 4-31 所示。

```
2020-10-04 22:06:32.917  INFO 6092 --- [nio-8080-exec-1] o.a.c.c.C.[Tomcat].[localhost].[/]
2020-10-04 22:06:32.917  INFO 6092 --- [nio-8080-exec-1] o.s.web.servlet.DispatcherServlet
2020-10-04 22:06:32.924  INFO 6092 --- [nio-8080-exec-1] o.s.web.servlet.DispatcherServlet
LoginInterceptor-执行了preHandle()方法!
SecondInterceptor-执行了preHandle()方法!
SecondInterceptor-执行了postHandle()方法!
LoginInterceptor-执行了postHandle()方法!
SecondInterceptor-执行了afterCompletion()方法!
LoginInterceptor-执行了afterCompletion()方法!
```

图 4-31

对于多个拦截器，正常流程下拦截器方法的执行顺序如图 4-32 所示。非正常情况下，第一个拦截器在 preHandle()处放行，第二个拦截器在 postHandle()处被拦截。

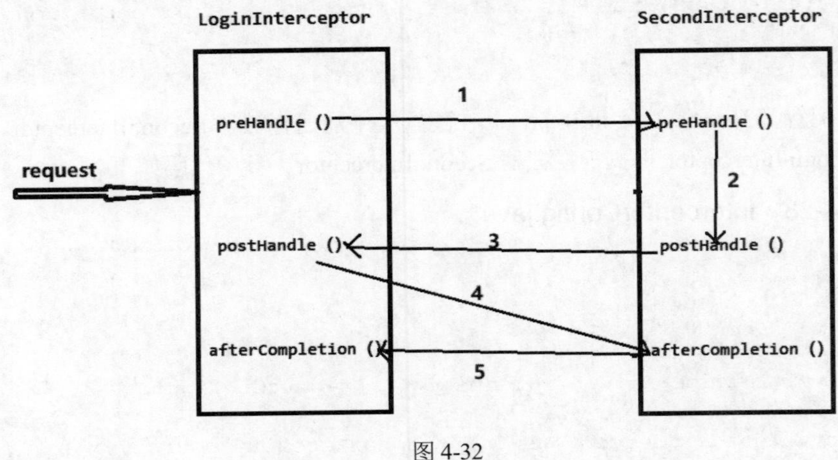

图 4-32

修改第一个拦截器 LoginInterceptor.java 的代码，参考代码如下。

示例代码 4-29　LoginInterceptor.java

```
public class LoginInterceptor implements HandlerInterceptor {
    @Override
    public boolean preHandle(HttpServletRequest request, HttpServletResponse response, Object handler) throws Exception {
        System.out.println("LoginInterceptor-执行了 preHandle()方法！");
        return true;
    }
}
```

修改第二个拦截器 SecondInterceptor.java 的代码，参考代码如下。

示例代码 4-30　SecondInterceptor.java

```
public class SecondInterceptor implements HandlerInterceptor {
    @Override
    public boolean preHandle(HttpServletRequest request, HttpServletResponse response, Object handler) throws Exception {
        System.out.println("SecondInterceptor-执行了 preHandle()方法！");
        return false;
    }
```

```
}
```

启动项目，在浏览器中访问路径 http://localhost:8080/interceptor/test。两个拦截器的执行结果如图 4-33 所示。

```
2020-10-04 22:35:22.100  INFO 18832 --- [nio-8080-exec-1] o.a.c.c.C.[Tomcat].[localhost].[/]
2020-10-04 22:35:22.101  INFO 18832 --- [nio-8080-exec-1] o.s.web.servlet.DispatcherServlet
2020-10-04 22:35:22.106  INFO 18832 --- [nio-8080-exec-1] o.s.web.servlet.DispatcherServlet
LoginInterceptor-执行了preHandle()方法!
SecondInterceptor-执行了preHandle()方法!
LoginInterceptor-执行了afterCompletion()方法!
```

图 4-33

非正常情况下的拦截器（第一个拦截器在 preHandle()处被拦截，在第二个拦截器处放行）。修改第一个拦截器 LoginInterceptor.java 的代码，参考代码如下。

示例代码 4-31 LoginInterceptor.java（部分代码）

```
public class LoginInterceptor implements HandlerInterceptor {
    @Override
    public boolean preHandle(HttpServletRequest request, HttpServletResponse response, Object handler) throws Exception {
        System.out.println("LoginInterceptor-执行了 preHandle()方法！");
        return false;
    }
}
```

修改第二个拦截器 SecondInterceptor.java 的代码，参考代码如下。

示例代码 4-32 SecondInterceptor.java（部分代码）

```
public class SecondInterceptor implements HandlerInterceptor {
    @Override
    public boolean preHandle(HttpServletRequest request, HttpServletResponse response, Object handler) throws Exception {
        System.out.println("SecondInterceptor-执行了 preHandle()方法！");
        return true;
    }
}
```

启动项目，在浏览器中访问路径 http://localhost:8080/interceptor/test。两个拦截器的执行结果如图 4-34 所示。

```
2020-10-04 22:37:00.362  INFO 18348 --- [nio-8080-exec-1] o.a.c.c.C.[Tomcat].[localhost].[/]
2020-10-04 22:37:00.362  INFO 18348 --- [nio-8080-exec-1] o.s.web.servlet.DispatcherServlet
2020-10-04 22:37:00.368  INFO 18348 --- [nio-8080-exec-1] o.s.web.servlet.DispatcherServlet
LoginInterceptor-执行了preHandle()方法!
```

图 4-34

对比以上两种非正常情况下执行的拦截器，我们可以得出三点结论：

（1）拦截器 1 放行，其后紧接着的拦截器 2 的 preHandle()方法才会执行。

（2）如果拦截器 1 放行，拦截器 2 的 preHandle()方法不放行，那么拦截器 2 的 postHandle()

和 afterCompletion()方法将不会执行。

（3）只要存在一个拦截器放行，那么此拦截器后的所有拦截器 preHandle()方法均不会被执行。

4.4.3 监听器

1. 概述

Web 监听器是 Servlet 中特殊的类，能帮助开发者监听 Web 中的一些特定事件，比如 ServletContext、HttpSession、ServletRequest 的创建和销毁，在某些动作前后增加处理，以及实现监控等。按照监听对象，Web 监听器可分为三类：Servlet 上下文监听器 ApplicationListener，Session 对象的监听器 HttpSessionListener，以及 Request 对象的监听器 ServletRequestListener。Web 监听器的使用场景很多，比如监听 Servlet 上下文来初始化一些数据，监听 Session 会话获取当前系统在线人数，监听客户端请求的 ServletRequest 来获取用户的一些访问信息。这里我们将通过实际例子和实际场景讲解这三类监听器以及自定义监听器。

2. 监听 ServletContext 对象

监听 ServletContext 对象主要用于初始化一些不经常改变的数据，比如网站首页的一些菜单信息、用户账号相关信息等，将这些数据存进 application 作用域中，在一定程度上会起到缓存的作用。页面上使用数据的时候直接从 application 作用域获取，而不是每次获取数据都要从数据库中获取。如果用户量大，那么每次首页菜单加载对于数据库来说都是很大的一笔开销。

下面将针对首页展示用户账号信息这个场景，来对 ServletContext 监听器做实际讲解。

（1）模拟数据库查询用户信息，参考代码如下。

示例代码 4-33　UserServiceImpl.java

```
@Service
public class UserServiceImpl implements UserService {
    @Override
    public User getUser() {
        return new User(111L,"tudoudou","aaa");
    }
}
```

（2）定义一个监听 ServletContext 的监听器，需要实现 ServletContextListener 接口，并且重写 onApplicationEvent 方法，参考代码如下。

示例代码 4-34　ServletContextListener.java

```
/**
 * Servlet 上下文监听器
 */
@Component
public class ServletContextListener implements
ApplicationListener<ContextRefreshedEvent> {
    @Override
    public void onApplicationEvent(ContextRefreshedEvent
contextRefreshedEvent) {
```

```
            //获取 application 上下文信息
            ApplicationContext applicationContext =
contextRefreshedEvent.getApplicationContext();
            //获取 userService 对象
            UserServiceImpl userServiceImpl =
applicationContext.getBean(UserServiceImpl.class);
            User user = userServiceImpl.getUser();
            //将获取的 user 对象放入 application 域中
            ServletContext applicationScope =
applicationContext.getBean(ServletContext.class);
            applicationScope.setAttribute("user",user);
        }
    }
```

（3）创建一个接口，从 application 作用域中获取初始化的 User 信息，返回到页面中，参考代码如下。

示例代码 4-35　UserController.java（部分代码）

```
@GetMapping("/user")
public User getUser(HttpServletRequest request){
    ServletContext applicationScope = request.getServletContext();
    User user = (User)applicationScope.getAttribute("user");
    return user;
}
```

接口请求结果如图 4-35 所示。

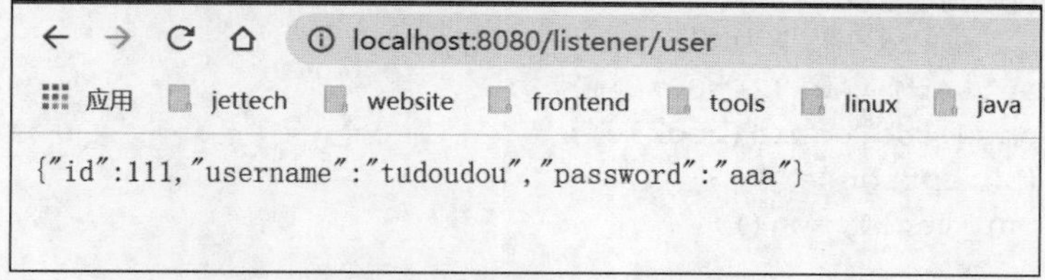

图 4-35

3. 监听 HTTP 会话 Session 对象

Web 监听器还有一个比较常用的作用就是监听 Session 对象，以统计当前 Web 应用在线用户总数量。下面将通过统计在线用户数量来具体讲解 Session 会话监听器。

（1）创建 Session 会话监听器，需要实现 HttpSessionListener 接口，并且按照自身需求重写 sessionCreated()方法和 sessionDestroyed()方法，参考代码如下。

示例代码 4-36　SessionListener.java

```
/**
 * httpSession 监听对象
 */
@Component
public class SessionListener implements HttpSessionListener {
    //用于记录在线用户数量
```

```
    public static Integer count = 0;
    @Override
    public void sessionCreated(HttpSessionEvent se) {
        System.out.println("新用户上线！");
        count++;
        se.getSession().getServletContext().setAttribute("count",count);
    }
    @Override
    public void sessionDestroyed(HttpSessionEvent se) {
        System.out.println("用户下线！");
        count--;
        se.getSession().getServletContext().setAttribute("count",count);
    }
}
```

（2）创建一个接口测试该监听器，参考代码如下。

示例代码 4-37　UserController.java（部分代码）

```
@GetMapping("/countOnlineUser")
public String getOnlineUserCount(HttpServletRequest request,
HttpServletResponse response){
    Cookie cookie;
    try {
        //将 sessionId 记录到浏览器中，防止在服务器找不到原有的用户信息，重新创建 session
        cookie = new Cookie("JSSIONID",
URLEncoder.encode(request.getSession().getId(), "utf-8"));
        cookie.setPath("/");
        //设置 cookie 有效期为 1 天
        cookie.setMaxAge(24*60*60);
        response.addCookie(cookie);
    }catch (Exception e){
        e.printStackTrace();
    }
    Integer count = (Integer)request.getSession().getServletContext().
getAttribute("count");
    return "total user online:"+count;
}
```

测试结果如图 4-36 所示。

图 4-36

4. 监听 ServletRequest 对象

监听客户端请求，主要是为了获取用户的一些请求信息，具体操作如下。

（1）创建监听 ServletRequest 对象的监听器，需要实现 ServletRequestListener 接口，按照自身

需求重写 requestInitialized()方法和 requestDestroyed()方法，参考代码如下。

示例代码 4-38 RequestListener.java（部分代码）

```
/**
 * request 请求 listener，获取请求访问信息
 */
@Component
public class RequestListener implements ServletRequestListener {
    @Override
    public void requestInitialized(ServletRequestEvent sre) {
        System.out.println("此次请求开始");
        HttpServletRequest request =
(HttpServletRequest)sre.getServletRequest();
        System.out.println("sessionId:"+request.getRequestedSessionId()+";
url:"+request.getRequestURL());
        String username = (String)request.getAttribute("username");
        if (username == null){
            request.setAttribute("username","tudoudou");
        }
    }
    @Override
    public void requestDestroyed(ServletRequestEvent sre) {
        System.out.println("此次请求结束");
        HttpServletRequest request = (HttpServletRequest)
sre.getServletRequest();
        String username = (String)request.getAttribute("username");
        System.out.println("requestScope 中 username 为:"+username);
    }
}
```

（2）新增一个接口，测试 ServletRequest 对象的监听器，参考代码如下。

示例代码 4-39 UserController.java（部分代码）

```
@GetMapping("/testRequest")
public String testRequest(HttpServletRequest request){
    String username = (String)request.getAttribute("username");
    System.out.println("init requestScope username:"+username);
    return "success";
}
```

测试结果如图 4-37 所示。

```
此次请求开始
sessionId:5B8AE8739B0F3D2730C1D7D8FA4FEC9A; url:http://localhost:8080/listener/testRequest
init requestScope username:tudoudou
此次请求结束
requestScope中username为:tudoudou
```

图 4-37

5. 自定义监听器

实现自定义监听器需要做三步操作：首先，自定义事件，即监听器用来监听的事件，该事件

需要继承 ApplicationEvent 对象；其次，自定义一个监听器加监听自定义的时间，该监听器需要实现 ApplicationListener 接口；最后，手动发布事件，发布之后监听器才能监听到，当然，发布事件可以根据实际业务场景进行操作。

（1）自定义事件，参考代码如下。

示例代码 4-40　MyEvent.java

```
/**
 * 自定义监听事件
 */
public class MyEvent extends ApplicationEvent {
    private User user;
    public MyEvent(Object source, User user) {
        super(source);
        this.user = user;
    }
    public User getUser() {
        return user;
    }
    public void setUser(User user) {
        this.user = user;
    }
}
```

（2）自定义监听器，参考代码如下。

示例代码 4-41　MyEventListener.java

```
/**
 * 自定义监听器
 */
@Component
public class MyEventListener implements ApplicationListener<MyEvent> {
    @Override
    public void onApplicationEvent(MyEvent myEvent) {
        //获取事件中的信息
        User user = myEvent.getUser();
        //打印获取到的信息
        System.out.println("username:"+user.getUsername());
    }
}
```

（3）发布事件，参考代码如下。

示例代码 4-42　UserServiceImpl.java（部分代码）

```
/**
 * User 用户 service 接口类实现类
 */
@Service
public class UserServiceImpl implements UserService {
    @Autowired
```

```
    private ApplicationContext applicationContext;
    @Override
    public User publishMyEvent() {
        User user = new User(100L,"doudou","111");
        //发布自定义事件
        MyEvent myEvent = new MyEvent(this, user);
        applicationContext.publishEvent(myEvent);
        return user;
    }
}
```

（4）创建接口测试自定义监听器，参考代码如下。

示例代码 4-43　UserController.java（部分代码）

```
@GetMapping("/publish")
public User publishMyEvent(HttpServletRequest request){
    return userService.publishMyEvent();
}
```

小　结

过滤器依赖于 Servlet 容器，在实现上是基于函数回调的，可以对几乎所有请求进行过滤，但是过滤器实例只能在容器初始化时调用一次。使用过滤器的时候主要用来做一些过滤操作，以获取有用的数据。比如，在过滤器中修改请求和响应的字符编码、过滤敏感词汇等。

拦截器依赖于 Web 框架，在 Spring MVC 中就是依赖于 Spring MVC 框架，在实现上基于 Java 反射机制，是面向切面编程（AOP）的一种应用。由于拦截器是基于 Spring 的依赖注入进行一些业务操作，因此拦截器实例在一个 Controller 的生命周期之内可以多次调用。但是拦截器只能拦截 Controller 相关的请求，对一些静态资源的访问是无法拦截的。

监听器是 Servlet 中特殊的类，主要用来监听 Web 中的一些特定事件，对监听到的事件做一些特殊的处理。按照监听对象不同，Web 监听器可分为三类：ApplicationListener、HttpSessionListener 和 ServletRequestListener。

过滤器和拦截器的区别如下：

（1）过滤器是基于函数回调的，拦截器是基于 Servlet 容器的。

（2）过滤器几乎可以过滤所有请求，但是拦截器只能拦截 Controller 的接口请求。

（3）拦截器可以访问 action 上下文、值栈里的对象，而过滤器不能访问。

（4）在 Controller 的生命周期中，过滤器只能在容器初始化时被调用一次，拦截器可以多次被调用。

（5）拦截器可以获取 IoC 容器中的各种 Bean，根据需求进行业务处理，但是过滤器不支持这一点。

4.5 前后端分离应用

4.5.1 前后端分离简介

1. 简介

前后端分离目前已经成为互联网项目开发的主流趋势，甚至是业内标准的使用方式，通过Nginx+Tomcat的方式（也可以加Node.js）有效地进行解耦，并且前后端分离为以后的大型分布式架构、弹性计算器、微服务架构、多端化服务打下坚实的基础，这是系统架构进化的必经之路。简而言之，前后端分离的核心思想就是前端HTML页面通过Ajax调用后端的RESTful API接口，并使用JSON数据进行交互。前后端分离架构如图4-38所示。

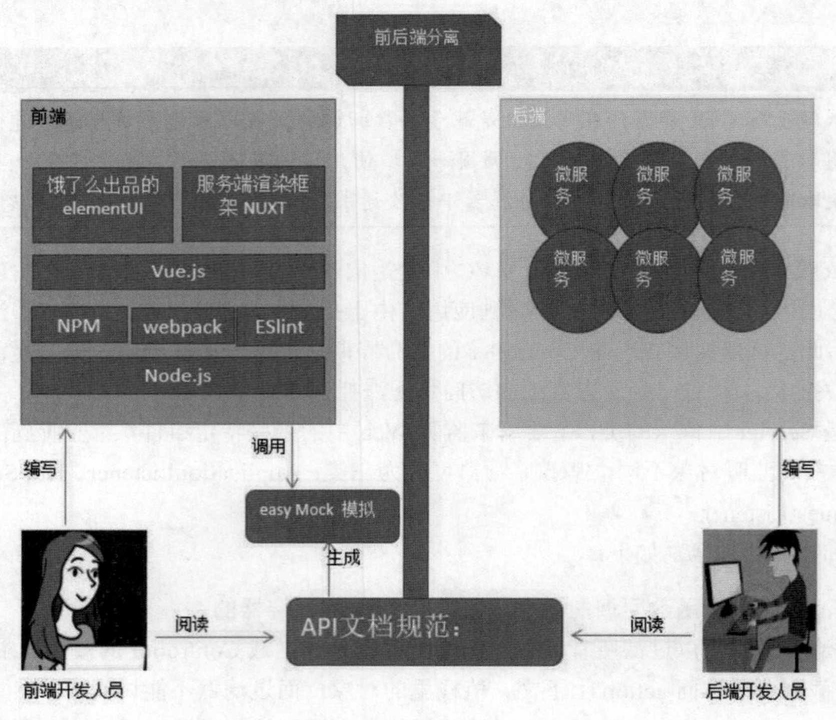

图4-38

在互联网架构中，Web服务器能被外网访问，一般是使用Nginx、Apache这类服务器，它们只解析静态资源。应用服务器只能在内网访问，一般是使用Tomcat、Jetty、Resin这类服务器，既可以解析动态资源，也可以解析静态资源，但是解析静态资源的能力没有Web服务器好。

2. 为什么要使用前后端分离

前后端分离是系统架构发展的必然，原因如下：

（1）彻底解放前端。前端不再需要向后台提供模板或是后台在前端 HTML 中嵌入后台代码。

（2）提高工作效率，分工更加明确。前后端分离的工作流程可以使前端只关注前端的事，后台只关心后台的活。前后端开发可以同时进行，在后台还没有时间提供接口的时候，前端可以先将数据写死或者调用本地的 JSON 文件，页面的增加和路由的修改也不必去麻烦后台，开发更加灵活。

（3）局部性能提升。通过前端路由的配置，我们可以实现页面的按需加载，无须一开始加载首页便加载网站的所有资源，服务器也不再需要解析前端页面，在页面交互及用户体验上有所提升。

（4）降低维护成本。前端与后端分两个项目，独立部署且维护更加方便。当后端服务不可用时，浏览器仍能访问前端页面，客户体验较好。

4.5.2 项目需求

本章讲的是 Spring Boot 的 Web 开发，首先在前面几节讲了一些 Spring Boot 在 Web 开发中的常用知识，比如 Servlet 容器的原理与配置、Spring MVC 支持、集成模板引擎 themeleaf 以及过滤器、拦截器、监听器的使用与区别。本节我们利用一个典型的前后端分离应用巩固加深 Spring Boot Web 开发的印象。

本书的内容大多数是后端知识，现在越来越多的后端开发者也能熟练掌握基本的前端知识了，比如 HTML、CSS、JS 等，甚至一些优秀的开发者还掌握了主流的前端框架，比如 BootStrap、Vue、Element-UI 等。

前端框架与技术层出不穷，这里采用当前流行的 Node.js+webpack+Vue+Element-UI 前端组合来开发一个后端管理系统的前端界面。后端则采用 Spring Boot+Spring MVC，接口 URL 统一采用 RESTful 风格（RESTful 相关的知识在后面讲解）。

员工管理是现代企业 ERP 系统中的一个模块，旨在方便员工信息的管理。ERP（Enterprise Resource Planning，企业资源计划）系统是一套软件系统，通常建立在信息技术基础上，采集需要的信息并结合系统化的管理思想，为公司决策层提供决策手段。这里只以员工管理模块的增删改查为例带大家入门前后端分离的相关技术栈与开发模式。

本节的例子只涉及员工表单表的增删改查，员工表有主键 ID、员工编号、员工姓名、部门、职位、生日、性别、月薪这些字段。由于 Spring Boot 数据访问的相关知识在后面才会讲到，因此这里在展示的时候并不会展示 Dao 层代码，重点关注前后端分离的开发模式，并从中学习主流的前端与后端技术。数据库表设计如图 4-39 所示。

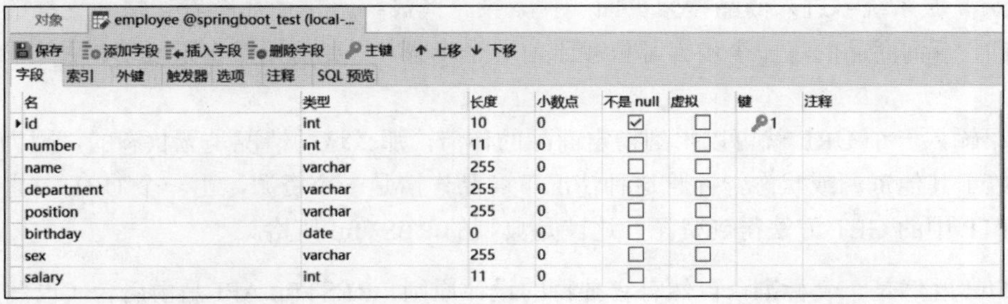

对象 | employee @springboot_test (local-...

保存 添加字段 插入字段 删除字段 主键 ↑上移 ↓下移

字段 索引 外键 触发器 选项 注释 SQL 预览

名	类型	长度	小数点	不是 null	虚拟	键	注释
id	int	10	0	☑	☐	1	
number	int	11	0	☐	☐		
name	varchar	255	0	☐	☐		
department	varchar	255	0	☐	☐		
position	varchar	255	0	☐	☐		
birthday	date	0	0	☐	☐		
sex	varchar	255	0	☐	☐		
salary	int	11	0	☐	☐		

图 4-39

4.5.3 后端开发

1. RESTful 设计原则

REST（Representational State Transfer，表征性状态转移）首次出现于 2000 年 Roy Fielding（HTTP 规范的主要编写者之一）的博士论文中。他在论文中提到：“我这篇文章的写作目的就是想在符合架构原理的前提下理解和评估以网络为基础的应用软件的架构设计，得到一个功能强、性能好、适宜通信的架构。REST 指的是一组架构约束条件和原则。”如果一个架构符合 REST 的约束条件和原则，就称之为 RESTful 架构，它是一套标准、一个规范，而不是一个框架。

REST 本身并没有创造新的技术、组件或服务，隐藏在 RESTful 背后的理念就是使用 Web 的现有特征和能力，更好地利用现有 Web 标准中的一些准则和约束。虽然 REST 本身受 Web 技术的影响很深，但是理论上 REST 架构风格并不是绑定在 HTTP 上的，只不过目前 HTTP 是唯一与 REST 相关的实例。所以，我们这里描述的 REST 是通过 HTTP 实现的 REST，可用一句话表示：URL 定位资源，用 HTTP 动词（GET、POST、PUT、PATCH、DELETE）描述操作。

首先，我们将从以下几个概念深入理解一下什么是 RESTful。

（1）资源：其实指的就是网络上的一个实体，或者说是网络上一个具体的信息，比如一张图片、一段文本、一个视频等。资源总是要通过一种载体来反映它的内容；文本既可以用 TXT，也可以用 HTML 或者 XML；图片可以用 JPG、PNG 等格式。在开发过程中，JSON 是常用的一种资源表现形式。

（2）统一入口：RESTful 风格的数据元操作 CRUD（CREATE、READ、UPDATE、DELETE）分别对应 HTTP 方法：GET 用来获取资源，POST 用来新建资源（也可以用于更新资源）。PUT 用来更新资源，DELETE 用来删除资源，这样就统一了数据操作的接口。

（3）URI：可以用一个 URI（统一资源定位符）指向资源，即每个 URI 都对应一个特定的资源。要获取这个资源，访问它的 URI 就可以，因此 URI 就成了一个资源的地址或识别符。一般情况下，每个资源至少有一个 URI 与之对应，最典型的 URI 就是 URL。

（4）无状态：所有的资源都可以用 URI 定位，而且这个定位与其他资源无关，也不会因为其他资源的变化而变化。

下面以查询商城中某个商品为例来说明有状态和无状态的区别：

①登录商城系统→进入商品搜索页面→搜索某个商品→点击商品查看详情。这样的操作流程是有状态的，查询商品的每一个步骤都依赖于前一个步骤，只要前置操作不成功，后续操作就无法执行。

②如果输入一个 URL 就可以得到指定商品的详情，那么这种情况是无状态的，因为获取商品信息不依赖于其他资源或状态，并且这种情况下商品详情是一个资源，由一个 URL 与之对应，可以通过 HTTP 中的 GET 方法得到资源，这就是典型的 RESTful 风格。

RESTful 既然是一套标准，自然有它独特的设计原则。RESTful API 是由后台（也就是 server）提供给前端调用的。前端调用 API 时向后台发起 HTTP 请求，后台响应请求将处理结果返回给前端，也就是说 RESTful 是典型的 HTTP 协议。下面我们将讲述一下 RESTful 架构的设计原则和常用的一些规范。

（1）在接口命名时应该用名词，不应该用动词，因为通过接口操作到的是资源。

（2）在 URL 中加入版本号，有利于版本的迭代管理，更加直观，参考代码如下。

```
//v1 表示版本号
https://ww.rest.com/v1
```

（3）对于资源的操作类型应该通过 HTTP 动词表示，如 POST、GET、PUT、DELETE。

```
GET /departments：列出公司所有部门
POST /departments：新建一个部门
GET /departments/id：获取某个指定部门的信息
PUT /departments/id：更新某个指定部门的信息（返回该部门的全部信息）
DELETE /departments/id：删除某个部门
GET /departments/id/jobs：列出某个指定部门的所有岗位信息
DELETE /departments/id/jobs/id：删除某个指定部门的指定岗位
```

（4）排序规则，默认是升序，“-”表示降序，多个排序规则以逗号间隔组合，使用 sort 查询参数限制。

```
//优先以 work_time 倒序，其次以 create_time 升序
GET /employees?sort=-work_time，create_time
```

（5）限制返回值的字段，明确指定输出字段列表，用于控制网络带宽和速度。使用 fileds 查询参数来限制。

```
//返回参数列表为 id,name,work_time,create_time，并且优先以 work_time 倒序，其次以
create_time 升序
GET /employees?fiiled=id,name，work_time，create_time &sort=-work_time，
create_time
```

（6）HTTP Method 分别对应资源的 CURD 操作。

- GET（SELECT）：从服务器取出资源（一项或多项）。
- POST（CREATE）：在服务器新建一个资源。
- PUT（UPDATE）：在服务器更新资源（客户端提供改变后的完整资源）。
- PATCH（UPDATE）：在服务器更新资源（客户端提供改变的属性）。
- DELETE（DELETE）：从服务器删除资源。

保证 POST、PUT、DELETE、PATCH、GET 操作幂等性。

（7）使用 SSL（Secure Sockets Layer，安全套接层）。

（8）参数和 URL 采用蛇形命名法（全由小写字母和下划线组成，在两个单词之间用下划线连接，比如 updated_name）。

（9）服务器请求和返回的数据格式应该尽量使用 JSON，避免使用 XML。设置 Response 的 Content-Type 为 application/json。

2. 编写 RESTful 接口

对于传统的单体应用架构来说，一般分为表现层（Controller）、业务逻辑层（service）和数

据访问层（dao），由于本节要讲的例子只涉及员工的增删改查，并不涉及复杂的业务处理，因此 service 层不予展示。同样地，数据访问层代码及连接数据库的配置也不会展示，重点关注表现层如何写好 RESTful 接口供前端调用。

在实际的工作开发中，可能一边写前端页面，一边开发后端接口。为了方便，这里先把所有需要用到的后端接口写好。对于员工的增删改查，这里设计了如下接口：

```
POST /employees：新增员工。
GET /employees/{id}：根据 id 获取员工。
GET /employees：获取所有员工。
PUT /employees/{id}：根据 id 修改员工。
DELETE /employees/{id}：根据 id 删除员工。
```

另外需要注意的是，如果 Controller 层的方法上有各种各样的返回值类型，比如 String、Employee、List<Employee>或者 Integer，那么前端代码接收到 response 后，由于没有一个统一的接口返回类型规范，对于 Controller 层的返回值解析不太方便。因此，我们需要设计一个统一的接口返回类型 Result 类，该类具有如下特征：

①具有三个字段，分别是状态码 code、信息 message、数据 data。

②由于返回的数据可能是各种各样的类型，因此数据的类型用泛型 T 表示。

③有两个 public 修改的方法 success、fail，分别在请求成功和失败时调用。

Result 类的详细代码如下。

示例代码 4-44　Result.java

```
package com.bobo.ems.util;
public class Result<T> {
    private Integer code;
    private String message;
    private T data;
    public Result<T> success(String message,T data){
        this.code = 200;
        this.message = message;
        this.data = data;
        return this;
    }
    public Result<T> failed(String message){
        this.code = 500;
        this.message = message;
        return this;
    }
    public Integer getCode() {
        return code;
    }
    public void setCode(Integer code) {
        this.code = code;
    }
    public String getMessage() {
        return message;
```

```
    }
    public void setMessage(String message) {
        this.message = message;
    }
    public T getData() {
        return data;
    }
    public void setData(T data) {
        this.data = data;
    }
}
```

application.yml 代码如下。

示例代码 4-45 application.yml

```
server:
  port: 8080
  servlet:
    context-path: /ems
```

编写 Employee 员工实体类，参考代码如下。

示例代码 4-46 Employee.java

```
package com.bobo.ems.entity;
import java.util.Date;
public class Employee {
    /**
     * 物理主键
     */
    private Integer id;
    /**
     * 员工编号，业务主键
     */
    private Integer number;
    /**
     * 员工姓名
     */
    private String name;
    /**
     * 部门
     */
    private String department;
    /**
     * 生日
     */
    private transient Date birthday;
    /**
     * 性别
     */
    private String sex;
    /**
```

```
     * 薪水
     */
    private Integer salary;
    /**
     * 职位
     */
    private String position;
    public Integer getId() {
        return id;
    }
    public void setId(Integer id) {
        this.id = id;
    }
    public Integer getNumber() {
        return number;
    }
    public void setNumber(Integer number) {
        this.number = number;
    }
    public String getName() {
        return name;
    }
    public void setName(String name) {
        this.name = name;
    }
    public String getDepartment() {
        return department;
    }
    public void setDepartment(String department) {
        this.department = department;
    }
    public Date getBirthday() {
        return birthday;
    }
    public void setBirthday(Date birthday) {
        this.birthday = birthday;
    }
    public String getSex() {
        return sex;
    }
    public void setSex(String sex) {
        this.sex = sex;
    }
    public Integer getSalary() {
        return salary;
    }
    public void setSalary(Integer salary) {
        this.salary = salary;
    }
    public String getPosition() {
```

```
        return position;
    }
    public void setPosition(String position) {
        this.position = position;
    }
}
```

编写 controller 层代码，参考代码如下。

示例代码 4-47　EmployeeController.java

```
@RestController
public class EmployeeController {
    @Autowired
    private EmployeeService employeeService;
    @GetMapping("/employees")
    public Result<List<Employee>> list(){
        List<Employee> employeeList = employeeService.list();
        Result<List<Employee>> r = new Result<>();
        return r.success("查询所有员工信息成功！",employeeList);
    }
    @PostMapping("/employees")
    public Result<Employee> addEmployee(@RequestBody Employee employee){
        Result<Employee> r = new Result();
        Employee e = employeeService.saveEmployee(employee);
        return r.success("新增员工信息成功！",e);
    }
    @GetMapping("/employees/{id}")
    public Result<Employee> getEmployee(@PathVariable("id") Integer id){
        Result<Employee> r = new Result();
        Employee e = employeeService.getById(id);
        return r.success("通过 id 获取员工信息成功！",e);
    }
    @PutMapping("/employees/{id}")
    public Result<Employee> updateEmployee(@RequestBody Employee employee){
        Result<Employee> r = new Result();
        Employee newEmployee = employeeService.updateEmployee(employee);
        return r.success("修改员工信息成功！",newEmployee);
    }
    @DeleteMapping("/employees/{id}")
    public Result<Employee> deleteEmployee(@PathVariable("id") Integer id){
        Result<Employee> result = new Result();
        Employee employee = employeeService.removeEmployee(id);
        return result.success("根据 id 删除员工信息成功！",employee);
    }
}
```

现在测试一下查询全部员工接口。启动项目，访问 http://localhost:8080/ems/employees，可以看到测试数据全都查出来了，如图 4-40 所示。

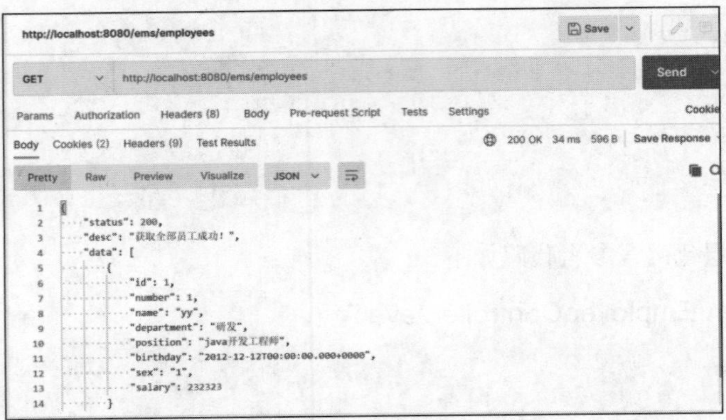

图 4-40

至此，后端 RESTful 接口全部编写完毕。接下来，我们需要从搭建环境开始，开发一个在 Node.js 上运行的前端项目。

4.5.4 前端开发

简单地说 Node.js 就是运行在服务端的 JavaScript。Node.js 是一个基于 Chrome JavaScript 运行时建立的一个平台。Node.js 是一个事件驱动 I/O 服务端 JavaScript 环境，基于 Google 的 V8 引擎（V8 引擎执行 JavaScript 的速度非常快，性能非常好）。

1. 安装 Node.js

接下来介绍在 Windows 平台上如何安装 Node.js。

访问 Node.js 的官方网站 https://node.js.org/en/download/，下载 LTS 版本对应的 Windows 64bit 的 msi 文件（计算机是 32 位的就下载 32 位），如图 4-41 所示。

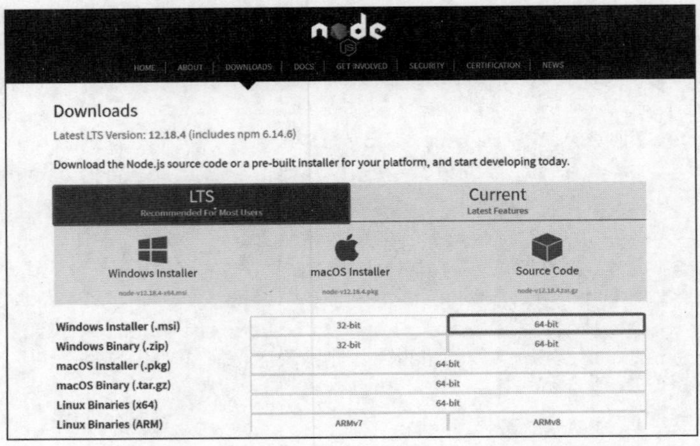

图 4-41

下载安装之后，双击 msi 文件执行安装过程。安装的过程很简单，一般一直单击 next 按钮，最后单击 Install 按钮确认安装即可。

安装完成之后，需要配置系统环境变量，具体的做法是：编辑 Path 变量，并将 Node.js 的安装根目录（比如 D:/dev/node.js/）添加上。

下面接着验证一下 Path 环境变量是否配置正确。打开 cmd，输入“node –v”命令，如果能正确输出 Node.js 的版本号，比如 v12.18.4，则说明 Path 环境变量配置正确。

接下来我们还需要配置 Node 缓存文件夹和 Node 全局安装目录。首先在 Node 安装根目录下分别创建 node_cache 和 node_global 文件夹，比如 D:/dev/node.js/node_cache 和 D:/dev/node.js/node_global，然后打开 cmd 窗口，分别执行如下命令：

```
npm config set prefix "D:/dev/node.js/node_global"
npm config set cache "D:/dev/node.js/node_cache"
```

上述命令执行完毕后，需要在系统环境变量的 Path 变量中添加上 Node 全局安装目录，即添加上 D:/dev/node.js/node_global。

接下来，我们验证一下 Node 全局安装目录的 Path 环境变量是否配置正确。首先打开 node_global 的根目录，比如 D:/dev/node.js/node_global（之前有安装过 cnpm 模块），如图 4-42 所示。

软件 (D:) > dev > nodejs > node_global

名称	修改日期	类型	大小
node_modules	2020/10/3 20:01	文件夹	
cnpm	2020/5/12 15:51	文件	1 KB
cnpm	2020/5/12 15:51	Windows 命令脚本	1 KB
cnpm	2020/5/12 15:51	Windows PowerShell ...	1 KB

图 4-42

打开 cmd 窗口，用 cnpm -v 来验证一下，如果执行该命令能够正确输出 cnpm 的版本号，则 Node 全局安装目录的 Path 环境变量配置正确。

另外补充一下 npm 和 cnpm 的区别：npm 是 Node 官方的包管理器。cnpm 是中国版的 npm，是用淘宝定制的 cnpm（gzip 压缩支持）命令行工具代替默认的 npm。大多数情况下用 npm 安装模块会出现超时问题，而用 cnpm 安装模块很难出现超时问题，因为 cnpm 默认用的是国内的淘宝镜像，网速是可以保证的。

安装 Node.js 后默认没有 cnpm，可以执行如下命令安装 cnpm（这里-g 参数表示全局安装，即安装至 node_global 文件夹中）：

```
npm install -g cnpm --registry=https://registry.npm.taobao.org
```

至此，Node 缓存文件夹和 Node 全局安装目录配置完毕。

如果上述配置都没有问题，那么 Node.js 的安装与配置就告一段落了。有了 Node.js 之后，前端项目运行的基础环境就有了。下面教大家如何搭建一个典型的 Vue 工程。

2. 项目搭建

一般情况下，Vue 项目都是用 vue-cli 搭建好的，会生成很多配置文件和基础文件，这样做的好处主要是不用自己新建文件和手动配置了，节省了很多工作量，提高了前端的开发效率。这就好比后端开发者利用 Spring initializr 去创建一个 Spring Boot 项目。

vue-cli 是 vue.js 的脚手架，用于自动生成 vue.js+webpack 的项目模板。webpack 是一个前端打包工具，我们暂不做深入研究。vue-cli 需要通过 Node.js 进行安装，首先打开 cmd 窗口，执行 npm install -g @vue/cli 命令，安装 vue-cli。

vue-cli 安装完成后，打开 cmd 命令窗口，先用 cd 命令进入想要存放项目的路径，然后执行 vue init webpack employee-manage-portal 命令初始化创建项目，此时控制台会显示一系列参数：

```
    ? Project name employee-manage-portal # 项目名称，直接回车，按照括号中默认名字
    #（注意这里的名字不能有大写字母，如果有就会报错：Sorry，name can no longer contain
    # capital letters）
    ? Project description A Vue.js project # 项目描述，随便写
    ? Author # 作者名称
    ? Vue build standalone # 这里选择的是运行加编译时
      Runtime + Compiler: recommended for most users
    ? Install vue-router? Yes # 是否需要 vue-router，路由肯定是要的
    ? Use ESLint to lint your code? Yes # 是否使用 ESLint 作为代码规范
    ? Pick an ESLint preset Standard # 一样的 ESlint 相关
    ? Set up unit tests Yes # 是否安装单元测试
    ? Pick a test runner # 是否需要测试模块，按需选择
    ? Setup e2e tests with Nightwatch? # 是否需要 e2e 测试，按需选择
    ? Should we run `npm install` for you after the project has been created?
(recommended) npm
```

以上参数都确认完毕后，等待项目初始化完成。前端开发工具推荐使用 HBuilder 或者微软的 VSCode。用 HBuilder 打开刚刚初始化好的项目，其项目的层次结构如图 4-43 所示。

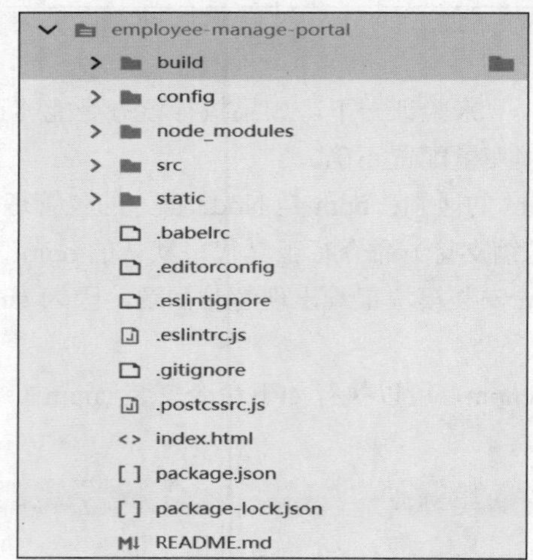

图 4-43

项目搭建完毕后还要做一步工作，即在 config/index.js 中配置前端服务的端口号，比如 8081，如图 4-44 所示。

3. 集成 Element-UI 框架

Element-UI 是饿了么前端团队推出的一款基于 Vue.js 2.0 的桌面端 UI 框架，手机端对应的框架是 Mint UI。前面提到过，这里将采用 Vue+Element-UI 开发前端，接下来将介绍如何集成 Element-UI 框架。

```
index.js

'use strict'
// Template version: 1.3.1
// see http://vuejs-templates.github.io/webpack for documentation.

const path = require('path')

module.exports = {
  dev: {

    // Paths
    assetsSubDirectory: 'static',
    assetsPublicPath: '/',
    proxyTable: {},

    // Various Dev Server settings
    host: 'localhost', // can be overwritten by process.env.HOST
    port: 8081, // can be overwritten by process.env.PORT, if port is in use, a free one will be determined
    autoOpenBrowser: true,
    errorOverlay: true,
    notifyOnErrors: true,
    poll: false, // https://webpack.js.org/configuration/dev-server/#devserver-watchoptions-
```

图 4-44

集成 Element-UI 的方式很简单，因为我们不需要像以前那样在 HTML 文件的 head 标签中通过 link 标签和 script 标签手动引入框架的样式文件和脚本文件。我们只需要在项目的某个文件中统一引入项目所依赖的模块即可。

具体引入方式为，在 src/main.js 中添加如下代码，表示在项目中引入 Element-UI 的相关组件及样式。

示例代码 4-48　main.js（部分代码）

```
import ElementUI from 'element-ui'
import 'element-ui/lib/theme-chalk/index.css'
Vue.use(ElementUI);
```

引用完成后 main.js 的内容如图 4-45 所示。在 main.js 中引用过的组件是全局生效的，不用再在每个模板页面中单独引用。

```
main.js

// The Vue build version to load with the `import` command
// (runtime-only or standalone) has been set in webpack.base.conf with an alias.
import Vue from 'vue'

import App from './App'
import router from './router'

//引入element-ui
import ElementUI from 'element-ui'
import 'element-ui/lib/theme-chalk/index.css'
Vue.use(ElementUI);
Vue.config.productionTip = false

/* eslint-disable */
new Vue({
  el: '#app',
  router,
  components: { App },
  template: '<App/>'
})
```

图 4-45

Element-UI 界面简洁大方又不失美观，包含丰富的组件和样式，适合快速开发后端管理系统。

本节中的前端项目例子可能不会运用很多 Element-UI 的组件，如果有兴趣学习更多关于 Element-UI 的知识，可以访问 Element-UI 的官方网站。

4. 集成 Axios 网络请求库

由于项目中需要请求后端接口获取员工信息，因此需要网络请求库的支持。Axios 是一个基于 Promise 的 HTTP 库，可以用在浏览器和 Node.js 中，具有以下特性：

（1）从浏览器中创建 XMLHttpRequests。
（2）从 Node.js 创建 HTTP 请求。
（3）支持 Promise API。
（4）拦截请求和响应。
（5）转换请求数据和响应数据。
（6）取消请求。
（7）自动转换 JSON 数据。
（8）客户端支持防御 XSRF。

Axios 和 Ajax 的使用方法基本一样，只有个别参数不同，如果用到了 vue-cli，则推荐使用 Axios，不建议使用 jQuery 的 Ajax。

集成与配置 Axios 的代码均在 src/api/api.js 中。

步骤 01 导入 Axios：

```
import axios from 'axios'
```

步骤 02 设置后端服务的 Web 上下文：

```
let base = '/ems'
```

步骤 03 设置 Axios 请求统一的 base 路径和超时时间：

```
axios.defaults.baseURL = 'http://localhost:8080';
axios.defaults.timeout = 10000;
```

步骤 04 定义一个获取所有员工信息的接口：

```
export const listEmployee = params => { return axios.get(`${base}/employees`); }
```

完整代码如图 4-46 所示。

```
api.js

import axios from 'axios'
//在api中导入axios
let base = '/ems'
axios.defaults.baseURL = 'http://localhost:8080';
axios.defaults.timeout = 10000;

export const listEmployee = params => { return axios.get(`${base}/employees`); }


```

图 4-46

5. 开发菜单页面

大多数系统都有菜单，所以这里首先要有一个菜单，而且这个菜单是固定的，不随页面的跳转而变化。

这里将菜单放在 App.vue（值得注意的是，Vue 项目中的页面都是以.vue 结尾的）中，然后通过路由组件 router-view 引用当前路由对应的模板页面。比如，创建四个菜单，分别是员工管理、考勤管理、薪酬管理、系统管理，然后每个菜单对应一个路由，每个路由对应一个模板页面，通过单击不同的菜单跳转到不同的页面。参考代码如下。

示例代码 4-49　App.vue（部分代码）

```
<template>
  <div id="app">
      <el-menu
        :default-active="activeIndex"
        class="el-menu-demo"
        mode="horizontal"
        background-color="#545c64"
        text-color="#fff"
        active-text-color="#ffd04b"
        router>
        <el-menu-item index="employee">员工管理</el-menu-item>
        <el-menu-item index="attendance">考勤管理</el-menu-item>
        <el-menu-item index="salary">薪酬管理</el-menu-item>
        <el-menu-item index="system">系统管理</el-menu-item>
      </el-menu>
    <router-view></router-view>
  </div>
</template>

<script>
export default {
  name: 'App',
  data () {
    return {
      activeIndex: 'employee'
    }
  }
}
</script>
<style>
</style>
```

Vue 项目中的每一个模板页面都包含三个标签，分别是 template、script、style。其中，template 放页面元素，script 和 style 分别存放脚本代码和样式代码。el-menu 是 Element-UI 提供的菜单组件，el-menu-item 标签的 index 属性值表示路由。

6. 路由配置

在创建项目时有一个 Install vue-router 选项，如果选择 Yes，就会默认引入路由，而不需要再手动引入。

传统的前后端不分离项目是通过服务端的 Spring MVC 的 ModeAndView 进行页面跳转的。现在的前后端分离项目中前端完全可以通过自己去控制路由来达到页面切换的目的，因此 vue-router 组件应景而生。

路由的配置在 src/router/index.js 中。在这个 js 中配置两个路由：一个是默认路由“/”，访问该路由跳到员工管理页面；另一个是员工管理页面的路由“/employee”，访问该路由也跳到员工管理页面，参考代码如下。

示例代码 4-50　index.js（部分代码）

```
import Vue from 'vue'
import Router from 'vue-router'
import employee from '@/components/employee'
Vue.use(Router)
export default new Router({
  routes: [
      {
          path: '/',
          redirect: '/employee',//设置默认指向的路径
          name: 'default'
        },
      {
      path: '/employee',
      name: 'employee',
      component: employee
      }
    ]
})
```

7. 开发员工管理页面

员工管理是核心业务页面，该页面中有员工新增、删除、修改按钮，并且要以一个表格的方式展示所有员工的信息。

首先在 src/components 目录下新建 employee.vue，参考代码如下。

示例代码 4-51　employee.vue

```
<template>
  <div>
      <el-button type="primary">新增员工</el-button>
      <el-table :data="tableData" style="width: 100%">
            <el-table-column prop="id" label="id"
width="80"></el-table-column>
            <el-table-column prop="number" label="员工编号"
width="100"></el-table-column>
            <el-table-column prop="name" label="姓名"
width="100"></el-table-column>
            <el-table-column prop="department" label="部门"
width="180"></el-table-column>
            <el-table-column prop="position" label="职位"
```

```
width="180"></el-table-column>
            <el-table-column prop="birthday" label="生日"
width="180" :formatter="formatDate"></el-table-column>
            <el-table-column prop="sex" label="性别"
width="80"></el-table-column>
            <el-table-column prop="salary" label="薪资"
width="180"></el-table-column>
                  <el-table-column label="操作" width="200">
                     <template slot-scope="scope">
                        <el-button type="warning" size="small">修改
</el-button>
                        <el-button type="danger" size="small">删除
</el-button>
                     </template>
                  </el-table-column>
            </el-table>
      </div>
    </template>
    <script>
    import { listEmployee } from '../api/api';
    export default {
      name: 'employee',
      data () {
        return {
          tableData : []
        }
      },
      methods: {
          loadData() {
            listEmployee({}).then(result=>{
              this.tableData = result.data.data;
            })
          },
          formatDate(row, column) {
              // 获取单元格数据
              let data = row[column.property]
              if(data == null) {
                  return null
              }
              let dt = new Date(data)
               return dt.getFullYear() + '-' + (dt.getMonth() + 1) + '-' +
dt.getDate();
          },
      },
      created(){
          this.loadData();
      }
    }
    </script>
    <style>
```

```
</style>
```

注 意

我们需要使用 import 关键字来导入 listEmployee 方法，否则无法使用该方法。

8. 运行测试

启动后端项目，如图 4-47 所示，若出现框中的内容，则表示启动成功，后端服务的 Web 端口采用的是默认 8080 端口。

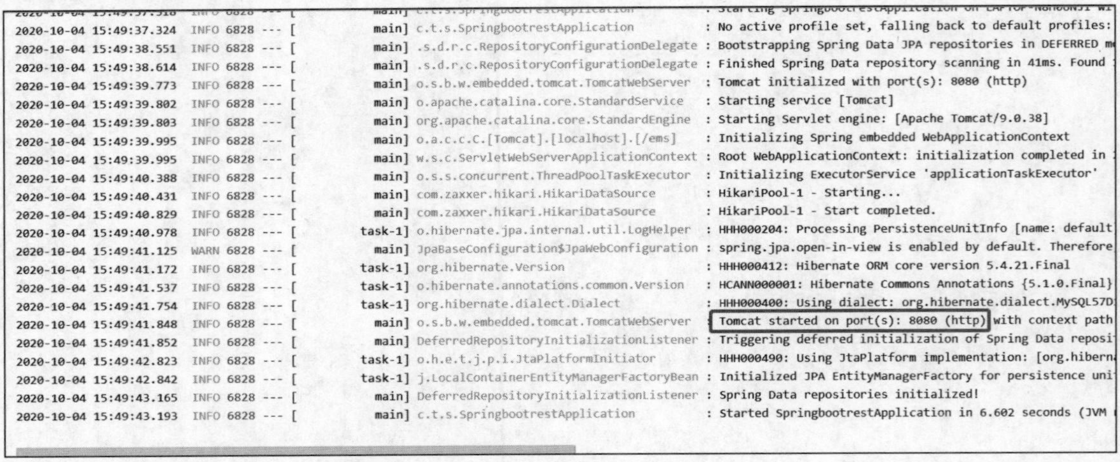

```
2020-10-04 15:49:37.324  INFO 6828 --- [           main] c.t.s.SpringbootrestApplication          : No active profile set, falling back to default profiles:
2020-10-04 15:49:38.551  INFO 6828 --- [           main] .s.d.r.c.RepositoryConfigurationDelegate : Bootstrapping Spring Data JPA repositories in DEFERRED m
2020-10-04 15:49:38.614  INFO 6828 --- [           main] .s.d.r.c.RepositoryConfigurationDelegate : Finished Spring Data repository scanning in 41ms. Found
2020-10-04 15:49:39.773  INFO 6828 --- [           main] o.s.b.w.embedded.tomcat.TomcatWebServer  : Tomcat initialized with port(s): 8080 (http)
2020-10-04 15:49:39.802  INFO 6828 --- [           main] o.apache.catalina.core.StandardService   : Starting service [Tomcat]
2020-10-04 15:49:39.803  INFO 6828 --- [           main] org.apache.catalina.core.StandardEngine  : Starting Servlet engine: [Apache Tomcat/9.0.38]
2020-10-04 15:49:39.995  INFO 6828 --- [           main] o.a.c.c.C.[Tomcat].[localhost].[/ems]    : Initializing Spring embedded WebApplicationContext
2020-10-04 15:49:39.995  INFO 6828 --- [           main] w.s.c.ServletWebServerApplicationContext : Root WebApplicationContext: initialization completed in
2020-10-04 15:49:40.388  INFO 6828 --- [           main] o.s.s.concurrent.ThreadPoolTaskExecutor  : Initializing ExecutorService 'applicationTaskExecutor'
2020-10-04 15:49:40.431  INFO 6828 --- [           main] com.zaxxer.hikari.HikariDataSource       : HikariPool-1 - Starting...
2020-10-04 15:49:40.829  INFO 6828 --- [           main] com.zaxxer.hikari.HikariDataSource       : HikariPool-1 - Start completed.
2020-10-04 15:49:40.978  INFO 6828 --- [         task-1] o.hibernate.jpa.internal.util.LogHelper  : HHH000204: Processing PersistenceUnitInfo [name: default
2020-10-04 15:49:41.125  WARN 6828 --- [           main] JpaBaseConfiguration$JpaWebConfiguration : spring.jpa.open-in-view is enabled by default. Therefore
2020-10-04 15:49:41.172  INFO 6828 --- [         task-1] org.hibernate.Version                    : HHH000412: Hibernate ORM core version 5.4.21.Final
2020-10-04 15:49:41.537  INFO 6828 --- [         task-1] o.hibernate.annotations.common.Version   : HCANN000001: Hibernate Commons Annotations {5.1.0.Final}
2020-10-04 15:49:41.754  INFO 6828 --- [         task-1] org.hibernate.dialect.Dialect            : HHH000400: Using dialect: org.hibernate.dialect.MySQL57D
2020-10-04 15:49:41.848  INFO 6828 --- [           main] o.s.b.w.embedded.tomcat.TomcatWebServer  : Tomcat started on port(s): 8080 (http) with context path
2020-10-04 15:49:41.852  INFO 6828 --- [           main] DeferredRepositoryInitializationListener : Triggering deferred initialization of Spring Data reposi
2020-10-04 15:49:42.823  INFO 6828 --- [         task-1] o.h.e.t.j.p.i.JtaPlatformInitiator       : HHH000490: Using JtaPlatform implementation: [org.hibern
2020-10-04 15:49:42.842  INFO 6828 --- [         task-1] j.LocalContainerEntityManagerFactoryBean : Initialized JPA EntityManagerFactory for persistence uni
2020-10-04 15:49:43.165  INFO 6828 --- [           main] DeferredRepositoryInitializationListener : Spring Data repositories initialized!
2020-10-04 15:49:43.193  INFO 6828 --- [           main] c.t.s.SpringbootrestApplication          : Started SpringbootrestApplication in 6.602 seconds (JVM
```

图 4-47

Node.js 是前端项目运行的基础环境。首先打开 cmd 窗口，利用 cd 命令进入前端项目的根目录下，然后执行 npm install 命令安装项目所需要的所有模块，安装完成后会在项目根目录下生成 node_modules 文件夹。需要了解的是，npm install 只需执行一次，再次运行项目就不需要执行了。

接着执行 npm run dev 命令运行项目，运行成功后访问 http://localhost:8081/#/employee，如图 4-48 所示，如果表格中有数据，就表示前端能成功请求后端服务，并且后端 RESTful 接口返回了正确的员工信息结果集。

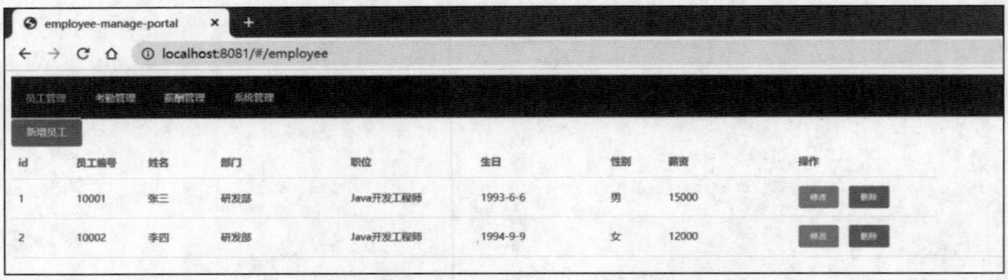

employee-manage-portal

localhost:8081/#/employee

员工管理 考勤管理 薪酬管理 系统管理

新增员工

id	员工编号	姓名	部门	职位	生日	性别	薪资	操作
1	10001	张三	研发部	Java开发工程师	1993-6-6	男	15000	修改 删除
2	10002	李四	研发部	Java开发工程师	1994-9-9	女	12000	修改 删除

图 4-48

这里通过真实的业务案例实现一个完整的前后端分离应用：后端采用 Spring Boot 框架，前端采用比较流行的 Node.js+Vue+Element-UI 技术组合；后端主要侧重 RESTful 风格 URL 的讲解，前端方面则对搭建 Node.js 环境、搭建项目以及集成各种框架等多方面进行渗透。

第 5 章

Spring Boot 原理解读

Spring Boot 简化了基于 Spring 的 Java 应用开发，降低了使用难度，从这个意义上讲 Spring Boot 是对 Spring 的进一步封装。这种封装使得很多使用者“只知其然，而不知其所以然”，在使用过程中出现问题时，不知道如何排错或者不能更好地使用 Spring Boot。本章将讲解 Spring Boot 的自动配置原理、启动流程、starter 和它的配置以及内嵌 Web 服务器原理等。

5.1 获取源代码

阅读源代码使得我们能快速、正确地理解 Spring Boot 的原理，但这要求阅读者具有较强的 Java 语言功底、精力、耐心和开发经验。虽然使用 IDEA 的反编译功能也能看到 Java 源代码，但是这种源代码没有注释，不利于理解。

目前 Java 项目大多基于 Maven 构建，使用 Git 分布式版本控制工具对项目进行管理。可以使用 Git 工具一次性把所有代码复制到本地，也可以使用 Maven 工具下载某个 jar 对应的源代码和 javadoc 文档。本节将讲述这两种获取源代码的方式。

5.1.1 使用 Git 复制

Spring Boot 的源代码托管在 https://github.com/spring-projects/spring-boot，可以在 cmd 中直接使用 Git 的 clone 命令（注意要安装 Git 客户端）复制。

```
git clone https://github.com/spring-projects/spring-boot.git
```

如果感觉输入命令复杂，也可以使用如图 5-1、图 5-2 所示的图形化界面，使用 Git 一次性下载 Spring Boot 到本地。

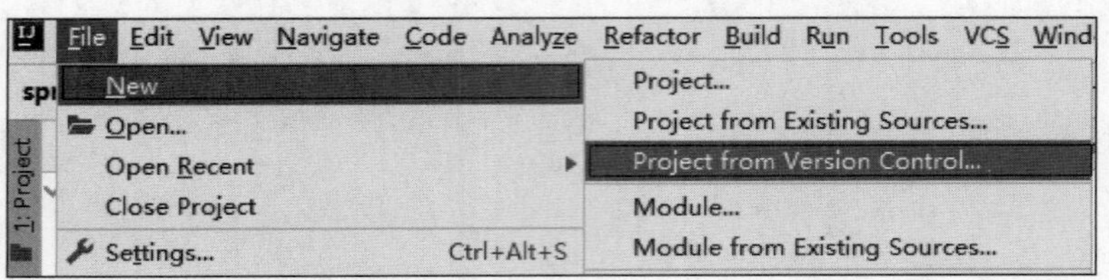

图 5-1

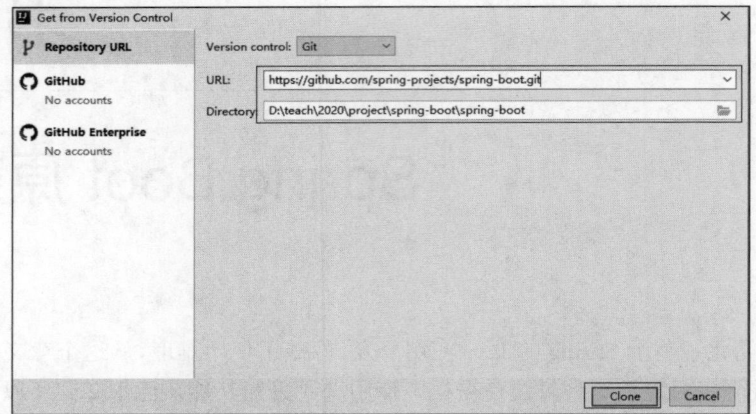

图 5-2

由于源代码比较大和 Github 国内访问速度慢的问题，下载需要一段时间，如图 5-3 所示。

图 5-3

5.1.2 使用 Maven 自动下载

我们可以直接通过 mvn 命令行（在环境变量中配置好 Maven 路径）下载 jar 的源代码，比如下载 spring-web 的 4.3.16 版本的源代码：

```
mvn org.apache.maven.plugins:maven-dependency-plugin:2.8:get
-Dartifact=org.springframework:spring-web:4.3.16.RELEASE:jar:sources
```

如果觉得命令不易使用，也可以使用如图 5-4 所示的图形化界面，在 IDEA 中使用 Maven 来下载 jar 源代码。

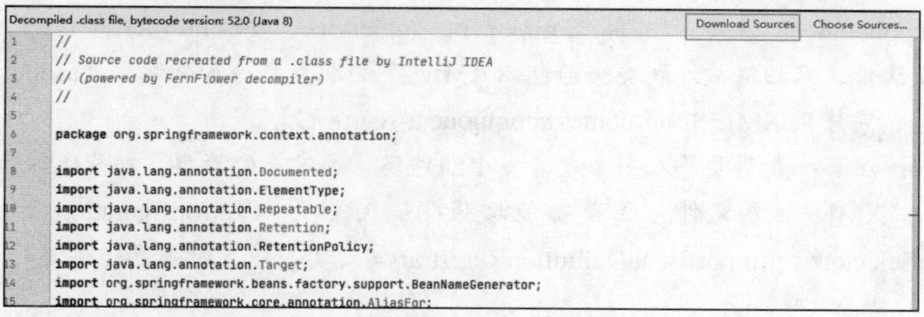

```
Decompiled .class file, bytecode version: 52.0 (Java 8)    Download Sources    Choose Sources...
//
// Source code recreated from a .class file by IntelliJ IDEA
// (powered by FernFlower decompiler)
//

package org.springframework.context.annotation;

import java.lang.annotation.Documented;
import java.lang.annotation.ElementType;
import java.lang.annotation.Repeatable;
import java.lang.annotation.Retention;
import java.lang.annotation.RetentionPolicy;
import java.lang.annotation.Target;
import org.springframework.beans.factory.support.BeanNameGenerator;
import org.springframework.core.annotation.AliasFor;
```

图 5-4

首先到jar中找到对应类，然后双击，IDEA将自动反编译。接着单击顶部的“Download Sources”，下载该类所在 jar 的对应源代码。下载完毕后，IDEA 会自动打开源代码（带有完整的注释），如图 5-5 所示。

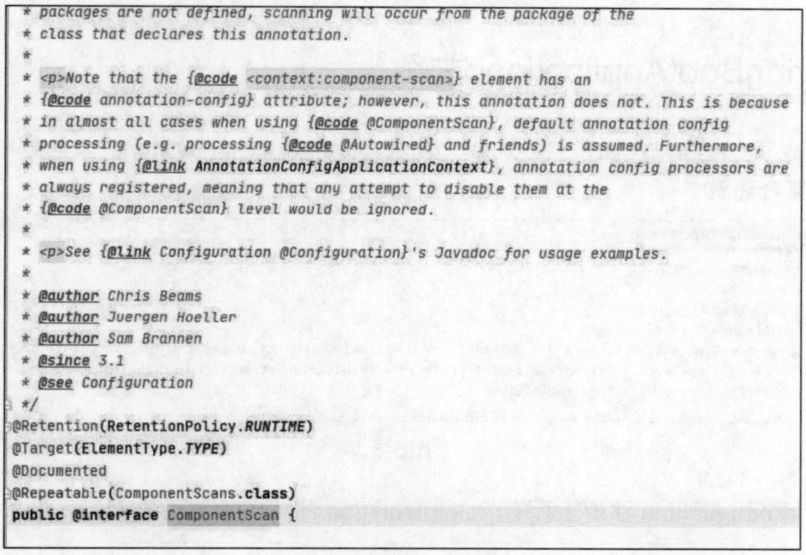

```
 * packages are not defined, scanning will occur from the package of the
 * class that declares this annotation.
 *
 * <p>Note that the {@code <context:component-scan>} element has an
 * {@code annotation-config} attribute; however, this annotation does not. This is because
 * in almost all cases when using {@code @ComponentScan}, default annotation config
 * processing (e.g. processing {@code @Autowired} and friends) is assumed. Furthermore,
 * when using {@link AnnotationConfigApplicationContext}, annotation config processors are
 * always registered, meaning that any attempt to disable them at the
 * {@code @ComponentScan} level would be ignored.
 *
 * <p>See {@link Configuration @Configuration}'s Javadoc for usage examples.
 *
 * @author Chris Beams
 * @author Juergen Hoeller
 * @author Sam Brannen
 * @since 3.1
 * @see Configuration
 */
@Retention(RetentionPolicy.RUNTIME)
@Target(ElementType.TYPE)
@Documented
@Repeatable(ComponentScans.class)
public @interface ComponentScan {
```

图 5-5

Maven 会把下载后的源代码安装到本地，后面再使用时就不需要下载 jar 源代码了，并且 IDEA 会根据源代码给出友好提示，因此推荐使用这种方式。

5.2 剖析自动配置原理

Spring Boot 是基于零 XML 配置的，使用“惯例优先于配置”策略。如果默认不满足要求，那么大部分情况下只需要在配置文件中配置即可。要完成这些工作，需要自动配置（Spring Boot 的主要特性之一）。要搞明白 Spring Boot 自动配置原理，需要了解 Spring 替代 XML 配置的注解：

- @Configuration 注解：用在某个类上，表明被注解类是一个配置类，对应 Spring 的 XML 配置文件。在该类某个方法上使用@Bean，完成 Spring 的 Bean 创建，以替代<bean>配置。

- @ComponentScan 注解：扫描@Component、@Service、@Controller 等注解。如果使用该注解的默认值，只扫描被注解类当前包以及子包下的配置类；如果配置 basePackages，则扫描指定包。它替代 XML 中的<context:component-scan>元素。
- @Import 注解：表明要导入一个或者多个组件类，通常是配置类。功能上等价于<import/>导入其他 XML 配置文件，但是功能增强了，可以导入@Configuration 注解的配置类、ImportSelector 和 ImportBeanDefinitionRegistrar 实现类，以及常规的组件类 。

通常 Spring Boot 启动类和 stater 不在同一个包下，而@ComponentScan 默认只能自动扫描当前包以及子包，在启动类中靠@Import 去导入 stater 配置类不算自动配置。

Spring Boot 要完成自动配置，需要有新机制来读取其他包下的配置类、事件监听器等。Spring Boot 依靠 SpringFactoriesLoader 类读取 META-INF/spring.factories 配置的 Spring 配置类等，实现了自动配置。

5.2.1 SpringBootApplication 注解

对于 Spring Boot 项目，通常要添加一个注解 SpringBootApplication。该注解源代码如图 5-6 所示，是一个复合注解。

```
SpringBootApplication.java
@Documented
@Inherited
@SpringBootConfiguration
@EnableAutoConfiguration
@ComponentScan(excludeFilters = { @Filter(type = FilterType.CUSTOM, classes = TypeExcludeFilter.class),
        @Filter(type = FilterType.CUSTOM, classes = AutoConfigurationExcludeFilter.class) })
public @interface SpringBootApplication {
```

图 5-6

SpringBootConfiguration 注解上面有@Configuration 注解，如图 5-7 所示。

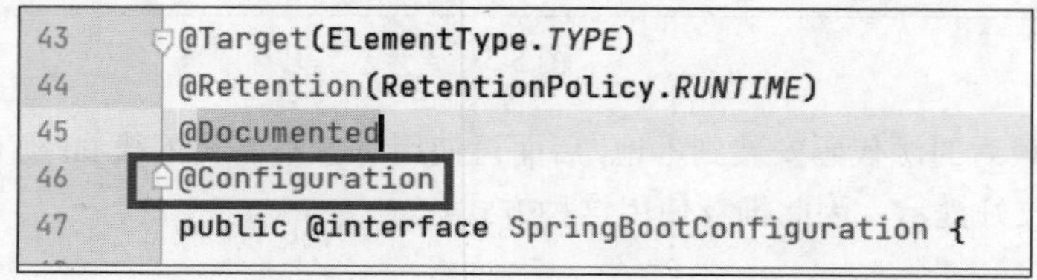

图 5-7

SpringBootConfiguration 注解是一个 Configuration 配置类，用于定义一个或者多个 Bean。

- EnableAutoConfiguration 触发自动配置。
- ComponentScan 扫描当前包以及子包下的组件。

这是一个简便的注解，等价于声明了@Configuration、@EnableAutoConfiguration 和@ComponentScan 三个注解。

5.2.2 EnableAutoConfiguration 注解

该注解触发配置类自动导入，源代码如图 5-8 所示。

```
EnableAutoConfiguration.java ×
    See Also: ConditionalOnBean, ConditionalO
              ConditionalOnClass, AutoConfigu
    Author:   Phillip Webb, Stephane Nicoll
@Target(ElementType.TYPE)
@Retention(RetentionPolicy.RUNTIME)
@Documented
@Inherited
@AutoConfigurationPackage
@Import(AutoConfigurationImportSelector.class)
public @interface EnableAutoConfiguration {
```

图 5-8

其上的@Import 注解用于导入其他配置类，配置的 value 属性值是一个 ImportSelector 接口（导入满足条件的@Configuration 配置类）的实现类。如图 5-9 所示，方法 selectImports 根据参数 AnnotationMetadata 值返回所有候选配置类全限定名，而 getExclusionFilter 从候选配置类中筛选满足条件的候选配置类。

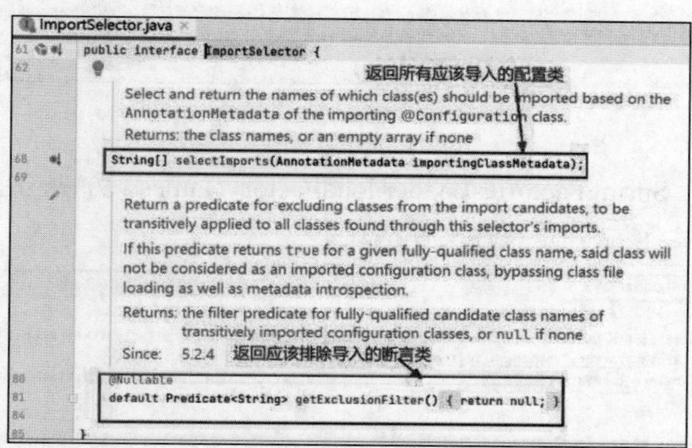

图 5-9

5.2.3 AutoConfigurationImportSelector 类

该类实现了接口 ImportSelector，其中的 selectImports 方法返回所有配置类，如图 5-10 所示。

该方法返回需要导入的配置类

读取自动配置类

```
AutoConfigurationImportSelector.java ×                                Reader Mode
@Override
public String[] selectImports(AnnotationMetadata annotationMetadata) {
    if (!isEnabled(annotationMetadata)) {
        return NO_IMPORTS;
    }
    AutoConfigurationMetadata autoConfigurationMetadata = AutoConfigurationMetadataLoader
            .loadMetadata(this.beanClassLoader);
    AutoConfigurationEntry autoConfigurationEntry = getAutoConfigurationEntry(autoConfigurationMetadata,
            annotationMetadata);
    return StringUtils.toStringArray(autoConfigurationEntry.getConfigurations());
}
```

图 5-10

在 getAutoConfigurationEntry 方法中调用 getCandidateConfigurations 来读取候选配置类，如图 5-11 所示。

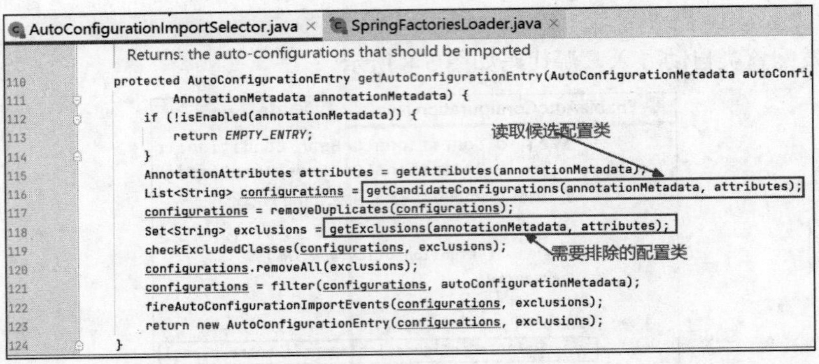

```
AutoConfigurationImportSelector.java × SpringFactoriesLoader.java ×
    Returns: the auto-configurations that should be imported
protected AutoConfigurationEntry getAutoConfigurationEntry(AutoConfigurationMetadata autoConfi
        AnnotationMetadata annotationMetadata) {
    if (!isEnabled(annotationMetadata)) {
        return EMPTY_ENTRY;
    }
    AnnotationAttributes attributes = getAttributes(annotationMetadata);
    List<String> configurations = getCandidateConfigurations(annotationMetadata, attributes);
    configurations = removeDuplicates(configurations);
    Set<String> exclusions = getExclusions(annotationMetadata, attributes);
    checkExcludedClasses(configurations, exclusions);
    configurations.removeAll(exclusions);
    configurations = filter(configurations, autoConfigurationMetadata);
    fireAutoConfigurationImportEvents(configurations, exclusions);
    return new AutoConfigurationEntry(configurations, exclusions);
}
```

读取候选配置类

需要排除的配置类

图 5-11

在 getCandidateConfigurations 方法中调用 SpringFactoriesLoader 的静态方法 loadFactoryNames，如图 5-12 所示。

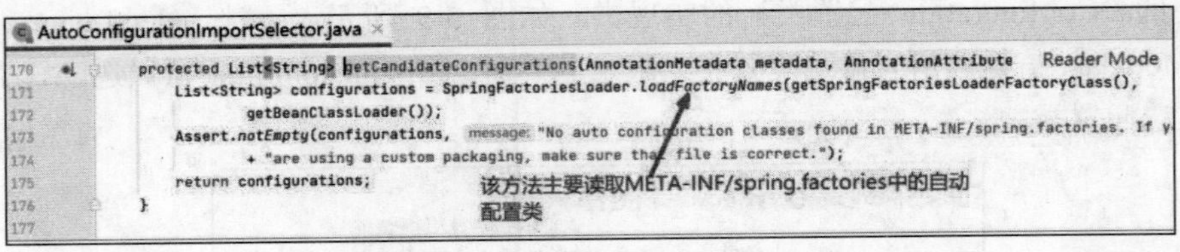

```
AutoConfigurationImportSelector.java ×
protected List<String> getCandidateConfigurations(AnnotationMetadata metadata, AnnotationAttribute    Reader Mode
    List<String> configurations = SpringFactoriesLoader.loadFactoryNames(getSpringFactoriesLoaderFactoryClass(),
            getBeanClassLoader());
    Assert.notEmpty(configurations, message: "No auto configuration classes found in META-INF/spring.factories. If y
            + "are using a custom packaging, make sure that file is correct.");
    return configurations;
}
```

该方法主要读取META-INF/spring.factories中的自动配置类

图 5-12

图 5-13 所示的 SpringFactoriesLoader.loadFactoryNames 方法用于读取 META-INF /spring.factories 文件内容，该文件是属性文件格式。

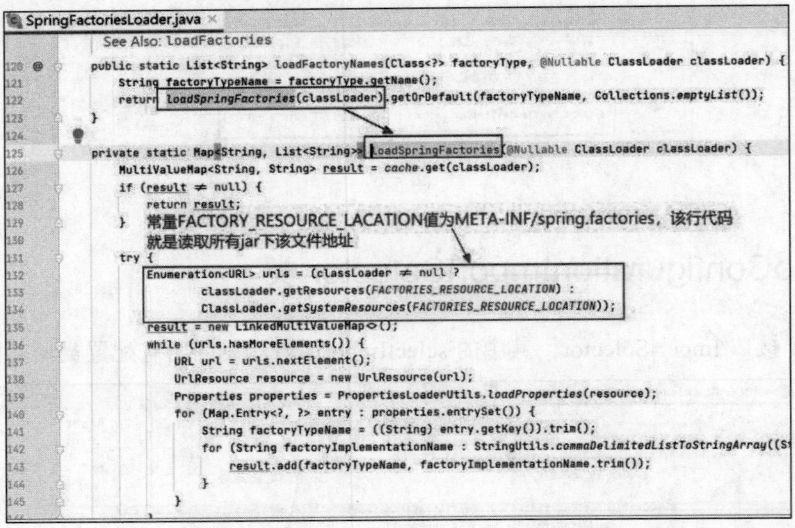

```
SpringFactoriesLoader.java ×
    See Also: loadFactories
public static List<String> loadFactoryNames(Class<?> factoryType, @Nullable ClassLoader classLoader) {
    String factoryTypeName = factoryType.getName();
    return loadSpringFactories(classLoader).getOrDefault(factoryTypeName, Collections.emptyList());
}

private static Map<String, List<String>> loadSpringFactories(@Nullable ClassLoader classLoader) {
    MultiValueMap<String, String> result = cache.get(classLoader);
    if (result ≠ null) {
        return result;
    }
    try {
        Enumeration<URL> urls = (classLoader ≠ null ?
                classLoader.getResources(FACTORIES_RESOURCE_LOCATION) :
                ClassLoader.getSystemResources(FACTORIES_RESOURCE_LOCATION));
        result = new LinkedMultiValueMap<>();
        while (urls.hasMoreElements()) {
            URL url = urls.nextElement();
            UrlResource resource = new UrlResource(url);
            Properties properties = PropertiesLoaderUtils.loadProperties(resource);
            for (Map.Entry<?, ?> entry : properties.entrySet()) {
                String factoryTypeName = ((String) entry.getKey()).trim();
                for (String factoryImplementationName : StringUtils.commaDelimitedListToStringArray((St
                    result.add(factoryTypeName, factoryImplementationName.trim());
                }
            }
```

常量FACTORY_RESOURCE_LACATION值为META-INF/spring.factories，该行代码就是读取所有jar下该文件地址

图 5-13

把自定义的配置类、监听器等配置到 META-INF/spring.factories 中，无论是否在 Spring Boot

启动类包以及子包下，类都会被 Spring Boot 自动扫描到，如图 5-14 所示。这就是自动配置的基本原理。

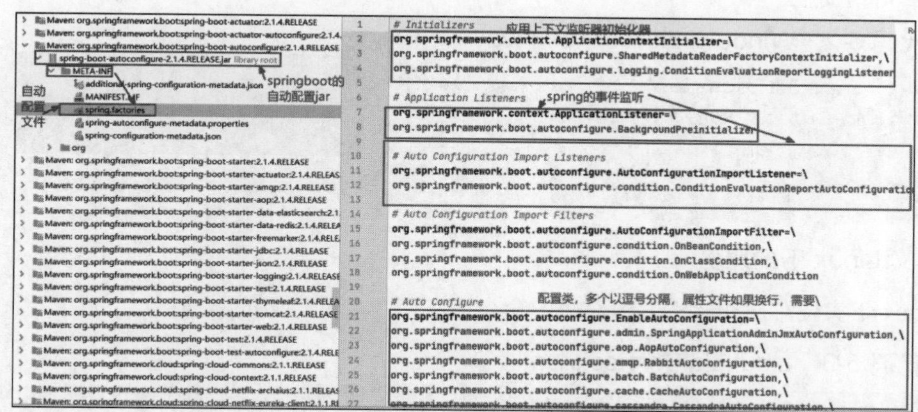

图 5-14

5.2.4　Conditional 注解

Conditional 注解从 Spring 4 开始引入，用于条件性启用或者禁用@Configuration 类或者@Bean 方法。Starter 配置的一些 Bean 可能需要修改，例如把默认数据源 HikariDataSource 换成 Druid，那么默认的数据源 HikariDataSource 对应的 Bean 就不能再配置，否则会存在两个数据源，因而某些 Bean 是否需要注册到 Spring 容器是有条件的。Spring Boot 使用@Conditional 来完成 Bean 的条件注册。下面通过一个案例讲述该注解的使用。

1. 需求

根据当前操作系统返回列举文件夹的命令：

- Windows：dir。
- Linux：ls。

2. 新建一个 Maven 项目

在 IDEA 中使用 Spring Initializr 向导创建一个名字为 condition-test 的 Spring Boot 项目，添加 Spring Web 依赖。

3. 定义接口

约定需要提供的功能方法，代码如下所示。

示例代码 5-1　ListService.java

```
public interface ListService {
    //显示查看文件夹的命令
    public String showCommand();
}
```

4. 定义 Windows 下的服务

在 Windows 下实现对应约定：显示对应的命令“dir”，如下代码所示。

示例代码 5-2　WindowsListService.java

```
public class WindowsListService implements ListService {
    public String showCommand() {
        return "dir";
    }
}
```

5. 定义 Linux 下的服务

在 Linux 下实现对应约定：显示对应的命令“ls”，如下代码所示。

示例代码 5-3　LinuxListService.java

```
public class LinuxListService implements ListService {
    public String showCommand() {
        return "ls";
    }
}
```

6. 定义 Controller

Spring MVC 前端控制器需要注入 ListService 接口实现类，这里仅仅演示能否注入成功，不需要提供 Controller 方法，如下代码所示。

示例代码 5-4　ListController.java

```
@Controller
public class ListController {
    @Autowired
    private ListService listService;
}
```

7. 定义配置类

Spring Boot 基于零 XML 配置 Spring 管理的 Bean，需要配置两个接口的实现类，以便能自动注入 ListController 中，如下代码所示。

示例代码 5-5　AppConfig.java

```
@Configuration
public class AppConfig {
    @Bean
    public ListService windowsListService() {
        return new WindowsListService();
    }
    @Bean
    public ListService linuxListService() {
        return new LinuxListService();
    }
}
```

8. 定义启动类

Spring Boot 从一个普通 main 方法开始执行，其上用注解@SpringBootApplication，代码如下所示。

示例代码 5-6　Application.java

```
@SpringBootApplication
public class Application {
    public static void main(String[] args) {
    ConfigurableApplicationContext
            context=SpringApplication.run(Application.class, args);
        //打印所有 bean 的名字，以便于测试
        String[] beanNames = context.getBeanDefinitionNames();
        for (String beanName : beanNames) {
            System.out.println(beanName);
        }
    }
}
```

9. 测试

项目启动失败，控制台输出如下所示。因为两个实现类满足条件，Spring 无法判断注入哪个实现类对象给控制器。

```
***************************
APPLICATION FAILED TO START
***************************
Description:
Field listService in com.guodexian.condition.controller.ListController
required a single bean, but 2 were found:
    - windowsListService: defined by method 'windowsListService' in class path
resource [com/guodexian/condition/config/AppConfig.class]
    - linuxListService: defined by method 'linuxListService' in class path
resource [com/guodexian/condition/config/AppConfig.class]
```

需要把什么条件下注册哪个 ListService 表达给 Spring。表达需要编写代码，代码不能脱离方法，方法不能脱离类。这里采用接口方式：使用者实现接口 Condition，Spring 按照 Condition 接口中定义的方法调用。

10. Windows 下的条件类

实现 Condition 接口：如果是在 Windows 下就返回 true，否则返回 false。

示例代码 5-7　WindowsConditional.java

```
public class WindowsConditional implements Condition{
    @Override
public boolean matches(ConditionContext context,AnnotatedTypeMetadata
    metadata) {
        //把条件表达在这里
    return
    context.getEnvironment().getProperty("os.name").toLowerCase().contains("
    windows");
    }
```

```
}
```

11. Linux 下的条件类

实现 Condition 接口：如果是在 Linux 下就返回 true，否则返回 false。

示例代码 5-8　LinuxConditional.java

```
public class LinuxConditional implements Condition{
   @Override
public boolean matches(ConditionContext context, AnnotatedTypeMetadata
                       metadata) {
       //把条件表达在这里
   return
          context.getEnvironment().getProperty("os.name").toLowerCase().con
          tains("linux");
    }
}
```

12. 修改配置类，添加条件

修改示例代码 5-5，添加上条件注解。这样 Spring 在扫描@Bean 时，如果@Conditional 注解的 value 属性值配置类的 matches 返回 true，该方法就会被 Spring 处理；反之，该方法被忽略。

示例代码 5-5　AppConfig.java

```
@Configuration
public class AppConfig {
   @Conditional(WindowsConditional.class)
   @Bean
   public ListService windowsListService() {
      return new WindowsListService();
   }
   @Bean
   @Conditional(LinuxConditional.class)
   public ListService linuxListService() {
      return new LinuxListService();
   }
}
```

13. 测试

运行启动类，项目启动成功，如下控制台输出所示。由于项目是运行在 Windows 系统下的，因此 WindowsConditional 的 matches 方法返回 true，而 LinuxConditional 返回 false，只有 WindowsListService 注册到 Spring 容器中。

```
......
appConfig
listController
windowsListService
org.springframework.boot.autoconfigure.AutoConfigurationPackages
......
```

14. 修改配置类，添加条件在类上

修改 AppConfig，把注解条件添加到类上面：为了验证@Conditional 在方法和类上哪个优先，让类上返回 false。

示例代码 5-5 AppConfig.java

```
@Configuration
@Conditional(LinuxConditional.class)
public class AppConfig {
```

15. 测试

项目启动失败，控制台输出如下所示。由于在类上添加了条件注解，而目前运行在 Windows 环境下，类上注解 value 属性值类的 matches 方法返回 false，因而整个类不再处理，没有一个实现 ListService 接口的 Bean 注册给 Spring 容器。这里先判断类是否需要处理，如果不需要处理就结束。

```
***************************
APPLICATION FAILED TO START
***************************
Description:
Field listService in com.guodexian.condition.controller.ListController
required a bean of type 'com.guodexian.condition.service.ListService' that could
not be found.
The injection point has the following annotations:
    - @org.springframework.beans.factory.annotation.Autowired(required=true)
```

16. 使用@Conditional 注解的常用注解

使用@Conditional 注解需要定义条件类，比较麻烦，如图 5-15 所示。为了使用方便，Spring Boot 提供一些常用条件性注解以及对应条件类。

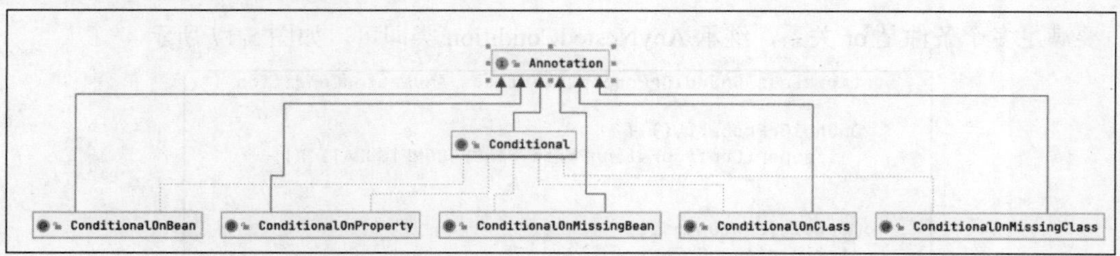

图 5-15

在 Spring Boot 中，被@Conditional 注解的常用注解需要满足如表 5-1 所示的条件。

表 5-1　使用@Conditional 注解的常用注解需要满足的条件

注解	功能
@ConditionalOnClass(name={"cn.edu.nyist.MyClass"})	指定类在类路径上
@ConditionalOnMissingClass	指定类不在类路径上
@ConditionalOnBean	指定的 Bean 类和/或名字已经包含在 BeanFactory。当该注解放置被@Bean 注解的方法上时，Bean 类默认为工厂方法返回值类型

（续表）

注解	功能
@ConditionalOnMissingBean	指定的 Bean 类和/或名字不在 BeanFactory。当该注解放置被@Bean 注解的方法上时，Bean 类默认为工厂方法返回值类型 Bean
@ConditionalOnProperty	指定的属性有指定的值时匹配，默认属性必须出现在 Environment 且不等于 false
@ConditionalOnNotWebApplication	不是 Web 环境时有效
@ConditionalOnWebApplication	是 Web 环境时有效
@OnJavaCondition	如果当前 Java 版本满足指定条件就返回 true，否则返回 false

17. @Conditional 注解组合

要满足多个条件是 and 关系，可以在类上使用多个条件注解，或者自定义类继承 org.springframework.boot.autoconfigure.condition.AllNestedConditions，如图 5-16 所示。

```
Condition that will match when all nested class conditions match. Can be used to create composite
conditions, for example:
   static class OnJndiAndProperty extends AllNestedConditions {

       OnJndiAndProperty() {
           super(ConfigurationPhase.PARSE_CONFIGURATION);
       }

       @ConditionalOnJndi()
       static class OnJndi {
       }

       @ConditionalOnProperty("something")
       static class OnProperty {
       }

   }
```

图 5-16

要满足多个条件是 or 关系，继承 AnyNestedCondition 类即可，如图 5-17 所示。

```
static class OnJndiOrProperty extends AnyNestedCondition {

    OnJndiOrProperty() {
        super(ConfigurationPhase.PARSE_CONFIGURATION);
    }

    @ConditionalOnJndi()
    static class OnJndi {
    }

    @ConditionalOnProperty("something")
    static class OnProperty {
    }
}
```

图 5-17

要满足多个条件是 not 关系，继承 NoneNestedConditions 类即可，如图 5-18 所示。

```
static class OnNeitherJndiNorProperty extends NoneOfNestedConditions {

    OnNeitherJndiNorProperty() {
        super(ConfigurationPhase.PARSE_CONFIGURATION);
    }

    @ConditionalOnJndi()
    static class OnJndi {
    }

    @ConditionalOnProperty("something")
    static class OnProperty {
    }

}
```

图 5-18

5.3　Spring Boot 启动流程

5.3.1　SpringApplication 初始化方法

我们在 Spring Boot 启动类中调用 SpringApplication 的静态方法 run，如下代码所示。

示例代码 5-9　Application.java

```
public static void main(String[] args) {
    SpringApplication.run(Application.class,args);
}
```

该 run 方法的源代码如下所示。

示例代码 5-10　SpringApplication.java（run 方法）

```
  public static ConfigurableApplicationContext run(Class<?> primarySource,
String... args) {
    return run(new Class<?>[] { primarySource }, args);
}
```

它又调用了另外一个重载 run 方法，首先创建一个 SpringApplication 对象，然后调用非静态 run 方法。

示例代码 5-10　SpringApplication.java（run 重载方法）

```
public static ConfigurableApplicationContext run(Class<?>[] primarySources,
        String[] args){
    return new SpringApplication(primarySources).run(args);
}
```

接下来看看构造函数完成的工作。

示例代码 5-10　SpringApplication.java 的构造函数

```
public SpringApplication(Class<?>... primarySources) {
    this(null, primarySources);
}
```

```
public SpringApplication(ResourceLoader resourceLoader,
 Class<?>...primarySources) {
   this.resourceLoader = resourceLoader;
   Assert.notNull(primarySources, "PrimarySources must not be null");
   this.primarySources = new LinkedHashSet<>(Arrays.asList(primarySources));
   this.webApplicationType = WebApplicationType.deduceFromClasspath();
   setInitializers((Collection)
getSpringFactoriesInstances(ApplicationContextInitializer.class));
   setListeners((Collection)
getSpringFactoriesInstances(ApplicationListener.class));
    this.mainApplicationClass = deduceMainApplicationClass();
}
```

该构造函数主要完成的工作是：

- 对 primarySources 初始化。
- 根据添加 jar 推断 webApplicationType 类型，进而创建对应类型的 ApplicationContext。
- 初始化 ApplicationContextInitializer 列表。
- 初始化 ApplicationListener 列表。
- 推断包含 main 方法的主类。

下面就各个步骤进行详细说明。

1. primarySources 初始化

Spring 现在提倡使用 Java 配置来代替 XML，配置信息可以来自多个类，这里我们指定一个主配置类。

2. 推断 webApplicationType 类型

示例代码 5-10　SpringApplication.java（常量部分）

```
   private static final String[] SERVLET_INDICATOR_CLASSES =
   { "javax.servlet.Servlet",
   "org.springframework.web.context.ConfigurableWebApplicationContext" };
private static final String WEBMVC_INDICATOR_CLASS =
   "org.springframework.web.servlet.DispatcherServlet";
private static final String WEBFLUX_INDICATOR_CLASS=
   "org.springframework.web.reactive.DispatcherHandler";
private static final String JERSEY_INDICATOR_CLASS =
   "org.glassfish.jersey.servlet.ServletContainer";
private static final String SERVLET_APPLICATION_CONTEXT_CLASS =
   "org.springframework.web.context.WebApplicationContext";
private static final String REACTIVE_APPLICATION_CONTEXT_CLASS =
   "org.springframework.boot.
    web.reactive.context.ReactiveWebApplicationContext";
    static WebApplicationType deduceFromClasspath() {
     if (ClassUtils.isPresent(WEBFLUX_INDICATOR_CLASS, null)
     && !ClassUtils.isPresent(WEBMVC_INDICATOR_CLASS, null)
                     && !ClassUtils.isPresent(JERSEY_INDICATOR_CLASS, null)) {
      return WebApplicationType.REACTIVE;
```

```
    }
    for (String className : SERVLET_INDICATOR_CLASSES) {
        if (!ClassUtils.isPresent(className, null)) {
            return WebApplicationType.NONE;
        }
    }
    return WebApplicationType.SERVLET;
}
```

方法 deduceFromClasspath 主要根据几个常量指定类是否在类路径上返回 WebApplicationType 的类型：

- NONE：不需要内嵌 Web 容器。
- SERVLET：一个基于 Servlet 的 Web 应用，应该启动内嵌 Servlet 容器。
- REACTIVE：一个基于 reactive 的 Web 应用，应该启动内嵌 reactive 容器。

3. 初始化 ApplicationContextInitializer 列表和 ApplicationListener

这两个初始化跟 5.2.3 小节读取自动配置类的原理一样，都是到 jar 的 META-INF/spring.factories 中读取。它们分别读取的 key 不同，如图 5-19 所示。

spring.factories

```
# Initializers
org.springframework.context.ApplicationContextInitializer=\
org.springframework.boot.autoconfigure.SharedMetadataReaderFactoryContextInitializer,\
org.springframework.boot.autoconfigure.logging.ConditionEvaluationReportLoggingListener

# Application Listeners
org.springframework.context.ApplicationListener=\
org.springframework.boot.autoconfigure.BackgroundPreinitializer

# Auto Configuration Import Listeners
org.springframework.boot.autoconfigure.AutoConfigurationImportListener=\
org.springframework.boot.autoconfigure.condition.ConditionEvaluationReportAutoConfigurationImportListener

# Auto Configuration Import Filters
org.springframework.boot.autoconfigure.AutoConfigurationImportFilter=\
org.springframework.boot.autoconfigure.condition.OnBeanCondition,\
org.springframework.boot.autoconfigure.condition.OnClassCondition,\
org.springframework.boot.autoconfigure.condition.OnWebApplicationCondition

# Auto Configure
org.springframework.boot.autoconfigure.EnableAutoConfiguration=\
org.springframework.boot.autoconfigure.admin.SpringApplicationAdminJmxAutoConfiguration,\
org.springframework.boot.autoconfigure.aop.AopAutoConfiguration,\
org.springframework.boot.autoconfigure.amqp.RabbitAutoConfiguration,\
org.springframework.boot.autoconfigure.batch.BatchAutoConfiguration,\
org.springframework.boot.autoconfigure.cache.CacheAutoConfiguration,\
org.springframework.boot.autoconfigure.cassandra.CassandraAutoConfiguration,\
```

图 5-19

- getSpringFactoriesInstances(ApplicationContextInitializer.class)：读取标号 1 部分。
- getSpringFactoriesInstances(ApplicationListener.class)：读取标号 2 部分。
- getSpringFactoriesInstances(EnableAutoConfiguration)：读取标号 3 部分。

4. 推断包含 main 方法的主类

示例代码 5-10　SpringApplication.java（推断主类部分）

```
private Class<?> deduceMainApplicationClass() {
    try {
        StackTraceElement[] stackTrace = new RuntimeException().
```

```
getStackTrace();
            for (StackTraceElement stackTraceElement : stackTrace) {
                if ("main".equals(stackTraceElement.getMethodName())) {
                    return Class.forName(stackTraceElement.getClassName());
                }
            }
        }catch (ClassNotFoundException ex) {
            // Swallow and continue
        }
        return null;
    }
```

以标准 Java 程序启动，从 main 方法开始执行，目前正在执行的方法通过调用栈可以找到 main 方法所在类，如上面的代码所示。

5.3.2 Spring Boot 启动流程

当 SpringApplication 创建完毕后，就开始执行 run 方法，如下代码所示。

示例代码 5-10 SpringApplication.java（run 部分）

```
public ConfigurableApplicationContext run(String... args) {
    StopWatch stopWatch = new StopWatch();
    stopWatch.start();
    ConfigurableApplicationContext context = null;
    Collection<SpringBootExceptionReporter> exceptionReporters = new
 ArrayList<>();
    configureHeadlessProperty();
    SpringApplicationRunListeners listeners = getRunListeners(args);
    listeners.starting();
    try {
    ApplicationArguments applicationArguments = new
 DefaultApplicationArguments(args);
    ConfigurableEnvironment environment = prepareEnvironment(listeners,
  applicationArguments);
        configureIgnoreBeanInfo(environment);
        Banner printedBanner = printBanner(environment);
        context = createApplicationContext();
        exceptionReporters = getSpringFactoriesInstances(
SpringBootExceptionReporter.class,
            new Class[] { ConfigurableApplicationContext.class }, context);
            prepareContext(context, environment, listeners,
      applicationArguments, printedBanner);
        refreshContext(context);
        afterRefresh(context, applicationArguments);
        stopWatch.stop();
        if (this.logStartupInfo) {
            new StartupInfoLogger(this.mainApplicationClass).
    logStarted(getApplicationLog(), stopWatch);
        }
```

```
            listeners.started(context);
            callRunners(context, applicationArguments);
        }catch (Throwable ex) {
            handleRunFailure(context, ex, exceptionReporters, listeners);
            throw new IllegalStateException(ex);
        }
        try {
            listeners.running(context);
        }
        catch (Throwable ex) {
            handleRunFailure(context, ex, exceptionReporters, null);
            throw new IllegalStateException(ex);
        }
        return context;
    }
```

该方法完成的工作如下：

（1）启动一个秒表（StopWatch）来统计启动时间。

（2）通过 SpringFactoriesLoader 获取所有 jar 目录下 META-INF/spring.factories 下配置的 SpringApplicationRunListener，该接口对 SpringApplication 的 run 方法不同阶段进行监听。

（3）listeners.starting()调用所有 SpringApplicationRunListener 的 starting()。

（4）根据 WebApplicationType 类型准备对应类型的类型 ConfigurableEnvironment，同时调用 listeners.environmentPrepared(environment)通知所有 SpringApplicationRunListener 环境准备完毕。

（5）打印 banner，如果 spring.main.banner-mode=off，就不打印；如果值是 console，就打印 banner 到控制台；如果值是 log，就输出到日志。我们可以在 resources 目录下新建 banner.txt 来修改默认的 banner。

（6）根据 WebApplicationType 类型，创建对一个类型的 ApplicationContext 对象。

（7）准备上下文。

示例代码 5-10　SpringApplication.java（准备上下文部分）

```
    private void prepareContext(ConfigurableApplicationContext context,
ConfigurableEnvironment environment,SpringApplicationRunListeners listeners,
ApplicationArguments applicationArguments, Banner printedBanner) {
        context.setEnvironment(environment);//设置运行环境
        postProcessApplicationContext(context);//ApplicationContext 进行后继处理
        /*调用所有 ApplicationContextInitializer 的 initialize*/
        applyInitializers(context);
        listeners.contextPrepared(context);//通知所有监听器上下文准备完毕
        if (this.logStartupInfo) {
            logStartupInfo(context.getParent() == null);
            logStartupProfileInfo(context);
        }
        // Add boot specific singleton beans
        ConfigurableListableBeanFactory beanFactory = context.getBeanFactory();
        beanFactory.registerSingleton("springApplicationArguments",
         applicationArguments);
        if (printedBanner != null) {
```

```
            beanFactory.registerSingleton("springBootBanner",
printedBanner);
        }
        if (beanFactory instanceof DefaultListableBeanFactory) {
            ((DefaultListableBeanFactory) beanFactory)

    .setAllowBeanDefinitionOverriding(this.allowBeanDefinitionOverriding);
        }
        if (this.lazyInitialization) {
            context.addBeanFactoryPostProcessor(new
        LazyInitializationBeanFactoryPostProcessor());
        }
        // 加载所有源
        Set<Object> sources = getAllSources();
        Assert.notEmpty(sources, "Sources must not be empty");
       //注册所有 bean 到 spring 容器
       load(context, sources.toArray(new Object[0]));
       //通知监听器上下文加载完毕
       listeners.contextLoaded(context);
    }
```

5.4 Spring Boot 的 starter

现在使用 Spring 流行零 XML 配置，也就是通过配置类和注解完成 Spring 配置。在搭建每个 Spring 项目过程中，需要反复完成如下操作。

（1）配置属性，如图 5-20 所示，例如 JDK 版本、类库的版本等。

```
<properties>
    <!--常用属性配置-->
    <project.build.sourceEncoding>UTF-8</project.build.sourceEncoding>
    <maven.compiler.source>1.8</maven.compiler.source>
    <maven.compiler.target>1.8</maven.compiler.target>
    <!--依赖版本控制-->
    <lombok.version>1.18.12</lombok.version>
    <commons.beanutils.version>1.9.4</commons.beanutils.version>
    <javax.servlet-api.version>3.1.0</javax.servlet-api.version>
    <javax.servlet.jsp-api.version>2.3.3</javax.servlet.jsp-api.version>
    <mysql-connector-java.version>5.1.49</mysql-connector-java.version>
</properties>
```

图 5-20

（2）添加多个依赖到 pom.xml，如图 5-21 所示，耗费大量精力解决依赖以及版本搭配问题。

```
<!--2 Servlet:这里主要是为了提示功能-->
<dependency>
    <groupId>javax.servlet</groupId>
    <artifactId>javax.servlet-api</artifactId>
    <version>${javax.servlet-api.version}</version>
    <scope>provided</scope>
</dependency>
<!--3 lombok:是为了快速编写javaBean-->
<dependency>
    <groupId>org.projectlombok</groupId>
    <artifactId>lombok</artifactId>
    <version>${lombok.version}</version>
</dependency>
<!--4jsp依赖：这里添加是为了提示-->
<dependency>
    <groupId>javax.servlet.jsp</groupId>
    <artifactId>javax.servlet.jsp-api</artifactId>
    <version>${javax.servlet.jsp-api.version}</version>
    <scope>provided</scope>
</dependency>
<!--帮我们提取参数-->
<dependency>
    <groupId>commons-beanutils</groupId>
    <artifactId>commons-beanutils</artifactId>
    <version>${commons.beanutils.version}</version>
</dependency>
```

图 5-21

（3）编写多个配置类，如图 5-22 所示。

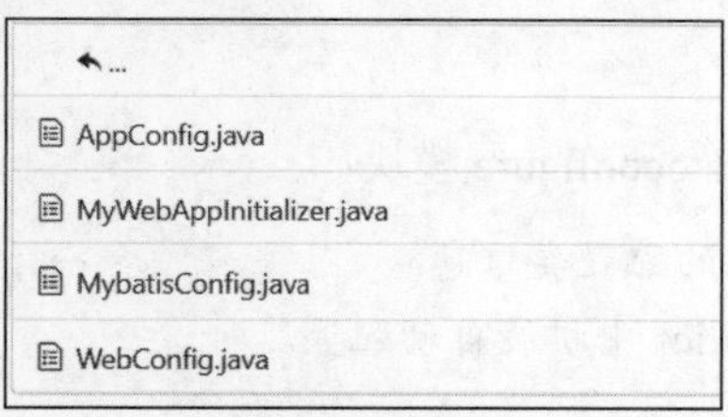

图 5-22

（4）把易变信息（例如数据库用户名和密码）写在属性文件中，如图 5-23 所示。

```
jdbc.url=jdbc:mysql://localhost:3306/just
jdbc.password=root
jdbc.username=root
jdbc.driverClassName=com.mysql.jdbc.Driver
```

图 5-23

针对这个情况，Spring Boot 使用 stater 来减少重复工作。一般 starter 由两个模块组成：

（1）stater 模块：完成版本控制、属性和依赖导入。

（2）自动配置模块：提供自动配置类以及自动配置类在 META-INF/spring.factories 中的配置。易变信息写在 Spring Boot 默认配置文件 application.properites/application.yml 中。

这样就可以做到开箱即用，不用再做很多重复工作，仅仅需要修改一些配置即可。

5.4.1 官方 starter

Spring Boot 预先定义好了很多 starter，需要使用某个功能时，导入对应 starter 即可。常用的 starter 如表 5-2 所示。

表 5-2 常用的 starter 及其作用

名称	作用
spring-boot-starter-web	构建 Web，包括 RESTful、Spring MVC 应用。默认使用 Tomcat 作为嵌入容器
spring-boot-starter-thymeleaf	支持 Thymeleaf 视图的 MVC Web 应用
spring-boot-starter-parent	它的父项目是 spring-boot-dependencies，用于项目属性配置和版本管理，Spring Boot 项目一般都使用它作为父项目
spring-boot-starter-test	对 Spring Boot 项目进行测试
spring-boot-starter-data-neo4j	提供 Neo4j 支持
spring-boot-starterdata-redis	提供 Redis 支持

5.4.2 自定义 starter

在 Web 应用中会使用验证码，一般使用 kaptcha 类库，需要很多配置。下面制作一个 Spring Boot 的 starter，简化 kaptcha 使用。

1. 步骤

①创建 kaptcha-spring-boot-autoconfigure 模块：

- 定义代码生成器核心对象的配置类信息。
- 在 META-INF/spring.factories 中扩展自动配置。

②创建 kaptcha-spring-boot-starter 模块。
③将自动配置模块及 starter 模块安装到本地仓库。
④测试模块 test-kaptcha-auto-configuration：

- 引入自定义的 kaptcha-spring-boot-starter 依赖坐标。
- 测试核心代码生成器对象是否自动配置成功了。

2. 新建一个空项目

在合适的目录下新建一个 custom-kaptcha-autoconfigure 目录，然后在 IDEA 中选择 File→open...，选中新建的目录。

3. 新建 module

创建 kaptcha-spring-boot-autoconfigure 模块，选中 Maven 项目，修改 pom.xml，代码如下。

示例代码 5-11 pom.xml（部分代码）

```
<groupId>com.guodexian</groupId>
<artifactId>kaptcha-spring-boot-autoconfigure</artifactId>
```

```
    <version>1.0-SNAPSHOT</version>
    <parent>
        <groupId>org.springframework.boot</groupId>
        <artifactId>spring-boot-starter-parent</artifactId>
        <version>2.2.11.RELEASE</version>
    </parent>
    <properties>
        <maven.compiler.source>8</maven.compiler.source>
        <maven.compiler.target>8</maven.compiler.target>
    </properties>
    <dependencies>
        <dependency>
            <groupId>org.springframework.boot</groupId>
            <artifactId>spring-boot-starter-web</artifactId>
        </dependency>
        <!--导入自动配置的相关依赖坐标-->
        <dependency>
            <groupId>org.springframework.boot</groupId>
            <artifactId>spring-boot-autoconfigure</artifactId>
        </dependency>
   <!--方便 IDE 能够检测到该依赖中用到的配置属性，能够自动补全，其实就是在编译的时候在
       META-INF 下面生成了一个 spring-configuration-metadata.json 文件
        -->
    <dependency>
        <groupId>org.springframework.boot</groupId>
        <artifactId>spring-boot-configuration-processor</artifactId>
        <optional>true</optional>
    </dependency>
    <!--验证码依赖-->
    <dependency>
            <groupId>com.github.penggle</groupId>
            <artifactId>kaptcha</artifactId>
            <version>2.3.2</version>
        </dependency>
        <dependency>
            <groupId>org.projectlombok</groupId>
            <artifactId>lombok</artifactId>
        </dependency>
    </dependencies>
```

4. 新建属性类

属性类接收配置文件中配置的属性，可以给属性提供默认值，如下代码所示。

示例代码 5-12 KaptchaProperties.java

```
@Data
@Component
@ConfigurationProperties(prefix = "kaptcha")
public class KaptchaProperties {
    private int imageWidth=100;
    private int imageHeigth=40;
```

```
    private int textproducerFontSize=32;
    private String textproducerFontColor="0,0,0";
    private String textproducerCharString=
"0123456789ABCDEFGHIJKLMNOPQRSTUVWXYAZ";
    private int textproducerCharLength=4;
    private String noiseImpl="com.google.code.kaptcha.impl.NoNoise";
}
```

5. 新建配置类

完成自动配置类，只有 Kaptcha 类库添加后才需要配置，因而使用注解@ConditionalOnClass，如下代码所示。

示例代码 5-13　KaptchaProperties.java

```
@Configuration
@ConditionalOnClass(KaptchaServlet.class)
@Import(KaptchaProperties.class)//导入属性文件接收类
public class KaptchaConfiguration {
    @Autowired
    private KaptchaProperties kaptchaProperties;
    @Bean
    public Producer  producer(){
        Properties properties = new Properties();
        properties.setProperty("kaptcha.image.width",
                        kaptchaProperties.getImageWidth()+"");
        properties.setProperty("kaptcha.image.height",
                        kaptchaProperties.getImageHeigth()+"");
        properties.setProperty("kaptcha.textproducer.font.size",
                      kaptchaProperties.getTextproducerFontSize()+"");
        properties.setProperty("kaptcha.textproducer.font.color",
                        kaptchaProperties.getTextproducerFontColor());
        properties.setProperty("kaptcha.textproducer.char.string",
                      kaptchaProperties.getTextproducerCharString());
        properties.setProperty("kaptcha.textproducer.char.length",
                kaptchaProperties.getTextproducerCharLength()+"");
        properties.setProperty("kaptcha.noise.impl",
                kaptchaProperties.getNoiseImpl());
        DefaultKaptcha kaptchaProducer = new DefaultKaptcha();
        Config config = new Config(properties);
        kaptchaProducer.setConfig(config);
        return  kaptchaProducer;
    }
}
```

6. 新建 spring.factories

在 resources 目录新建 META-INF 目录，然后新建 spring.factories，内容如下所示。

示例代码 5-14　spring.factories

```
# 注册自定义自动配置
org.springframework.boot.autoconfigure.EnableAutoConfiguration=com.guodexi
```

```
an.kaptcha.config.KaptchaConfiguration
```

7. 新建模块 kaptcha-spring-boot-starter

新建一个 Maven 模块后删除 src，对应 starter 项目不需要有内容，修改 pom.xml。

示例代码 5-15　pom.xml

```
<groupId>com.guodexian</groupId>
<artifactId>kaptcha-spring-boot-starter</artifactId>
<version>1.0-SNAPSHOT</version>
<properties>
    <maven.compiler.source>8</maven.compiler.source>
    <maven.compiler.target>8</maven.compiler.target>
</properties>
<dependencies>
    <dependency>
        <groupId>com.guodexian</groupId>
        <artifactId>kaptcha-spring-boot-autoconfigure</artifactId>
        <version>1.0-SNAPSHOT</version>
    </dependency>
    <dependency>
        <groupId>com.github.penggle</groupId>
        <artifactId>kaptcha</artifactId>
        <version>2.3.2</version>
    </dependency>
</dependencies>
```

8. 安装两个模块到本地库（见图 5-24）

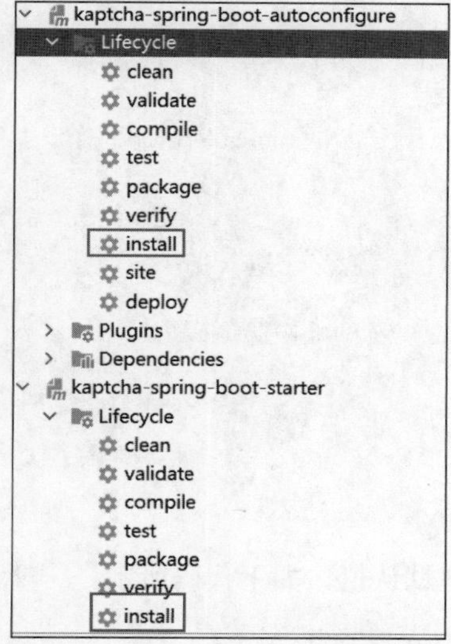

图 5-24

单击后安装到本地库，以便其他项目可以引用。

9. 新建测试 Maven 项目

新建一个基于 Spring Boot 名称为 test-kaptcha-auto-configuration 的项目，修改 pom.xml 导入自

定义的 stater。

示例代码 5-16　pom.xml（部分代码）

```
<dependency>
    <groupId>com.guodexian</groupId>
    <artifactId>kaptcha-spring-boot-starter</artifactId>
    <version>1.0-SNAPSHOT</version>
</dependency>
```

10. 新建启动类

定义一个 Controller 方法，返回验证码，代码如下所示。

示例代码 5-17　Application.java

```
@SpringBootApplication
@Controller
public class Application {
    @Autowired
    private Producer producer;
    public static void main(String[] args) {
        SpringApplication.run(Application.class,args);
    }
    @GetMapping(path = "/kaptcha")
    public void getKaptcha(HttpServletResponse response, HttpSession session)
{
        // 生成验证码
        String text = producer.createText();
        BufferedImage image = producer.createImage(text);
        // 将验证码存入 session
        session.setAttribute("kaptcha", text);
        // 将图片输出给浏览器
        response.setContentType("image/png");
        try {
            OutputStream os = response.getOutputStream();
            ImageIO.write(image, "png", os);
        } catch (IOException e) {
            System.out.println("响应验证码失败:" + e.getMessage());
        }
    }
}
```

11. 配置 application.yml

如果不满足需求，就修改某些默认值，如下代码所示。

示例代码 5-18　application.yml 文件

```
kaptcha:
  textproducer-char-length: 5
```

12. 测试

启动测试项目，通过浏览器访问，如果成功，就会看到如图 5-25 所示的结果。

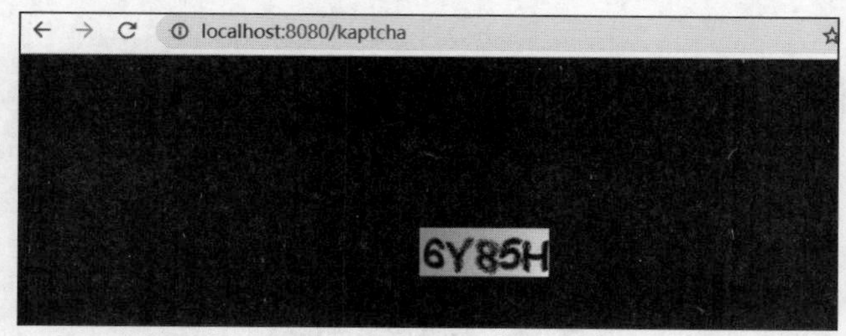

图 5-25

5.5　Spring Boot 配置详解

在使用 Spring Boot 的 stater 过程中，虽然 Spring Boot 使用“惯例优先于配置”，但是一些易变信息和 starter 特性仍然需要用户配置。Spring Boot 支持从多个位置读取，按照优先级从以下位置读取（优先级依次降低，优先级高的覆盖优先级低的）：

- 当前目录下的/config 子目录。
- 当前目录。
- 类路径下的/config 包。
- 类路径根目录。

配置文件支持的两种常用格式为 .properties 和.yml，如果同一个位置存在两种格式的配置文件，那么属性文件优先级高。Spring Boot 官方配置很多，我们可以查看官方文档。对于 Spring Boot 标准配置，请看文档：https://docs.spring.io/spring-boot/docs/current/reference/html/appendix-application-properties.html#common-application-properties。

5.5.1　配置的两种文件格式

JDK 提供对属性文件读写的相关类，需要使用工具 native2ascii 把属性文件内容转为 ascii 或者 unicode 编码，否则会出现乱码。YML 是 JSON 的超集，后缀名通常是.yml，适合表达具有层次结构的配置信息，需要额外添加类库 snakeyaml。两者的对比如表 5-3 所示。

表 5-3　属性文件与 YAML 文件的对比

属性文件	YAML 文件

jdbc.url=jdbc:mysql://localhost:3306/just
jdbc.password=root
jdbc.username=root
jdbc.driverClassName=com.mysql.jdbc.Driver
	jdbc:
	  driverClassName:
	com.mysql.jdbc.Driver
	  password: root
	  url:
	jdbc:mysql://localhost:3306/just
	  username: root

5.5.2 数据源配置

Spring Boot 内置 HikariCP 数据源，如果要切换为其他数据源，就需要在 Maven 中添加对应依赖，并在配置类添加对应的 Bean。数据源的常用配置如表 5-4 所示。

表 5-4 数据源常用配置

属性名	作用
spring.datasource.username	数据库的登录用户名
spring.datasource.password	数据库的登录密码
spring.datasource.url	数据库的 JDBC URL
spring.datasource.driver-class-name	JDBC 驱动类名

5.5.3 Web 配置

Web 配置包括前端控制器 DispatcherServlet、上下文路径、视图解析路径等，如表 5-5 所示。

表 5-5 Web 配置信息

属性名	作用
spring.mvc.servlet.load-on-startup	dispatcher servlet 启动优先级，默认值为－1
spring.mvc.servlet.path	dispatcher servlet 的路径，默认值为“/”
spring.mvc.static-path-pattern	静态资源路径模式，默认值为“/**”
spring.mvc.view.prefix	Spring MVC 视图的前缀
spring.mvc.view.suffix	Spring MVC 视图的后缀
server.port	修改内置服务器端口，默认值是 8080
server.servlet.context-path	应用的上下文路径，默认是“”

5.5.4 日志配置

如果在运行过程中出现错误或者想知道工作原理，就可以借助日志来输出更多有用的信息，如表 5-6 所示。

表 5-6 日志配置信息

属性名	作用
logging.level.*	设置某个包或者类日志级别，例如 logging.level.org.springframework=DEBUG
logging.path	设置日志文件路径
logging.file.max-size	设置日志文件最大容量，默认值为 10MB

5.5.5 自定义配置

首先编写一个 JavaBean 风格的属性类，然后在注解 ConfigurationProperties 中指定一个唯一前缀，如图 5-26 所示。

```
@ConfigurationProperties(prefix = "kaptcha")
public class KaptchaProperties {
    private int imageWidth=100;
    private int imageHeight =40;
    private int textproducerFontSize=32;
    private String textproducerFontColor="0,0,0";
    private String textproducerCharString="0123456789ABCDEFGHIJKLMNOPQRSTUVWXYAZ";
    private int textproducerCharLength=4;
    private String noiseImpl="com.google.code.kaptcha.impl.NoNoise";
}
```

使用注解配置一个唯一前缀

图 5-26

然后使用该属性类，如图 5-27 所示。Spring Boot 会自动把配置文件中的值注入如图 5-28 所示的属性类对象中。

```
@Import(KaptchaProperties.class)
public class KaptchaConfiguration {
    @Autowired
    private KaptchaProperties kaptchaProperties;
```

图 5-27

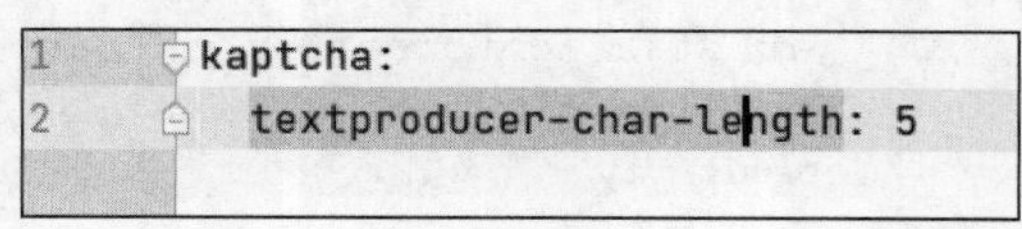

图 5-28

5.6　内置 Web 容器原理

Spring Boot 对于 Web 项目默认内置一个服务器，把项目打包成一个可执行的 jar 文件，这样项目发布后就不需要额外安装 Web 服务器了。Spring Boot 默认内置的服务器是 Tomcat，下面我们通过代码展示如何内置服务器。

5.6.1　内嵌 Tomcat

在 Tomcat 官网下载页面中提供了 Tomcat 的嵌入版本，如图 5-29 所示。

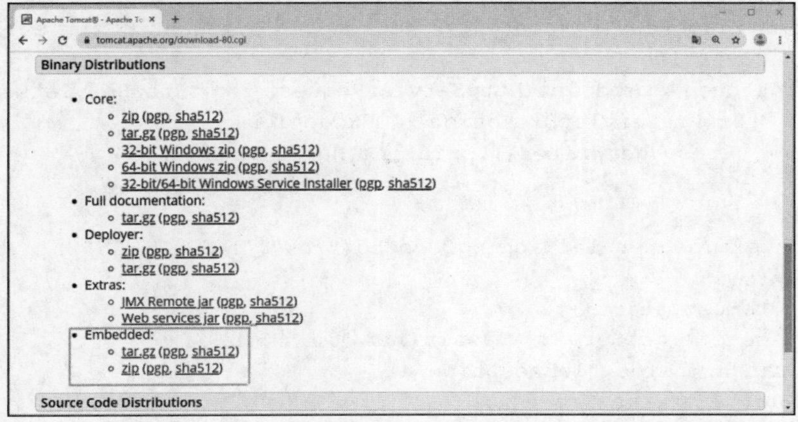

图 5-29

使用嵌入式版本可以把 Tomcat 嵌入 Java 程序中成为一个独立应用，不再需要部署到独立的外部服务器中。下面我们展示内置服务器的基本使用。

（1）新建一个 Maven 项目，添加依赖，代码如下所示。

示例代码 5-19　pom.xml（部分代码）

```
<!--提供 Tomcat 核心功能-->
<dependency>
    <groupId>org.apache.tomcat.embed</groupId>
    <artifactId>tomcat-embed-core</artifactId>
    <version>8.5.58</version>
</dependency>
<!--提供 JSP 翻译、编译等功能-->
<dependency>
    <groupId>org.apache.tomcat.embed</groupId>
    <artifactId>tomcat-embed-jasper</artifactId>
    <version>8.5.58</version>
</dependency>
```

（2）编写启动类，代码如下所示。

示例代码 5-20　EmbedServer.java（部分代码）

```
public class EmbedServer {
    public static void main(String[] args) {
        Tomcat tomcat=new Tomcat();
        ///修改默认端口
        tomcat.setPort(9999);
        //设置应用的根目录
        tomcat.getHost().setAppBase(System.getProperty("user.dir")+"/.");
        /**
         * 添加一个 webapp 应用到服务器
         * contextPath:用于上下文映射
         * docBase:上下文的根目录
         */
        Context context = tomcat.addWebapp("/test", "src/main/webapp");
        //以编程方式添加一个 Servlet
        Tomcat.addServlet(context, "helloWorld", new HttpServlet() {
            @Override
        protected void doGet(HttpServletRequest req, HttpServletResponse resp)
             throws ServletException, IOException {
                resp.getWriter().println("Hello World");
            }
        });
        context.addServletMappingDecoded("/hw","helloWorld");
        try {
            tomcat.start();
            tomcat.getServer().await();//防止程序立即结束
        } catch (LifecycleException e) {
            e.printStackTrace();
        }
```

```
    }
}
```

（3）在 webapp 目录中新建一个 index.jsp，代码如下所示。

示例代码 5-21 index.jsp

```
<!DOCTYPE html>
<html lang="en">
    <head>
        <meta charset="UTF-8">
        <title>hello world</title>
    </head>
    <body>
        <h1>Hello World!</h1>
        <%=1+2%>
    </body>
</html>
```

（4）跟执行普通 Java 程序一样执行 EmbedServer 中的 main 方法，发现服务器启动了，然后在浏览器中访问，结果如图 5-30、图 5-31 所示。

图 5-30

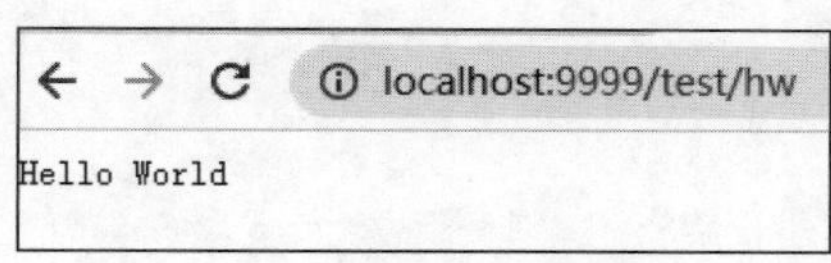

图 5-31

5.6.2 Spring Boot 内嵌 Tomcat 原理

Spring Boot 默认采用内置 Web 服务器的方式，可扩展性较好，并提供多种 Web 服务器（例如 Tomcat、Jetty）的支持，类关系如下：

静态内部类 TomcatWebServerFactoryCustomizer 上的@ConditionalClass 注解，判断当前项目类路径下是否有 Tomcat 类，如图 5-32 所示。如果存在，就创建 TomcatWebServerFactoryCustomizer 对象，用 customize 方法完成 Tomcat 服务器配置，如图 5-33 所示。

```
EmbeddedWebServerFactoryCustomizerAutoConfiguration.java
 * @since 2.0.0
 */
@Configuration(proxyBeanMethods = false)
@ConditionalOnWebApplication
@EnableConfigurationProperties(ServerProperties.class)
public class EmbeddedWebServerFactoryCustomizerAutoConfiguration {

    /**
     * Nested configuration if Tomcat is being used.
     */
    @Configuration(proxyBeanMethods = false)
    @ConditionalOnClass({ Tomcat.class, UpgradeProtocol.class })
    public static class TomcatWebServerFactoryCustomizerConfiguration {

        @Bean
        public TomcatWebServerFactoryCustomizer tomcatWebServerFactoryCustomizer(Environment environment,
                ServerProperties serverProperties) {
            return new TomcatWebServerFactoryCustomizer(environment, serverProperties);
        }

    }
```

图 5-32

```java
public class TomcatWebServerFactoryCustomizer
        implements WebServerFactoryCustomizer<ConfigurableTomcatWebServerFactory>, Ordered {

    private final Environment environment;

    private final ServerProperties serverProperties;

    public TomcatWebServerFactoryCustomizer(Environment environment, ServerProperties serverProperties) {
        this.environment = environment;
        this.serverProperties = serverProperties;
    }

    @Override
    public int getOrder() { return 0; }

    @Override
    public void customize(ConfigurableTomcatWebServerFactory factory) {
        ServerProperties properties = this.serverProperties;
        ServerProperties.Tomcat tomcatProperties = properties.getTomcat();
        PropertyMapper propertyMapper = PropertyMapper.get();
        propertyMapper.from(tomcatProperties::getBasedir).whenNonNull().to(factory::setBaseDirectory);
        propertyMapper.from(tomcatProperties::getBackgroundProcessorDelay).whenNonNull().as(Duration::getSeconds)
                .as(Long::intValue).to(factory::setBackgroundProcessorDelay);
        customizeRemoteIpValve(factory);
        ServerProperties.Tomcat.Threads threadProperties = tomcatProperties.getThreads();
        propertyMapper.from(threadProperties::getMax).when(this::isPositive)
                .to((maxThreads) -> customizeMaxThreads(factory, threadProperties.getMax()));
        propertyMapper.from(threadProperties::getMinSpare).when(this::isPositive)
                .to((minSpareThreads) -> customizeMinThreads(factory, minSpareThreads));
```

图 5-33

Spring Boot 应用（无论是否是 Web 应用）可以打包为一个 jar 文件，从一个标准 Java 程序 main 方法执行。在 main 方法中调用 SpringApplicaiton 的 run 方法，并在 main 方法所在类中配置注解 @SpringBootApplication，触发自动配置，默认扫描所在包以及子包下的组件。

第6章

Spring Boot 数据访问与事务

本章将讲解 Spring Boot 数据访问的几种方式，包括 Spring Data JDBC、Spring Data JPA 和集成 MyBatis 框架，以及事务的类型、特性、并发问题等。

6.1 Spring Data JDBC

6.1.1 数据访问简介

JDBC（Java Database Connectivity）是 Java 连接数据库的原生接口。JDBC 对实现与数据库连接的服务提供商而言，就是接口模型第三方中间件厂商实现与数据的连接提供了标准的方法，对 Java 开发工程师来说就是 API，标准的接口。JDBC 是所有操作数据库的框架必须使用的，所以为了方便 Java 工程师调用数据库，数据库厂商通常都会实现 JDBC 接口。Spring 为了支持各种持久化技术，提供了简单的模板以及回调方法。

1. JdbcTemplate 模板

Spring Data JDBC 抽象框架 core 包提供了 JDBC 模板类，是 Spring 中最基本的 JDBC 模板，是 core 包中的核心类，其他模板都是基于 JdbcTemplate 完成的。JdbcTemplate 利用 JDBC 和简单的索引参数查询对数据库进行简单的访问。

2. NamedParameterJdbcTemplate

能够在查询时把值绑定到 SQL 里的命名参数中，而不是索引参数，其内部包含了一个 JdbcTemplate，所以它是基于 JdbcTemplate 的功能又额外增加了参数命名的功能。

3. SimpleJdbcTemplate

利用 Java 5 的特性（比如自动装箱、通用和可变参数列表）来简化 JDBC 模板的使用，相较于 NamedParameterJdbcTemplate 主要增加了 JDK 5.0 的反省和可变长度参数的支持。

Spring Data JDBC 相较于传统的 JDBC 而言，只需要在配置文件中配置数据库驱动、连接等相关信息，Java 代码中只需要通过注入模板调用方法执行 SQL 即可。下面我们主要讲解 JdbcTemplate。其实 Spring 中的 JdbcTemplate 就是对 JDBC 进行了一层封装，将 DataSource 注册到 JdbcTemplate 中，处理了资源的建立和释放，我们只需要提供 SQL 语句和提取结果即可，使得 JDBC 更加易于使用。

（1）JdbcTemplate 中主要提供了以下 4 类方法：

- execute()方法：可以执行 SQL 语句，一般用于执行 DDL（Data Definition Languages，数据定义语言）语句，常用的关键字包括 create、drop、alter 等。
- update()方法及 batchUpdate()方法：update()方法主要用于单条数据的新增、修改和删除；batchUpdate()方法用于执行新增、修改、删除批处理的相关语句。
- query()方法及 queryForXXX()方法：用于执行查询相关的方法。
- call()方法：用于执行存储过程、函数相关语句。

（2）JdbcTemplate 类中支持的回调类：

- 预编译语句及存储过程创建回调：用于根据 JdbcTemplate 提供的连接创建相应的语句。
  - PreparedStatementCreator：通过回调获取 JdbcTemplate 提供的 Connection，由用户使用该 Conncetion 创建相关的 PreparedStatement。
  - CallableStatementCreator：通过回调获取 JdbcTemplate 提供的 Connection，由用户使用该 Conncetion 创建相关的 CallableStatement。
- 预编译语句设值回调：用于预编译语句相应参数设值。
  - PreparedStatementSetter：通过回调获取 JdbcTemplate 提供的 PreparedStatement，由用户来对预编译语句相应参数设值。
  - BatchPreparedStatementSetter：类似于 PreparedStatementSetter，但用于批处理，需要指定批处理大小。
- 自定义功能回调：
  - ConnectionCallback：通过回调获取 JdbcTemplate 提供的 Connection，用户可在该 Connection 执行任何数量的操作。
  - StatementCallback：通过回调获取 JdbcTemplate 提供的 Statement，用户可以在该 Statement 执行任何数量的操作。
  - PreparedStatementCallback：通过回调获取 JdbcTemplate 提供的 PreparedStatement，用户可以在该 PreparedStatement 执行任何数量的操作。
  - CallableStatementCallback：通过回调获取 JdbcTemplate 提供的 CallableStatement，用户可以在该 CallableStatement 执行任何数量的操作。
- 结果集处理回调：
  - RowMapper：用于将结果集每行数据转换为需要的类型，用户需实现方法 mapRow(ResultSet rs, int rowNum)来完成将每行数据转换为相应的类型。
  - RowCallbackHandler：用于处理 ResultSet 的每一行结果，用户需实现方法 processRow(ResultSet rs)来完成处理，在该回调方法中无须执行 rs.next()，该操作由

JdbcTemplate 来执行，用户只需按行获取数据然后处理即可。

- ResultSetExtractor：用于结果集数据提取，用户需实现方法 extractData(ResultSet rs)来处理结果集（必须处理整个结果集）。

6.1.2　实战

我们主要讲一下 Spring Boot 项目使用 Spring Data JDBC 进行数据库操作。核心依赖如代码 6-1 所示。

示例代码 6-1　pom.xml（核心代码）

```
<dependency>
    <groupId>org.springframework.boot</groupId>
    <artifactId>spring-boot-starter-data-jdbc</artifactId>
</dependency>
```

通过 JdbcTemplate 添加学生信息功能，如示例代码 6-2 所示。

示例代码 6-2　StudentDaoImpl.java 文件的核心代码

```
@Repository
public class StudentDaoImpl implements StudentDao {
    @Autowired
    private JdbcTemplate jdbcTemplate;
    @Override
    public int addStudent(Student student) {
        String sql = "insert into student
(username,password,birthday,sex,address) values (?,?,?,?,?)";
        int row = jdbcTemplate.update(sql, student.getUsername(), student.
getPassword(), student.getBirthday(), student.getSex(), student.getAddress());
        return row;
    }
}
```

业务层代码如下。

示例代码 6-3　StudentServiceImpl.java

```
@Service
public class StudentServiceImpl implements StudentService {
    @Autowired
    private StudentDao studentDao;
    @Override
    public int addStudent(Student student) {
        return studentDao.addStudent(student);
    }
}
```

测试接口代码如下。

示例代码 6-4　StudentController.java

```
/**
 * 用于测试 spring-boot-starter-jdbc 的接口
*/
@RestController
@RequestMapping("/student")
public class StudentController {
    @Autowired
    private StudentService studentService;
    @PostMapping("/add")
    public int addStudent(@RequestBody Student student){
        return studentService.addStudent(student);
    }
}
```

采用 Postman 解耦测试工具来测试 post 请求，如图 6-1 所示，在编号 5 的区域设置请求参数。

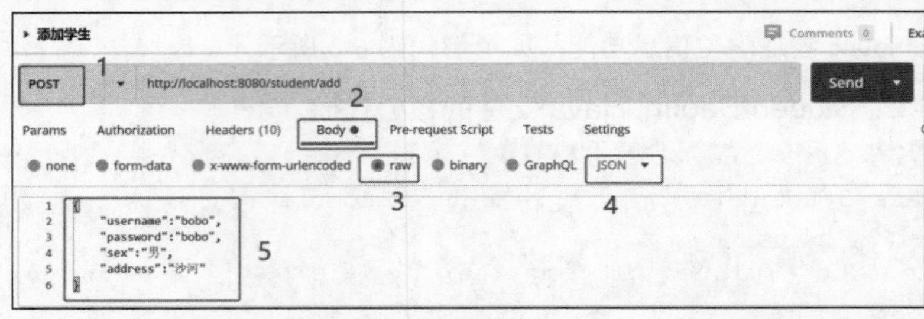

图 6-1

查询学生列表，代码如下。

示例代码 6-5　StudentDaoImpl.java（部分代码）

```
@Override
public List<Map<String, Object>> listStudent() {
    return jdbcTemplate.queryForList("select * from student");
}
```

根据 id 查询学生信息，代码如下。

示例代码 6-6　StudentDaoImpl.java（部分代码）

```
@Override
public Student getStudentById(Long id) {
    String sql = "select * from student where id = ?";
    return jdbcTemplate.queryForObject(sql, new RowMapper<Student>() {
        @Override
        public Student mapRow(ResultSet resultSet, int row) throws SQLException
{
            Student student = new Student();
            student.setId(Long.valueOf(resultSet.getInt(1)));
            student.setUsername(resultSet.getString(2));
            student.setPassword(resultSet.getString(3));
            student.setBirthday(resultSet.getDate(4));
```

```
            student.setSex(resultSet.getString(5));
            student.setAddress(resultSet.getString(6));
            return student;
        }
    },id);
}
```

修改学生信息，代码如下。

示例代码 6-7　StudentDaoImpl.java（部分代码）

```
    @Override
    public int updateStudent(Student student) {
        String sql = "update student set username = ?,password = ?, address = ? where
id = ?";
        int row = jdbcTemplate.update(sql, student.getUsername(),
student.getPassword(), student.getAddress(), student.getId());
        return row;
    }
```

删除学生信息，代码如下。

示例代码 6-8　StudentDaoImpl.java（部分代码）

```
    @Override
    public int deleteStudent(Long id) {
        String sql = "delete from student where id = ?";
        int row = jdbcTemplate.update(sql, id);
        return row;
    }
```

小　结

本小节主要讲解 Spring Boot 项目基于 JDBC 的数据库访问。此处的 JDBC 并非原生 JDBC，而是基于 JDBC 做了深层次的封装，其最核心的 API 是基于 JdbcTemplate 对象的相关方法。springboot-data-jdbc 所有数据库访问几乎都是围绕这个类去工作的，通过依赖注入，把数据源相关配置信息装配到 JdbcTemplate 中，由 JdbcTemplate 进行具体的数据库访问。

6.2　Spring Data JPA

6.2.1　JPA 简介

JPA（Java Persistence API）是由 Sun 官方提出的 Java 持久化规范，并不是一套产品。它为 Java 开发人员提供一种对象/关联映射工具来管理 Java 应用中的关系数据。它的出现主要是为了简化现有的持久化开发工作和整合 ORM 技术，结束现在 Hibernate、TopLink、JDO 等 ORM 框架各自为营的局面。值得注意的是，JPA 是在充分吸收现有的 Hibernate、TopLink、JDO 等 ORM 框架的基础上发展而来的，具有易于使用、伸缩性强等优点。从目前开发社区的反应上来看，JPA 受到了极大的支持和赞扬，其中包括 Spring 与 EJB 3.0 的开发团队。

Spring Data JPA 是 Spring 基于 PRM 框架、JPA 规范基础上封装的一套 JPA 应用框架，底层使用 Hibernate 的 JPA 技术实现，可使开发者用极简的代码实现对数据库的访问和操作。它提供了包括增删改查等在内的常用功能，且易于扩展。Spring Data JPA 让我们从 dao 层的操作中解放出来，基本上所有的 CRUD 都可以依赖它来实现，如图 6-2 所示。

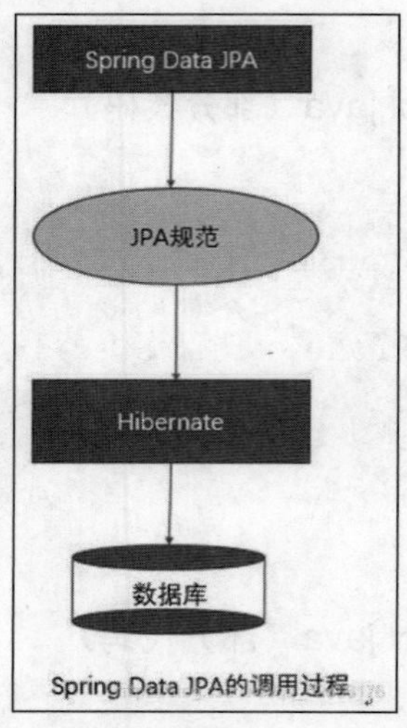

图 6-2

JPA 的完整调用过程如图 6-3 所示。

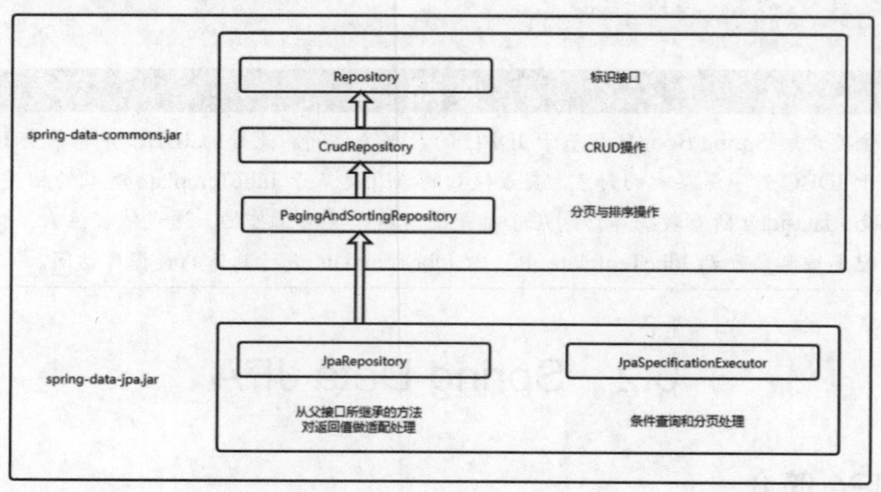

图 6-3

6.2.2 实战

Spring Data JPA 的基本配置如下。

示例代码 6-9　pom.xml 中的相关依赖

```
<dependency>
    <groupId>org.springframework.boot</groupId>
    <artifactId>spring-boot-starter-data-jdbc</artifactId>
</dependency>
<dependency>
    <groupId>org.springframework.boot</groupId>
    <artifactId>spring-boot-starter-data-jpa</artifactId>
</dependency>
```

由于部分 jar 包的版本已经在 Spring Boot 的 pom 依赖中的 parent 指定了，因此上述 JPA 的依赖不用显式指定，如下代码所示。

示例代码 6-10　pom.xml 中的相关依赖

```
<parent>
    <groupId>org.springframework.boot</groupId>
    <artifactId>spring-boot-starter-parent</artifactId>
    <version>2.2.6.RELEASE</version>
    <relativePath/> <!-- lookup parent from repository -->
</parent>
```

示例代码 6-11　application.properties（核心代码）

```
spring.datasource.url=jdbc:mysql://*:*/test?useUnicode=true&characterEncod
ing=utf-8&serverTimezone=UTC&useSSL=true
spring.datasource.username=
spring.datasource.password=
spring.datasource.driver-class-name=com.mysql.jdbc.Driver
spring.jpa.properties.hibernate.hbm2ddl.auto=update
spring.jpa.properties.hibernate.dialect=org.hibernate.dialect.MySQL5InnoDB
Dialect
spring.jpa.show-sql=true
```

spring.jpa.properties.hibernate.hbm2ddl.auto 的相关配置如下：

（1）create：每次加载 Hibernate 时都会删除上一次生成的表（包括数据），然后重新生成新表，即使两次没有任何修改也会这样执行，适用于每次执行单测前清空数据库。

（2）create-drop：每次加载 Hibernate 时都会生成表，但当 SessionFactory 关闭时所生成的表将自动删除。

（3）update：常用的属性值，第一次加载 Hibernate 时创建数据表（前提是需要先有数据库），以后加载 Hibernate 时不会删除上一次生成的表，而是会根据实体更新，只新增字段，不会删除字段（即使实体中已经删除）。

（4）validate：每次加载 Hibernate 时都会验证数据表结构，只会和已经存在的数据表进行比较，根据 model 修改表结构，但不会创建新表。

（5）不配置此项，表示禁用自动建表功能。

User 实体代码如下所示。

示例代码 6-12　User.java

```
@Entity
@Table(name="user")
public class User {
    @Id
    @GeneratedValue(strategy = GenerationType.IDENTITY)
    @Column(name = "id")
    private Integer id;
    @Column(name = "username", nullable = false)
    private String username;
    @Column(name = "password", nullable = false)
    private String password;
    @Column(name = "phone")
    private String phone;
    public Integer getId() {
        return id;
    }
    public void setId(Integer id) {
        this.id = id;
    }
    public String getUsername() {
        return username;

    public void setUsername(String username) {
        this.username = username;
    }
    public String getPassword() {
        return password;
    }
    public void setPassword(String password) {
        this.password = password;
    }
    public String getPhone() {
        return phone;
    }
    public void setPhone(String phone) {
        this.phone = phone;
    }
}
```

dao 层代码如下所示。

示例代码 6-13　UserDao.java

```
public interface UserDao extends JpaRepository<User,Integer> {

}
```

ServiceImpl 层代码如下所示。

示例代码 6-14 UserServiceImpl.java

```
@Service
public class UserServiceImpl implements UserService {
    @Autowired
    private UserDao userDao;
    @Override
    public List<User> findAll() {
        return userDao.findAll();
    }
    @Override
    public Optional<User> findById(Integer id) {
        Optional<User> user = userDao.findById(id);
        return user;
    }
    @Override
    public void saveUser(User user) {
        userDao.save(user);
    }
    @Override
    public void delete(Integer id) {
        User user = new User();
        user.setId(id);
        userDao.delete(user);
    }
}
```

Controller 层代码如下。

示例代码 6-15 UserController.java

```
/**
 * 用于测试spring-boot-starter-jdbc的接口
 */
@RestController
@RequestMapping("/user")
public class UserController {
    @Autowired
    private UserService userService;
    @GetMapping("/list")
    public List<User> listUser(){
        return userService.findAll();
    }
    @GetMapping("/{id}")
    public Optional<User> findById(@PathVariable("id") Integer id){
        return userService.findById(id);
    }
    @PostMapping("/add")
    public String add(@RequestBody() User user){
        userService.saveUser(user);
        return "success!";
    }
```

```
    @DeleteMapping("/{id}")
    public String delete(@PathVariable("id") Integer id){
        userService.delete(id);
        return "success!";
    }
}
```

小 结

JPA 是 Sun 官方在 JDX 1.5 之后提出的 Java 持久化规范，主要是为了简化持久层开发以及整合 ORM 技术，结束 Hibernate、TopLink、JDO 等 ORM 各自为营的局面。JPA 易于使用，伸缩性强，提供了一系列数据库 CRUD 操作的接口，简化了操作持久层的代码。

6.3 Spring Boot 集成 MyBatis-Plus

MyBatis 是一流的持久性框架，支持自定义 SQL、存储过程和高级映射。MyBatis 消除了几乎所有的 JDBC 代码以及参数的手动设置和结果检索。MyBatis 可以使用简单的 XML 或注释进行配置，并将图元、映射接口和 Java POJO（普通的旧 Java 对象）映射到数据库记录。虽然 MyBatis 直接在项目中通过 SQL 语句操作数据库很灵活，但是这也意味要写大量的 XML 文件，在 XML 中配置所需的 SQL 很麻烦。MyBatis-Plus 就很好地解决了这个问题。

6.3.1 MyBatis-Plus 简介

MyBatis-Plus（MP）是一个 MyBatis 的增强工具，在 MyBatis 的基础上只做增强不做改变，为简化开发、提高效率而生。这是官方给的定义，那么它是怎么增强的呢？其实就是它已经在 service 层封装好了一些 CRUD 方法。我们自定义的 service 层业务类只需要继承 com.baomidou.mybatisplus.extension.service.impl.ServiceImpl 类即可，我们不需要再写 XML 配置 SQL，直接调用这些方法就可以实现 CURD 操作，类似于 JPA。MyBatis-Plus 核心功能包括代码生成器、CRUD 接口、条件构造器、分页插件、Sequence 主键、自定义 ID 生成器等。

1. MyBatis-Plus 的特性

（1）无侵入。只做增强不做改变，引入它不会对现有工程产生影响，如丝般顺滑。

（2）损耗小。启动即会自动注入基本 CURD，性能基本无损耗，直接面向对象操作。

（3）强大的 CRUD 操作。内置通用 Mapper、通用 Service，仅仅通过少量配置即可实现单表大部分 CRUD 操作，更有强大的条件构造器，满足各类使用需求。

（4）支持 Lambda 形式调用。通过 Lambda 表达式，方便地编写各类查询条件，无须再担心字段写错。

（5）支持主键自动生成。支持多达 4 种主键策略（内含分布式唯一 ID 生成器 Sequence），可自由配置，完美解决主键问题。

（6）支持 ActiveRecord 模式。支持 ActiveRecord 形式调用，实体类只需继承 Model 类即可进行强大的 CRUD 操作。

（7）支持自定义全局通用操作。支持全局通用方法注入（write once, use anywhere）。

（8）内置代码生成器。采用代码或者 Maven 插件可快速生成 Mapper、Model、Service、Controller 层代码，支持模板引擎，更有超多自定义配置。

（9）内置分页插件。基于 MyBatis 物理分页，开发者无须关心具体操作。配置好插件之后，写分页等同于普通 List 查询。

（10）分页插件支持多种数据库。支持 MySQL、MariaDB、Oracle、DB2、H2、HSQL、SQLite、PostgreSQL、SQL Server 等多种数据库。

（11）内置性能分析插件。可输出 SQL 语句及其执行时间，建议开发测试时启用该功能，能快速揪出慢查询。

（12）内置全局拦截插件。提供全表 delete、update 操作智能分析阻断，也可自定义拦截规则，预防误操作。

2. 支持的数据库

（1）MySQL、SQL Server、Oracle、DB2、H2、MariaDB、HSQL、SQLite、PostgreSQL、Presto、Gauss、Firebird。

（2）Phoenix、Clickhouse、Sybase ASE、OceanBase、达梦数据库、虚谷数据库、人大金仓数据库、南大通用数据库。

3. 框架结构（见图 6-4）

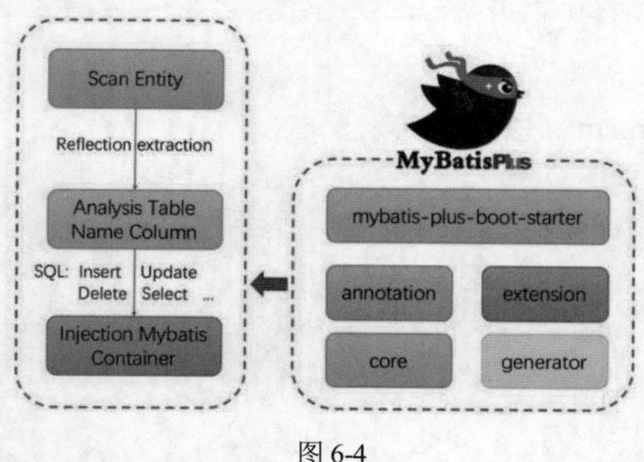

图 6-4

6.3.2 MyBatis-Plus 实战

下面通过一个简单的示例来讲述 MyBatis-Plus 的强大功能。

（1）初始化工程和数据库表。新建 Spring Boot 项目的步骤在此不再赘述。另外，Spring Boot 版本为 2.2.6.RELEASE，JDK 版本为 1.8。创建一张数据库表脚本如下：

```
DROP TABLE IF EXISTS student;
CREATE TABLE student  (
  id int(11) NOT NULL AUTO_INCREMENT COMMENT '主键 ID',
  username varchar(255) CHARACTER SET utf8mb4 COLLATE utf8mb4_general_ci NULL
```

```
DEFAULT NULL COMMENT '用户名',
    password varchar(255) CHARACTER SET utf8mb4 COLLATE utf8mb4_general_ci NULL
DEFAULT NULL COMMENT '密码',
    phone varchar(255) CHARACTER SET utf8mb4 COLLATE utf8mb4_general_ci NULL
DEFAULT NULL COMMENT '电话',
    PRIMARY KEY (id)
  );
```

该数据库表 student 的数据如表 6-1 所示。

表 6-1　student 表数据

id	username	password	phone
1	Tom	111	15611111111
2	Apple	111	15622222222
3	Lady	111	15633333333
4	BoBo	111	15644444444
5	Tiya	111	15655555555

student 表的数据脚本如下：

```
INSERT INTO student VALUES (1, 'Tom', '111', '15611111111');
INSERT INTO student VALUES (2, 'Apple', '111', '15622222222');
INSERT INTO student VALUES (3, 'Lady', '111', '15633333333');
INSERT INTO student VALUES (4, 'BoBo', '111', '15644444444');
INSERT INTO student VALUES (5, 'Tiya', '111', '15655555555');
```

（2）添加 pom 依赖，参考代码如下。

示例代码 6-16　pom.xml（核心依赖）

```
<dependencies>
    <!--Spring Boot Starter 父工程-->
    <dependency>
        <groupId>org.springframework.boot</groupId>
        <artifactId>spring-boot-starter-web</artifactId>
    </dependency>
    <!--MySQL 数据库驱动-->
    <dependency>
        <groupId>mysql</groupId>
        <artifactId>mysql-connector-java</artifactId>
        <scope>runtime</scope>
    </dependency>
    <!--mybatis plus-->
    <dependency>
        <groupId>com.baomidou</groupId>
        <artifactId>mybatis-plus-boot-starter</artifactId>
        <version>3.4.0</version>
    </dependency>

    <dependency>
        <groupId>org.springframework.boot</groupId>
```

```
            <artifactId>spring-boot-starter-test</artifactId>
            <scope>test</scope>
            <exclusions>
                <exclusion>
                    <groupId>org.junit.vintage</groupId>
                    <artifactId>junit-vintage-engine</artifactId>
                </exclusion>
            </exclusions>
        </dependency>
    </dependencies>
```

（3）在 application.properties 配置文件中添加数据库相关配置。

示例代码 6-17　application.properties（核心代码）

```
    spring.datasource.driver-class-name=com.mysql.cj.jdbc.Driver
    spring.datasource.url=jdbc:mysql://localhost:3306/mybatis-plus-test?useUni
code=true&characterEncoding=utf-8&zeroDateTimeBehavior=convertToNull&serverTim
ezone=UTC
    spring.datasource.username=root
    spring.datasource.password=ok
```

在 Spring Boot 启动类中添加@MapperScan 注解，扫描 Mapper 文件夹，代码如下所示。

示例代码 6-18　SpringbootMybatisApplication.java

```
    @SpringBootApplication
    @MapperScan("com.tudou.springbootmybatis.mapper")
    public class SpringbootMybatisApplication {
        public static void main(String[] args) {
            SpringApplication.run(SpringbootMybatisApplication.class, args);
        }
    }
```

（4）编写测试样例。创建 Student.java 实体类，代码如下。

示例代码 6-19　Student.java

```
    public class Student {
        private Integer id;
        private String username;
        private String password;
        private String phone;
        public Integer getId() {
            return id;
        }
        public void setId(Integer id) {
            this.id = id;
        }
        public String getUsername() {
            return username;
        }
        public void setUsername(String username) {
            this.username = username;
```

```
    }
    public String getPassword() {
        return password;
    }
    public void setPassword(String password) {
        this.password = password;
    }
    public String getPhone() {
        return phone;
    }
    public void setPhone(String phone) {
        this.phone = phone;
    }
}
```

（5）创建 StudentMapper.java，代码如下。

示例代码 6-20　StudentMapper.java

```
public interface StudentMapper extends BaseMapper<Student> {
}
```

（6）测试 CRUD 操作，代码如下。

示例代码 6-21　SimpleDemoTest.java

```
@SpringBootTest
public class SimpleDemoTest {
    @Autowired
    private StudentMapper studentMapper;
    @Test
    public void testSelect() {
        //参数 Wrapper<T> queryWrapper 用于构造 SQL 的查询参数，此处查询所有学生信息，
        //故为 null
        List<Student> studentList = studentMapper.selectList(null);
        for (Student student : studentList) {
            System.out.println(student.getUsername());
        }
    }
}
```

6.3.3　代码生成器

AutoGenerator 是 MyBatis-Plus 的代码生成器，通过 AutoGenerator 可以快速生成 Entity、Mapper、Mapper XML、Service、Controller 等各个模块的代码，极大地提升了开发效率。

（1）添加代码生成器相关依赖，核心代码如下。

示例代码 6-22　pom.xml（核心代码）

```
<dependency>
    <groupId>com.baomidou</groupId>
```

```
    <artifactId>mybatis-plus-generator</artifactId>
    <version>3.4.0</version>
</dependency>
```

（2）MyBatis-Plus 从 3.0.3 之后移除了代码生成器与模板引擎的默认依赖，需要手动添加相关依赖。MyBatis-Plus 支持 Velocity（默认）、Freemarker、Beetl，用户可以选择自己熟悉的模板引擎，如果都不满足你的要求，可以采用自定义模板引擎。

Velocity（默认）：

```
<dependency>
    <groupId>org.apache.velocity</groupId>
    <artifactId>velocity-engine-core</artifactId>
    <version>2.2</version>
</dependency>
```

Freemarker:

```
<dependency>
<groupId>org.freemarker</groupId>
<artifactId>freemarker</artifactId>
<version>2.3.30</version>
</dependency>
```

Beetl:

```
<dependency>
<groupId>com.ibeetl</groupId>
<artifactId>beetl</artifactId>
<version>3.2.2.RELEASE</version>
</dependency>
```

注意，选择了非默认引擎时，需要在 AutoGenerator 中设置模板引擎。

```
AutoGenerator autoGenerator = new AutoGenerator();
// set freemarker engine
autoGenerator.setTemplateEngine(new FreemarkerTemplateEngine());
// set beetl engine
autoGenerator.setTemplateEngine(new BeetlTemplateEngine());
// set custom engine (reference class is your custom engine class)
autoGenerator.setTemplateEngine(new CustomTemplateEngine());
```

配置 ClobalConfig:

```
//1.全局配置
GlobalConfig globalConfig = new GlobalConfig();
//项目绝对路径
String projectPath = "your project dir";
globalConfig.setOutputDir(projectPath + "/src/main/java");
globalConfig.setAuthor("username");
globalConfig.setOpen(false);
// gc.setSwagger2(true); 实体属性 Swagger2 注解
```

配置 DataSourceConfig:

```
    //2. 数据源配置
    DataSourceConfig dataSourceConfig = new DataSourceConfig();
    dataSourceConfig.setUrl("jdbc:mysql://localhost:3306/mybatis-plus-test?use
Unicode=true&characterEncoding=utf-8&zeroDateTimeBehavior=convertToNull&server
Timezone=UTC");
    // dataSourceConfig.setSchemaName("public");
    dataSourceConfig.setDriverName("com.mysql.cj.jdbc.Driver");
    dataSourceConfig.setUsername("database username");
    dataSourceConfig.setPassword("database password");
```

配置 PackageConfig：

```
    PackageConfig packageConfig = new PackageConfig();
    //在 controller service mapper mapper.xml 等文件外层创建一个文件夹
    //packageConfig.setModuleName("address");
    packageConfig.setModuleName("");
    packageConfig.setParent("com.tudou.springbootmybatis");
```

自定义配置：

```
    // 自定义新建文件部分配置
    InjectionConfig injectionConfig = new InjectionConfig() {
        @Override
        public void initMap() {
            // to do nothing
        }
    };
    // 如果模板引擎是 freemarker
    //String templatePath = "/templates/mapper.xml.ftl";
    // 默认模板引擎是 velocity
    String templatePath = "/templates/mapper.xml.vm";
    // 自定义输出配置
    List<FileOutConfig> fileOutConfigList = new ArrayList<>();
    // 自定义配置会被优先输出
    fileOutConfigList.add(new FileOutConfig(templatePath) {
        @Override
        public String outputFile(TableInfo tableInfo) {
            // 自定义输出文件名，如果 Entity 设置了前后缀，那么此处 XML 的名称会跟着发生变化
            return projectPath + "/src/main/resources/mapper/" +
packageConfig.getModuleName()
                    + "/" + tableInfo.getEntityName() + "Mapper" +
StringPool.DOT_XML;
        }
    });
    injectionConfig.setFileOutConfigList(fileOutConfigList);
```

策略配置：

```
    // 策略配置
    StrategyConfig strategy = new StrategyConfig();
    strategy.setNaming(NamingStrategy.underline_to_camel);
    strategy.setColumnNaming(NamingStrategy.underline_to_camel);
```

```
//strategy.setSuperEntityClass("你自己的父类实体，没有则不用设置!");
strategy.setEntityLombokModel(true);
strategy.setRestControllerStyle(true);
// 公共父类
//strategy.setSuperControllerClass("你自己的父类控制器，没有则不用设置!");
// 写于父类中的公共字段
//strategy.setSuperEntityColumns("id");
//strategy.setInclude(scanner("表名，多个英文逗号分割").split(","));
String tableNames = "address,order,order_item";
//表名，逗号分隔
strategy.setInclude(tableNames.split(","));
strategy.setControllerMappingHyphenStyle(true);
strategy.setTablePrefix(packageConfig.getModuleName() + "_");
autoGenerator.setStrategy(strategy);
//设置模板引擎
autoGenerator.setTemplateEngine(new VelocityTemplateEngine());
```

MyBatis-Plus 代码生成器的完整代码参见第 6 章 springboot-mybatis 项目的 MybatisPlusGenerator.java 类，直接运行 main 方法即可。

生成代码的架构如图 6-5 所示。

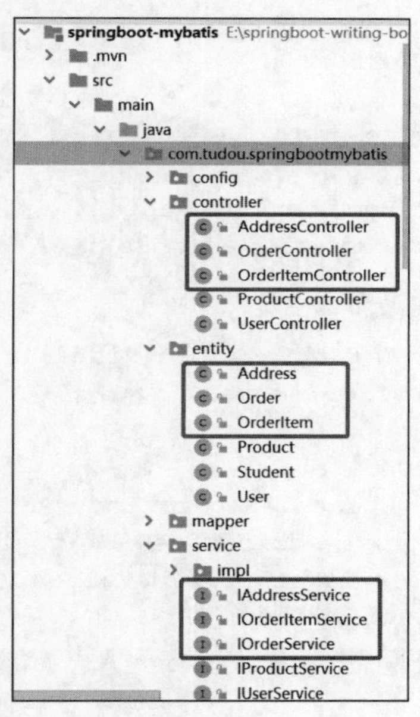

图 6-5

6.3.4　CRUD 接口

通用 Service CRUD 封装 IService 接口，进一步封装 CRUD，采用 get 查询单行、remove 删除、list 查询集合、page 分页前缀命名方式区分 mapper 层，避免混淆。如果存在自定义的 Service 方法，需要创建自己的 XXXService 层，继承 com.baomidou.mybatisplus.extension.service.IService 接口即可使用 MyBatis-Plus 提供的 CRUD 方法，也可以自定义 service 层方法。业务接口层的代码如下所示。

示例代码 6-23　IAddressService.java

```
public interface IAddressService extends IService<Address> {
}
```

业务实现层的代码如下所示。

示例代码 6-24　AddressServiceImpl.java

```
@Service
public class AddressServiceImpl extends ServiceImpl<AddressMapper, Address>
implements IAddressService {
}
```

以 AddressController.java 为例，注入 IAddresServic，调用 com.baomidou.mybatisplus.extension.service.IService 接口提供的一系列 CRUD 方法进行一个简单的测试，代码如下。

示例代码 6-25　AddressController.java

```
/**
 * 前端控制器
*/
@RestController
@RequestMapping("/address")
public class AddressController {
    @Autowired
    private IAddressService addressService;
    @PostMapping("/insert")
    public boolean save(@RequestBody Address address){
        return addressService.save(address);
    }
    @DeleteMapping("/remove/{id}")
    public boolean remove(@PathVariable("id") Integer id){
        return addressService.removeById(id);
    }
    @PutMapping("/saveOrUpdate")
    public boolean saveOrUpdate(@RequestBody Address address){
        QueryWrapper<Address> addressQueryWrapper = new QueryWrapper<>();
        addressQueryWrapper.isNotNull("detailed_address")
                .eq("province", "湖北省");
        return addressService.saveOrUpdate(address, addressQueryWrapper);
    }
    @GetMapping("/list")
    public List<Address> remove(){
        QueryWrapper<Address> addressQueryWrapper = new QueryWrapper<>();
        addressQueryWrapper.eq("province", "湖北省");
        return addressService.list(addressQueryWrapper);
    }
    @GetMapping("/page")
    public IPage<Address> pageAddress(){
        Page<Address> page = new Page<>(1,3);
        QueryWrapper<Address> addressQueryWrapper = new QueryWrapper<>();
        //构造查询条件
```

```
        addressQueryWrapper.eq("province", "湖北省");
        IPage<Address> addressPage = addressService.page(page,
addressQueryWrapper);
        return addressPage;
    }
}
```

通用 Mapper CRUD 封装 com.baomidou.mybatisplus.core.mapper.BaseMapper 接口，BaseMapper<T>泛型 T 表示任意实体，当项目启动时会自动解析实体表关系映射转换为 MyBatis 内部对象注入容器。如果需要自定义 SQL 操作，创建的 XXXMapper 需要继承 BaseMapper<T>接口，参考如下代码片段：

```
public interface AddressMapper extends BaseMapper<Address> {
}
```

6.3.5　分页插件

MyBatis-Plus 提供了分页插件 PaginationInterceptor，可以通过插件拦截四大对象 Executor、StatementHandler、ParameterHandler、ResultSetHandler 相关方法的执行，根据需求完成数据动态改变。所以，MyBatis-Plus 的分页是基于拦截器的，这个拦截器拦截的是方法和方法中的参数，先判断是否是查询操作，如果是查询操作就会进入分页逻辑，拦截器会通过反射获取该方法的参数，判断是否存在 IPage 对象的实现类。如果不存在就不进行分页，如果存在则将该参数复制给 IPage 对象，最后凭借 SQL 的处理完成分页操作。使用 IPage 时需要配置一个 PaginationInterceptor 类型的 Bean 交给 Spring 进行管理，在 Spring Boot 项目中配置分页插件的代码如下所示。

示例代码 6-26　MybatisPlusConfig.java

```
@EnableTransactionManagement
@Configuration
public class MybatisPlusConfig {
    /**
     * 配置分页
     * @return com.baomidou.mybatisplus.extension.plugins.
PaginationInterceptor
     * @throws
     * @update
     * @see  MybatisPlusConfig paginationInterceptor()
     * @since V1.0
     */
    @Bean
    public PaginationInterceptor paginationInterceptor() {
        PaginationInterceptor paginationInterceptor = new
PaginationInterceptor();
        // 设置请求的页面大于最大页后的操作，为 true 时调回首页，为 false 时继续请求，默认
为 false
        // paginationInterceptor.setOverflow(false);
        // 设置最大单页限制数量，默认 500 条，-1 表示不受限制
        // paginationInterceptor.setLimit(500);
```

```
        // 开启 count 的 join 优化，只针对部分 left join
        paginationInterceptor.setCountSqlParser(new
JsqlParserCountOptimize(true));
        return paginationInterceptor;
    }
}
```

com.baomidou.mybatisplus.extension.service.IService 接口中提供了许多分页查询的方法，非常便利。

IService 接口中提供的分页方法如下代码所示。

```
// 无条件分页查询
IPage<T> page(IPage<T> page);
// 条件分页查询
IPage<T> page(IPage<T> page, Wrapper<T> queryWrapper);
// 无条件分页查询
IPage<Map<String, Object>> pageMaps(IPage<T> page);
// 条件分页查询
IPage<Map<String, Object>> pageMaps(IPage<T> page, Wrapper<T> queryWrapper);
```

pageMaps 参数说明如表 6-2 所示。

表 6-2　pageMaps 参数说明

参数类型	说明
IPage<T> page	分页对象
Wrapper<T> queryWrapper	实体对象封装操作类 queryWrapper，用于设置查询参数
T	任意实体对象

如果是简单的查询分页，即单表查询，那么我们可以直接使用 IService 提供的 CRUD 接口中现有的 selectPage()或者 selectMapsPage()方法，源码如下所示。

示例代码 6-27　BaseMapper.java 部分代码

```
<E extends IPage<T>> E selectPage(E page,@Param("ew") Wrapper<T> queryWrapper);
<E extends IPage<Map<String, Object>>> E selectMapsPage(E page, @Param("ew")
Wrapper<T> queryWrapper);
```

其中，Page 实现了 IPage 接口。常用的构造方法如下所示。其中，参数 current 表示当前页，size 表示分页的页面大小。除此之外，Page.java 还提供了其他构造方法，可根据自身需要选择，在此不再赘述。

```
public Page(long current, long size) {
    this(current, size, 0L);
}
```

如果是复杂的分页查询，即多表关联查询，就需要用到自定义的 mapper，此时的分页操作也很简单，只需要将 mapper 的第一个参数设置为 Page 对象即可。当执行到此方法时，MyBatis-Plus 的分页插件将会自动拦截，自动在 SQL 后拼接分页条件，并且把分页查询到的结果返回这个 Page 对象中，如下代码所示。

示例代码 6-28　ProductController.java（部分代码）

```
@GetMapping("/list")
public Page<Product> listProduct(){
    Page<Product> page = new Page<>(1,5);
    Page<Product> productPage = productService.listProductPage(page);
    return productPage;
}
```

示例代码 6-29　ProductServiceImpl.java

```
@Override
public Page<Product> listProductPage(Page<Product> page){
    List<Product> products = productMapper.listProduct(page);
    page.setRecords(products);
    return page;
}
```

示例代码 6-30　ProductMapper.xml

```
<?xml version="1.0" encoding="UTF-8"?>
<!DOCTYPE mapper PUBLIC "-//mybatis.org//DTD Mapper 3.0//EN"
"http://mybatis.org/dtd/mybatis-3-mapper.dtd">
<mapper namespace="com.tudou.springbootmybatis.mapper.ProductMapper">
    <resultMap id="BaseResultMap"
type="com.tudou.springbootmybatis.entity.Product">
        <id property="id" column="id"></id>
        <result property="name" column="name"></result>
        <result property="productTime" column="product_time"></result>
        <result property="category" column="category"></result>
    </resultMap>
    <select id="listProduct" resultMap="BaseResultMap">
        select * from product
    </select>
</mapper>
```

请求 http://localhost:8080/product/list，测试结果如图 6-6 所示。

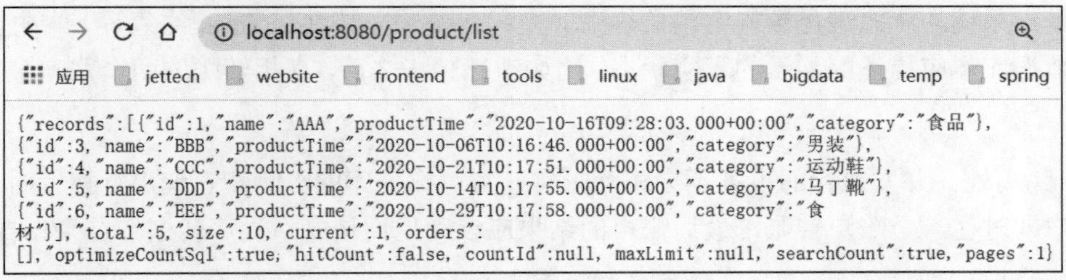

localhost:8080/product/list

应用　jettech　website　frontend　tools　linux　java　bigdata　temp　spring

{"records":[{"id":1,"name":"AAA","productTime":"2020-10-16T09:28:03.000+00:00","category":"食品"},
{"id":3,"name":"BBB","productTime":"2020-10-06T10:16:46.000+00:00","category":"男装"},
{"id":4,"name":"CCC","productTime":"2020-10-21T10:17:51.000+00:00","category":"运动鞋"},
{"id":5,"name":"DDD","productTime":"2020-10-14T10:17:55.000+00:00","category":"马丁靴"},
{"id":6,"name":"EEE","productTime":"2020-10-29T10:17:58.000+00:00","category":"食
材"}],"total":5,"size":10,"current":1,"orders":
[],"optimizeCountSql":true,"hitCount":false,"countId":null,"maxLimit":null,"searchCount":true,"pages":1}

图 6-6

小 结

本小节主要讲述了数据库访问的增强工具 MyBatis-Plus 的使用方法，其核心功能包括代码生成器、CRUD 接口、条件构造器、分页插件、Sequence 主键、自定义 ID 生成器等。Spring Boot 项目引入 MyBatis-Plus 相关 Maven 依赖后，不但可以通过 AutoGenerator 代码生成器快速生成 Entity、Mapper、Mapper XML、Service、Controller 等各个模块的代码，而且通过 Mapper 接口类注入之后就能对对应的数据库表进行 CRUD 操作，不需要编写 XML 文件，极大地提升了开发效率。

6.4 事 务

6.4.1 事务的定义与特性

1. 事务的定义

数据库事务是构成单一逻辑工作单元的操作集合。数据库事务可以包含一个或多个数据库操作，但这些操作构成一个逻辑上的整体。这个逻辑整体中的数据库操作，要么全部执行成功，要么全部不执行。也就是说，构成事务的所有操作，要么全都对数据库产生影响，要么全都不产生影响，不管事务是否执行成功，数据库总是保持一致性状态。

2. 事务的特性

事务具有 4 个特性，即原子性（Atomicity）、一致性（Consistency）、隔离性（Isolation）和持久性（Durability），通常简称为 ACID。

（1）原子性。事务中的所有操作作为一个整体，像原子一样不可分割，要么全部成功，要么全部失败。

（2）一致性。事务的执行结果必须使数据库从一个一致性的状态到另一个一致性的状态。一致性状态是指系统的状态满足数据完整性约束（主码、参照完整性、check 约束等），并且系统的状态反应数据库本应描述的显示的真实状态，比如银行转账之后，互相转账的两个账户金额总和保持不变。

（3）隔离性。并发执行的事务不会相互影响，其对数据库的影响和它们串行执行时一样。比如多个用户同时往一个账户转账，最后账户的结果应该和他们按先后次序转账的结果一样。

（4）持久性。事务一旦提交，其对数据库的更新就是持久的，任何事务或系统故障都不会导致数据丢失，不会因为系统故障或者断电造成数据不一致或者丢失。

6.4.2　事务的并发问题

1. 事务的并发问题

一个数据库中的同一份数据，由于被多客户端并发式访问，或者被多线程并发式访问，会导致多个事务同时访问同一份数据，如果没有采取必要的隔离措施，就会导致各种并发问题，破坏数据的完整性。这种数据库的并发问题可以大致归结为 4 类：脏读、不可重复读、幻读、丢失更新。下面我们将根据实际场景讲解这 4 类事务的并发问题。

（1）脏读。事务 A 读取了事务 B 更新的数据，然后 B 回滚操作，那么 A 读取到的数据是脏数据，如表 6-3 所示。

表 6-3　脏读例子

时间节点	转账事务 A	取款事务 B
T1		开始事务
T2	开始事务	
T3		查询账户余额为 500 元
T4		取出现金 200 元，余额更新为 300 元
T5	查询账户余额为 300 元（**脏读**）	
T6		**事务回滚**（余额回复为 500 元）
T7	汇款 100 元（余额更新为 400 元）	
T8	**提交事务**	
余额	应有余额	实际余额
	600	400

在这个场景中，取款事务 B 希望从账户中取出 200 元现金，与此同时转账事务 A 向账户中汇款 100 元，因为转账事务 A 在汇款之前读取了 B 事务尚未提交的数据，产生了脏读，导致账户丢失了 200 元。

（2）不可重复读。一个事务中两次（不同时间点）读取同一行数据，但是这两次读取到的数据不一致，如表 6-4 所示。

表 6-4　不可重复读例子

时间节点	取款事务 A	查询事务 B
T1	**开始事务**	
T2		**开始事务**
T3	查询账户余额为 500 元	
T4		查询账户余额为 500 元
T5	取出 200 元（余额更新为 300 元）	
T6	**提交事务**	
T7		查询账户余额为 300 元
T8		**提交事务**

在这个场景中，事务 B 在事务取款前后读取的账户余额不一致，T4 节点查询的余额为 500 元，T7 节点查询的余额为 300 元。

（3）幻读。事务 A 表示系统管理员将数据库中某张存放成绩表的所有成绩都更新为 ABCDEF 等级，此时事务 B 插入了一条具体分数的记录，当事务 A 结束之后发现有一条数据学生成绩没有更新为 ABCDEF 等级制的，就好像发生了幻觉一样，此种情况称为幻读。

在这个场景中，我们会发现当事务 A 和事务 B 都结束后数据库表会存在一条学生成绩信息是具体分数的形式，而非 ABCDEF 等级制的形式，此时就出现了幻读（见表 6-5）。

表 6-5　幻读例子

时间节点	事务 A	事务 B
T1		**开始事务**
T2	**开始事务**	
T3	把所有学生成绩由具体分数更新为 ABCDEF 等级	
T4		新增一条学生成绩信息记录
T5		**提交事务**
T6	**提交事务**	

（4）丢失更新。丢失更新又可细分为第一类丢失更新和第二类丢失更新。事务 A 覆盖了事务 B 已提交的更新数据，导致事务 B 的更新数据好像丢失了一样，称为丢失更新。

①第一类丢失更新，如表 6-6 所示。

表 6-6　第一类丢失更新例子

时间节点	存款事务 A	取款事务 B
T1		**开始事务**
T2	**开始事务**	
T3		查询账户余额为 500 元
T4	查询账户余额为 500 元	
T5		取款 100 元（余额 400 元）
T6	存款 200 元（余额 700 元）	
T7		**提交事务**
T8	**事务回滚**	
余额	应有余额	实际余额
	500 元	400 元

在这个场景中，我们会发现事务 A 发生回滚之后事务 B 的操作丢失了，这种数据丢失会导致严重的问题，比如上述场景中个人账户就损失了 100 元。

②第二类丢失更新，如表 6-7 所示。

表 6-7　第二类丢失更新例子

时间节点	存款事务 A	取款事务 B
T1		**开始事务**
T2	**开始事务**	
T3		查询账户余额为 500 元
T4	查询账户余额为 500 元	
T5		取款 100 元（余额 400 元）
T6	存款 200 元（余额 700 元）	
T7		**提交事务**
T8	**提交事务**	
余额	应有余额	实际余额
	600 元	700 元

在这个场景中，我们会发现事务 A 提交之后事务 B 的操作丢失了。这种数据丢失也会导致严重的问题，比如上述场景银行层面损失 100 元。

通过以上场景对比，我们会发现幻读和不可重复读很类似，都是读取到不一致的数据，当然本质上也是有区别的。幻读的侧重点在于插入和删除，即第二次查询数据会比第一次查询的数据变多了或者变少了。不可重复读的侧重点在于修改，即会出现第二次查询与第一次查询的同一条记录中某些字段值不一致的情况。

2. 事务的隔离级别

事务具有隔离性，从理论上讲，事务与事务之间都是独立存在的，不应该互相影响；事务的执行也应该是完全串行化的，即先执行完事务 1，再执行事务 2，以此类推。然而，完全的隔离性会导致系统的并发性很低，从而降低了资源的利用。所以，数据库对事务隔离性的要求有所放宽，从而在一定程度上造成了数据库数据的不一致性。SQL 标准为事务定义了不同的隔离级别，从低到高依次是：

（1）读未提交的（ISOLATION_READ_UNCOMMITTED）：最低的隔离级别，即事务未提交前就可以被其他事务读取，易导致幻读、脏读和不可重复读。

（2）读已提交的（ISOLATION_READ_COMMITTED）：表示一个事务提交之后才能被其他事务读取，即禁止其他事物读取到未提交的事务数据，易导致幻读和不可重复读。

（3）可重复读（ISOLATION_REPEATABLE_READ）：表示多次读取到同一个数据时，其值都与事务开始时读取到的内容一致，即禁止读取其他事务未提交的数据，可防止脏读和不可重复读，但易导致幻读。

（4）序列化（ISOLATION_SERIALIZABLE）：最可靠的隔离级别，也是最耗费效率、代价最大的隔离级别，该隔离级别可有效防止脏读、不可重复读和幻读。

四种隔离级别易导致的并发异常如表 6-8 所示。

表 6-8　四种隔离级别易导致的并发异常

事务的隔离级别	可能导致的并发异常			
	脏读	不可重复读	幻读	丢失更新
读未提交的	可能	可能	可能	可能
读已提交的		可能	可能	可能
可重复读			可能	
串行化				

当然，每个数据库的默认隔离级别是不一样的。常用的关系型数据库 MySQL、Oracle、SQL Server 支持的隔离级别以及默认的隔离级别如表 6-9 所示。

表 6-9　三种数据库支持的隔离级别及默认的隔离级别

数据库	读未提交的	读已提交的	可重复读	序列化	默认隔离级别
MySQL	√	√	√	√	可重复读
Oracle		√		√	读已提交的
SQL Server	√	√	√	√	读已提交的

3. 以 MySQL 为例测试隔离级别

下面我们将以 MySQL 做隔离级别的演示说明。

（1）准备工作

①首先在“任务管理器”→“服务列表”中启动 MySQL 服务，如图 6-7 所示。

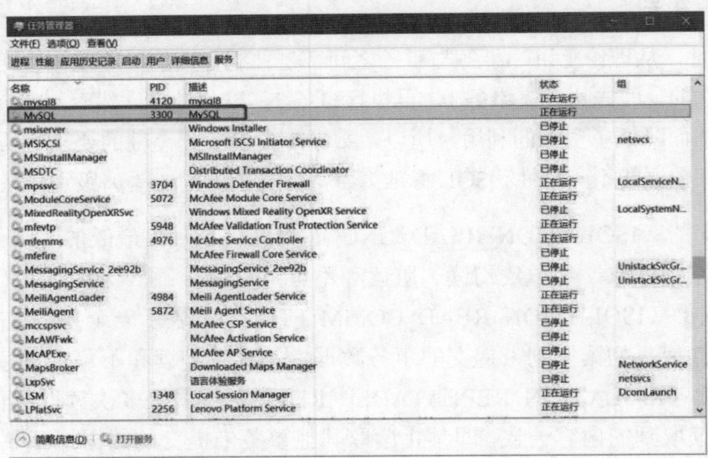

图 6-7

②打开命令窗口，切换到 MySQL 的 bin 目录，运行命令 mysql.exe -h 127.0.0.1 -uroot -pok（root 为数据库用户名，ok 为数据库密码），打开一个 MySQL 客户端，如图 6-8 所示。

```
C:\Windows\System32\cmd.exe - mysql -h 127.0.0.1 -uroot -pok
Microsoft Windows [版本 10.0.18362.1016]
(c) 2019 Microsoft Corporation。保留所有权利。

D:\dev\mysql\mysql-5.7.27-winx64\bin>mysql -h 127.0.0.1 -uroot -pok
mysql: [Warning] Using a password on the command line interface can be insecure.
Welcome to the MySQL monitor.  Commands end with ; or \g.
Your MySQL connection id is 16
Server version: 5.7.27 MySQL Community Server (GPL)

Copyright (c) 2000, 2019, Oracle and/or its affiliates. All rights reserved.

Oracle is a registered trademark of Oracle Corporation and/or its
affiliates. Other names may be trademarks of their respective
owners.

Type 'help;' or '\h' for help. Type '\c' to clear the current input statement.

mysql>
```

图 6-8

③执行命令“select @@tx_isolation;”查看 MySQL 数据库默认隔离级别，如图 6-9 所示。

④执行命令“create database transation_test;”创建一个数据库 transation_test，再创建一个表 user_account，进行后续的测试。

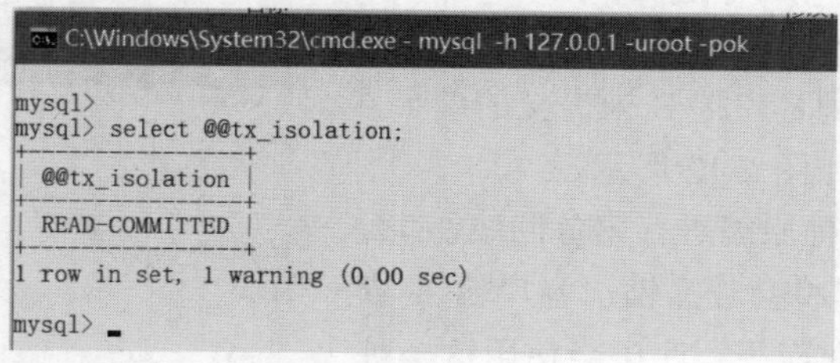

```
C:\Windows\System32\cmd.exe - mysql -h 127.0.0.1 -uroot -pok
mysql>
mysql> select @@tx_isolation;
+----------------+
| @@tx_isolation |
+----------------+
| READ-COMMITTED |
+----------------+
1 row in set, 1 warning (0.00 sec)

mysql> _
```

图 6-9

执行命令“use transation_test;”切换至刚创建的数据库 transation_test，并创建表 user_account，语句如下：

```
create table user_account(id int primary key,name varchar(255),balance number);
```

插入测试语句，具体如下：

```
insert into user_account(id,name,balance) values (1,'tudou',500);
```

执行结果如图 6-10 所示。

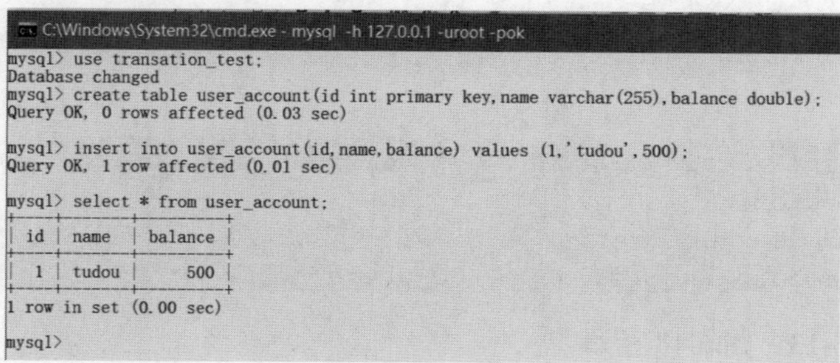

```
C:\Windows\System32\cmd.exe - mysql -h 127.0.0.1 -uroot -pok
mysql> use transation_test;
Database changed
mysql> create table user_account(id int primary key,name varchar(255),balance double);
Query OK, 0 rows affected (0.03 sec)

mysql> insert into user_account(id,name,balance) values (1,'tudou',500);
Query OK, 1 row affected (0.01 sec)

mysql> select * from user_account;
+----+-------+---------+
| id | name  | balance |
+----+-------+---------+
|  1 | tudou |     500 |
+----+-------+---------+
1 row in set (0.00 sec)

mysql>
```

图 6-10

（2）测试隔离级别——读取未提交

打开一个 MySQL 客户端 A，设置当前事务隔离级别为 read uncommitted（读取未提交），并开启事务。

执行结果如图 6-11 所示。

```
选择C:\Windows\System32\cmd.exe - mysql -h 127.0.0.1 -uroot -pok
mysql> set session transaction isolation level read uncommitted;
Query OK, 0 rows affected (0.00 sec)

mysql> start transaction;
Query OK, 0 rows affected (0.00 sec)
```

图 6-11

在客户端 A 提交事务之前，打开另外一个客户端 B，更新表 user_account 中的数据，并将用户 tudou 的账户余额由 500 元更新为 300 元，核心命令如下：

```
mysql.exe -h 127.0.0.1 -uroot -pok
show databases;
use transation_test;
select * from user_account;
update user_account set balance=300 where id=1;
```

此时客户端 B 并未提交更新结果，我们在客户端 A 进行查询，发现能查询到客户端 B 未提交的结果，即余额被更改为 300 元，如图 6-12 所示。

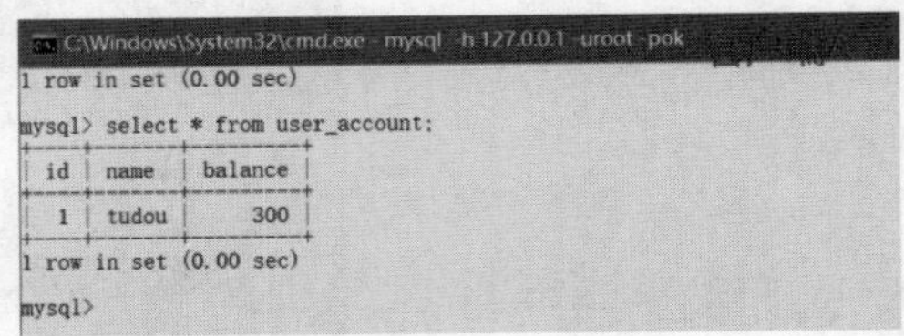

图 6-12

如果客户端 B 因为某种原因进行事务回滚，那么客户端 B 做的 update 操作将会被撤销，客户端 A 查询到的数据就是脏数据，在应用程序中会使用这个脏数据进行显示或计算等。这就容易导致某应用程序中的某个功能出现数据不一致的情况。

（3）测试隔离级别——读取已提交

①设置隔离级别为读取已提交，并将用户 id=1 的账户余额恢复至 500 元。

②客户端 A 查询到的余额为 500 元。

③客户端 B 执行更新操作，将余额更新为 600 元，未提交事务。

④客户端 A 再次查询，余额仍为 500 元。

⑤客户端 B 提交事务。

⑥客户端 A 再次查询余额，余额变为 600。

核心命令如下：

```
set session transaction isolation level read committed;
start transaction;
select * from user_account;
update user_account set balance=600 where id=1;
```

执行结果如图 6-13、图 6-14 所示。

客户端 A 在同一个事务中，不同阶段查询到的 tudou 账户余额数据不一致（初次查询到的账户余额为 500，当客户端 B 提交更新结果后，查询到的账户余额为 600），所以“读取已提交”的隔离级别易导致不可重复读问题。

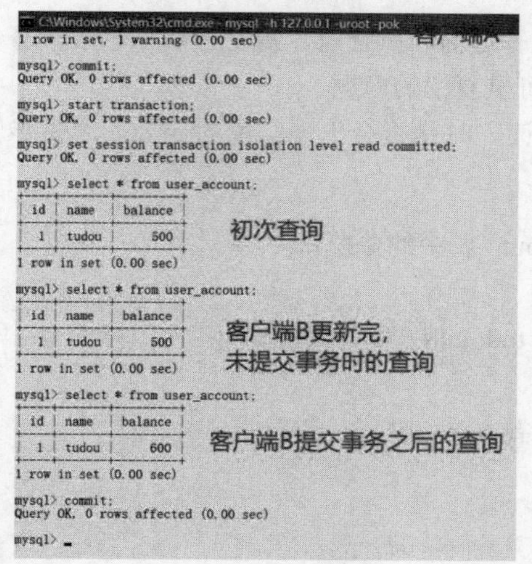

图 6-13

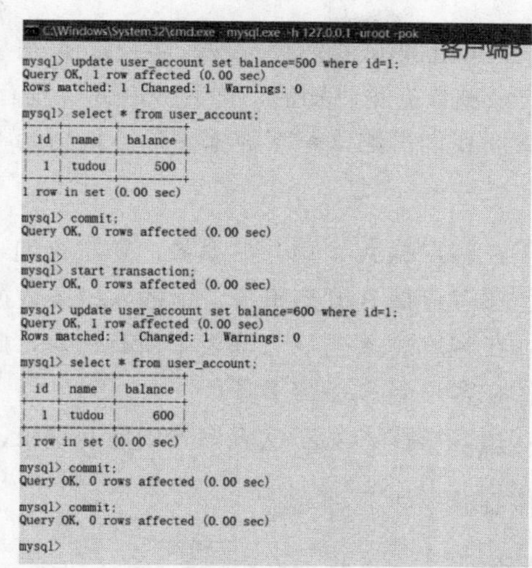

图 6-14

（4）测试隔离级别——可重复读

①设置隔离级别为可重复读。
②客户端 A 查询到的余额为 600 元。
③客户端 B 执行更新操作，将余额更新为 500 元，未提交事务。
④客户端 A 再次查询，余额仍未 600 元。
⑤客户端 B 提交事务。
⑥客户端 A 再次查询余额时，余额仍然为 600。

核心命令如下：

```
set session transaction isolation level repeatable read;
start transaction;
select * from user_account;
update user_account set balance=500 where id=1;
```

执行结果如图 6-15、图 6-16 所示。

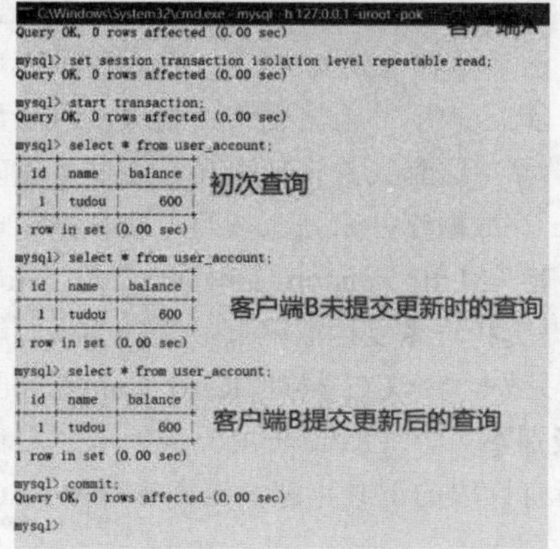

图 6-15

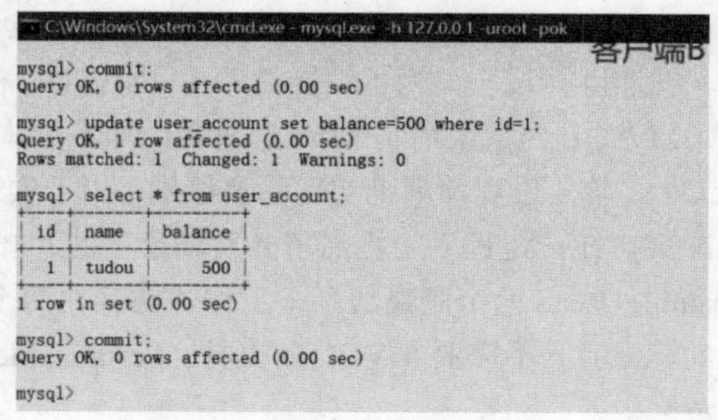

图 6-16

经过以上测试我们会发现，客户端 A 在事务 B 执行之前、执行时、提交之后三个阶段查询到的账户余额数据是已知的，均为 600 元，没有出现不可重复读的问题。

下面我们再测试一下数据库隔离机制为“可重复读”时是否会出现幻读的现象。测试步骤如下：

①在客户端 A 窗口开启事务，初次查询 user_account 表中的数据。

②在客户端 B 开启事务，新插入一条数据。

③在客户端 A 再次查询表中所有记录，此时只有 tudou 唯一一个用户。

④在客户端 B 提交事务。

⑤在客户端 A 第三次查询表中所有记录，仍然只有 tudou 唯一一个用户。

核心命令如下：

```
start transaction;
set session transaction isolation level repeatable read;
select * from user_account;
insert into user_account (id,name,balance) values (2,'bobo',1000);
```

具体测试结果如图 6-17、图 6-18 所示。

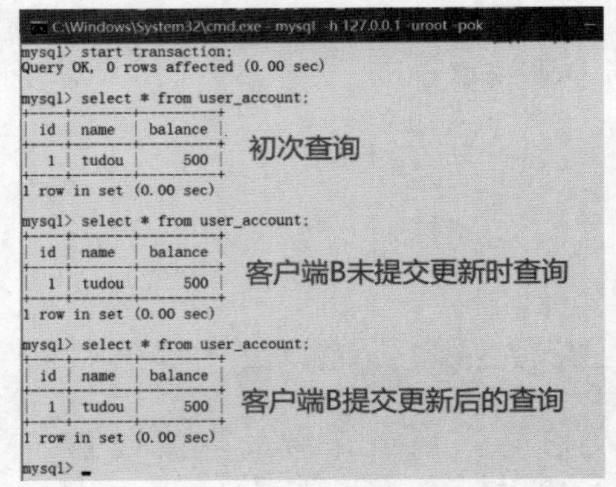

图 6-17

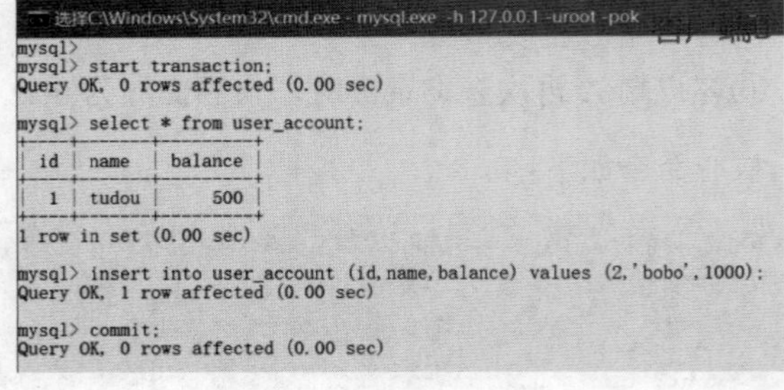

图 6-18

根据以上流程测试，从预期来说，客户端 A 在第一次和第二次查询时均应该查询到一条记录，当客户端 B 提交事务之后，客户端 A 进行第三次查询时，应该是查询到 tudou 和 bobo 两个账户信息；然而，实际测试中发现客户端 A 三次的查询结果均一致，表示并未出现幻读现象。这与数据库理论中隔离级别为“可重复读”时仍会出现幻读现象违背了。这到底是为什么呢？经过多番研究发现，在实际情况中，为了性能的考虑，MySQL 和 Oracle 等数据库并未完全按照上述数据库理论实现。MySQL 内部采用了 MVCC 机制，即多版本并发控制（Multi-Version Concurrency Control，MVCC），基于乐观锁理论实现隔离级别，用于实现读已提交和可重复读取隔离级别的实现，解决读写冲突，在一定程度上提高了并发场景下的吞吐性能。另外，MVCC 只在 Read Committed 和 Repeatable Read 两个隔离级别下工作，与其他两个隔离级别不兼容（Read Uncommitted 总是读取最新的记录行，不需要 MVCC 的支持；Serializable 会对所有读取的记录行都加锁，单靠 MVCC 无法完成）。

在讲解 MVCC 机制之前，我们引入系统版本号和事务版本号。系统版本号和事务版本号的基本概念如表 6-10 所示。

表 6-10　版本号的定义和用法

版本号	定义
系统版本号	一个递增数字，每开始一个新事务，系统版本号就会自动递增
事务版本号	事务开始时的系统版本号

并且在 InnoDB 存储引擎 MVCC 的实现策略中，会在每张表的每条数据后面添加创建版本号和删除版本号两个字段。在创建一行数据时，将当前系统版本号作为创建版本号；删除一行数据时，将当前系统版本号作为删除版本号。

对于数据库增删改查操作，对应的创建版本号和删除版本号底层原理如表 6-11 所示。

表 6-11　数据库增删改查操作与版本号

操作	创建版本号	删除版本号	备注
insert	插入数据时，当前系统版本号赋值给创建版本号	空	
delete	删除数据时，当前系统版本号赋值给删除版本号字段，标识该行数据是在哪一个事务中被删除的	空	
update	插入一条新记录，保存当前事务版本号为创建版本号	同时保存当前事务版本号到原来删除的行	update 实际上是通过 delete 和 insert 实现的
select	创建版本号小于等于当前事务版本号，保证取出的数据不会有后启动的事务创建的数据	删除版本号为空或大于当前事务版本号。 好整好了至少在该事务开启之前数据没有被其他事务删除	（1）创建版本号小于等于当前事务版本号。 （2）删除版本号为空或大于当前事务版本号时，select 语句才会读到数据

我们将用以下示例复现可重复读隔离机制下的幻读现象，具体操作步骤如下：

①在客户端 A 窗口开启事务，设置隔离级别为可重复读。

②在客户端 A 初次查询 user_account 表中的数据，含有两条记录。

③在客户端 B 开启事务，新插入一条数据。

④在客户端 A 再次查询表中的数据，仍含有两条记录。

⑤在客户端 B 提交事务。

⑥在客户端 A 第三次查询表中所有记录，仍含两条记录。

⑦在客户端 A 进行 update 操作，将 id 为 1 和 2 的两个用户余额更新为 50。

⑧在客户端 A 第四次查询表中所有记录。

核心命令如下：

```
start transaction;
set session transaction isolation level repeatable read;
select * from user_account;
insert into user_account (id,name,balance) values (3,'zhangsan',750);
update user_account set balance=50;
```

测试结果如图 6-19、图 6-20 所示。

通过以上测试结果我们会发现，当客户端 A 第四次查询表 user_account 的数据时，变成了 3 条记录，并且 3 个用户的余额均被改为 50，而预期只是把前两条数据余额改为 50，测试结果与预期不符，出现了幻读现象。MySQL 中的 MVCC 机制在可重复读隔离机制下并未完全解决幻读问题，只是解决了读取数据情况下的幻读，对于修改操作仍然存在幻读问题。这证明一个问题：在可重复读机制下，select 操作读取的是历史数据（也称为快照），update 操作读取的是数据库最新版本的数据。

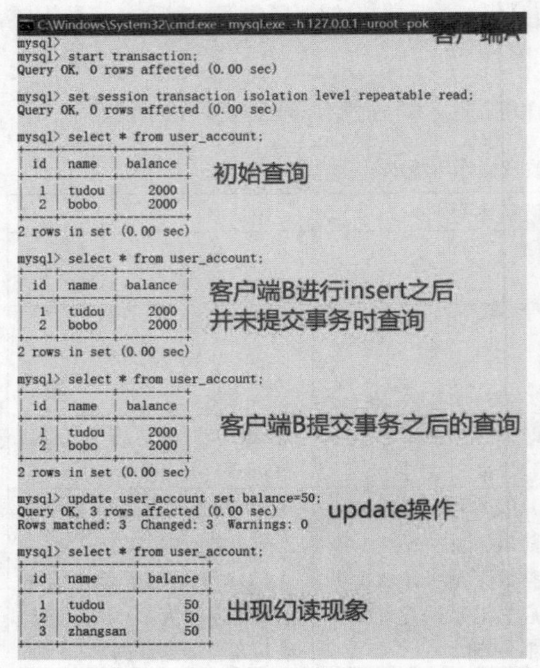
```
mysql>
mysql> start transaction;
Query OK, 0 rows affected (0.00 sec)

mysql> set session transaction isolation level repeatable read;
Query OK, 0 rows affected (0.00 sec)

mysql> select * from user_account;
+----+-------+---------+
| id | name  | balance |
+----+-------+---------+
|  1 | tudou |    2000 |
|  2 | bobo  |    2000 |
+----+-------+---------+
2 rows in set (0.00 sec)

mysql> select * from user_account;
+----+-------+---------+
| id | name  | balance |
+----+-------+---------+
|  1 | tudou |    2000 |
|  2 | bobo  |    2000 |
+----+-------+---------+
2 rows in set (0.00 sec)

mysql> select * from user_account;
+----+-------+---------+
| id | name  | balance |
+----+-------+---------+
|  1 | tudou |    2000 |
|  2 | bobo  |    2000 |
+----+-------+---------+
2 rows in set (0.00 sec)

mysql> update user_account set balance=50;
Query OK, 3 rows affected (0.00 sec)
Rows matched: 3  Changed: 3  Warnings: 0

mysql> select * from user_account;
+----+----------+---------+
| id | name     | balance |
+----+----------+---------+
|  1 | tudou    |      50 |
|  2 | bobo     |      50 |
|  3 | zhangsan |      50 |
+----+----------+---------+
```


图 6-19

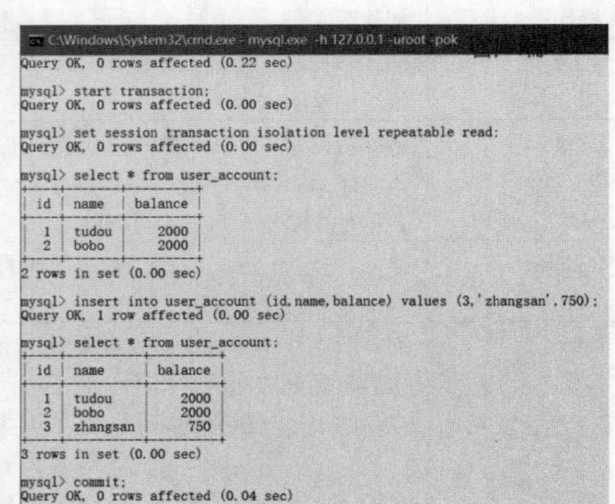
```
Query OK, 0 rows affected (0.22 sec)

mysql> start transaction;
Query OK, 0 rows affected (0.00 sec)

mysql> set session transaction isolation level repeatable read;
Query OK, 0 rows affected (0.00 sec)

mysql> select * from user_account;
+----+-------+---------+
| id | name  | balance |
+----+-------+---------+
|  1 | tudou |    2000 |
|  2 | bobo  |    2000 |
+----+-------+---------+
2 rows in set (0.00 sec)

mysql> insert into user_account (id,name,balance) values (3,'zhangsan',750);
Query OK, 1 row affected (0.00 sec)

mysql> select * from user_account;
+----+----------+---------+
| id | name     | balance |
+----+----------+---------+
|  1 | tudou    |    2000 |
|  2 | bobo     |    2000 |
|  3 | zhangsan |     750 |
+----+----------+---------+
3 rows in set (0.00 sec)

mysql> commit;
Query OK, 0 rows affected (0.04 sec)
```


图 6-20

至此，我们知道设置可重复读隔离级别和单纯的 MVCC 机制是无法解决幻读的。在项目中解决幻读通常有两种方式：

- 使用串行化读的隔离级别。
- MVCC+next-key locks：next-key locks 有 record locks（索引加锁）和 gap locks（间隙锁，每次锁住的不仅是需要使用的数据，还有这些数据附近的数据）。

实际上，很多项目是不会使用到上面两种方法的，串行化读的性能太差，而且很多时候我们是完全可以接受幻读的。

（5）测试隔离级别——串行化

串行化隔离级别是最严格的隔离级别。所谓串行化，指的是事务 A 和事务 B 需要按照开启事务的顺序执行，并且执行当前事务时会锁表，直到当前事务进行 commit 操作提交事务时才解锁，并开始下一个事务，因此不会出现脏读、不可重复读以及幻读的情况。由于这种隔离级别并发性极低，会导致数据库性能极低，因此在实际开发过程中很少会用到。

我们将通过以下案例进行串行化隔离级别的测试，具体步骤如下：

①开启数据库客户端 A 和 B，并设置客户端数据库隔离级别为串行化。
②客户端 A 开启事务，查询 user_account 中所有记录数。
③客户端 B 开启事务，向 user_account 表中插入一条新记录。
④客户端 A 提交事务。
⑤客户端 B 提交事务。

核心命令如下：

```
mysql.exe -h 127.0.0.1 -uroot -pok
show databases;
use transation_test;
select * from user_account;
update user_account set balance=300 where id=1;
insert into user_account (id,name,balance) values (4,'lisi',500);
```

测试结果如图 6-21、图 6-22 所示，客户端 B 的 insert 操作需要等待客户端 A 的事务执行完，表解锁之后才进行 insert 操作，事务 B 的执行等待了约 46 秒。

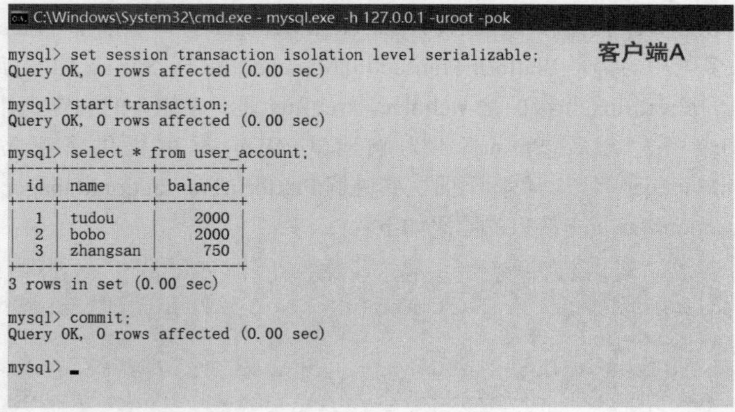

图 6-21

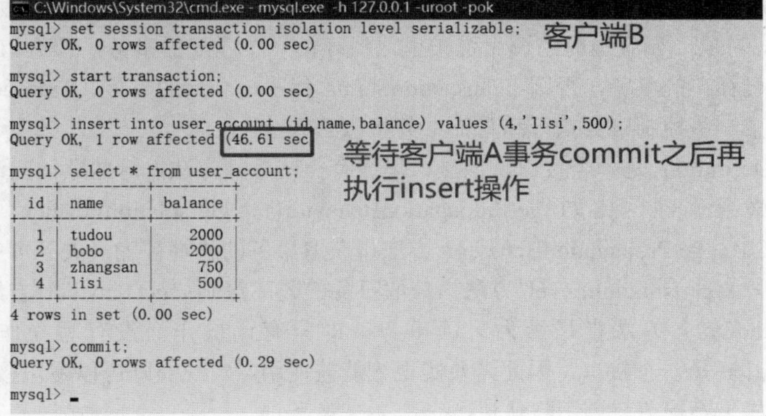

图 6-22

经过以上所有隔离级别的测试，我们了解了事务隔离级别的一系列特性。另外，我们对事务并发问题再做一些补充和总结：

- 事务隔离级别为“读已提交”时，写入数据时只会锁住对应的行。
- 事务隔离级别为“可重复读”时，如果检索条件有索引（包括主键索引），默认加锁的方式是 next-key 锁；如果检索条件没有索引，更新数据时会锁住整张表。一个间隙被事务加了锁，其他事务是不能在这个间隙插入记录的，这样可以防止幻读。
- 事务隔离级别为“串行化”时，读写数据都会锁住整张表。
- 事务隔离级别越高，越能保证数据的完整性和一致性，但是执行代价越高，并发性能越低。因此实际项目开发使用时需要综合考虑。

6.4.3 编程式事务和声明式事务

事务管理对于企业应用来说至关重要，当出现异常情况时，它可以保证数据的一致性。Spring 支持编程式事务管理和声明式事务管理两种方式。

首先我们来看看 Spring 框架的事物抽象。Spring 的事务策略由 TransactionManager 接口定义，PlatformTransactionManager 接口和 ReactiveTransactionManager 接口继承了 TransactionManager 接口。我们的程序大多数用的都是 PlatformTransactionManager 接口。Spring 5.0 之后引入了 Reactive Web 框架 webflux，与 webflux 平级的是 webmvc，webflux 是一个完全的响应式并且非阻塞的 Web 框架，因此 Spring 5.2 之后 Spring 还为响应式 Web 框架提供了事务管理抽象，即 ReactiveTransactionManager 接口。我们下面主要讲解 PlatformTransactionManager 事务管理抽象。

PlatformTransactionManager 接口的源码如下：

```
    public interface PlatformTransactionManager extends TransactionManager {
        TransactionStatus getTransaction(@Nullable TransactionDefinition var1)
throws TransactionException;
        void commit(TransactionStatus var1) throws TransactionException;
        void rollback(TransactionStatus var1) throws TransactionException;
    }
```

getTransaction 方法通过传入一个 TransactionDefinition 类型的参数来获取一个 TransactionStatus 对象，如果当前的调用堆栈里已经存在一个匹配的事务，那么 TransactionStatus 代表的就是这个已存在的事务，否则 TransactionStatus 代表一个新的事务。TransactionStatus 接口为事务代码提供了一些控制事务执行和查询事务状态的方法。

TransactionDefinition 是一个接口，该接口里面有一些默认方法，这些默认方法的返回值是声明一个事务的必要属性，比如 getPropagationBehavior()、getIsolationLevel()、getTimeout()、isReadOnly()。其中，getPropagationBehavior 方法指获取事务的传播行为（关于事务的传播行为下一节会专门讲解），getIsolationLevel 方法指获取事务的隔离级别；getTimeout 方法指获取事务的超时时间；isReadOnly 方法指是否为只读事务，即只有读操作而没有写操作的事务。由于 TransactionDefinition 是一个接口，因此需要实现才能被使用。实现类既可以覆盖接口的默认方法，也可以不覆盖，如果不覆盖就使用默认值。

commit 方法用于提交事务，rollback 方法用于回滚事务。

1. 声明式事务

声明式事务一般使用@Transactional 注解，并且使用@EnableTransactionManagement 开启事务管理，接下来我们讲解其背后的工作原理。

声明式事务是建立在 AOP 之上的，首先我们的应用程序会通过 XML 的方式或者注解的方式提供元数据，AOP 与事务元数据结合产生一个代理。当执行目标方法时拦截器 TransactionInterceptor 会对目标方法进行拦截，然后在目标方法的前后调用代理。其本质是在目标方法开始之前创建或者加入一个事务，在执行完目标方法之后根据执行情况提交或者回滚事务。

拦截器 TransactionInterceptor 通过检查方法返回类型来检测是哪种事务管理风格。如果返回的是响应式类型（例如 Publisher）的方法，就符合响应式事务管理的条件，其他返回类型（包括 void）则使用 PlatformTransactionManager。

@Transactional 注解是基于注解的声明式事务，当然基于 XMl 配置的也可以，但是要在 Spring Boot 应用中使用，所以一律使用基于注解的声明式事务。@Transactional 可以作用于接口定义、接口方法、类定义和类的公共方法上，如果作用于私有方法或者包可见的方法上，虽然不会引发错误，但是并不会激活事务的一些行为。另外，将@Transactional 作用于类上要比作用于接口上更好，下面说明一下原因。在 Spring AOP 框架中有两种模式，分别是基于代理和基于 AspectJ。基于代理又分为两种，一种是基于接口的，一种是基于类的。如果是基于类的代理或是基于 AspectJ，那么将@Transactional 作用于接口上不会起到任何作用。上面说到了事务是基于 AOP 的，那么如何让事务支持多种模式呢？ @EnableTransactionManagement 注解提供了相关的支持，@EnableTransactionManagement 注解的源码如下：

```
public @interface EnableTransactionManagement {
    boolean proxyTargetClass() default false;
    AdviceMode mode() default AdviceMode.PROXY;
    int order() default 2147483647;
}
```

事务管理默认模式为基于代理，即 mode=AdviceMode.PROXY，且默认的代理方式为基于接口的，因为 proxyTargetClass=false。

@Transactional 注解的属性如表 6-12 所示。

表 6-12　@Transactional 注解的属性

属性	类型	描述
value	String	指定事务管理器
propagation	enum: Propagation	事务传播行为
isolation	enum: Isolation	事务隔离级别
timeout	Int(单位为秒)	事务超时时间
readOnly	boolean	是否只读
rollbackFor	Class[]	引起回滚的异常类数组
rollbackForClassName	String[]	引起回滚的异常类名称数组
noRollbackFor	Class[]	不会引起回滚的异常类数组
noRollbackForClassName	String[]	不会引起回滚的异常类名称数组

关于声明式事务就先介绍这些，下面用代码的方式了解事务的行为。

以银行转账业务为例，现有四张表，分别为银行工作人员表、银行客户表、银行账户表和操作记录表。为了简便起见，客户表与账户表为一对一关系，即每位客户都只有一张银行卡。业务场景为 A 客户通过银行工作人员给 B 客户转账，如果转账成功，则账户表中对应 A 和 B 的两条记录要进行账户余额的修改，操作记录表中会新增两条记录，分别是 A 出账的记录和 B 入账的记录。因为转账业务必须是一个事务，要么都成功，要么都失败，所以两条 update 语句和两条 insert 语句必须被事务包围，如果发生异常，就必须回滚。

数据库采用 MySQL，表的结构如图 6-23 所示。

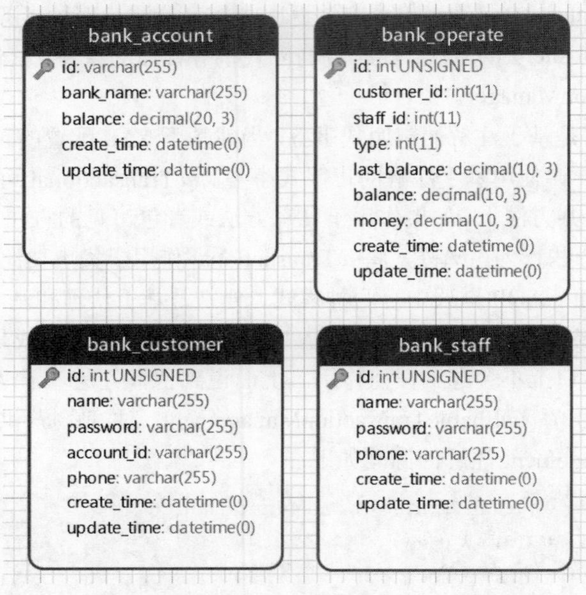

图 6-23

创建表的脚本如下：

```
-- 银行账户表
CREATE TABLE bank_account (
  id varchar(255) NOT NULL COMMENT '银行卡号',
  bank_name varchar(255) DEFAULT NULL COMMENT '银行名称',
  balance decimal(20,3) DEFAULT NULL COMMENT '账户余额',
  create_time datetime DEFAULT NULL COMMENT '创建时间',
  update_time datetime DEFAULT NULL COMMENT '修改时间',
  PRIMARY KEY (id)
);
-- 银行工作人员表
CREATE TABLE bank_staff (
  id int(10) unsigned NOT NULL AUTO_INCREMENT COMMENT '员工 ID',
  name varchar(255) DEFAULT NULL COMMENT '员工姓名',
  password varchar(255) DEFAULT NULL COMMENT '密码',
  phone varchar(255) DEFAULT NULL COMMENT '手机号',
  create_time datetime DEFAULT NULL COMMENT '创建时间',
  update_time datetime DEFAULT NULL COMMENT '修改时间',
  PRIMARY KEY (id)
```

```
);
-- 银行客户表
CREATE TABLE bank_customer (
  id int(10) unsigned NOT NULL AUTO_INCREMENT COMMENT '客户 ID',
  name varchar(255) DEFAULT NULL COMMENT '客户姓名',
  password varchar(255) DEFAULT NULL COMMENT '密码',
  account_id varchar(255) DEFAULT NULL COMMENT '银行卡号',
  phone varchar(255) DEFAULT NULL COMMENT '手机号',
  create_time datetime DEFAULT NULL COMMENT '创建时间',
  update_time datetime DEFAULT NULL COMMENT '修改时间',
  PRIMARY KEY (id)
);
-- 操作表
CREATE TABLE bank_operate (
  id int(10) unsigned NOT NULL AUTO_INCREMENT COMMENT '流水号',
  customer_id int(11) DEFAULT NULL COMMENT '客户 ID',
  staff_id int(11) DEFAULT NULL COMMENT '员工 ID',
  type int(11) DEFAULT NULL COMMENT '操作类型,0 表示入账,1 表示出账',
  last_balance decimal(10,3) DEFAULT NULL COMMENT '上次余额',
  balance decimal(10,3) DEFAULT NULL COMMENT '当前余额',
  money decimal(10,3) DEFAULT NULL COMMENT '交易金额,且(上次余额-当前余额)的绝对
值=交易金额',
  create_time datetime DEFAULT NULL COMMENT '创建时间',
  update_time datetime DEFAULT NULL COMMENT '修改时间',
  PRIMARY KEY (id)
);
```

现在往表里插入初始化数据，与前面说过的业务场景对应起来，分别是一条工作人员记录、两条客户记录、两条账户记录，操作记录表则不插入初始化数据。insert 脚本如下：

```
INSERT INTO bank_account(id,bank_name,balance,create_time,update_time) VALUES
('6217000011112222333','中国建设银行',10000.000,'2020-10-31 17:10:10','2020-10-31
17:10:17');
INSERT INTO bank_account(id,bank_name,balance,create_time,update_time) VALUES
('6217000011112222444','中国建设银行',20000.000,'2020-10-31 17:11:25','2020-10-31
17:11:27');
INSERT INTO bank_customer(id, name, password, account_id, phone, create_time,
update_time) VALUES (1, '张三', 'abcdef', '6217000011112222333', '13712345678',
'2020-10-31 17:13:55', '2020-10-31 17:13:58');
INSERT INTO bank_customer(id, name, password, account_id, phone, create_time,
update_time) VALUES (2, '李四', '123456', '6217000011112222444', '13712340000',
'2020-10-31 17:14:28', '2020-10-31 17:14:30');
INSERT INTO bank_staff(id,name,password,phone,create_time,update_time) VALUES
(1,'员工一号','admin','13700001111','2020-10-31 17:15:07','2020-10-31 17:15:10');
```

框架主要为 Spring Boot+MyBatis 组合，Spring Boot 集成 MyBatis 的配置前面已经学过，下面只展示一些关键代码。

示例代码 6-31 BankOperateController.java（核心代码）

```
@RestController
public class BankOperateController {
```

```
    @Autowired
    private BankOperateService bankOperateService;
    /**
     * @param staffId 工作人员 ID
     * @param fromCustomerId 转账客户 ID
     * @param toCustomerId 被转账客户 ID
     * @param money 转账金额
     */
    @GetMapping("/transfer")
    public void transfer(Integer staffId,Integer fromCustomerId,Integer
toCustomerId, BigDecimal money){
        bankOperateService.transfer(staffId,fromCustomerId,
toCustomerId,money);
    }
}
```

示例代码 6-32　BankOperateServiceImpl.java

```
@Service
public class BankOperateServiceImpl implements BankOperateService {
    @Autowired
    private BankMapper bankMapper;
    @Autowired
    private TransactionTemplate transactionTemplate;
    @Override
    @Transactional
    public boolean transfer(Integer staffId, Integer fromCustomerId, Integer
toCustomerId, BigDecimal money) {
        BankCustomer fromCustomer =
bankMapper.selectBankCustomerById(fromCustomerId);
        BankCustomer toCustomer =
bankMapper.selectBankCustomerById(toCustomerId);
        //第一步：给转账客户扣钱
        bankMapper.updateBankAccountBalance(fromCustomer.getBankAccount().
getId(),money.negate());
        //模拟交易过程中发生异常
        //int a = 1/0;
        //第二步：记录转账客户流水
        BankOperate fromOperate = new BankOperate();
        fromOperate.setCustomerId(fromCustomerId);
        fromOperate.setStaffId(staffId);
        fromOperate.setType(1);
        fromOperate.setLastBalance(fromCustomer.getBankAccount().
getBalance());
        fromOperate.setBalance(fromCustomer.getBankAccount().getBalance().
subtract(money));
        fromOperate.setMoney(money);
        Date date1 = new Date();
        fromOperate.setCreateTime(date1);
        fromOperate.setUpdateTime(date1);
        bankMapper.insertBankOperate(fromOperate);
```

```
        //第三步：给被转账客户加钱
        bankMapper.updateBankAccountBalance(toCustomer.getBankAccount().
getId(),money);
        //第四步：记录被转账客户流水
        BankOperate toOperate = new BankOperate();
        toOperate.setCustomerId(toCustomerId);
        toOperate.setStaffId(staffId);
        toOperate.setType(0);
        toOperate.setLastBalance(toCustomer.getBankAccount().getBalance());
        toOperate.setBalance(toCustomer.getBankAccount().getBalance().
add(money));
        toOperate.setMoney(money);
        Date date2 = new Date();
        toOperate.setCreateTime(date2);
        toOperate.setUpdateTime(date2);
        bankMapper.insertBankOperate(toOperate);
        return true;
    }
}
```

在 transfer 方法中，主要做了四件事：给转账客户扣钱、记录扣钱流水、给被转账客户加钱、记录加钱流水。这四步操作密不可分，必须在同一个事务内进行，因此在 transfer 方法上加了@Transactional 注解，通过声明式事务的方式来保障转账业务的安全性。

在第一步操作后，通过模拟交易过程中发送异常来检验@Transactional 注解是否生效。注意，只有 service 方法将异常抛出来，Spring 才会知道发生了异常，并且抛出来的异常必须是 RuntimeException 或者其子类（比如 java.lang. NullPointerException、java.lang.ArithmeticException 等）才会回滚事务；如果抛出的是其他异常，则不会回滚。

下面通过访问接口来模拟转账业务。转账前张三和李四的账户余额如图 6-24 所示。

id	bank_name	balance	create_time	update_time
6217000011112222333	中国建设银行	10000.000	2020-10-31 17:10:10	2020-10-31 17:10:17
6217000011112222444	中国建设银行	20000.000	2020-10-31 17:11:25	2020-10-31 17:11:27

图 6-24

启动项目后，访问接口 http://localhost:8080/transfer?staffId=1&fromCustomerId=1&toCustomerId=2&money=1000 模拟张三给李四转 1000 块钱。访问后张三和李四的账户余额如图 6-25 所示。

id	bank_name	balance	create_time	update_time
6217000011112222333	中国建设银行	10000.000	2020-10-31 17:10:10	2020-10-31 17:10:17
6217000011112222444	中国建设银行	20000.000	2020-10-31 17:11:25	2020-10-31 17:11:27

图 6-25

这是因为第一步操作成功了，但是紧接着发生了异常，所以将第一步给张三扣钱的 update 语句回滚了。

将@Transactional 注解注释掉，重启项目，再次访问一遍那个接口。访问后张三和李四的账户余额如图 6-26 所示。

id	bank_name	balance	create_time	update_time
6217000011112222333	中国建设银行	9000.000	2020-10-31 17:10:10	2020-10-31 17:10:17
6217000011112222444	中国建设银行	20000.000	2020-10-31 17:11:25	2020-10-31 17:11:27

图 6-26

张三的账户扣钱了，李四的账户却没有加钱。另外，bank_operate 表也没有新增流水记录。显然，张三的钱白白减少了 1000 块且没有任何记录。由此可见，没有事务管理的业务是非常危险的。

2. 编程式事务

Spring 提供了以下 3 种支持编程式事务的方式：

- 使用 TransactionTemplate。
- 使用 TransactionalOperator。
- 直接实现 TransactionManager。

下面主要介绍 TransactionTemplate。这种方式用于响应式 Web 应用中，是最常用的。

TransactionTemplate 与 Spring 提供的其他模板相似，如 JdbcTemplate。TransactionTemplate 使用回调方法来让应用程序代码从获取和释放事务性资源的样板操作中解放出来，也就意味着使用 TransactionTemplate 不用手动开启事务和提交事务。应用程序代码只用关注业务，其他的交给 TransactionTemplate 即可。

TransactionTemplate 使用 execute 方法传入一个 TransactionCallback 类型的参数来实现编程式事务。TransactionCallback 类型的参数通常可以通过匿名内部类来做，需要重写它的 doInTransaction 方法。将业务逻辑放到 doInTransaction 方法中（其中方法有一个 TransactionStatus 类型的参数），并将整个 doInTransaction 方法的代码块用 try...catch 包围，再在 catch 语句中调用 TransactionStatus 对应的 setRollbackOnly 方法来进行事务回滚。如果没有发生异常，事务就会正常提交。TransactionTemplate 的 execute 方法声明如下：

```
    public <T> T execute(TransactionCallback<T> action) throws
TransactionException;
```

TransactionTemplate 还提供了一些 API，用于设置事务的相关属性。

```
setTransactionManager(PlatformTransactionManager transactionManager);
setIsolationLevel(int isolationLevel);
setPropagationBehavior(int propagationBehavior);
setTimeout(int timeout);
setReadOnly(boolean readOnly);
```

下面对前面展示的由声明式事务所控制的转账方法进行改造。为了检验编程式事务是否有效，一定不要在 service 接口或实现类上使用@Transational 注解。

示例代码 6-33　BankOperateServiceImpl.java

```
@Service
public class BankOperateServiceImpl implements BankOperateService {
    @Autowired
    private BankMapper bankMapper;
    @Autowired
    private TransactionTemplate transactionTemplate;
```

```
        @Override
        public boolean transfer(Integer staffId, Integer fromCustomerId, Integer
toCustomerId, BigDecimal money) {
            BankCustomer fromCustomer =
bankMapper.selectBankCustomerById(fromCustomerId);
            BankCustomer toCustomer =
bankMapper.selectBankCustomerById(toCustomerId);
            //编程式事务
            return transactionTemplate.execute(new TransactionCallback<Boolean>() {
                @Override
                public Boolean doInTransaction(TransactionStatus transactionStatus) {
                    try {
                        //第一步：给转账客户扣钱
                        bankMapper.updateBankAccountBalance(fromCustomer.
getBankAccount().getId(),money.negate());
                        //模拟交易过程中发生异常
                        //int a = 1/0;
                        //第二步：记录转账客户流水
                        BankOperate fromOperate = new BankOperate();
                        fromOperate.setCustomerId(fromCustomerId);
                        fromOperate.setStaffId(staffId);
                        fromOperate.setType(1);
                        fromOperate.setLastBalance(fromCustomer.
getBankAccount().getBalance());
                        fromOperate.setBalance(fromCustomer.getBankAccount().
getBalance().subtract(money));
                        fromOperate.setMoney(money);
                        Date date1 = new Date();
                        fromOperate.setCreateTime(date1);
                        fromOperate.setUpdateTime(date1);
                        bankMapper.insertBankOperate(fromOperate);
                        //第三步：给被转账客户加钱
                        bankMapper.updateBankAccountBalance(toCustomer.
getBankAccount().getId(),money);
                        //第四步：记录被转账客户流水
                        BankOperate toOperate = new BankOperate();
                        toOperate.setCustomerId(toCustomerId);
                        toOperate.setStaffId(staffId);
                        toOperate.setType(0);
                        toOperate.setLastBalance(toCustomer.getBankAccount().
getBalance());
                        toOperate.setBalance(toCustomer.getBankAccount().
getBalance().add(money));
                        toOperate.setMoney(money);
                        Date date2 = new Date();
                        toOperate.setCreateTime(date2);
                        toOperate.setUpdateTime(date2);
                        bankMapper.insertBankOperate(toOperate);
                        return true;
                    } catch (RuntimeException e) {
```

```
                        // 遇到运行时异常就回滚
                        transactionStatus.setRollbackOnly();
                        return false;
                    }
                }
            });
        }
    }
```

下面通过访问接口来测试编程式事务是否生效。转账前张三和李四的账户余额如图 6-27 所示。

id	bank_name	balance	create_time	update_time
6217000011112222333	中国建设银行	10000.000	2020-10-31 17:10:10	2020-10-31 17:10:17
6217000011112222444	中国建设银行	20000.000	2020-10-31 17:11:25	2020-10-31 17:11:27

图 6-27

启动项目后，访问接口 http://localhost:8080/transfer?staffId=1&fromCustomerId=1&toCustomerId=2&money=1000 模拟张三给李四转 1000 块钱。访问后张三和李四的账户余额如图 6-28 所示。

id	bank_name	balance	create_time	update_time
6217000011112222333	中国建设银行	10000.000	2020-10-31 17:10:10	2020-10-31 17:10:17
6217000011112222444	中国建设银行	20000.000	2020-10-31 17:11:25	2020-10-31 17:11:27

图 6-28

由此可见，虽然声明式事务和编程式事务的事务管理实现方式略有不同，但是其本质是没有区别的，最终呈现出来的效果也是相同的。

3. 声明式事务和编程式事务对比

声明式事务最大的优点就是不需要通过编程的方式管理事务，不需要在业务逻辑代码中掺杂事务管理的代码，只需在配置文件中做相关的事务规则声明（或通过基于@Transactional 注解的方式）便可以将事务规则应用到业务逻辑中。

声明式事务管理优于编程式事务管理，是 Spring 倡导的非侵入式的开发方式。声明式事务管理使业务代码不受污染，一个普通的 POJO 对象只要加上注解，就可以获得完全的事务支持。和编程式事务相比，声明式事务唯一不足的是，它的最细粒度只能作用到方法级别，无法像编程式事务那样作用到代码块级别。即便如此，也存在很多变通的方法，比如将需要进行事务管理的代码块独立为方法等。

声明式事务管理也有两种常用的方式，一种是基于 tx 和 aop 名字空间的 XML 配置文件，另一种是基于@Transactional 注解。基于注解的方式更简单易用、更清爽。

6.4.4 Spring 事务的传播行为

事务是逻辑处理原子性的保证手段，通过使用事务控制可以极大地避免出现逻辑处理失败导致的脏数据等问题。事务最重要的两个特性就是事务隔离级别和事务传播类型。在前面讲解事务隔离级别时，我们知道了事务隔离级别是对数据库读写层面的控制范围，而事务的传播性是对事务层面的范围控制。

Spring 对事务的控制是使用 aop 切面实现的，我们不用关心事务的开始、提交、回滚，只需要加上开启事务的注解@Transcation 即可。不过这种便利会导致一些事务嵌套方面的问题，比如 Service1 和 Service2 均开启了事务，当在 Service1 中调用 Service2 的方法时，如果 Service2 的方法出现了异常，那么只回滚 Service2 事务还是将 Service1 和 Service2 一起回滚呢？

Spring 事务是在数据库事务的基础上进行封装和扩展的，主要特性如下：

①支持原有事务隔离级别。

②加入了事务传播的概念，提供多个事务合并和隔离的功能。

③提供声明式事务，使业务代码与事务分离，事务更易使用。

Spring 提供了以下 3 个事务接口：

①TransactionDefinition：事务定义。

②PlatformTransactionManager：事务管理。

③TransactionStatus：事务运行状态。

Spring 事务传播行为主要有 7 种类型，如表 6-13 所示。

表 6-13　Spring 事务传播行为的类型及说明

事务传播行为类型	说明
Propagation.REQUIRED	如果当前存在事务，则加入该事务；如果当前不存在事务，则创建一个新的事务（默认的事务传播行为）
Propagation.SUPPORTS	如果当前存在事务，则加入该事务；如果当前不存在事务，则以非事务的方式继续运行
Propagation.MANDATORY	如果当前存在事务，则加入该事务；如果当前不存在事务，则抛出异常
Propagation.REQUIRES_NEW	重新创建一个新的事务，如果当前存在事务，则延缓当前的事务
Propagation.NOT_SUPPORTED	以非事务的方式运行，如果当前存在事务，则暂停当前的事务
Propagation.NEVER	以非事务的方式运行，如果当前存在事务，则抛出异常
Propagation.NESTED	如果没有事务，就新建一个事务；如果有事务，就在当前事务中嵌套其他事务

下面对事务传播类型 Propagation.REQUIRED 进行详细测试。

（1）准备工作

①创建数据库表 user，字段包含 id、username、password。

②通过 MyBatis-Plus 代码生成器生成 User 对应的 UserController.java、IUserService.java、UserServiceImpl.java、UserMapper.java、UserMapper.xml。代码生成方式查看 6.3.3 小节，此处不再赘述。

③手动生成 IUserService2.java、UserServiceImpl2.java。

核心命令及密码说明如下。

创建数据库表 user，代码如下：

```
CREATE TABLE user  (
  id int(11) NOT NULL AUTO_INCREMENT,
```

```
    username varchar(255) CHARACTER DEFAULT NULL,
    password varchar(255) CHARACTER DEFAULT NULL,
    PRIMARY KEY (`id`)
)
```

示例代码 6-34　IuserService2.java

```
public interface IuserService2 extends Iservice<User> {
    void insertByRequired(User user);
    void insertByRequiredForException(User user);
}
```

示例代码 6-35　UserServiceImpl2.java

```
@Service
public class UserServiceImpl2 extends ServiceImpl<UserMapper, User> implements
IuserService2 {
    @Autowired
    private UserMapper userMapper;
    @Override
    @Transactional(propagation = Propagation.REQUIRED)
    public void insertByRequired(User user) {
        userMapper.insert(user);
    }
    @Override
    @Transactional(propagation = Propagation.REQUIRED)
    public void insertByRequiredForException(User user) {
        userMapper.insert(user);
        throw new RuntimeException("新用户插入异常!");
    }
}
```

示例代码 6-36　IuserService.java

```
public interface IuserService extends Iservice<User> {
    void insertByRequired(User user);
}
```

示例代码 6-37　UserServiceImpl.java

```
@Service
public class UserServiceImpl extends ServiceImpl<UserMapper, User> implements
IuserService {
    @Autowired
    private UserMapper userMapper;
    @Autowired
    private IuserService2 userService2;
    @Override
    @Transactional(propagation = Propagation.REQUIRED)
    public void insertByRequired(User user) {
        userMapper.insert(user);
    }
}
```

（2）测试案例

①场景一：外围方法并未开启事务。

外围方法抛出异常：

示例代码 6-38　UserServiceImpl.java（部分代码）

```
@Override
public void testRequiredForNoTransaction() {
    User user1 = new User();
    user1.setUsername("Tom");
    insertByRequired(user1);
    User user2 = new User();
    user2.setUsername("Lili");
    userService2.insertByRequired(user2);
    throw new RuntimeException("测试 required 异常！");
}
```

内部事务抛出异常：

示例代码 6-39　UserServiceImpl.java（部分代码）

```
@Override
public void testRequiredForNoTransactionException() {
    User user1 = new User();
    user1.setUsername("Tom");
    insertByRequired(user1);
    User user2 = new User();
    user2.setUsername("Lili");
    userService2.insertByRequiredForException(user2);
}
```

②场景二：外围方法开启事务。

外围方法抛出异常：

示例代码 6-40　UserServiceImpl.java（部分代码）

```
@Override
@Transactional(propagation = Propagation.REQUIRED)
public void testRequiredForTransaction() {
    User user1 = new User();
    user1.setUsername("Tom");
    insertByRequired(user1);
    User user2 = new User();
    user2.setUsername("Lili");
    userService2.insertByRequired(user2);
    throw new RuntimeException("测试 required 异常！");
}
```

内部事务抛出异常：

示例代码 6-41　UserServiceImpl.java（部分代码）

```
@Override
```

```
@Transactional(propagation = Propagation.REQUIRED)
public void testRequiredForTransactionException() {
    User user1 = new User();
    user1.setUsername("Tom");
    insertByRequired(user1);ohk[/'pyi0u;lj
    User user2 = new User();
    user2.setUsername("Lili");
    userService2.insertByRequiredForException(user2);
}
```

内部事务抛出异常被捕获：

示例代码 6-42　UserServiceImpl.java（部分代码）

```
@Override
@Transactional(propagation = Propagation.REQUIRED)
public void testRequiredForTransactionCatchException() {
    User user1 = new User();
    user1.setUsername("Tom");
    insertByRequired(user1);
    User user2 = new User();
    user2.setUsername("Lili");
    try {
        userService2.insertByRequiredForException(user2);
    }catch (Exception e){
        System.out.println("userService2 中事务异常！");
    }
}
```

Controller 层测试接口代码如下：

示例代码 6-43　UserController.java

```
@RestController
@RequestMapping("/user")
public class UserController {
    @Autowired
    private IUserService userService;
    /**
     * 测试 1：外部方法不开启事务，外围方法抛出异常
     */
    @GetMapping("/testRequiredForNoTransaction")
    public void testRequiredForNoTransaction(){
        userService.testRequiredForNoTransaction();
    }
    /**
     * 测试 2：外部方法不开启事务，内部事务抛出异常
     */
    @GetMapping("/testRequiredForNoTransactionException")
    public void testRequiredForNoTransactionException(){
        userService.testRequiredForNoTransactionException();
    }
    /**
```

```
     * 测试 3：外部方法开启事务，外围方法抛出异常
     */
    @GetMapping("/testRequiredForTransaction")
    public void testRequiredForTransaction(){
        userService.testRequiredForTransaction();
    }
    /**
     * 测试 4：外部方法开启事务，内部事务抛出异常
     */
    @GetMapping("/testRequiredForTransactionException")
    public void testRequiredForTransactionException(){
        userService.testRequiredForTransactionException();
    }
    /**
     * 测试 5：外部方法开启事务，内部事务抛出异常但是被捕获
     */
    @GetMapping("/testRequiredForTransactionCatchException")
    public void testRequiredForTransactionCatchException(){
        userService.testRequiredForTransactionCatchException();
    }
}
```

测试结果如表 6-14 所示。

表 6-14 测试结果

测试实例编号	测试结果	外围方法是否开启事务	说明
1	用户“Tom”“Lili”均成功插入数据库	否	外围方法 testRequiredForNoTransaction()未开启事务，插入“Tom”“Lili”的方法在自己的事务中独立运行，外围方法的异常并不影响其内部事务
2	用户“Tom”成功插入，“Lili”插入失败	否	外围方法 testRequiredForNoTransactionException ()未开启事务，插入“Tom”“Lili”的方法在自己的事务中独立运行，所以插入“Lili”方法抛出异常时，只会回滚自己本身的事务，并不影响数据库插入“Tom”的事务
3	用户“Tom”“Lili”均未插入数据库	是	外围方法 testRequiredForTransaction()开启事务，内部方法事务加入外围方法的事务，所以当外围方法发生异常回滚时，内部事务也需要回滚
4	用户“Tom”“Lili”均未插入数据库	是	外围方法 testRequiredForTransactionException()开启事务，内部方法事务加入外围方法的事务，所以内部事务发生异常回滚时外围事务感知到异常，导致整体事务全部回滚
5	用户“Tom”“Lili”均未插入数据库	是	外围方法 testRequiredForTransactionCatchException()开启事务，内部方法事务加入外围方法的事务，所以内部事务发生异常回滚时，即使异常被 catch 到不被外围事务感知，但是整体事务依然全部回滚

通过以上测试，对于 Spring 事务传播类型 Propagation.REQUIRED 我们可以得出以下两点结论：

- 当外围方法未开启事务并且其内部事务传播类型设置为 Propagation.REQUIRED 时，内部事务都会开启自身的事务，并且事务之间相互独立、互不干扰。
- 外围方法开启事务、事务传播类型设置为 Propagation.REQUIRED 并且其内部事务也设置为 Propagation.REQUIRED 时，所有 Propagation.REQUIRED 修饰的内部事务和外围事务属于同一事务，只要有一个方法回滚，整个事务均回滚。

另外，Spring 事务传播类型可以根据上述测试案例自行测试，此处不再赘述。

本小节主要从事务的特性、并发问题、编程式事务和声明式事务、Spring 事务的传播行为四个方面讲解事务。在企业级应用程序开发中，事务管理是必不可少的技术，主要用来确保数据的完整性和一致性。Spring 事务管理解决了全局事务和本地事务的缺陷，允许应用开发者在任何环境下使用一致的编程模型。Spring 同时支持编程式事务和声明式事务，使应用程序可以运行在任何具体的底层事务基础之上。

第7章 Spring Boot高并发

本章将讲解Spring Boot高并发处理，首先引入高并发中常用的缓存技术和消息队列技术，然后通过模拟两个高并发场景来达到Spring Boot应用处理高并发实战的目的。

7.1 Spring Boot缓存技术

7.1.1 Spring缓存抽象简介

这里要讲的是Spring的缓存抽象，而不是Spring Boot。缓存抽象的底层核心是Spring框架的一部分，如何将缓存技术封装起来并自动化配置才是Spring Boot框架要做的事。

从Spring 3.1版本开始，Spring支持向一个已存在的Spring应用在不影响其他功能的前提下透明地添加缓存，和Spring对于事务的支持相似，Spring的缓存抽象允许各种各样的缓存解决方案，并且对代码的影响最小化。

从Spring 4.1版本开始，在JSR-107注解的支持下，缓存抽象得到了显著扩展并且有更多的自定义选项。

（1）缓存与缓冲区

术语“缓冲”和“缓存”往往可以互换使用。但是请注意，它们代表不同的事物。

传统上，缓冲区用作快速实体和慢速实体之间数据的中间临时存储。由于快的一方必须等待慢的一方，势必会影响性能，缓冲区通过一次性移动整个数据块而不是小块数据来缓解这种情况。

根据定义，缓存是隐藏的，任何一方都不知道发生了缓存。缓存同样可以提高性能，但是是通过让相同的数据以快速方式多次读取来实现的。

（2）缓存抽象

缓存抽象的核心是将缓存应用于 Java 方法。一种通用的场景是，在每次调用方法时，如果缓存已存在，则返回缓存的结果，而不必调用实际的方法；如果缓存不存在，则执行方法的逻辑，并将方法的返回值缓存起来，以便下次调用该方法时直接返回。

与 Spring 中的其他服务一样，缓存服务是一种抽象，不是缓存实现，所以需要使用实际的存储器来存储缓存数据。也就是说，抽象不必编写缓存逻辑，但是也不提供实际的数据存储器。org.springframework.cache.Cache 和 org.springframework.cache.CacheManager 两个接口代表了 Spring 的缓存抽象。

Cache 是一个缓存单位，每个 Cache 对象都有一个 name 属性，也可以称为 cacheName，用于唯一标识一个 Cache。每个 Cache 内部维护了 n 个 key-value 对，因为缓存的本质就是 key-value 对，这些 key-value 对就是缓存的实际数据。

CacheManager 意为缓存管理器，用来管理多个 Cache，其内部维护了一个 Map 结构，用于管理多个 Cache 对象。Map 的 key 存储的是 Cache 对象的 name 属性，Map 的 value 存储的是 Cache 对象本身。

Spring 为了简化使用，顺便还提供了 CacheManager 接口的一些实现：

①基于 JDK ConcurrentMap 的缓存 SimpleCacheManager。

②基于 Ehcache 2.x 的缓存 EhCacheCacheManager。

③基于 Caffeine 的缓存 CaffeineCacheManager。

④基于 Gemfire 的缓存。

⑤基于 JSR-107 的缓存，例如 Ehcache 3.x。

也就是说，这些实现可以直接拿来用，而不用每次都自定义缓存实现。不过这些实现可以不用一一了解，因为我们重点介绍的是缓存抽象。

如果想要自定义缓存实现，需要实现 Cache 和 CacheManager 接口。如果要将缓存抽象应用于项目中，总的来说需要做两件事情：

①缓存声明：确定需要缓存的方法及其策略。

②缓存配置：数据存储和读取的缓存存储器。

Spring 的缓存抽象是基于注解的，下面介绍一下有哪些可用的注解。

- @Cacheable：触发缓存一个实体。
- @CacheEvict：清除缓存。
- @CachePut：在不影响方法执行的情况下更新缓存。
- @Caching：组合多个注解一起使用。
- @CacheConfig：在类级别共享一些常见的缓存相关设置。
- @EnableCaching：开启缓存。

下面重点介绍一下常用的@Cacheable、@CachePut 和@CacheEvict 注解。

@Cacheable 注解是最重要的一个注解，作用于方法上，缓存方法的返回值。当调用使用了@Cacheable 注解的某个方法时，首先会判断缓存数据是否存储，如果存在，则不执行方法本体，直接返回缓存数据；如果不存在，则执行方法本体，并且将方法的缓存值缓存起来。@Cacheable 的常见属性如下：

- value：cacheNames 的 string 数组。
- cacheNames：与 value 属性互为别名，是一回事。value 和 cacheNames 不要同时使用。
- key：定义缓存数据的键值。key 属性的值可以包含 spel 表达式，与缓存相关的 spel 元数据如表 7-1 所示。
- keyGenerator：自定义 keyGenerator，用于生成 Key。key 属性和 keyGenerator 不要同时使用。

表 7-1　与缓存相关的 spel 元数据

属性名	定位	说明	示例
methodName	Root object	方法名	#root.methodName
method	Root object	方法	#root.method.name
target	Root object	目标对象	#root.target
targetClass	Root object	目标类	#root.targetClass
args	Root object	参数，数据类型	#root.args[0]
caches	Root object	当前方法所依据的缓存的集合	#root.caches[0].name
argument name	Evaluation context	参数名称，如果不知道参数名称，则可以利用索引	#p0
result	Evaluation context	方法返回值	#result

@CachePut 注解也作用于方法上，可以触发缓存，和@Cacheable 的作用几乎一模一样，和@Cacheable 注解的区别是： @CachePut 总是会执行方法本体，并且更新缓存。

@CacheEvict 注解用于清除一个缓存。@CacheEvict 的 allEntries 如果设为 true，则清除的是该 cacheName 包含所有的缓存对；如果设为 false，则只清除当前 key 对应的缓存对。当前 key 可以通过@CacheEvict 注解的 key 或 keyGenerator 属性定义。

7.1.2　Ehcache 缓存实战

在上一节简单介绍了 Spring 缓存抽象的基本概念和一些常用注解，本节将以 EhCache 缓存框架为缓存存储器，完成一个项目中使用缓存的案例。

EhCache 是一个纯 Java 的进程内缓存框架，效率高，主要用于单体应用的缓存，对分布式支持不够好。一般在第三方库中用到的比较多，比如 Hibernate 的默认缓存就是 EhCache。

首先新建一个 Spring Boot 项目，在 pom.xml 中引入 MyBatis、EhCache 等相关依赖，如下代码所示。

示例代码 7-1　pom.xml 部分代码

```
<dependency>
    <groupId>org.springframework.boot</groupId>
    <artifactId>spring-boot-starter-web</artifactId>
```

```
    </dependency>
    <dependency>
        <groupId>org.mybatis.spring.boot</groupId>
        <artifactId>mybatis-spring-boot-starter</artifactId>
        <version>2.1.3</version>
    </dependency>
    <dependency>
        <groupId>mysql</groupId>
        <artifactId>mysql-connector-java</artifactId>
        <scope>runtime</scope>
    </dependency>
    <dependency>
        <groupId>org.springframework.boot</groupId>
        <artifactId>spring-boot-starter-cache</artifactId>
    </dependency>
    <dependency>
        <groupId>net.sf.ehcache</groupId>
        <artifactId>ehcache</artifactId>
        <version>2.10.4</version>
    </dependency>
```

关于 MyBatis 和数据源的配置这里不再赘述，下面主要展示一下 EhCache 缓存的配置和业务逻辑代码。

在 src/main/resources 目录下新建一个 mycache.xml，该 XML 文件是 EhCache 框架的核心配置文件，主要内容如下。

示例代码 7-2 mycache.xml

```
    <?xml version="1.0" encoding="UTF-8"?>
    <ehcache xmlns:xsi="http://www.w3.org/2001/XMLSchema-instance"
            xsi:noNamespaceSchemaLocation="http://ehcache.org/ehcache.xsd"
            updateCheck="false">
        <diskStore path="F:/test/"/>
        <defaultCache eternal="false" maxElementsInMemory="1000"
timeToIdleSeconds="0" timeToLiveSeconds="600"
                    memoryStoreEvictionPolicy="LRU"/>
        <cache
            name="allbooks"
            eternal="false"
            maxElementsInMemory="200"
            timeToIdleSeconds="0"
            timeToLiveSeconds="300"
            memoryStoreEvictionPolicy="LRU"/>
    </ehcache>
```

其中，diskStore 元素用于指定一个文件目录，当 EhCache 把数据写到硬盘上时，将把数据写到这个文件目录下。

defaultCache 元素用于指定默认缓存策略，当 EhCache 找不到定义的缓存时，就使用默认缓存策略。

cache 元素用来定义要缓存的实体，比较常用的属性如表 7-2 所示。

表 7-2 cache 元素常用的属性

属性	说明
name	定义缓存名称
eternal	true 表示对象永不过期，此时会忽略 timeToIdleSeconds 和 timeToLiveSeconds 属性，默认为 false
timeToIdleSeconds	设定允许对象处于空闲状态的最长时间，以秒为单位。当对象自从最近一次被访问后，如果处于空闲状态的时间超过了 timeToIdleSeconds 属性值，这个对象就会过期，EhCache 将把它从缓存中清空。只有当 eternal 属性为 false 时该属性才有效。如果该属性值为 0，则表示对象可以无限期地处于空闲状态
timeToLiveSeconds	设定对象允许存在于缓存中的最长时间，以秒为单位。当对象自从被存放到缓存中后，如果处于缓存中的时间超过了 timeToLiveSeconds 属性值，这个对象就会过期，EhCache 将把它从缓存中清除。只有当 eternal 属性为 false 时该属性才有效。如果该属性值为 0，则表示对象可以无限期地存在于缓存中。timeToLiveSeconds 必须大于 timeToIdleSeconds 属性才有意义
maxElementsInMemory	内存中最大缓存对象数。maxElementsInMemory 界限后，会把溢出的对象写入硬盘缓存中。注意：如果缓存的对象要写入硬盘中，那么该对象必须实现了 Serializable 接口
memoryStoreEvictionPolicy	当达到 maxElementsInMemory 限制时，EhCache 将会根据指定的策略去清理内存，可选策略有 LRU（最近最少使用，默认策略）、FIFO（先进先出）、LFU（最少访问次数）。
maxElementsOnDisk	硬盘中最大缓存对象数，若是 0 则表示无穷大
overflowToDisk	当系统宕机时是否保存到磁盘
diskPersistent	是否缓存虚拟机重启期数据，是否持久化磁盘缓存，当这个属性的值为 true 时，系统在初始化时会在磁盘中查找文件名为 cache、后缀名为 index 的文件，这个文件中存放了已经持久化在磁盘中的 cache 的 index，找到后会把 cache 加载到内存，要想把 cache 真正持久化到磁盘，写程序时注意执行 net.sf.ehcache.Cache.put(Element element)后调用 flush()方法
diskSpoolBufferSizeMB	这个参数设置 DiskStore（磁盘缓存）的缓存区大小，默认是 30MB。每个 Cache 都应该有自己的一个缓冲区
diskExpiryThreadIntervalSeconds	磁盘失效线程运行时间间隔，默认为 120 秒
clearOnFlush	内存数量最大时是否清除

前面说过要使用 Spring 的缓存抽象就必须提供一个 CacheManager，下面配置一下 EhCache 框架的实现 EhCacheCacheManager。

示例代码 7-3 EhCacheCacheManagerConfig.java

```
public class EhCacheCacheManagerConfig {
    @Bean
    public EhCacheCacheManager ehCacheCacheManager(EhCacheManagerFactoryBean
bean) {
        return new EhCacheCacheManager(bean.getObject());
    }
    @Bean
```

```
    public EhCacheManagerFactoryBean ehCacheManagerFactoryBean() {
        EhCacheManagerFactoryBean cacheManagerFactoryBean = new
EhCacheManagerFactoryBean();
        cacheManagerFactoryBean.setConfigLocation(new
ClassPathResource("mycache.xml"));
        cacheManagerFactoryBean.setShared(true);
        return cacheManagerFactoryBean;
    }
}
```

然后定义业务实体类 Book。

示例代码 7-4　Book.java

```
public class Book implements Serializable {
    private static final long serialVersionUID = -5095285519170350640L;
    private Long id;
    private String name;
    private BigDecimal price;
    private Integer amount;
    public Long getId() {
        return id;
    }
    public void setId(Long id) {
        this.id = id;
    }
    public String getName() {
        return name;
    }
    public void setName(String name) {
        this.name = name;
    }
    public BigDecimal getPrice() {
        return price;
    }
    public void setPrice(BigDecimal price) {
        this.price = price;
    }
    public Integer getAmount() {
        return amount;
    }
    public void setAmount(Integer amount) {
        this.amount = amount;
    }
}
```

BookServiceImpl 定义两个接口，分别是查询所有 Book 的接口和更新一个 Book 的接口，详细代码如下所示。

示例代码 7-5　BookServiceImpl.java

```
@Service
```

```
public class BookServiceImpl implements BookService {
    @Autowired
    private BookMapper bookMapper;
    @Override
    @CacheEvict(cacheNames = {"allbooks"},allEntries = true)
    public Book updateBook(Book book) {
        bookMapper.updateBook(book);
        return book;
    }
    @Override
    @Cacheable(cacheNames = {"allbooks"},key =
"#root.targetClass+'.listBooks'")
    public List<Book> listBooks() {
        return bookMapper.selectBooks(null);
    }
}
```

listBooks 方法查询所有的 Book。updateBook 方法用于更新一个 Book，所以必须清除缓存来保证数据库与缓存数据的一致性。

Controller 类的详细代码如下所示。

示例代码 7-6　BookController.java

```
@RestController
public class BookController {
    @Autowired
    private BookService bookService;
    @GetMapping("/listall")
    public List<Book> listBooks(){
        return bookService.listBooks();
    }
    @PostMapping("/update")
    public String updateBook(Book book){
        bookService.updateBook(book);
        return "更新成功";
    }
}
```

注意，程序主类必须用@EnableCaching 注解才能使缓存生效。启动 Spring Boot 程序，验证一下这两个接口是否生效。连续访问两次 http://localhost:8080/listall 接口，DEBUG 日志如图 7-1 所示。

从图 7-1 中可以看出来，第一次访问该接口时还没有缓存，所以是数据库查询；第二次查询时缓存已经存在，所以直接返回缓存数据，并没有打印 SQL 语句，没有用数据库查询。

下面用接口测试工具（如 Postman）访问一次更新 Book 接口，接口 URL 可以是 http://localhost:8080/update?id=1&price=200，意思就是将 id 为 1 的 Book 的 price 更新为 200 元，注意是 post 请求；update 接口请求成功后，再请求一次 listall 接口，DEBUG 日志如图 7-2 所示。

```
o.s.web.servlet.DispatcherServlet              : GET "/listall", parameters={}
s.w.s.m.m.a.RequestMappingHandlerMapping       : Mapped to com.bobo.group.springbootmybatisbase.book.
com.zaxxer.hikari.HikariDataSource             : HikariPool-1 - Starting...
com.zaxxer.hikari.HikariDataSource             : HikariPool-1 - Start completed.
c.b.g.s.b.mapper.BookMapper.selectBooks        : ==>  Preparing: select * from book
c.b.g.s.b.mapper.BookMapper.selectBooks        : ==> Parameters:
c.b.g.s.b.mapper.BookMapper.selectBooks        : <==      Total: 14
m.m.a.RequestResponseBodyMethodProcessor       : Using 'application/json;q=0.8', given [text/html, ap
m.m.a.RequestResponseBodyMethodProcessor       : Writing [[com.bobo.group.springbootmybatisbase.book.
o.s.web.servlet.DispatcherServlet              : Completed 200 OK

o.s.web.servlet.DispatcherServlet              : GET "/listall", parameters={}
s.w.s.m.m.a.RequestMappingHandlerMapping       : Mapped to com.bobo.group.springbootmybatisbase.book.
m.m.a.RequestResponseBodyMethodProcessor       : Using 'application/json;q=0.8', given [text/html, ap
m.m.a.RequestResponseBodyMethodProcessor       : Writing [[com.bobo.group.springbootmybatisbase.book.
o.s.web.servlet.DispatcherServlet              : Completed 200 OK
```

图 7-1

```
o.s.web.servlet.DispatcherServlet              : POST "/update", parameters={masked}
s.w.s.m.m.a.RequestMappingHandlerMapping       : Mapped to com.bobo.group.springbootmybatisbase.book.controller.Bo
c.b.g.s.b.mapper.BookMapper.updateBook         : ==>  Preparing: update book SET price=? where id=?
c.b.g.s.b.mapper.BookMapper.updateBook         : ==> Parameters: 200(BigDecimal), 1(Long)
c.b.g.s.b.mapper.BookMapper.updateBook         : <==    Updates: 1
m.m.a.RequestResponseBodyMethodProcessor       : Using 'text/plain', given [*/*] and supported [text/plain, */*,
m.m.a.RequestResponseBodyMethodProcessor       : Writing ["更新成功"]
o.s.web.servlet.DispatcherServlet              : Completed 200 OK

o.s.web.servlet.DispatcherServlet              : GET "/listall", parameters={}
s.w.s.m.m.a.RequestMappingHandlerMapping       : Mapped to com.bobo.group.springbootmybatisbase.book.controller.Bo
c.b.g.s.b.mapper.BookMapper.selectBooks        : ==>  Preparing: select * from book
c.b.g.s.b.mapper.BookMapper.selectBooks        : ==> Parameters:
c.b.g.s.b.mapper.BookMapper.selectBooks        : <==      Total: 14
m.m.a.RequestResponseBodyMethodProcessor       : Using 'application/json;q=0.8', given [text/html, application/xht
m.m.a.RequestResponseBodyMethodProcessor       : Writing [[com.bobo.group.springbootmybatisbase.book.entity.Book@
o.s.web.servlet.DispatcherServlet              : Completed 200 OK
```

图 7-2

从图 7-2 可以看出，调用了更新方法后，为了保证缓存一致性，清空了缓存，接着调用查询接口时使用数据库查询。

小　结

本节首先讲述了 Spring 框架如何通过 Cache 和 CacheManager 接口定义 Spring 统一的缓存抽象层，紧接着介绍了各个缓存注解的作用及使用方式，最后通过 Spring Boot 集成第三方缓存框架 EhCache 认识了在实际项目中该如何使用 Spring 的缓存抽象。

7.2 分布式缓存 Redis

7.2.1 Redis 简介

Redis 是一个开源的高性能非关系型数据库，更确切一点来说它是一个基于键值对的缓存系统。Redis 基于内存存储，并提供多种持久化策略，保证在宕机情况下能根据磁盘进行数据恢复；

提供了多种丰富的数据类型，足够应对不同场景的需求；同时还是一个单线程的服务端程序，内部采用 IO 多路复用机制，极大地提高了 IO 的利用率，单线程避免了频繁的线程上下文切换，这使得基于内存的单线程的 Redis 在性能上不弱于大多数的多线程服务端程序。另外，Redis 还支持主从复制、哨兵、集群等多种高可用技术，避免了单点故障。由于现代计算机内存比磁盘昂贵得多，因此一台计算机的内存往往是很有限的，Redis 集群技术解决了单个 Redis 的内存容易成为瓶颈的问题。

为什么要用 Redis 呢？

Redis 与一些进程内缓存（比如 EhCache）不同，它是一个独立的进程，也就是我们常说的分布式缓存。分布式缓存易于水平扩展，而传统的进程内缓存比较困难。Redis 采用 socket 协议与客户端进行通信。

另外，与同样是分布式缓存的 Memecache 相比，Memecache 不支持持久化，只支持 string 类型，而 Redis 则支持 string、list、set、zset、hash 五种数据类型，Redis 的 Value 值最大可以达到 512MB，Memcache 只有可怜的 1MB，另外，Redis 的速度比 Memecache 快很多。

在一些互联网应用中，特别是淘宝京东这样的电商平台或者是 12306 售票系统等，经常会有一些高并发、秒杀场景，当瞬时流量突然增大时，如果让大流量直接到达数据库服务器，就会导致数据库服务器不堪重负，比如 CPU 使用率百分之百，用户执行一次普通查询速度异常缓慢，严重时会造成数据库服务器直接宕机从而导致整个服务不可用。这时架构师通常会有针对性地对系统架构采取一些优化措施，而分布式缓存就是其中之一。通过将热点数据加载到 Redis 中，流量首先到达 Redis 而不是数据库服务器，这样就能有效减轻数据库服务器的负担。

7.2.2 Redis 安装及基本命令

本小节介绍在 Linux 上如何安装 Redis，首先下载、解压、编译 Redis：

```
$ wget http://download.redis.io/releases/redis-6.0.6.tar.gz
$ tar xzf redis-6.0.6.tar.gz
$ cd redis-6.0.6
$ make
```

如果执行 make 命令后有报错，一般是环境问题，可采用如下解决方案。

（1）安装 gcc 套装：

```
yum install cpp
yum install binutils
yum install glibc
yum install glibc-kernheaders
yum install glibc-common
yum install glibc-devel
yum install gcc
yum install make
```

（2）升级 gcc：

```
yum -y install centos-release-scl
```

```
yum -y install devtoolset-9-gcc devtoolset-9-gcc-c++ devtoolset-9-binutils
scl enable devtoolset-9 bash
```

（3）设置永久升级：

```
echo "source /opt/rh/devtoolset-9/enable" >>/etc/profile
```

gcc 安装成功后，再执行 make 编译，编译成功后，进入解压后的 src 目录，通过如下命令启动 Redis：

```
$ src/redis-server
```

Redis 进程启动后，我们可以使用内置的客户端 redis-cli 与 Redis 进行交互。下面介绍一些常用的命令。

前面已经说到 Redis 至少支持 string、list、set、zset、hash 五种最基本的数据类型，下面通过命令熟悉一下这五种数据类型。

字符串类型是其他四种类型的基础，这是因为其他四种类型的值也是字符串，只不过是形式不同而已。

set 和 get 命令用于操作 string 类型的 key。添加一个 string 类型的键值对用 set 命令，获取一个 string 类型的键值对的值用 get 命令。

```
127.0.0.1:6379> set name redis
OK
127.0.0.1:6379> get name
"redis"
```

list 类型按元素的插入顺序存储，并且支持在 list 的左右两端插入/删除元素，元素可重复。list 类型可以实现简单消息队列的功能。

lpush、lpop、rpush、rpop 命令用于操作 list 类型的 key：在 list 的左侧插入和删除元素分别使用 lpush 和 lpop 命令，右侧则使用 rpush 和 rpop 命令。如果不想删除元素，只想查看 list 中有哪些元素，则可以使用 lrange 命令。

```
127.0.0.1:6379> lpush cacheList redis memecache
(integer) 2
127.0.0.1:6379> rpush cacheList memecache ehcache
(integer) 4
127.0.0.1:6379> lrange cacheList 0 3
1) "memecache"
2) "redis"
3) "memecache"
4) "ehcache"
127.0.0.1:6379> lpop cacheList
"memecache"
127.0.0.1:6379> rpop cacheList
"ehcache"
```

set 类型是元素无序不重复的集合。添加元素使用 sadd 命令，删除元素使用 srem 命令，查看集合中的所有元素使用 smembers 命令。

```
127.0.0.1:6379> sadd cacheSet redis redis ehcache memecache
```

```
(integer) 2
127.0.0.1:6379> smembers cacheSet
1) "ehcache"
2) "redis"
3) "memecache"
127.0.0.1:6379> srem cacheSet memecache
(integer) 1
```

zset是有序不重复集合，每个元素都有一个分数，集合按分数排序。添加元素时如果元素已存在，则覆盖之前的分数。zset类型非常适合用于实现一些网站的排行榜功能。

添加元素使用zadd命令，zrange命令会按从小到大的顺序排序，再返回start到end之间的元素，传入withscores会返回分数。

```
127.0.0.1:6379> zadd cacheZSet 80 ehcache 90 redis 70 memecache 85 mysql
(integer) 4
127.0.0.1:6379> zrange cacheZSet 0 3 withscores
1) "memecache"
2) "70"
3) "ehcache"
4) "80"
5) "mysql"
6) "85"
7) "redis"
8) "90"
```

hash类型适合存储对象。在下面的代码中，user称为一级key，username、age、sex称为二级key。一级key是Redis数据的key，可以用来存放对象名称；二级key是hash的key，可以用来存储对象的属性名。

```
127.0.0.1:6379> hset user username zhangsan age 20 sex nan
(integer) 3
127.0.0.1:6379> hget user username
"zhangsan"
127.0.0.1:6379> hget user age
"20"
127.0.0.1:6379> hget user sex
"nan"
```

下面再介绍几种通用的命令。

查询所有key使用keys *命令，判断某个key是否存在使用exists命令，查看某个key的数据类型使用type命令，删除一个key使用del命令。

```
127.0.0.1:6379> keys *
1) "cacheZSet"
2) "cacheSet"
3) "cacheList"
4) "user"
5) "name"
127.0.0.1:6379> exists user
(integer) 1
```

```
127.0.0.1:6379> type cacheZSet
zset
127.0.0.1:6379> del name
(integer) 1
```

Redis 提供了大量丰富的命令，上述命令只是其中的一部分，这里就不一一列举其他命令了，有兴趣的读者可以访问 Redis 官网进一步学习。

7.2.3 Redis 缓存实战

下面来看一下在实际项目中使用缓存的经典案例——缓存穿透。

缓存穿透是指用户查询数据，该数据在数据库压根就没有，在缓存中自然也不会有。这样就会导致用户查询的时候在缓存中找不到，每次都要去数据库再查询一遍，然后返回空，相当于进行了两次无用的查询。这时的用户很可能是攻击者，攻击会导致数据库压力过大。

解决方案有两种：

（1）如果一个查询返回的数据为空（不管是数据不存在还是系统故障），就把这个空结果进行缓存，但它的过期时间会很短，最长不超过五分钟。然后设置一个默认值存放到缓存，这样第二次到缓存中获取就有值了，而不会继续访问数据库，这种办法最简单粗暴。

（2）使用布隆过滤器。布隆过滤器的巨大用处就是，能够迅速判断一个元素是否在一个集合中，对程序性能的影响可以忽略不计。因此，我们将所有可能存在的数据缓存到布隆过滤器中，当黑客访问不存在的缓存时迅速返回，避免缓存及数据库挂掉。

下面实现布隆过滤器的解决方案。我们可以使用 Google 的 guava 包中的布隆过滤器。项目在 7.1.2 小节中搭建过，只不过将 EhCache 的依赖换成了 Redis 的依赖。pom 依赖如下所示。

示例代码 7-7　pom.xml

```xml
<dependencies>
    <dependency>
        <groupId>org.springframework.boot</groupId>
        <artifactId>spring-boot-starter-web</artifactId>
    </dependency>
    <dependency>
        <groupId>org.mybatis.spring.boot</groupId>
        <artifactId>mybatis-spring-boot-starter</artifactId>
        <version>2.1.3</version>
    </dependency>
    <dependency>
        <groupId>mysql</groupId>
        <artifactId>mysql-connector-java</artifactId>
        <scope>runtime</scope>
    </dependency>
    <dependency>
        <groupId>org.springframework.boot</groupId>
        <artifactId>spring-boot-starter-data-redis</artifactId>
    </dependency>
```

```
    <dependency>
        <groupId>com.google.guava</groupId>
        <artifactId>guava</artifactId>
        <version>22.0</version>
    </dependency>
</dependencies>
```

示例代码 7-8 application.yml

```
spring:
  datasource:
    driver-class-name: com.mysql.cj.jdbc.Driver
    url:
jdbc:mysql://localhost:3306/springboot-ehcache-redis-secondskill?useUnicode=true&characterEncoding=utf-8&serverTimezone=Asia/Shanghai
    username: root
    password: ok
server:
  servlet:
    context-path: /
  port: 8080
  redis:
    host: 127.0.0.1
    port: 6379
mybatis:
  type-aliases-package: com.bobo.springbootredis.entity
  mapper-locations: classpath:mapper/*Mapper.xml
```

引入 Redis 相关依赖之后，还需要配置 Redis 的 CacheManager。另外，还可以配置 RedisTemplate（Spring 提供的一个针对 Redis 的模板操作类）。Spring 的缓存抽象虽然方便，但是局限性很大，缓存的最小粒度是方法，不够灵活，使用 RedisTemplate 的 API 则灵活得多。Redis 的配置如下所示。

示例代码 7-9 RedisCacheManagerConfig.java

```
@Configuration
public class RedisCacheManagerConfig {
    @Bean(name = "template")
    public RedisTemplate<String, Object> template(RedisConnectionFactory
factory) {
        // 创建 RedisTemplate<String, Object>对象
        RedisTemplate<String, Object> template = new RedisTemplate<>();
        // 配置连接工厂
        template.setConnectionFactory(factory);
        // 定义 Jackson2JsonRedisSerializer 序列化对象
        Jackson2JsonRedisSerializer<Object> jacksonSeial = new
Jackson2JsonRedisSerializer<>(Object.class);
        ObjectMapper om = new ObjectMapper();
        // 指定要序列化的域，field、get/set 以及修饰符范围，ANY 是都有（包括 private 和
public）
        om.setVisibility(PropertyAccessor.CREATOR,
```

```
JsonAutoDetect.Visibility.ANY);
            // 指定序列化输入的类型，类必须是非 final 修饰的，final 修饰的类（比如 String,
            // Integer 等）会报异常
            om.enableDefaultTyping(ObjectMapper.DefaultTyping.NON_FINAL);
            jacksonSeial.setObjectMapper(om);
            StringRedisSerializer stringSerial = new StringRedisSerializer();
            // redis key 序列化方式使用 stringSerial
            template.setKeySerializer(stringSerial);
            // redis value 序列化方式使用 jackson
            template.setValueSerializer(jacksonSeial);
            // redis hash key 序列化方式使用 stringSerial
            template.setHashKeySerializer(stringSerial);
            // redis hash value 序列化方式使用 jackson
            template.setHashValueSerializer(jacksonSeial);
            template.afterPropertiesSet();
            return template;
        }
        @Bean
        public CacheManager cacheManager(RedisTemplate<String, Object> template)
{
            RedisCacheConfiguration defaultCacheConfiguration =
                    RedisCacheConfiguration
                            .defaultCacheConfig()
                            // 设置 key 为 String
                            .serializeKeysWith(RedisSerializationContext.Serializa
tionPair.fromSerializer(template.getStringSerializer()))
                            // 设置 value 为自动转 Json 的 Object
                            .serializeValuesWith(RedisSerializationContext.Seriali
zationPair.fromSerializer(template.getValueSerializer()))
                            // 不缓存 null
                            .disableCachingNullValues()
                            // 缓存数据保存 1 小时
                            .entryTtl(Duration.ofDays(30));
            RedisCacheManager redisCacheManager =
                    RedisCacheManager.RedisCacheManagerBuilder
                            // Redis 连接工厂
                            .fromConnectionFactory(template.getConnectionFactory()
)
                            // 缓存配置
                            .cacheDefaults(defaultCacheConfiguration)
                            // 配置同步修改或删除 put/evict
                            .transactionAware()
                            .build();
            return redisCacheManager;
        }
    }
```

在 BookService 中，定义一个根据 ID 查询 Book 的方法，如下所示。

示例代码 7-10 BookService.java

```
public interface BookService {
    /**
     * 根据 ID 获取 book
     */
    Book getBook(Long id);
}
```

示例代码 7-11 BookServiceImpl.java（部分代码）

```
@Service
public class BookServiceImpl implements BookService {
    @Autowired
    private BookMapper bookMapper;
    @Override
    @Cacheable(cacheNames = {"single_book"},key =
"#root.targetClass+'.'+#root.methodName+'.'+#p0",
            unless = "#result == null")
    public Book getBook(Long id){
        return bookMapper.selectBookById(id);
    }
}
```

下面定义一个布隆过滤器。可以通过 Spring 事件机制来监听上下文刷新事件，假设 Book 的 ID 范围是 1 到 1000000，就将 1 到 1000000 全部放到布隆过滤器中。此时，如果用户查询的 book 的 ID 在 1 到 1000000 之间，就进行正常查询，否则，直接返回 null，既不用 Redis 查询也不用数据库查询，有效避免了两次无用查询。

布隆过滤器的代码如下。

示例代码 7-12 BookBloomFilter.java

```
@Component
public class BookBloomFilter {
    public BloomFilter<Long> bloomFilter;
    private static final long SIZE = 1000000;
    @EventListener
    public void contextRefreshedEventListener(ContextRefreshedEvent
contextRefreshedEvent) {
        bloomFilter = BloomFilter.create(Funnels.longFunnel(), SIZE);
        for (long i = 1; i <= SIZE; i++) {
            bloomFilter.put(i);
        }
    }
}
```

Controller 的代码如下。

示例代码 7-13 BookController.java

```
@RestController
public class BookController {
    @Autowired
```

```
    private BookService bookService;
    @Autowired
    private BookBloomFilter bookBloomFilter;
    @GetMapping("/get")
    public Book getBook(Long id){
        if(bookBloomFilter.bloomFilter.mightContain(id)){
            System.out.println("id为"+id+"的book在数据库中存在");
            return bookService.getBook(id);
        }else{
            System.out.println("id为"+id+"的book在数据库中不存在");
            return null;
        }
    }
}
```

分别请求 http://localhost:8080/get?id=10 和 http://localhost:8080/get?id=1001000，后台打印的日志如图 7-3 所示，布隆过滤器的作用显现出来。

```
id为10的book在数据库中存在
2021-01-17 19:12:38.520 DEBUG 1772 --- [nio-8080-exec-3] m.m.a.RequestResponseBodyMethodProcessor
2021-01-17 19:12:38.520 DEBUG 1772 --- [nio-8080-exec-3] m.m.a.RequestResponseBodyMethodProcessor
2021-01-17 19:12:38.520 DEBUG 1772 --- [nio-8080-exec-3] o.s.web.servlet.DispatcherServlet
2021-01-17 19:12:47.397 DEBUG 1772 --- [nio-8080-exec-4] o.s.web.servlet.DispatcherServlet
2021-01-17 19:12:47.397 DEBUG 1772 --- [nio-8080-exec-4] s.w.s.m.m.a.RequestMappingHandlerMapping
id为1001000的book在数据库中不存在
```

图 7-3

另外，布隆过滤器有一定的概率会出现误判，比如 ID 的范围是 1 到 1000000，当输入不在此范围的 ID 时，布隆过滤器可能认为该 ID 在此范围。误判是由于布隆过滤器的内部机制所造成的，布隆过滤器占用的空间越大，误判率越低，反之则越高。下面通过一个简单的例子来测试一下布隆过滤器的默认误判率有多大。

示例代码 7-14　BloomFilterWuPanTest.java

```
public class BloomFilterWuPanTest {
    private static int size = 1000000;
    private static BloomFilter<Integer> bloomFilter =
BloomFilter.create(Funnels.integerFunnel(), size);
    public static void main(String[] args) {
        for (int i = 1; i <= size; i++) {
            bloomFilter.put(i);
        }
        int count = 0;
        //故意取10000个不在过滤器里的值，看看有多少个会被认为是在过滤器里的
        for (int i = size + 1; i <= size*2; i++) {
            if (bloomFilter.mightContain(i)) {
                count++;
            }
        }
        System.out.println("误判的数量：" + count);
```

```
    }
}
```

这段代码先将 1 到 1000000 之间的数字加入布隆过滤器中，然后故意取 1000001 到 2000000 之间的数字让布隆过滤器判断，程序运行后的打印结果如图 7-4 所示。

```
"C:\Program Files\Java\jdk1.8.0_212\bin\java.exe" ...
误判的数量：30155

Process finished with exit code 0
```

图 7-4

这说明误判率大概在 3.1%。如果网站要求不高，那么还是能接受的。

小 结

本小节重点介绍分布式缓存 Redis，首先对什么是 Redis、Redis 有哪些特性、为什么要用 Redis 做了一个全面的概述，随后又介绍了 Redis 的安装与基本命令的使用，最后使用布隆过滤器来解决高并发场景下 Redis 缓存穿透的问题。

7.3 消息中间件

7.3.1 消息中间件简介

MQ（Message Queue）是应用程序与应用程序之间的通信方法，应用程序通过一定规则进行写入队列和读取某队列的消息来实现它们之间的通信。消息是消息队列中的最小概念，本质上就是一段数据，被一个或者多个应用程序读取，是应用程序之间传递信息的载体。消息队列中涉及 AMQP（Advanced Message Queuing Protocol，高级消息队列协议），AMQF 是一个应用层协议的开发标准，主要是面向消息中间件设计的，具有可靠性和安全性。消息中间件是分布式系统中的重要组件，主要解决应用解耦、异步消息、流量削峰等问题，实现高性能、高可用、可伸缩和最终一致性架构，主要具有以下特性：

①可靠传输：对于应用系统来说，只要成功把消息提交给消息中间件，关于数据可靠传输的问题就可以交给消息中间件来解决。

②不重复传输：也就是断点续传的功能，特别适合网络不稳定的环境，节约网络资源。

③异步传输：发布和接收消息的双方应用不必同时在线，具有脱机能力和安全性。

④消息驱动：当队列接收到消息后，会主动通知消息接收方。

⑤支持事务：目前消息队列也被广泛应用于分布式事务的同步。

消息中间件何时何地使用、如何使用都是值得思考和实践的，胡乱使用消息中间件不仅会增加系统的复杂度，出问题时还可能导致整个分布式系统瘫痪。下面将从异步处理、应用解耦、流量削峰和消息通信四个方面介绍消息队列在实际应用中的常用场景。

（1）异步处理

场景：一个用户需要注册某应用的会员，当注册成功后发送应用的系统消息和邮件告知用户已经注册成功。传统做法有串行和并行两种。

①串行方式：将用户注册会员信息成功写入数据库之后，发送系统消息提醒，再发送邮件告知，以上三个步骤全部完成之后将注册结果返回给客户端，如图 7-5 所示。

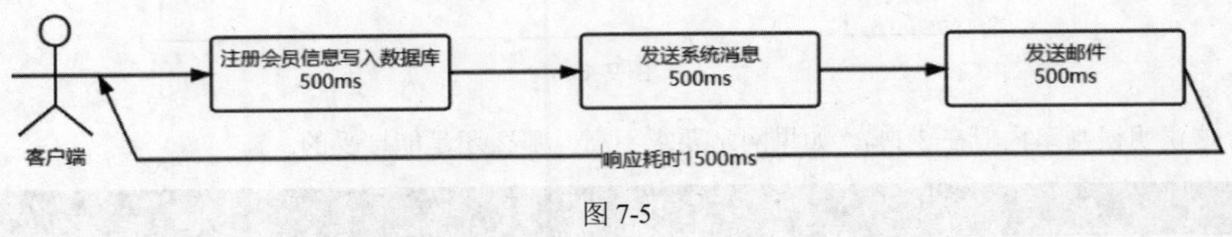

图 7-5

②并行方式：将用户注册会员信息成功写入数据库之后，发送系统消息的同时发送邮件告知，以上三个任务完成之后返回给客户端，如图 7-6 所示。

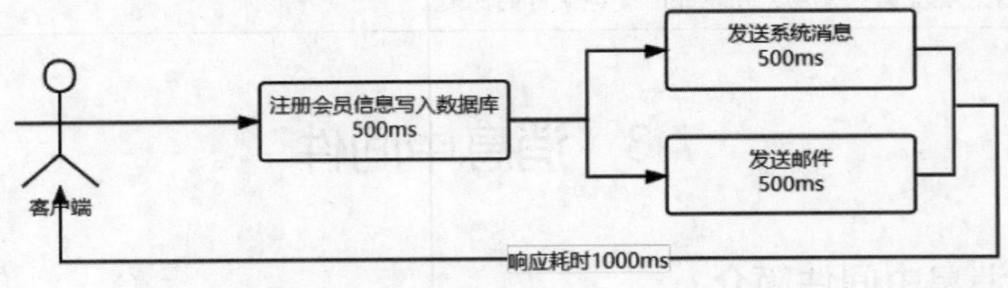

图 7-6

③引入消息队列中间件：引入消息中间件之后，用户响应时间约为注册会员信息写入数据库的时间，也就是 500ms。因为会员信息入库成功后，将发送系统消息和发送邮件两个任务写入消息队列即可直接返回，写入消息队列速度很快（5ms），基本可忽略，因此用户响应时间约为 500ms，如图 7-7 所示。

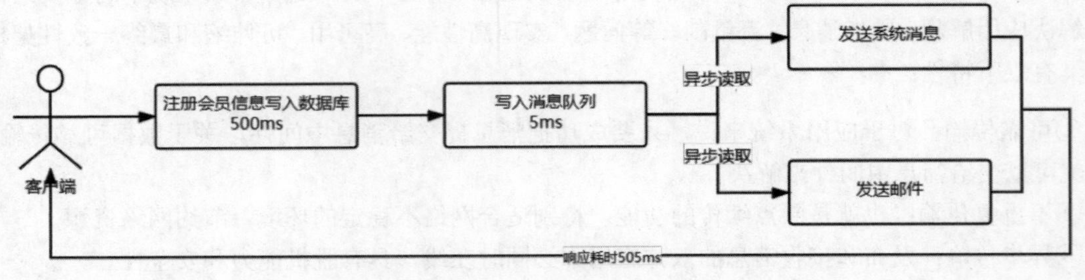

图 7-7

通过以上分析可知，串行方式完成会员注册功能耗时 1500ms，并行方式耗时 1000ms，所以从一定程度上讲并行方式可以提高任务处理速度。引入中间件之后，系统架构发生改变，响应速度直接提升为 500ms（约数），即业务处理速率大概比串行快 3 倍、比并行快 2 倍。

（2）应用解耦

场景：一个商城系统采用分布式架构，即订单、库存、会员等模块为单独系统，当用户下单后会更新订单系统，订单系统再通知库存系统更新该物品的库存信息。

①传统方式：传统方式导致订单系统和库存系统耦合，如果库存系统异常，导致请求无法访问，则库存更新失败，从而导致用户下单失败，如图 7-8 所示。

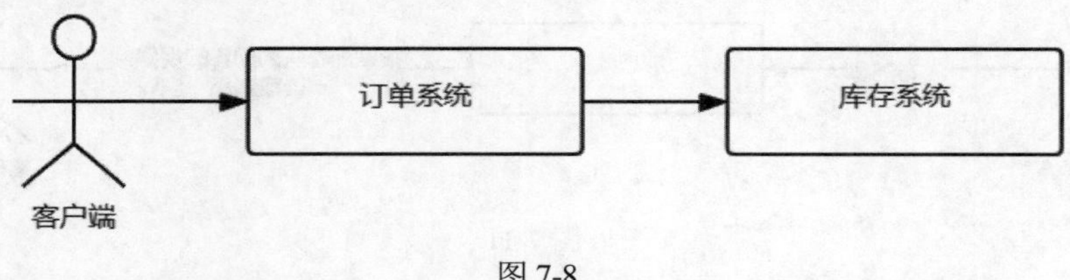

图 7-8

②加入消息中间件：当系统中引入消息中间件，用户下单，将订单信息写入数据库后，会将该订单任务写入消息队列中，直接返回给用户下单成功。至于库存系统，根据设定的规则从消息队列中读取用户订单信息，然后进行库存更新的业务操作。

即使用户在下单时库存系统异常，无法正常使用，也不会影响用户下单，因为订单的信息已经写入消息队列中，当被对应的消费者消费时该消息才会被移除队列，这样就实现了订单系统和库存系统之间的解耦，如图 7-9 所示。

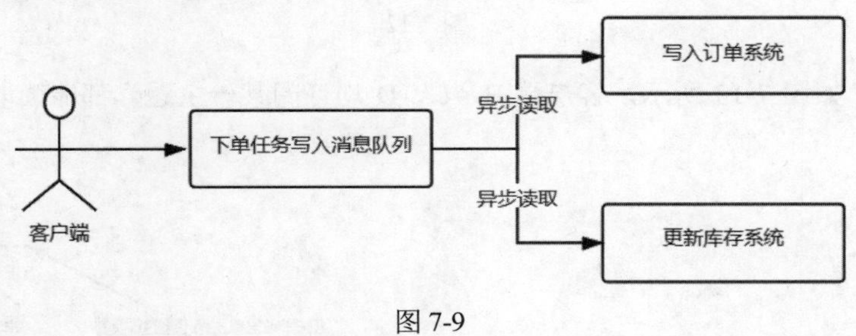

图 7-9

（3）流量削峰

流量削峰是消息中间件的常用场景，一般用来削减大量请求给系统造成的压力，起到缓存的作用，减轻下游业务系统的压力，比如我们熟知的直播间秒杀活动或双十一淘宝的秒杀活动等。当大量用户在同一时间下单时，用户请求数量短时间内剧增，应用系统很容易挂掉。

为了解决这个问题，一般会选择在决堤的业务应用之前增加一个消息中间件。当服务器接收到用户下单请求时先将这些请求写入消息队列中，订单系统再从消息队列中读取用户请求信息，有条不紊地进行业务处理，这样就可以控制同时访问订单系统的人数，起到缓存削峰的作用，如图 7-10 所示。

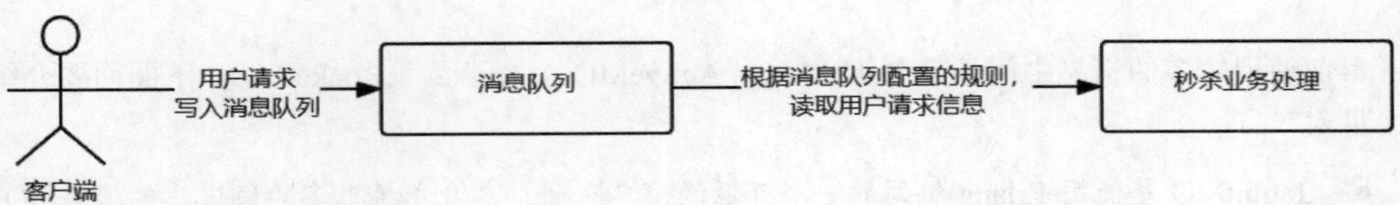

图 7-10

（4）消息通信

消息中间件一般都内置了高效的通信机制，因此可以实现点对点、聊天室和广播等消息通信。

①点对点通信。客户端 A 和客户端 B 使用同一队列，进行消息通信。如图 7-11 所示，客户端 A 向消息队列发布消息，客户端 B 可以直接根据配置的规则获取到客户端 A 发布的消息。

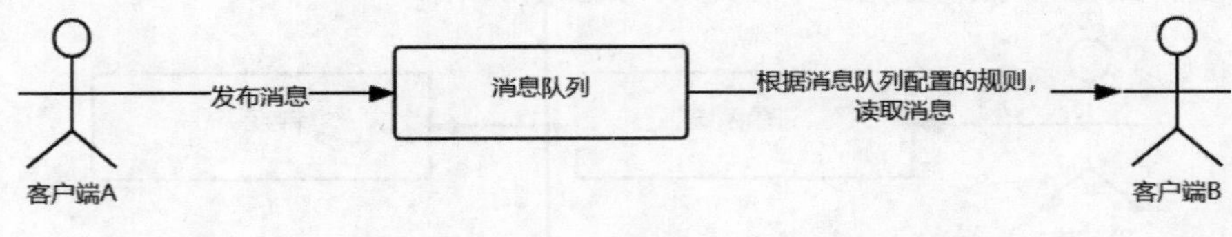

图 7-11

②聊天室通信。如图 7-12 所示，客户端 A 和客户端 B 同时订阅一个主题，进行消息的发布和接收，实现聊天室的效果。

图 7-12

③广播通信。如图 7-13 所示，客户端 B、C、D 均订阅某一主题，都能读取到客户端 A 发布的消息。

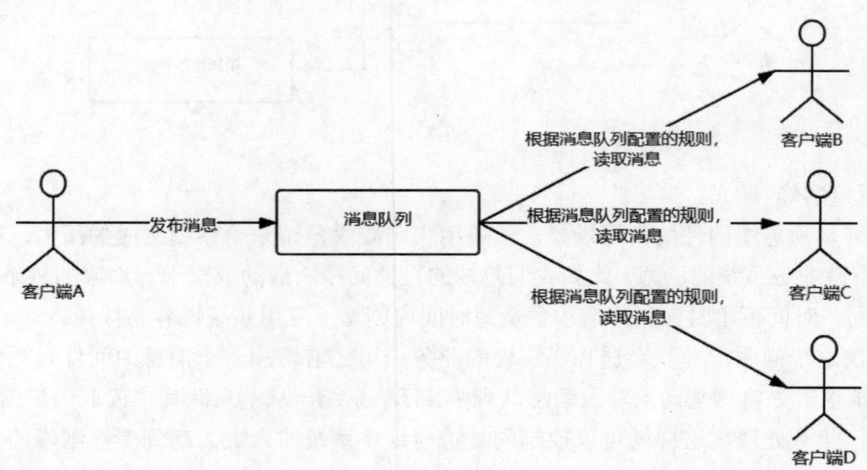

图 7-13

目前使用较多的消息中间件有 RabbitMQ、ActiveMQ 、Kafka 、RocketMQ，下面简要介绍一下这四款产品。

- RabbitMQ 是使用 Erlang 编写的一个开源的消息队列，本身支持很多的协议，包括 AMQP、XMPP、SMTP 和 STOMP。正是如此，它的重量级很大，更适合于企业级的开发。同时，它实现了一个经纪人（Broker）构架，意味着消息在发送给客户端时先在中心队列排队，对路由（Routing）、负载均衡（Load balance）或者数据持久化都有很好的支持，是高可用的主从架构。另外，它还支持常用的多种语言客户端，比如C++、Java、.Net、Python、PHP、Ruby等。

- ActiviteMQ是Apache下的一个子项目，类似于ZeroMQ，能够以代理人和点对点的技术实现队列；同时也类似于 RabbitMQ，用少量的代码就可以高效地实现高级应用场景。它也支持常用的多种语言客户端，比如C++、Java、.Net、Python、PHP、Ruby等。
- Kafka是Apache下的一个子项目，是一个高性能跨语言分布式Publish/Subscribe消息队列系统。Jafka是在Kafka之上孵化而来的，即Kafka的一个升级版，具有快速持久化、高吞吐、分布式系统自动实现负载均衡、支持Hadoop数据并行加载等特性。
- RcoketMQ 是一款低延迟、高可靠、可伸缩、易于使用的消息中间件，支持多种消息协议，如JMS、MQTT等，而且支持发布/订阅（Pub/Sub）和点对点（P2P）消息模型。单一的消息队列具备支持百万消息的堆积能力。另外，其还提供Docker镜像，用于隔离测试和云集群部署。 RcoketMQ是一款具备非常高的可用性的分布式架构产品。

7.3.2 RabbitMQ 简介

1. RabbitMQ 基础概念

消息队列通常会涉及三个概念：生产者（发消息者）、消息队列、消费者（收消息者）。RabbitMQ在这三个基本概念之上又加入了路由的概念，即在生产者和消息队列之间增加了一个交换器（Exchange），生产者把消息发给交换器，交换器再根据调度策略把消息转发给消息队列。因此，生产者和消息队列之间没有直接的关系，消息队列起到储存消息的作用，等待消费者取出消息。RabbitMQ的具体执行过程如图7-14所示。

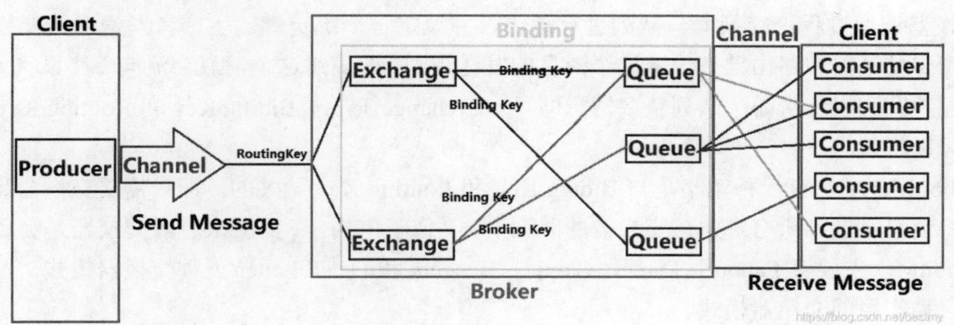

图 7-14

RabbitMQ运作过程中涉及如下几个概念：

- Channel（信道）：多路复用连接中的一条独立的双向数据流通道。信道是建立在真实的TCP连接内的虚拟连接，复用TCP连接的通道。
- Producer（消息的生产者）：向消息队列发布消息的客户端应用程序。
- Consumer（消息的消费者）：从消息队列取得消息的客户端应用程序。

- Message（消息）：消息由消息头和消息体组成。消息体是不透明的，而消息头则由一系列的可选属性组成，这些属性包括 Routing Key（路由键）、priority（消息优先权）和 delivery-mode（是否持久性存储）等。
- Routing Key（路由键）：消息头的一个属性，用于标记消息的路由规则，决定了交换器的转发路径，最大长度为 255 字节。
- Queue（消息队列）：存储消息的一种数据结构，用来保存消息，直到消息发送给消费者。它是消息的容器，也是消息的终点。一个消息可投入一个或多个队列。消息一直在队列里面，等待消费者连接到这个队列将消息取走。需要注意的是，当多个消费者订阅同一个 Queue 时，Queue 中的消息会被平均分摊给多个消费者进行处理，而不是每个消费者都收到所有的消息并处理，每一条消息只能被一个订阅者接收。
- Exchange（交换器|路由器）：提供 Producer 到 Queue 之间的匹配，接收生产者发送的消息并将这些消息按照路由规则转发到消息队列。交换器用于转发消息，不会存储消息，如果没有 Queue 绑定到 Exchange，它会直接丢弃 Producer 发送过来的消息。
- Binding（绑定）：用于建立 Exchange 和 Queue 之间的关联。一个绑定就是基于 Binding Key 将 Exchange 和 Queue 连接起来的路由规则，所以可以将交换器理解成一个由 Binding 构成的路由表。
- Binding Key（绑定键）：Exchange 与 Queue 的绑定关系，用于匹配 Routing Key，最大长度为 255 字节。
- Broker：RabbitMQ Server，服务器实体。

2. Exchange（交换器）调度模式

交换器消息调度模式指的是交换器在收到生产者发送的消息后，依据某种规则把消息转发到一个或多个队列中保存。交换器的调度策略与三个因素有关：Exchange Type（交换器类型）、Biding Key（绑定键）、Routing Key（路由键）。

交换器调度过程：交换器与队列进行绑定，并设定一个绑定键。生产者发消息给交换器时，还需要指定交换器和路由键来设定这个消息的发送规则，即最终这个消息会流向哪个队列，消费者只需要配置队列表示从哪个队列获取消息即可。Exchange Type、Biding Key 和 Routing Key 要联合使用才可以生效。

RabbitMQ 中也有一种策略是与 Biding Key 和 Routing Key 无关的，即只需要交换器和队列绑定就会将消息发送到所有绑定该交换器的队列上，类似于广播。下面就来看看交换器是哪种调度模式。RabbitMQ 主要有 Fanout、Direct、Topic、Headers 和 Dead Letter 五种交换器模式，这里主要讲解前三种常用的交换器模式。

（1）Fanout（订阅模式/广播模式），如图 7-15 所示。

扇形交换器会把所有接收到的消息都分发到与之绑定的队列中。此种模式是没有 Biding Key 和 Routing Key 概念的，即与 Biding Key 和 Routing Key 无关，类似于子网广播，子网内的每台主机都能获得一份复制的消息。相较于其他策略，Fanout Exchange 转发消息的效率最高，工作模式如图 7-16 所示。

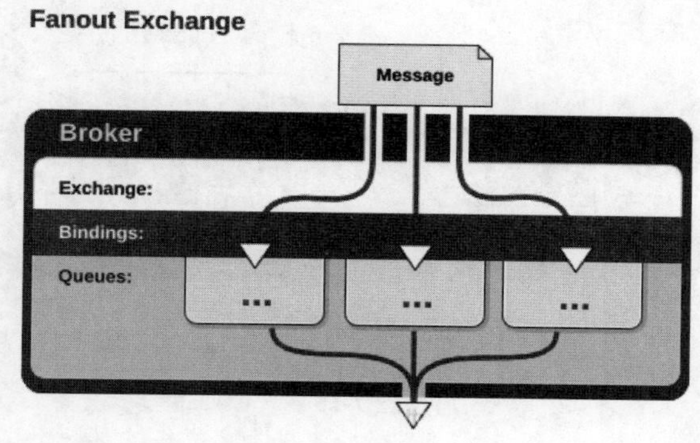

图 7-15

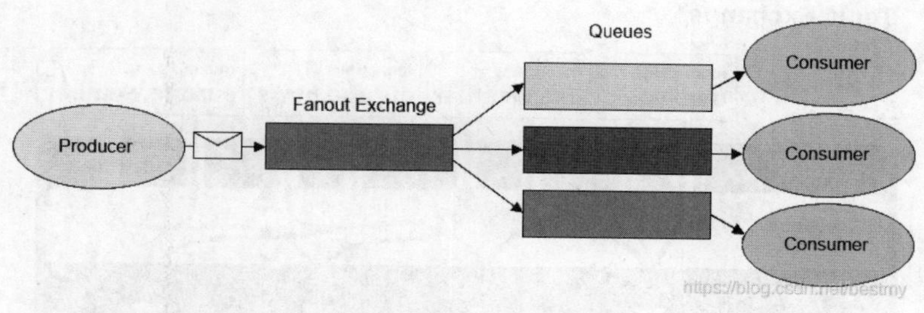

图 7-16

（2）Direct（点对点模式），如图 7-17 所示。

直连型交换器是一种精确匹配模式，即需要生产者发送消息时绑定的 Routing Key，需要和绑定 Exchange、Queue 的 Biding Key 完全匹配，才能将消息分发到对应的队列上。

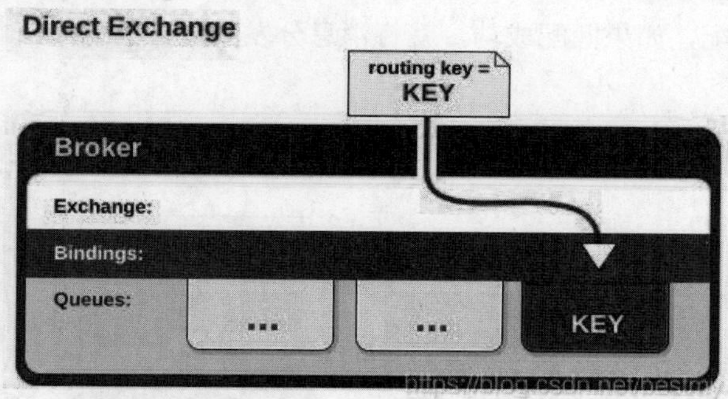

图 7-17

另外，RabbitMQ 默认提供了一个 Exchange，交换器名字是空字符串，交换器类型是 Direct，绑定到所有的 Queue，每一个 Queue 和这个无名 Exchange 之间的 Binding Key 是 Queue 的名字，所以有时我们感觉不需要交换器也可以发送和接收消息，实际上是使用了 RabbitMQ 默认提供的 Exchange，工作模式如图 7-18 所示。

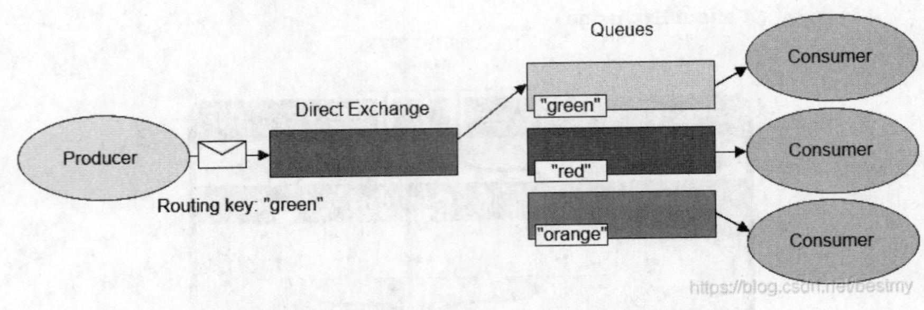

图 7-18

（3）Topic（通配符模式），如图 7-19 所示。

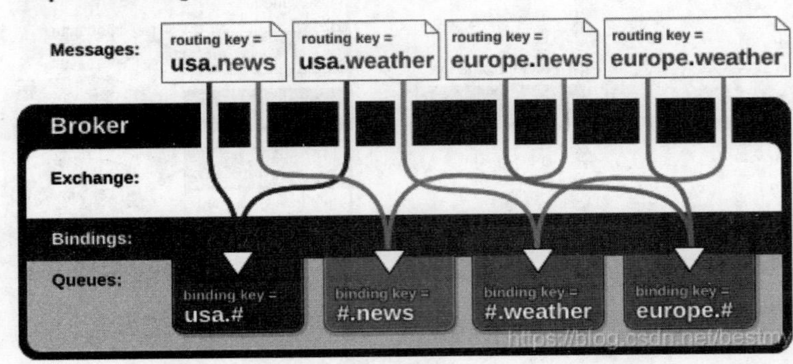

图 7-19

主题交换器是一种模糊匹配模式，是生产者发消息 Routing Key 与交换器、队列之间绑定的 Binding Key 进行模糊匹配，如果匹配成功，就将消息分发到对应的 Queue。工作模式如图 7-20 所示。

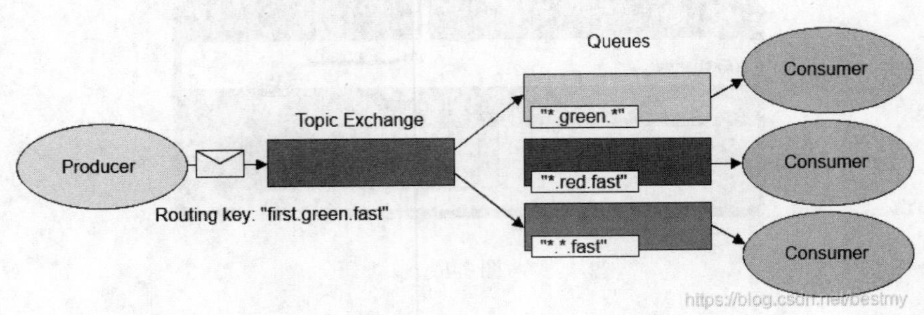

图 7-20

Routing Key 和 Binding Key 是一段字符串，用“.”进行分隔，分隔开的每段独立字符串称为一个单词。通配符构成的 key 一般是用来和队列进行绑定的，即用于 Binding Key 的设定。通配键可以用“#”和“*”两种特殊字符作为通配符，其中“#”可以匹配任意数量（零个或多个），“*”可以匹配一个单词，是必须出现的。下面通过例子讲解一下这两种通配符。

场景：

①一个主题交换器：TopicExchange。

②三个队列：Queue1、Queue2、Queue3。

③TopicExchange 和 Queue1 的绑定键为 TT.*，和 Queue2 的绑定键为 TT.#，和 Queue3 的绑定键为*.TT.*。

测试：

①如果消息 Message1 携带的路由键 Routing Key 为 TT.A，那么这条消息将会被 Queue1 和 Queue2 收到。

②如果消息 Message2 携带的路由键 Routing Key 为 TT，那么这条消息将只会被 Queue2 收到。

③如果消息 Message3 携带的路由键 Routing Key 为 TT.AA.BB，那么这条消息将只会被 Queue2 收到。

④如果消息 Message4 携带的路由键 Routing Key 为 A.TT.B，那么这条消息将只会被 Queue3 收到。

7.3.3　实战

由于 RabbitMQ 是以 Erlang 语言开发的，因此在安装 RabbitMQ 之前需要先安装 Erlang 语言开发包。本小节我们主要从安装 Erlang 语言开发包、安装 RabbitMQ、Spring Boot 集成 RabbitMQ 代码演示这三部分着手实操。

1. Erlang 语言开发包安装

（1）根据计算机的配置从 Erlang 官网（https://www.erlang.org/downloads/20.0）下载对应规格的语言开发包，现在计算机的一般都是 64 位的，所以以下示例均以 64 位进行演示，具体操作如图 7-21 所示。

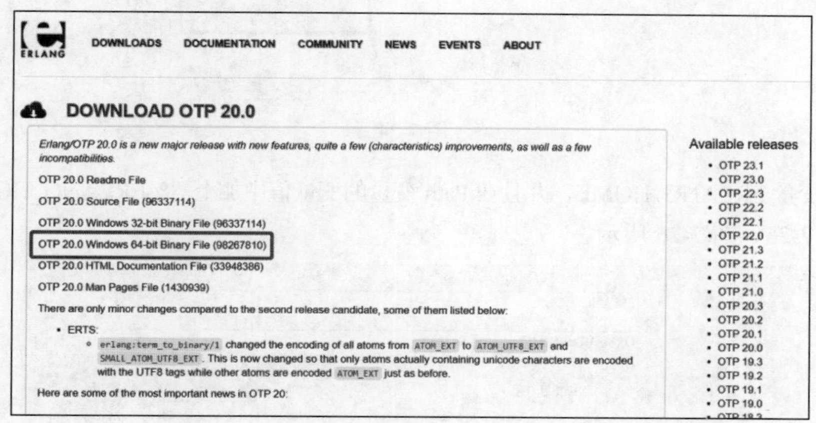

图 7-21

（2）下载完成后，双击安装，选择默认配置即可，既可自定义安装目录，也可以按照默认目录安装，具体操作如图 7-22、图 7-23 所示。

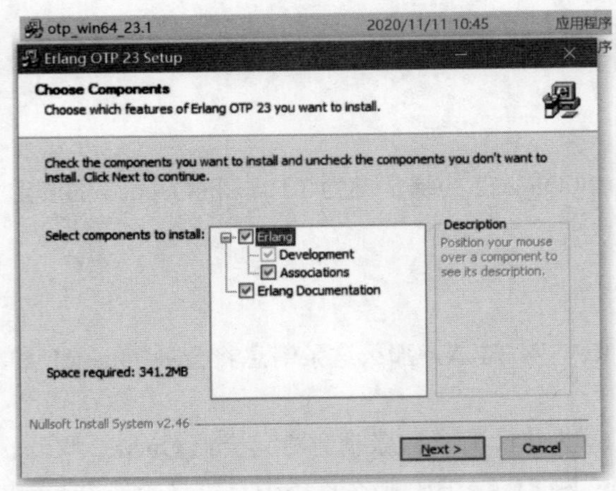

图 7-22

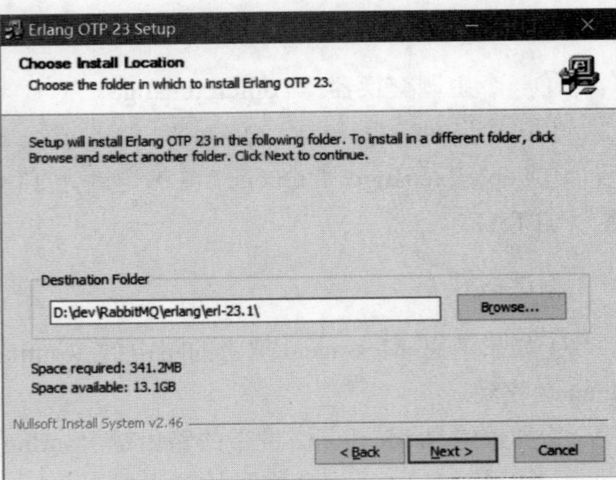

图 7-23

（3）配置高级环境变量（我的电脑→属性→高级系统设置），具体操作如图 7-24 所示。

图 7-24

配置环境变量 ERLANG_HOME，并且在 Path 变量的变量值中加上“%ERLANG_HOME%\bin;”，具体操作如图 7-25、图 7-26 所示。

图 7-25

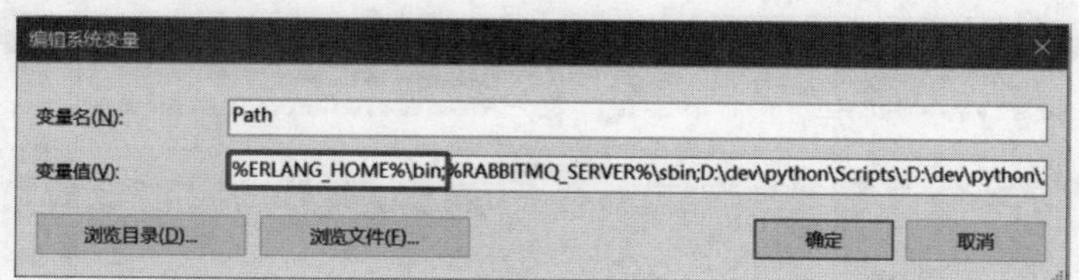

图 7-26

2. RabbitMQ 安装

（1）从 RabbitMQ 官网（https://www.rabbitmq.com/install-windows.html）下载应用程序，具体步骤如图 7-27~图 7-29 所示。

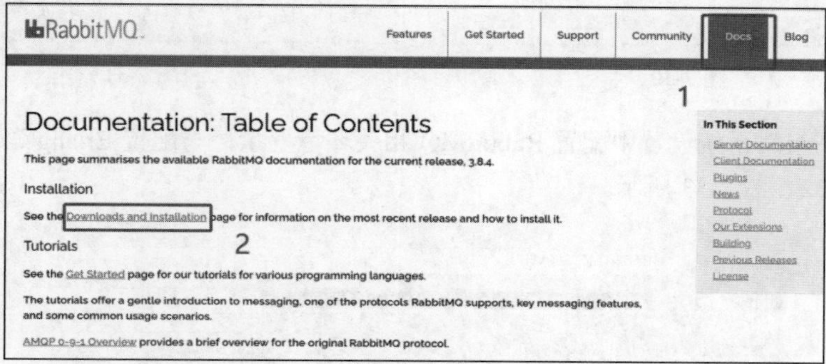

图 7-27

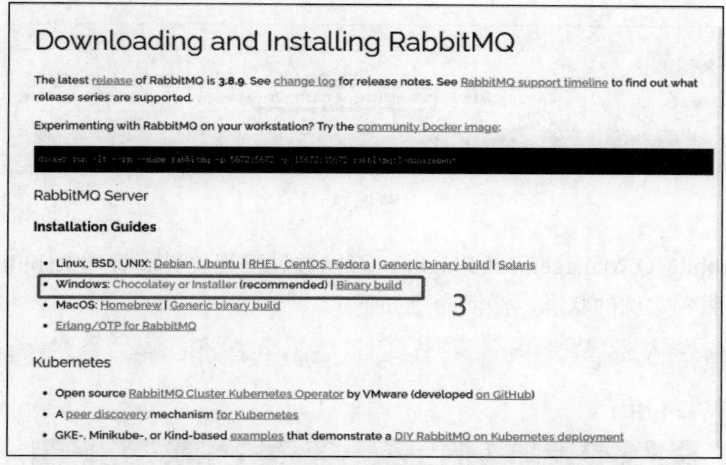

图 7-28

Description	Download	Signature
Installer for Windows systems (from GitHub, recommended)　4	rabbitmq-server-3.8.9.exe	Signature
Alternative download location (from Bintray)	rabbitmq-server-3.8.9.exe	

图 7-29

（2）双击 rabbitmq-server-3.8.9.exe，选择安装目录/默认目录安装，默认安装的 RabbitMQ 监听端口是 5672，如图 7-30、图 7-31 所示。

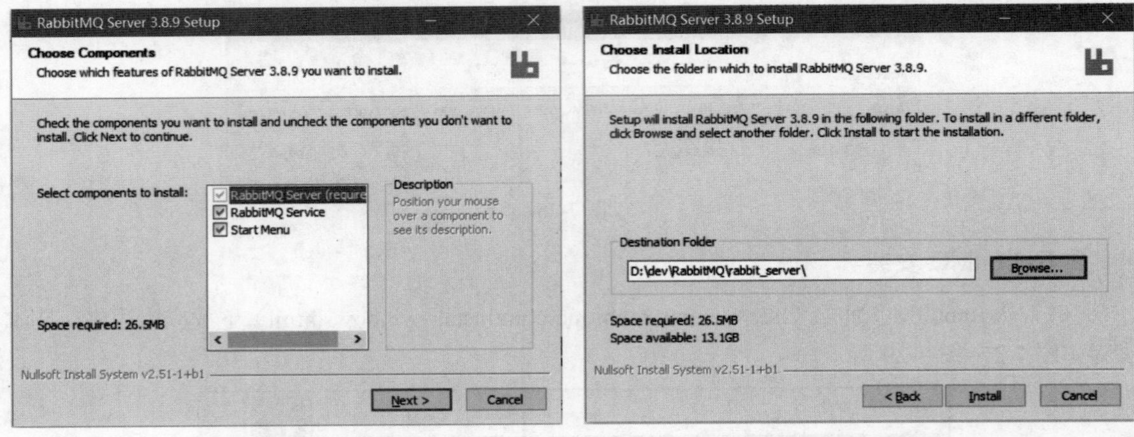

图 7-30　　　　图 7-31

（3）在“高级”系统设置中配置 RabbitMQ 相关环境变量，与配置 Erlang 语言包配置相似，具体步骤如图 7-32、图 7-33 所示。

变量名(N): RABBITMQ_SERVER
变量值(V): D:\dev\RabbitMQ\rabbitmq_server\rabbitmq_server-3.8.9
浏览目录(D)... 浏览文件(F)... 确定 取消

图 7-32

变量名(N): Path
变量值(V): %ERLANG_HOME%\bin;%RABBITMQ_SERVER%\sbin;D:\dev\python\Scripts\;D:\dev\python\;
浏览目录(D)... 浏览文件(F)... 确定 取消

图 7-33

（4）激活 RabbitMQ Management Plugin，可以通过可视化界面管理 RabbitMQ，并且能清晰地查看 RabbitMQ 服务实例的状态，激活命令如下：

```
rabbitmq-plugins.bat  enable  rabbitmq_management
```

执行结果如图 7-34 所示。

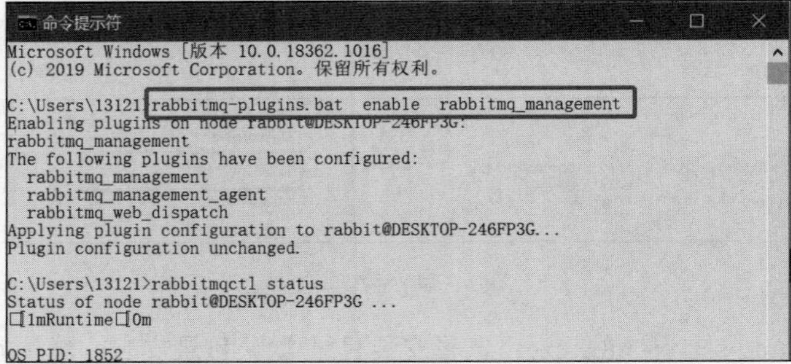

图 7-34

查看 RabbitMQ 管理插件是否安装成功，命令如下：

```
rabbitmqctl status
```

如果执行该命令有报错 Error: unable to perform an operation on node，则可能是 RabbitMQ 后台服务没有启动，执行 net start RabbitMQ 命令启动。

查看 RabbitMQ 已有用户以及用户对应的角色信息，命令如下：

```
Rabbitmqctl.bat list_users
```

RabbitMQ 的默认用户名密码是 guest/guest。

新增一个用户，并赋予超级管理员的角色，命令如下：

```
rabbitmqctl.bat add_user username password
rabbitmqctl.bat set_user_tags username administrator
```

执行结果如图 7-35 所示。

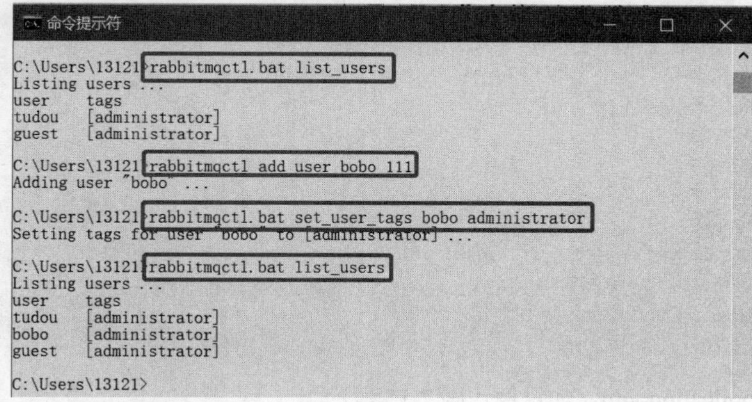

图 7-35

（5）访问网址 http://localhost:15672，并使用刚才创建的账号登录系统，可以查看 RabbitMQ 的管理控制台，操作如图 7-36 所示。

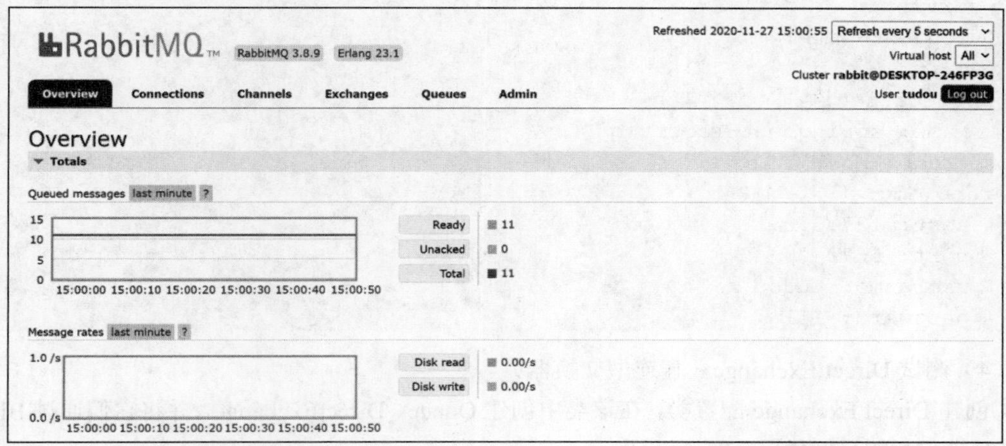

图 7-36

3. Spring Boot 集成 RabbitMQ

（1）新建一个 Spring Boot 项目 springboot-rabbitmq，创建方式在此不再赘述。

（2）在 pom.xml 中添加 RabbitMQ 相关依赖，如示例代码 7-14 所示。

示例代码 7-14　pom.xml（核心代码）

```
<dependencies>
    <dependency>
        <groupId>org.springframework.boot</groupId>
        <artifactId>spring-boot-starter-web</artifactId>
    </dependency>
    <dependency>
        <groupId>org.springframework.boot</groupId>
        <artifactId>spring-boot-starter-test</artifactId>
        <scope>test</scope>
        <exclusions>
            <exclusion>
                <groupId>org.junit.vintage</groupId>
                <artifactId>junit-vintage-engine</artifactId>
            </exclusion>
        </exclusions>
    </dependency>
  <!—rabbitmq→
    <dependency>
        <groupId>org.springframework.boot</groupId>
        <artifactId>spring-boot-starter-amqp</artifactId>
    </dependency>
</dependencies>
```

（3）配置 application.yml，如示例代码 7-15 所示。

示例代码 7-15　application.yml

```
server:
   port: 8081
   servlet:
    context-path: /mq
spring:
  application:
    name: springboot-rabbitmq
    #rabbitMq 相关设置
  rabbitmq:
    host: 127.0.0.1
    port: 5672
    username: tudou
    password: 111
```

（4）测试 Direct Exchange（直连型交换器）。

①创建 Direct Exchange 配置类，在该类中创建 Queue、DirectExchange，并将它们通过 Binding Key 进行绑定，如示例代码 7-16 所示。

示例代码 7-16　DirectExchangeConfig.java

```
/**
 * 直连型交换器配置类
 */
@Configuration
public class DirectExchangeConfig {
    /**
     * 创建队列
     * Queue(String name, boolean durable)
     *  name: 队列名字
     *  durable: 是否持久化,默认是 false
     *  持久化队列: 会被存储在磁盘上，当消息代理重启时仍然存在
     *  暂存队列: 当前连接有效
     */
    @Bean
    public Queue queueDemo(){
        return new Que“e("queue1-direct-t”st",true);
    }
    /**
     *  创建交换器
     *  DirectExchange(String name, boolean durable, boolean autoDelete)
     *      name: 交换器名称
     *      durable: 是否持久化,默认是 false,表示暂存队列,只有当前链接有效;如果为 true,
     *      表示持久化队列，会被存储在磁盘上
     *      autoDelete: 是否自动删除，默认是 false。当没有生产者或者消费者使用
     *      这个交换器时将会被删除
     */
    @Bean
    DirectExchange directExchangeDemo(){
        return new DirectExchan“e("direct-exchang”-1",true,false);
    }
    /**
     * 将队列和直连型交换器绑定，并设置匹配键为 direct-queue-routing
     */
    @Bean
    Binding bindingDirectQueue(){
        Binding binding = BindingBuilder.bind(queueDemo())
                .to(directExchangeDemo())
                .wi“h("direct-queue-”ey");
        return binding;
    }
}
```

②创建生产者，向 DirectExchange 交换器发送消息，如示例代码 7-17 所示。

示例代码 7-17　ProviderController.java

```
/**
 * 生产者
 */
@RestController
```

```
    public class ProviderController {
        @Autowired
        private RabbitTemplate rabbitTemplate;
        @GetMapping("/sendDirectMessage")
        public String sendDirectMessage(){
            String message = "direct message!!!";
            rabbitTemplate.convertAndSend("direct-exchange-1",
"direct-queue-key",message);
            return "successful";
        }
    }
```

③请求地址 http://localhost:8081/mq/sendDirectMessage，成功向交换器发送消息，此时还没有消费者消费该消息，我们可以在 RabbitMQ 管理界面查看消息是否发送成功，如图 7-37 所示。

④创建消费者，读取队列 queue1-direct-test 的消息，如示例代码 7-18 所示。

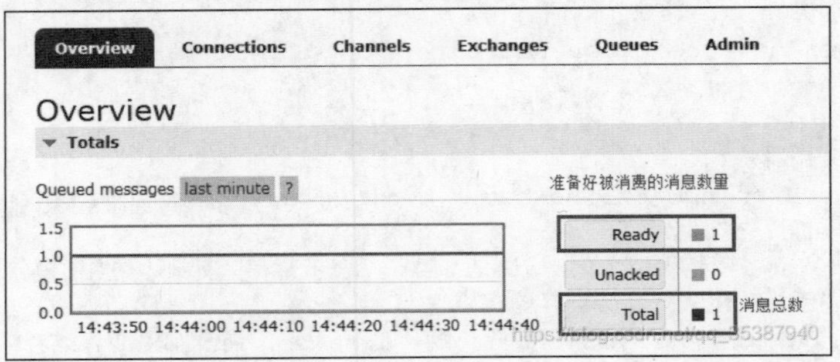

图 7-37

示例代码 7-18　DirectExchangeConsumer.java

```
/**
 * 测试直连型交换器
 */
@Component
@RabbitListener(queues = "queue1-direct-test")
public class DirectExchangeConsumer {
    @RabbitHandler
    public void consumerDirectMsg(String meaasge){
        System.out.println("DirectExchangeConsumer 收到消息: "+meaasge);
    }
}
```

重启项目，我们会发现，消费者收到了名为 queue1-direct-test 队列中的消息，测试结果如图 7-38 所示。

```
2020-11-27 16:23:24.175  INFO 15192 --- [           main] o.s.web.context.ContextLoader
2020-11-27 16:23:25.378  INFO 15192 --- [           main] o.s.s.concurrent.ThreadPoolTaskExecutor
2020-11-27 16:23:25.694  INFO 15192 --- [           main] o.s.a.r.c.CachingConnectionFactory
2020-11-27 16:23:25.748  INFO 15192 --- [           main] o.s.a.r.c.CachingConnectionFactory
DirectExchangeConsumer收到消息: direct message!!!
```

图 7-38

⑤既然 Direct Exchange 是一对一的消费，那么如果同时存在两个消费者监听同一个队列时，是否会存在重复消费呢？换句话说，两个消费者是否会收到相同的消息呢？下面我们创建一个消费者测试一下，如示例代码 7-19 所示。

示例代码 7-19　DirectExchangeConsumer2.yml

```
@Component
@RabbitListener(queues = "queue1-direct-test")
public class DirectExchangeConsumer2 {
    @RabbitHandler
    public void consumerDirectMsg(String meaasge){
        System.out.println("DirectExchangeConsumer2 收到消息: "+meaasge);
    }
}
```

⑥修改生产者 ProviderController.java 中的 sendDirectMessage()方法，如示例代码 7-20 所示。

示例代码 7-20　ProviderController.java（核心代码）

```
@GetMapping("/sendDirectMessage")
public String sendDirectMessage(){
   /* String message = "direct message!!!";
    rabbitTemplate.convertAndSend("direct-exchange-1","direct-queue-key",m
essage);*/
    for (int i=0; i<=10;i++){
        String message = "第"+i+"次, direct message!!!";

rabbitTemplate.convertAndSend("direct-exchange-1","direct-queue-key",messa
ge);
        System.out.println("sendDirectMessage 发送消息: "+message);
        try{
            Thread.sleep(500);
        }catch (Exception e){ }
    }
    return "successful";
}
```

⑦重启项目，请求地址 http://localhost:8081/mq/sendDirectMessage 进行测试，测试结果如图 7-39 所示。观察数据会发现，RabbitMQ 对于多个消费者消费同一个队列实行了轮询的方式，并不存在消息被重复消费的情况。

```
2020-11-27 16:35:04.076  INFO 6256 --- [nio-8081-exec-1] o.a.c.c.C.[Tomcat].[localhost].[/mq]
2020-11-27 16:35:04.076  INFO 6256 --- [nio-8081-exec-1] o.s.web.servlet.DispatcherServlet
2020-11-27 16:35:04.082  INFO 6256 --- [nio-8081-exec-1] o.s.web.servlet.DispatcherServlet
sendDirectMessage发送消息：第0次，direct message!!!
DirectExchangeConsumer收到消息：第0次，direct message!!!
sendDirectMessage发送消息：第1次，direct message!!!
DirectExchangeConsumer2收到消息：第1次，direct message!!!
sendDirectMessage发送消息：第2次，direct message!!!
DirectExchangeConsumer收到消息：第2次，direct message!!!
sendDirectMessage发送消息：第3次，direct message!!!
DirectExchangeConsumer2收到消息：第3次，direct message!!!
sendDirectMessage发送消息：第4次，direct message!!!
DirectExchangeConsumer收到消息：第4次，direct message!!!
sendDirectMessage发送消息：第5次，direct message!!!
DirectExchangeConsumer2收到消息：第5次，direct message!!!
sendDirectMessage发送消息：第6次，direct message!!!
DirectExchangeConsumer收到消息：第6次，direct message!!!
sendDirectMessage发送消息：第7次，direct message!!!
DirectExchangeConsumer2收到消息：第7次，direct message!!!
sendDirectMessage发送消息：第8次，direct message!!!
DirectExchangeConsumer收到消息：第8次，direct message!!!
sendDirectMessage发送消息：第9次，direct message!!!
DirectExchangeConsumer2收到消息：第9次，direct message!!!
sendDirectMessage发送消息：第10次，direct message!!!
DirectExchangeConsumer收到消息：第10次，direct message!!!
```

图 7-39

（5）测试 Topic Exchange（主题交换器）。

①创建主题交换器配置类，如示例代码 7-21 所示。

示例代码 7-21　TopicExchangeConfig.java（核心代码）

```
/**
 * 主题交换器配置类
 */
@Configuration
public class TopicExchangeConfig {
    public final static String KEY1 = "topic.key.A";
    public final static String KEY2 = "topic.key.B";
    public final static String KEY3 = "topic.key.*";
    public final static String KEY4 = "topic.key.#";
    public final static String KEY5 = "topic.key.CC.DD";
    public final static String TOPIC_EXCHANGE_NAME = "topicExchange";
    /**
     *  创建队列 A
     */
    @Bean
    public Queue queueA(){
        return new Queue("queueA");
    }
    /**
     *  创建队列 B
     */
    @Bean
    public Queue queueB(){
        return new Queue("queueB");
    }
    /**
     *  创建队列 c
     */
```

```
        @Bean
        public Queue queueC(){
            return new Queue("queueC");
        }
        /**
         * 创建交换器
         */
        @Bean
        TopicExchange getTopicExchange(){
            return new TopicExchange(TOPIC_EXCHANGE_NAME);
        }
        /**
         * 队列 A 和交换器绑定，绑定的 key 为 topic.key.A，即只有消息携带的路由 key 为
topic.key.A 才能发送到 queueA
         */
        @Bean
        Binding bindingExchangeQueueA(){
            return
BindingBuilder.bind(queueA()).to(getTopicExchange()).with(KEY1);
        }
        /**
         * 队列 B 和交换器绑定，绑定的 key 为 topic.key.*，即只有消息携带的路由 key 以
topic.key 开头才能发送到 queueB
         */
        @Bean
        Binding bindingExchangeQueueB(){
            return
BindingBuilder.bind(queueB()).to(getTopicExchange()).with(KEY3);
        }
        /**
         * 队列 C 和交换器绑定，绑定的 key 为 topic.key.#，即只有消息携带的路由 key 以
topic.key 开头才能发送到 queueC，#表示一个或多个单词，比如 topic.key.CC.DD
         */
        @Bean
        Binding bindingExchangeQueueC(){
            return BindingBuilder.bind(queueC()).
to(getTopicExchange()).with(KEY4);
        }
    }
```

②修改 ProviderController.java，新增方法 sendTopicMessage()，向主题交换器发送消息，如示例代码 7-22 所示。

示例代码 7-22　ProviderController.java（核心代码）

```
    @GetMapping("/sendTopicMessage")
    public String sendTopicMessage(){
        String key = TopicExchangeConfig.KEY1;
        String message = "主题交换器测试，路由键为:"+ key;
        //预期：队列 A、B、C 均可收到消息
```

```
rabbitTemplate.convertAndSend(TopicExchangeConfig.TOPIC_EXCHANGE_NAME,key,mess
age);
        return "successful";
    }
    @GetMapping("/sendTopicMessage2")
    public String sendTopicMessage2(){
        String key = TopicExchangeConfig.KEY3;
        String message = "主题交换器测试，路由键为:"+ key;
        //预期：队列 B、C 均可收到消息
        rabbitTemplate.convertAndSend(TopicExchangeConfig.
TOPIC_EXCHANGE_NAME,key,message);

        return "successful";
    }
    @GetMapping("/sendTopicMessage3")
    public String sendTopicMessage3(){
        //String key = TopicExchangeConfig.KEY4;
        String key = TopicExchangeConfig.KEY5;
        String message = "主题交换器测试，路由键为:"+ key;
        //预期：队列 C 可收到消息
        rabbitTemplate.convertAndSend(TopicExchangeConfig.TOPIC_EXCHANGE_NAME,
key,message);
        return "successful";
    }
```

③创建消费者，如下代码所示。

示例代码 7-23 TopicExchangeQueueABConsumer.java（核心代码）

```
    /**
     * 测试主题交换器——带通配符*的路由键
     */
    @Component
    @RabbitListener(queues = "queueB")
    public class TopicExchangeQueueABConsumer {
        @RabbitHandler
        public void consumerA(String message){
            System.out.println("queueB 收到消息为："+message);
        }
    }
```

示例代码 7-24 TopicExchangeQueueAConsumer.java（核心代码）

```
    /**
     * 主题交换器——无匹配符的路由 key
     */
    @Component
    @RabbitListener(queues = "queueA")
    public class TopicExchangeQueueAConsumer {
        @RabbitHandler
        public void consumerA(String message){
            System.out.println("queueA 收到消息为："+message);
```

```
    }
}
```

示例代码 7-25 TopicExchangeQueueCConsumer.java（核心代码）

```
/**
 * 主题交换器——测试带#的路由 key
 */
@Component
@RabbitListener(queues = "queueC")
public class TopicExchangeQueueCConsumer {
    @RabbitHandler
    public void consumerA(String message){
        System.out.println("queueC 收到消息为: "+message);
    }
}
```

④重启项目，请求地址 http://localhost:8081/mq/sendTopicMessage 进行测试，测试结果如图 7-40 所示。观察数据会发现，当生产者发消息带的路由键为 topic.key.A 时，队列 A、B、C 均能收到消息，因为该主题交换器与 QueueA、QueueB、QueueC 的 Binding Key 分别为 topic.key.A、topic.key.*、topic.key.#，根据前一小节分析的匹配策略，这三个 Binding Key 均能与 topic.key.A 这个 Routing Key 匹配上。

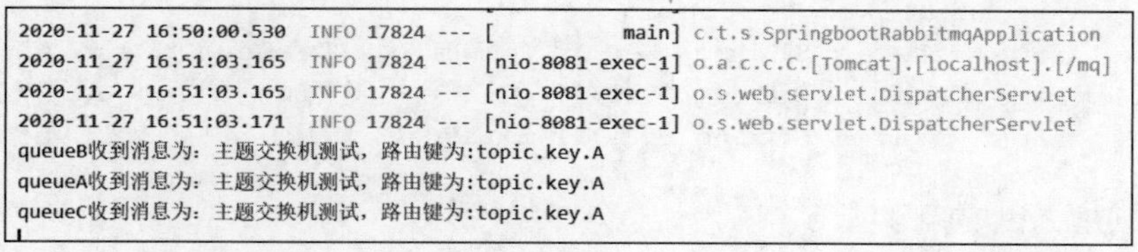

```
2020-11-27 16:50:00.530  INFO 17824 --- [           main] c.t.s.SpringbootRabbitmqApplication
2020-11-27 16:51:03.165  INFO 17824 --- [nio-8081-exec-1] o.a.c.c.C.[Tomcat].[localhost].[/mq]
2020-11-27 16:51:03.165  INFO 17824 --- [nio-8081-exec-1] o.s.web.servlet.DispatcherServlet
2020-11-27 16:51:03.171  INFO 17824 --- [nio-8081-exec-1] o.s.web.servlet.DispatcherServlet
queueB收到消息为：主题交换机测试，路由键为:topic.key.A
queueA收到消息为：主题交换机测试，路由键为:topic.key.A
queueC收到消息为：主题交换机测试，路由键为:topic.key.A
```

图 7-40

请求地址 http://localhost:8081/mq/sendTopicMessage1 进行测试，测试结果如图 7-41 所示。我们观察数据会发现，当生产者发消息带的路由键为 topic.key.*时，只有队列 B、C 均能收到消息，因为该主题交换器与 QueueA、QueueB、QueueC 的 Binding Key 分别为 topic.key.A、topic.key.*、topic.key.#，根据前一小节分析的匹配策略，只有 topic.key.#、topic.key.*能与 topic.key.*这个 Routing Key 匹配上。

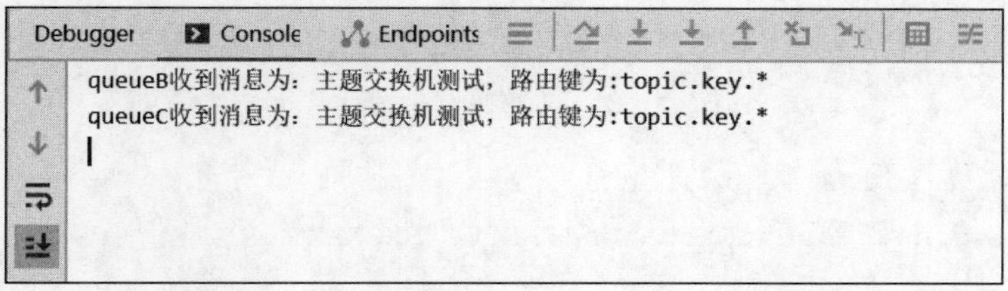

```
Debugger  Console  Endpoints
queueB收到消息为：主题交换机测试，路由键为:topic.key.*
queueC收到消息为：主题交换机测试，路由键为:topic.key.*
```

图 7-41

请求地址 http://localhost:8081/mq/sendTopicMessage2 进行测试，测试结果如图 7-42 所示。观察数据会发现，当生产者发消息带的路由键为 topic.key.CC.DD 时，只有队列 C 能收到消息，因为该主题交换器与 QueueA、QueueB、QueueC 的 Binding Key 分别为 topic.key.A、topic.key.*、 topic.key.#，根据前一小节分析的匹配策略，只有 topic.key.#能与 topic.key.CC.DD 这个 Routing Key 匹配上。

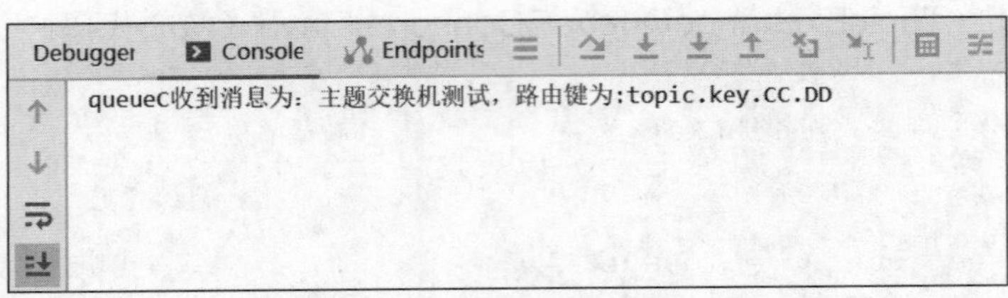

图 7-42

（6）测试 Fanout Exchange（扇形交换器）。

①创建扇形交换器配置类，如示例代码 7-26 所示。

示例代码 7-26 FanoutExchangeConfig.java（核心代码）

```
/**
 * 扇形交换器配置类
 * 扇形交换器没有路由的概念，就算绑定了路由键也是无视的，这个交换器接收到消息后会直接转发到与其绑定的所有队列上，类似于广播的性质
 */
@Configuration
public class FanoutExchangeConfig {
    public final static String FANOUT_QUEUE_A = "fanout.queue.A";
    public final static String FANOUT_QUEUE_B = "fanout.queue.B";
    public final static String FANOUT_EXCHANGE_NAME = "fanoutExchange";
    /**
     * 创建三个队列 A、B、C
     */
    @Bean
    public Queue fanoutQueueA(){
        return new Queue(FANOUT_QUEUE_A);
    }
    @Bean
    public Queue fanoutQueueB(){
        return new Queue(FANOUT_QUEUE_B);
    }
    /**
     * 创建交换器
     */
    @Bean
    FanoutExchange fanoutExchange() {
        return new FanoutExchange(FANOUT_EXCHANGE_NAME);
    }
    /**
     * 绑定队列和扇形交换器
     */
    @Bean
    Binding bindingFanoutExchangeA() {
        return BindingBuilder.bind(fanoutQueueA()).to(fanoutExchange());
```

```
    }
    @Bean
    Binding bindingFanoutExchangeB() {
        return BindingBuilder.bind(fanoutQueueB()).to(fanoutExchange());
    }
}
```

②修改 ProviderController.java，新增方法 sendFanoutMessage()，向扇形交换器发送消息，如下代码所示。

示例代码 7-27　ProviderController.java（核心代码）

```
@GetMapping("/sendFanoutMessage")
public String sendFanoutMessage(){
    String message = "send message to fanout exchange!!";
    rabbitTemplate.convertAndSend(FanoutExchangeConfig.
FANOUT_EXCHANGE_NAME,null,message);
    return "successful";
}
```

③创建消费者，如下代码所示。

示例代码 7-28　FanoutExchangeConsumerA.java（核心代码）

```
/**
 * 消费者——测试扇形交换器
 */
@Component
@RabbitListener(queues = "fanout.queue.A")
public class FanoutExchangeConsumerA {
    @RabbitHandler
    public void consume(String message){
        System.out.println("收到来自 fanout 交换器-queueA 的消息："+message);
    }
}
```

示例代码 7-29　FanoutExchangeConsumerB.java（核心代码）

```
/**
 * 消费者——测试扇形交换器
 */
@Component
@RabbitListener(queues = "fanout.queue.B")
public class FanoutExchangeConsumerB {
    @RabbitHandler
    public void consume(String message){
        System.out.println("收到来自 fanout 交换器-queueB 的消息："+message);
    }
}
```

④重启项目，请求地址 http://localhost:8081/mq/sendDirectMessage 进行测试，测试结果如图 7-43 所示。观察数据会发现，两个队列都绑定了该交换器，所以两个队列的消费者均能监听到这条消息。

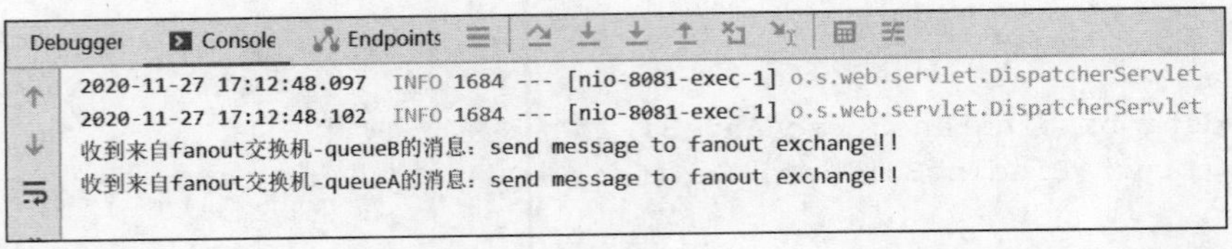

```
2020-11-27 17:12:48.097  INFO 1684 --- [nio-8081-exec-1] o.s.web.servlet.DispatcherServlet
2020-11-27 17:12:48.102  INFO 1684 --- [nio-8081-exec-1] o.s.web.servlet.DispatcherServlet
收到来自fanout交换机-queueB的消息：send message to fanout exchange!!
收到来自fanout交换机-queueA的消息：send message to fanout exchange!!
```

图 7-43

小 结

本小节主要讲解了消息中间件的概念，以及为什么要用消息中间件、消息中间件应用场景等，其中主要详细讲解了 RabbitMQ。RabbitMQ 在消息中间件涉及的三个概念（生产者、消息队列、消费者）以外又增加了路由的概念，即在生产者和消息队列之间增加了一个交换器（Exchange），生产者把消息发给交换器，交换器再根据调度策略把消息转发给消息队列。RabbitMQ 主要有 Fanout、Direct、Topic、Headers 和 Dead Letter 五种交换器（路由）模式。

消息中间件是分布式系统中的重要组件，主要解决应用解耦、异步消息、流量削峰等问题，实现高性能、高可用、可伸缩和最终一致性架构，为业务的实现提供了极大的便利。

7.4 高并发实战

7.4.1 分布式系统生成唯一 ID 方案

在高并发项目中，我们经常涉及系统唯一 ID 的生成，比如常用的订单号 ID。生成唯一 ID 的策略有很多种，当然我们需要找到适合当下场景、需求以及性能要求的 ID 生成策略。分布式高并发系统下生成唯一 ID 在一般情况下都会有以下要求：

①全局唯一性，不能重复。

②信息安全，不能包含公司运营和用户账号相关数据。

③ID 长度固定，且长度适宜。

④ID 需要保证递增顺序。

⑤ID 生成耗时较短。

下面介绍几种 ID 生成策略。

（1）数据库自增长序列或字段

数据库命令：

```
show VARIABLES;
set GLOBAL auto_increment_increment = 100;
set GLOBAL auto_increment_offset = 1;
```

优点：代码实现简单，ID 可按顺序排列，可读性强。

缺点：此种 ID 获取方式需要依赖于数据库，扩容麻烦，而且需要数据插入成功之后才能从数据库查询到产生的 ID，如果是数据库集群，主从数据库同步延迟时还必须从 master 数据库查询。

（2）UUID

UUID 是唯一标识码，一共 16 个字节（128 位），组成部分包括当前日期、时间、时钟序列、MAC 地址。UUID 有多种实现版本，实用且简单的方式就是 JDK 自带的 UUID。

具体实现如下代码所示。

示例代码 7-30　UUIDUtil.java

```
public class UUIDUtil {
    public static String generateUUID(){
        UUID uuid = UUID.randomUUID();
        return uuid.toString();
    }
    public static void main(String[] args) {
        System.out.println(generateUUID());
    }
}
```

优点：UUID 生成非常简单，并且是通过本地生成的，不占用宽带，不影响数据迁移。

缺点：UUID 格式类似于“e87235eb-48c1-49dd-be55-440b78519a92”，是无序字母样式，无法保证有序，即 ID 递增，并且其查询也慢，UUID 不可读。

（3）Redis 生成唯一 ID

Redis 生成 ID 格式：时间 14 位（年月日时分秒）+ Redis 自增（从 1 开始自增）。

具体实现代码如下。

示例代码 7-31　RedisService.java（部分代码）

```
public Long getIncrValue(String key){
    ValueOperations valueOperations = redisTemplate.opsForValue();
    redisTemplate.setValueSerializer(new StringRedisSerializer());
    Long incrementId = valueOperations.increment(key, 1);
    return  incrementId;
}
```

测试代码如下。

示例代码 7-32　RedisController.java（部分代码）

```
@RequestMapping("/getIncrementId")
public String getIncrementId(){
    String currentTime = DateUtil.format(new Date(), "yyyyMMddHHmmss");
    String key = "id_"+ currentTime;
    Long incrValue = redisService.getIncrValue(key);
    String id =currentTime+incrValue;
    return id;
}
```

优点：使用灵活，不依赖于数据库，并且性能优于数据库。

缺点：系统本身未使用 Redis，就需要额外集成 Redis 服务。

（4）Twitter 的 snowflake 算法生成唯一 ID

Twitter 的 snowflake（雪花）是一种开源算法，用于分布式系统生成唯一 ID，即 8 个字节（64bit）大小的整数。其生成 ID 组成部分包括 1bit 符号位、41bit 时间戳、10bit 工作机器 ID、12bit 序列号，如图 7-44 所示。

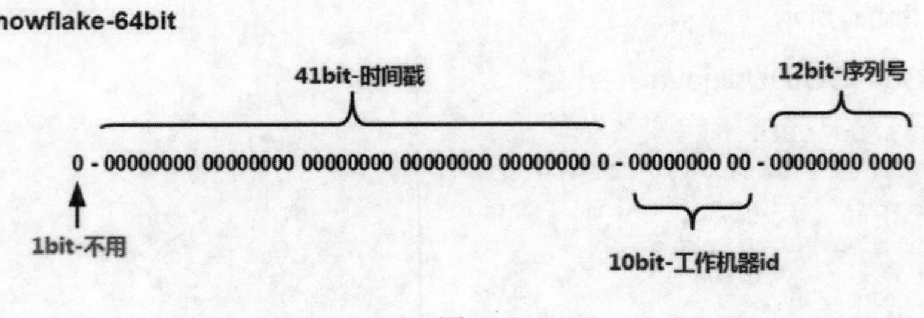

图 7-44

①1bit 标识符：二进制中最高位为符号位，0 表示正数，1 表示负数。由于生成的 id 通常为 long 类型的正数，因此最高位为 0。

②41bit 时间戳：用来记录时间戳（毫秒级），41bit 时间戳并不是存储的当前时间的时间戳，而是存储的时间戳的差值，即 41bit 时间戳 = 当前时间戳 - 开始时间戳（开始时间戳指的是 ID 生成器开始使用的时间，由我们的程序来指定）。

③10bit 工作机器 id：包括 5bit 机器 id（workId）和 5bit 数据中心 id（dataCenterId）。

④12bit 序列号：毫秒内的计数，计数序列号支持每个节点每毫秒，即同一台机器同一时间戳可产生 4096 个 ID 序列号。

雪花算法具体实现见第 7 章 springboot-order-id 项目中 SnowflakeIdWorker.java 文件。

优点：snowFlake 算法产生的 ID，整体上是按照时间自增排序的，并且整个分布式系统内不会产生 ID 碰撞，因为其中包含了数据中心 ID 和机器 ID。另外，雪花算法生成 ID 的效率较高，经测试每秒大概能产生 26 万 ID。

缺点：机器 ID 和数据中心 ID 在分布式系统部署中通常会使用相同的配置，会导致 ID 重复的风险。我们可以通过 ip 生成机器 ID，通过 hostName 生成数据中心 ID 的方式，最大限度防止 10 位机器码重复（上述代码示例已给出解决方案）。

总　结

比较以上四种 ID 生成方式，雪花算法生成的 ID 满足分布式系统 ID 生成的所有要求，相较而言是最合适的方案之一，推荐使用。

7.4.2　秒杀场景实战

这一小节里将在前面介绍的 Redis 缓存、消息队列以及订单号生成策略结合在一起，做一个实战案例——秒杀系统。我们知道每年的淘宝、京东双 11 或是 12306 抢票软件都会面临超大流量，当系统流量激增时，如何让系统稳定、快速地运行是技术人员的一大难题。淘宝在发展初期有很多问题，经常出现服务器崩溃，特别是 21 世纪 10 年代智能手机普及后，网上购物越来越受网友欢迎，因此每一年双 11 的流量都很大，这时淘宝系统问题充分暴露了出来，大大小小无数的难题被淘宝技术人员攻克，时至今日，淘宝已能够扛住亿级别的流量。

电商系统的技术含金量十足，本章前两节提到的缓存与消息队列技术在电商系统中用得很多，因为电商系统流量多，流量多服务器压力就大，就需要借助缓存来减轻服务器压力。瞬时压力大可以用消息队列起到削峰的作用，因为消息队列的显著特点就是异步、排队，并且消息可以持久化，不用担心消息丢失。缓存服务与消息队列服务都可以搭建高可用集群。不过不要为了使用消息队列而使用消息队列，如何构建一个快速稳定运行的业务系统才是重中之重，离开了业务，技术只是一个空壳。

一个秒杀系统的架构远不止使用缓存和消息队列那么简单，下面来谈一谈秒杀系统的架构思想。

（1）接口限流

秒杀最终的本质是数据库的更新，但是有很多大量无效的请求，我们最终要做的就是把这些无效的请求过滤掉，防止渗透到数据库。限流需要入手的方面很多，前端限流的做法是当页面上的秒杀按钮点击一次之后，将其设为 disable 状态，防止用户不停地点击秒杀按钮请求后端。有一些聪明的用户知道如何利用浏览器的开发者控制台的 network 找到秒杀 URL，然后利用一些 HTTP 请求工具无限循环请求。对于这种情况我们需要做接口防刷，主要有以下三种方案：

①为了避免有程序访问经验的人通过下单页面 URL 直接访问后台接口来秒杀货品，我们需要将秒杀的 URL 实现动态化，即使是开发整个系统的人都无法在秒杀开始前知道秒杀的 URL。

②在网关层做限流，避免流量渗透到下游系统，让同一个用户 xx 秒内重复请求直接拒绝，具体多少秒需要根据实际业务和秒杀的人数而定，一般限定为 10 秒。具体的做法就是通过 Redis 的键过期策略，键可以设置为 userId，让每个请求都从 Redis 中根据键获取一个 value，value 可以是任意值，一般放业务的属性好一点，比如 username。如果获取到这个 value 为空或者为 null，就表示它是有效的请求，然后放行这个请求；如果不为空，就表示它是重复性请求，直接丢掉这个请求。如果请求有效，则设置键，并对该键设置一个 10 秒的过期时间。

③接口限流的策略有很多，令牌桶算法是其中一种，在网关层做限制。令牌桶算法的基本思路是每个请求尝试获取一个令牌，后端只处理持有令牌的请求。生产令牌的速度和效率都可以自己限定，guava 提供了 RateLimter 的 API 供我们使用，具体的代码实现这里就不说了。

（2）高性能

秒杀场景伴随着高并发大流量，而 CPU、内存等系统资源整个过程的瓶颈十分珍贵，我们应该将有限的系统资源分配给主要的请求，而额外的不是必需的请求，在系统架构设计的时候就应其消除，比如将页面做动静分离、前端页面静态化，数据则通过异步 Ajax 请求从后端获取，尽量减少 HTTP 请求，合并 CSS 或 JS，启用浏览器缓存和文件压缩，CDN 加速，建立独立图片服务器等。

服务器的数据如果是热点数据，比如商品信息，则可以将它们缓存到 Redis 中，从而快速访问。简单来说，热点数据就是那些访问频率高、读多写少的数据，一般情况下热点数据才适合缓存，与热点数据相反的冷数据不建议缓存。

（3）服务降级

假如在秒杀过程中出现了某个服务器宕机，或者服务不可用，应该做好后备工作。通过 Spring

Cloud Netflix 项目中的 Hystrix 组件进行服务熔断和降级，可以开发一个备用服务。假如服务器真的宕机了，直接返回给用户一个友好的提示，而不是显示直接卡死、服务器错误等生硬的反馈。

（4）水平扩展

除了接口限流、静态缓存、热点数据缓存等方案，我们还需要对系统架构进行水平扩展。首先，对整个大的系统按业务垂直划分为多个子系统，每个子系统又进行水平扩展，利用 Nginx 做反向代理与负载均衡，Nginx 也可以进行水平扩展，是通过 DNS 轮询实现的。DNS 服务器对于一个域名配置了多个解析 ip，每次 DNS 解析请求会访问 DNS 服务器，DNS 服务器轮询返回这些 ip。因此，当 Nginx 成为瓶颈的时候，新增一个 Nginx 结点和对应外网 ip 就能扩展反向代理层的性能，做到理论上的无限高并发。Redis 构建高可用集群，可以防止单机 Redis 的内存成为系统瓶颈，数据库 MySQL 做分库分表、读写分离等。

（5）Redis 预扣库存

预扣库存就是用户抢到商品后库存减 1 的操作，为什么要预扣库存呢？一般减库存有下单减库存和付款减库存两种方案，Redis 预扣库存是介于这两者之间的折中方案。很多请求进来都需要后台查询库存，这是一个频繁读的场景。可以使用 Redis 来缓存商品库存，具体做法是在秒杀活动开始之前将秒杀商品的库存信息加载到 Redis 中，秒杀活动开始后，请求进来，先从 Redis 中查询商品库存，如果大于 0，则代表抢到商品，Redis 中库存减 1。这里需要注意的是，虽然 Redis 的单线程特性保证命令都是原子性操作，但是查询库存和扣除库存是两个命令，命令之间不具有原子性，如果不采取任何措施，在高并发的情况下就有可能出现超卖问题。我们可以利用 Lua 脚本将两个命令合并到一起，最后让 Redis 执行 Lua 脚本来保持原子性。另外，超卖问题的本质会导致 MySQL 的库存小于 0，因此我们可以利用如下 SQL 语句防止库存小于 0：update goods set amount=amount-1 where id=? and amount>0。由于 update 语句会给该记录加上排它锁，因此不会存在事务并发问题。还可以将数据库的 amount 字段设为无符号，如果插入负数，就会抛出异常，service 捕获异常后做容错处理。

（6）异步下单

上面说到在 Redis 中预扣库存，那么如何保证 Redis 库存与 MySQL 库存的一致性呢？答案是使用消息队列 rabbitmq 异步下单，做法是将生成的订单发送到队列里，随后立即向用户响应下单成功的消息。一方面，消费者收到消息后再在 MySQL 中执行修改库存的 update 操作；另一方面，用户接收到下单成功的响应后会跳转到一个带有计时的支付页面，一般是 15 分钟，如果超时或者是用户主动取消支付，则后台需要相应增加库存。

基于上述理论，下面我们来动手实现一个高并发的案例，主要实现的技术有：

- Redis 预扣库存。
- Redis 通过执行 Lua 脚本保证原子性。
- 使用消息队列 RabbitMQ 异步下单。
- 使用雪花算法生成不重复订单号。
- 项目整体框架采用 Spring Boot+MyBatis。

因为这里只是为了演示，没有必要搞得那么复杂，像接口限流、水平扩展等方案都没有加进去，重点掌握 Redis 预扣库存、异步下单等方案即可。

如何搭建项目就不说了，下面从 pom.xml 文件说起。在 pom.xml 中引入 Spring Boot、MyBatis、Redis、RabbitMQ、commons-lang3 等相关依赖，如下代码所示。

示例代码 7-33 pom.xml（部分代码）

```
<dependency>
    <groupId>org.springframework.boot</groupId>
    <artifactId>spring-boot-starter-web</artifactId>
</dependency>
<dependency>
    <groupId>org.mybatis.spring.boot</groupId>
    <artifactId>mybatis-spring-boot-starter</artifactId>
    <version>2.1.3</version>
</dependency>
<dependency>
    <groupId>mysql</groupId>
    <artifactId>mysql-connector-java</artifactId>
    <scope>runtime</scope>
</dependency>
<dependency>
    <groupId>org.springframework.boot</groupId>
    <artifactId>spring-boot-starter-data-redis</artifactId>
</dependency>
<dependency>
    <groupId>org.springframework.boot</groupId>
    <artifactId>spring-boot-starter-amqp</artifactId>
</dependency>
<dependency>
<groupId>org.apache.commons</groupId>
    <artifactId>commons-lang3</artifactId>
    <version>3.8</version>
</dependency>
```

为了演示，我们在数据库中设计了两张表，分别是 book 和 book_order，顾名思义，设计的是一个秒杀图书的项目，如果用户秒杀成功，则用户的订单会插入 book_order 这张订单表里，如图 7-45 所示。

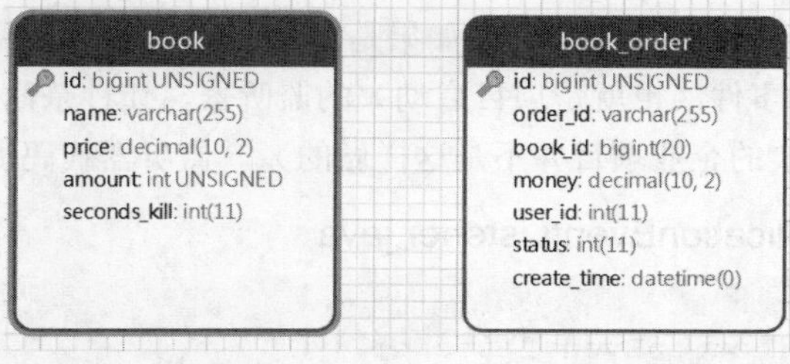

图 7-45

创建数据库表的命令如下：

book 表：

```
CREATE TABLE book (
```

```
  id    bigint(20) UNSIGNED NOT NULL AUTO_INCREMENT,
  name    varchar(255) CHARACTER SET utf8mb4 COLLATE utf8mb4_general_ci NULL
DEFAULT NULL,
  price    decimal(10, 2) NULL DEFAULT NULL,
  amount    int(11) UNSIGNED NULL DEFAULT NULL,
  seconds_kill    int(11) NULL DEFAULT NULL,
  PRIMARY KEY (   id   ) USING BTREE
);
```

book_order 表：

```
CREATE TABLE book_order (
  id    bigint(20) UNSIGNED NOT NULL AUTO_INCREMENT,
  order_id    varchar(255) CHARACTER SET utf8mb4 COLLATE utf8mb4_general_ci
NULL DEFAULT NULL,
  book_id    bigint(20) NULL DEFAULT NULL,
  money    decimal(10, 2) NULL DEFAULT NULL,
  user_id    int(11) NULL DEFAULT NULL,
  status    int(11) NULL DEFAULT NULL,
  create_time    datetime(0) NULL DEFAULT NULL,
  PRIMARY KEY (   id   ) USING BTREE
);
```

book 表记录如图 7-46 所示。

id	name	price	amount	seconds_kill
1	三国演义	200.00	206666	(Null)
2	水浒传	125.90	185869	(Null)
3	西游记	130.00	8899	(Null)
4	红楼梦	69.88	289	(Null)
5	数据结构与算法	79.50	1200	(Null)
6	重新定义springcloud实战	188.99	55700	(Null)
7	Hadoop权威指南	99.00	338	(Null)
8	mysql技术内幕	85.00	1208	(Null)
9	kafka权威指南	120.00	2865	(Null)
10	大话数据结构	89.00	6868	(Null)
11	三国演义	69.00	10	1
12	水浒传	59.00	10	1
13	西游记	59.00	10	1
14	红楼梦	59.00	10	1

图 7-46

amount 字段表示库存，seconds_kill 字段如果是 1，就表示秒杀商品。为了方便，这里写了一个监听 Spring 上下文刷新事件（也就是项目启动）的监听器，让秒杀商品在项目启动的时候提前加载到 Redis 缓存中（真实的企业项目并不是这样做的）。监听器代码如下。

示例代码 7-34　ApplicationEventListener.java

```
@Component
public class ApplicationEventListener {
   @Autowired
   private RedisTemplate<String,Object> redisTemplate;
   @Autowired
   private BookMapper bookMapper;
   public static final String SECONDS_BOOK_KEY = "secondsKillBooks";
   @EventListener
```

```
    public void contextRefreshedEventListener(ContextRefreshedEvent
contextRefreshedEvent) {
        if(redisTemplate == null){return;}
        System.out.println("往 redis 中加载秒杀商品开始! ");
        Book params = new Book();
        params.setSecondsKill(1);
        // 获取操作 redis hash 类型的操作类
        HashOperations<String, Object, Object> hashOperations =
redisTemplate.opsForHash();
        // 查询秒杀商品，并通过流的方式转化为 Map，Key 为商品 ID，value 为商品库存
        Map<String,Integer> map =
bookMapper.selectBooks(params).stream().collect(
                Collectors.toMap(t->
String.valueOf(t.getId()),Book::getAmount));
        // 往 redis 中加载数据
        hashOperations.putAll(SECONDS_BOOK_KEY,map);
        System.out.println("往 redis 中加载秒杀商品成功! ");
    }
}
```

秒杀商品采用 hash 类型存储，该 hash 的 Redis Key 为 secondsKillBooks，而该 hash 的 field 为书的 ID，该 hash 的 Value 为书的库存。

另外，Spring Boot 集成 Redis 的配置类、集成 RabbitMQ 的配置类，在本章的前面已经介绍过了，这里不再粘贴相关代码。生成订单号所采用的雪花算法相关代码这里也不再粘贴。准备工作做完之后，下面正式进入业务环节。

秒杀的核心操作就是下单，下面先定义一个下单的 Controller，具体代码如下。

示例代码 7-35 OrderController.java

```
@RestController
public class OrderController {
    @Autowired
    private OrderService orderService;
    @PostMapping("/order/add")
    public String addOrder(@RequestParam(value = "bookId") Long bookId,
                        @RequestParam(value = "userId") Integer userId){
        boolean success = orderService.addOrder(bookId,userId);
        if(success){
            return "下单成功! ";
        }
        return "下单失败";
    }
}
```

这个 Controller 只有一个 addOrder 方法，接收秒杀请求。下单请求的两个必选参数分别是 bookId 和 userId，也就是说，用户下单必须告诉后台服务下单的是哪件商品、下单人是谁。

注意，Controller 的 addOrder 方法中又调用了 OrderService 的 addOrder 方法，这里即将要展示的 OrderServiceImpl 代码中有两个 addOrder 方法，这两个 addOrder 的业务含义是异步下单和真实下单。毫无疑问，这里的 Controller 调用的是异步下单方法。

真正的下单业务代码如下。

示例代码 7-36　OrderServiceImpl.java

```
    @Service
    @Transactional
    public class OrderServiceImpl implements OrderService {
        @Autowired
        private OrderProducer orderProducer;
        @Autowired
        private RedisTemplate<String,Object> redisTemplate;
        @Autowired
        private BookMapper bookMapper;
        @Autowired
        private OrderMapper orderMapper;
        @Override
        public boolean addOrder(Long bookId,Integer userId) {
            DefaultRedisScript<String> redisScript = new DefaultRedisScript();
            // 设置 lua 脚本
            redisScript.setScriptSource(new ResourceScriptSource(new
ClassPathResource("stock_simple.lua")));
            redisScript.setResultType(String.class);
            List<String> keys = new ArrayList<>();
            keys.add(ApplicationEventListener.SECONDS_BOOK_KEY);
            StringRedisSerializer serializer = new StringRedisSerializer();
            // 执行 lua 脚本
            String res =
redisTemplate.execute(redisScript,serializer,serializer,keys,
bookId.toString());
            if(!"1".equals(res)){
                return false;
            }
            Order order = new Order();
            order.setOrderId(String.valueOf(SnowflakeIdWorker.generateId()));
            order.setBookId(bookId);
            order.setMoney(bookMapper.selectBookById(bookId).getPrice());
            order.setUserId(userId);
            order.setStatus(0);
            order.setCreateTime(new Date());
            orderProducer.sendOrder(order);
            return true;
        }
        @Override
        public boolean addOrder(Order order) {
            int result;
            result= orderMapper.insertOrder(order);
            if(result > 0){
                Book params = new Book();
                params.setId(order.getBookId());
                params.setAmount(-1);
                try {
```

```
                result = bookMapper.updateBook(params);
                if(result > 0){
                    return true;
                }
            } catch (Exception e) {
                System.out.println("库存不足，无法下单!");
            }

        }
        throw new RuntimeException("更新库存失败! ");
    }
}
```

注意看一下异步下单方法，前面讲过会通过 Redis 执行 Lua 脚本来保证多个 Redis 命令的原子性，而使用 Redis 主要是为了预扣库存，因此 Lua 脚本中势必包含了查询库存与扣减库存两步操作，下面来看一下 Lua 脚本。

Stock_simple.lua 脚本如下所示。

```
local val=redis.call('hget',KEYS[1],ARGV[1]);
-- 判断是否为空
if(val == false) then
-- ker 或 field 不存在
return '0'
elseif(tonumber(val) <= 0)
then
-- 库存为 0
redis.call('hdel',KEYS[1],ARGV[1]);
return '0'
else
-- 下单成功
redis.call('hincrby',KEYS[1],ARGV[1],-1);
return '1'
end
```

redis.call 函数是用来调用 Redis 命令的，local 表示本地变量，而 KEYS 和 ARGV 两个变量是全局变量，其实是两个数组，数组的下标从 1 开始，数组的值是外部传进来的。为什么说是外部传进来的呢？可以查看 redisTemplate 的 execute 方法：

```
public <T> T execute(RedisScript<T> script, List<K> keys, Object... args){
    return this.scriptExecutor.execute(script, keys, args);
}
```

keys 参数对应的是 Lua 脚本中的 KEYS 变量，args 可变参数对应的是 Lua 脚本中的 ARGV 变量。Redis 执行 Lua 脚本后会返回 0 或 1，如果是 0，就代表秒杀，直接退出整个方法；如果是 1，就代表秒杀成功，紧接着新建一个 Order 对象，由于是异步下单，因此会将 Order 对象作为消息通过生产者发送到 RabbitMQ 中。

生产者代码：

示例代码 7-37 OrderProducer.java

```
    @Service
    public class OrderProducer {
        @Autowired
        private RabbitTemplate rabbitTemplate;
        public boolean sendOrder(Order order){
            try {
rabbitTemplate.convertAndSend(OrderDirectExchangeConfig.EXCHANGE_NAME,
                    OrderDirectExchangeConfig.QUEUE_EXCHANGE_KEY,new
ObjectMapper().writeValueAsString(order));
            } catch (JsonProcessingException e) {
                e.printStackTrace();
            }
            return true;
        }
    }
```

消费者代码：

示例代码 7-38 OrderConsumer.java

```
    @Service
    public class OrderConsumer {
        @Autowired
        private OrderService orderService;
        @RabbitListener(queues = OrderDirectExchangeConfig.QUEUE_NAME)
        public void consumerOrder(Message meaasge, Channel channel){
            System.out.println("订单消费者收到消息："+meaasge);
            ObjectMapper mapper = new ObjectMapper();
            Order order = null;
            try {
                order = mapper.readValue(meaasge.getBody(), Order.class);
            } catch (IOException e) {
                e.printStackTrace();
                return;
            }
            //库存同步到mysql
            try {
                boolean result = orderService.addOrder(order);
                if(result){
                    System.out.println("下单成功["+new
String(meaasge.getBody())+"]!");
                }else{
                    System.out.println("下单失败["+new
String(meaasge.getBody())+"]!");
                }
            } catch (Exception e) {
                e.printStackTrace();
                System.out.println("下单失败["+new
String(meaasge.getBody())+"]!");
```

```
            }
        }
    }
```

消费者收到消息后会调用 OrderServiceImpl 的真实下单方法。我们可以回头看看 OrderServiceImpl 真实下单的 addOrder 方法，里面包含了两步操作：插入订单记录和修改库存。注意，这两步操作要保证原子性，所以要在 OrderServiceImpl 类上打上@Transactional 注解，从而让 Spring 接管事务，并且可通过主动抛出 RuntimeException 使事务回滚。

至此，业务代码就写完了。为了模拟高并发场景，这里定义了一个秒杀客户端类，通过 Apache 的 HttpClient 发出秒杀请求。SecondsKillClient 类代码如下所示。

示例代码 7-39　SecondsKillClient.java

```
    public class SecondsKillClient {
        // 请求数量
        private static final int HTTP_REQUEST_COUNT = 1000;
        // 线程唤醒时间
        private static final long UNTIL_TIME = System.currentTimeMillis()+5000;
        private static final String REQUEST_URL = 
"http://localhost:8080/order/add";
        public void execute(){
            for (int i = 0; i < HTTP_REQUEST_COUNT; i++) {
                Thread t = new Thread(() -> {
                    // 阻塞线程，设置指定的时间唤醒
                    LockSupport.parkUntil(UNTIL_TIME);
                    // 生成随机的 bookId
                    Long bookId = buildRandomBookId();
                    // 生成随机的 userId
                    Integer userId = buildRandomUserId();
                    // 构造 post 请求
                    HttpPost httpPost = new HttpPost(REQUEST_URL);
                    List<NameValuePair> params = new ArrayList<>();
                    params.add(new BasicNameValuePair("bookId", 
String.valueOf(bookId)));
                    params.add(new BasicNameValuePair("userId", 
String.valueOf(userId)));

                    // 发起 http 请求
                    try(CloseableHttpClient httpclient = 
HttpClients.createDefault()) {
                        UrlEncodedFormEntity entity = new 
UrlEncodedFormEntity(params, "utf-8");
                        httpPost.setEntity(entity);
                        httpclient.execute(httpPost);
                    } catch (IOException e) {
                        e.printStackTrace();
                        System.out.println("post 请求发送失败");
                    }
                },"线程"+i+"号！");
                t.start();
```

```
        }
    }
    private Long buildRandomBookId(){
        int a = (ThreadLocalRandom.current().nextInt(1000) + 1) % 4;
        switch (a) {
            case 0: return 11L;
            case 1: return 12L;
            case 2: return 13L;
            case 3: return 14L;
            default: return null;
        }
    }
    private Integer buildRandomUserId(){
        return ThreadLocalRandom.current().nextInt(100000) + 50;
    }
    public static void main(String[] args) {
        new SecondsKillClient().execute();
    }
}
```

这里 5000 个线程同时发起请求，四件商品随机下单。过当 Redis 预扣库存是非原子操作时的情况，MySQL 库存出现了负值，也就是商品超卖了。这是由于高并发导致的，因此 5000 个线程足矣。

下面进入正式的测试环节。首先启动项目，Redis 库存如图 7-47 所示。然后，运行 SecondsKillClient 的 main 方法。MySQL 中的 book 表数据如图 7-48 所示。

local::db0::secondsKillBooks

HASH: secondsKillBooks

row	key	value
1	11	10
2	12	10
3	13	10
4	14	10

图 7-47

id	name	price	amount	seconds_kill
1	三国演义	200.00	206666	(Null)
2	水浒传	125.90	185869	(Null)
3	西游记	130.00	8899	(Null)
4	红楼梦	69.88	289	(Null)
5	数据结构与算法	79.50	1200	(Null)
6	重新定义springcloud实战	188.99	55700	(Null)
7	Hadoop权威指南	99.00	338	(Null)
8	mysql技术内幕	85.00	1208	(Null)
9	kafka权威指南	120.00	2865	(Null)
10	大话数据结构	89.00	6868	(Null)
11	三国演义	69.00	0	1
12	水浒传	59.00	0	1
13	西游记	59.00	0	1
14	红楼梦	59.00	0	1

图 7-48

book_order 表数据如图 7-49 所示。

id	order_id	book_id	money	user_id	status	create_time
1	510380425322147840	13	59.00	548	0	2021-01-16 19:16:26
2	510380425334730756	11	69.00	91852	0	2021-01-16 19:16:26
3	510380425334730753	14	59.00	22868	0	2021-01-16 19:16:26
4	510380425309564928	14	59.00	1833	0	2021-01-16 19:16:26
5	510380430414032896	12	59.00	25691	0	2021-01-16 19:16:27
6	510380430493724672	13	59.00	85862	0	2021-01-16 19:16:27
7	510380425334730754	14	59.00	33339	0	2021-01-16 19:16:26
8	510380425334730752	13	59.00	41474	0	2021-01-16 19:16:26
9	510380425334730755	12	59.00	11655	0	2021-01-16 19:16:26
10	510380430539862016	14	59.00	49032	0	2021-01-16 19:16:27
11	510380425334730758	14	59.00	14593	0	2021-01-16 19:16:26
12	510380425330536448	14	59.00	88223	0	2021-01-16 19:16:26
13	510380430598582272	12	59.00	85229	0	2021-01-16 19:16:27
14	510380430695051264	13	59.00	27653	0	2021-01-16 19:16:27
15	510380425334730757	11	69.00	1912	0	2021-01-16 19:16:26
16	510380430816686080	14	59.00	48675	0	2021-01-16 19:16:27
17	510380430963486720	11	69.00	70139	0	2021-01-16 19:16:27
18	510380430883794944	12	59.00	77092	0	2021-01-16 19:16:27
19	510380431043178496	13	59.00	31904	0	2021-01-16 19:16:27
20	510380431072538625	12	59.00	1419	0	2021-01-16 19:16:27
21	510380431072538624	11	69.00	76654	0	2021-01-16 19:16:27
22	510380431139647488	11	69.00	4646	0	2021-01-16 19:16:27
23	510380431068344320	12	59.00	51210	0	2021-01-16 19:16:27

SELECT * FROM `springboot-mybatis-base`.`book_order` LIMIT 0,1000

第 1 条记录（共 40 条）于第 1 页

图 7-49

四件商品的库存都为 0，并且恰好生成了 40 个订单。由此可见，这个秒杀程序的业务功能是没有线程安全问题的。

本节介绍企业级高并发场景下的一些解决方案。首先讲述了在分布式系统中高并发下唯一 ID 的生成方案，以及不同方案的优劣对比；然后论述了电商项目中常见的秒杀场景的一些解决方案；最后结合前面讲的 Redis 和 RabbitMQ 技术解决秒杀时的扣减库存和下单问题。

第 8 章

Spring Boot 构建企业级应用

本章将讲解怎样构建一个企业级应用，在这一章引入了权限认证框架，讲述 Spring Security 和 Shiro 两种不同的权限框架，同时还介绍了怎样实现单点登录，其中涉及 3 种方式。然后讲解第三方登录（比如 QQ、微信、微博等）的认证方式。在讲解完认证部分的知识后，将详细介绍如何优雅地生成接口文档、集成日志框架等。

8.1　集成权限认证框架

登录和鉴权几乎是所有业务系统中必不可少的部分，权限管理框架属于系统安全的范畴，权限管理实现对用户访问系统的控制，按照安全规则或者安全策略控制用户可以访问而且只能访问自己被授权的资源，其中包括用户身份认证和授权两部分，简称认证授权。接下来讲述基本的身份认证过程和授权管理的集中方式，然后引入两大比较流行的开源框架 Spring Security 和 Shiro 集成到 Spring Boot 项目中，从而快速实现身份认证和授权。

8.1.1　权限认证基础知识

1. 身份认证

身份认证是计算机及软件系统用于确认用户具有合法身份的一种技术手段。众所周知，计算机系统和计算机网络是一个虚拟的数字世界。在这个数字世界里，一切信息（包括用户的身份信息）都是用一组特定的数据来表示的，计算机只能识别用户的数字身份，所有对用户的授权也是针对用户数字身份的授权。我们生活的现实世界是一个真实的物理世界，每个人都拥有独一无二的物理身份。如何保证以数字身份进行操作的操作者就是这个数字身份的合法拥有者，也就是说保证操作者的物理身份与数字身份相对应成为一个很重要的问题。身份认证技术的诞生就是为了解决这个问题。

在真实世界中，验证一个人的身份主要通过三种方式判定：一是根据你所知道的信息来证明你的身份（what you know），假设某些信息只有某个人知道，比如暗号等，通过询问这个信息就可以确认这个人的身份；二是根据你所拥有的东西来证明你的身份（what you have），假设某一个东西只有某个人有，比如印章等；三是直接根据你独一无二的身体特征来证明你的身份（who you are），比如指纹、面貌等。

在信息系统中，对用户的身份认证手段大体可以分为这三种，仅通过一个条件的符合来证明一个人的身份称之为单因子认证。仅使用一种条件判断用户的身份容易被仿冒，也可以通过组合两种不同条件来证明一个人的身份，称之为双因子认证。

身份认证技术从是否使用硬件方面可以分为软件认证和硬件认证。从认证需要验证的条件来看，可以分为单因子认证和双因子认证。从认证信息来看，可以分为静态认证和动态认证。身份认证技术的发展经历了从软件认证到硬件认证、从单因子认证到双因子认证、从静态认证到动态认证的过程。

目前 Web 应用的身份认证方式比较流行的有以下几种：

- HTTP Form 认证：通过表单方式提交用户名和密码的认证方式。
- HTTP Basic 认证：通过 HTTP header 提交用户名+密码的认证方式。
- HTTP Token 认证：通过一个 Token 字符串来识别用户合法身份的认证方式。
- HTTPS 证书认证：基于 CA 根证书签名的双向数字认证方式。

第一种 HTTP Form 认证方式是最常见的，我们在使用各大网站（比如微博、知乎、微信、QQ、Bilibili 等）时，都是通过用户名+密码的方式来进行登录认证的。由于这种方式比较简单，能够被广大用户所接受，因此这是一个最常见而且通用的认证方式。在这种方式下，用户先在客户端填写自己的账号名称和密码，然后通过单击登录方式将账号和密码从网络发送至服务器应用程序，服务器程序通过核对系统中存储的用户名和密码是否和客户端发送过来的一致（一致表明用户身份认证通过，不一致就会拒绝用户登录），整个流程如图 8-1 所示。

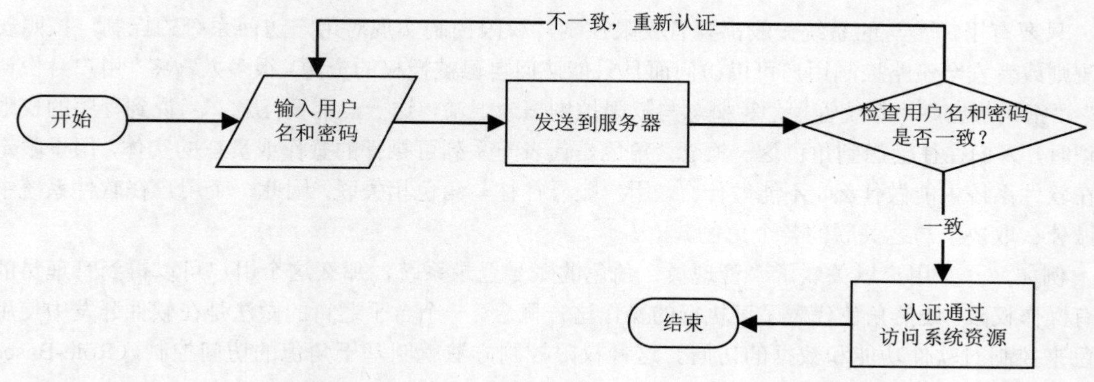

图 8-1

2. 授权管理

授权管理的目的是为了控制资源的访问（哪些用户能够访问哪些资源，什么资源不能被访问），从而控制用户能够使用哪些系统资源。在用户通过系统的身份认证之后，系统就要为用户分配系统资源访问权限了，而这个访问权限是由系统管理员预先设定好的，在用户登录之后，系统根据预先分配的规则去匹配，判断用户是否具有访问资源的权限，只有具有相关权限的用户才可以被放行，否则会被拒绝。授权实际就是 who、what、how 三者之间的关系，即 who 对 what 进行 how 的操作。who，权限的拥有者或主体（如 Principal、User、Group、Role、Actor 等）；what，权限针对的对象或资源（Object、Resource）；how，具体的权限（Privilege，正向授权与负向授权）。简单一点说就是通过给角色授权，将赋有权限的角色施加到某个用户身上，这样用户就可以实施相应的权力了。通过角色的划分，使权限管理更加灵活：角色的权力可以灵活改变，用户的角色身份可以随着场所的不同而发生改变等。

结合前面的身份认证过程，整个认证授权流程如图 8-2 所示。

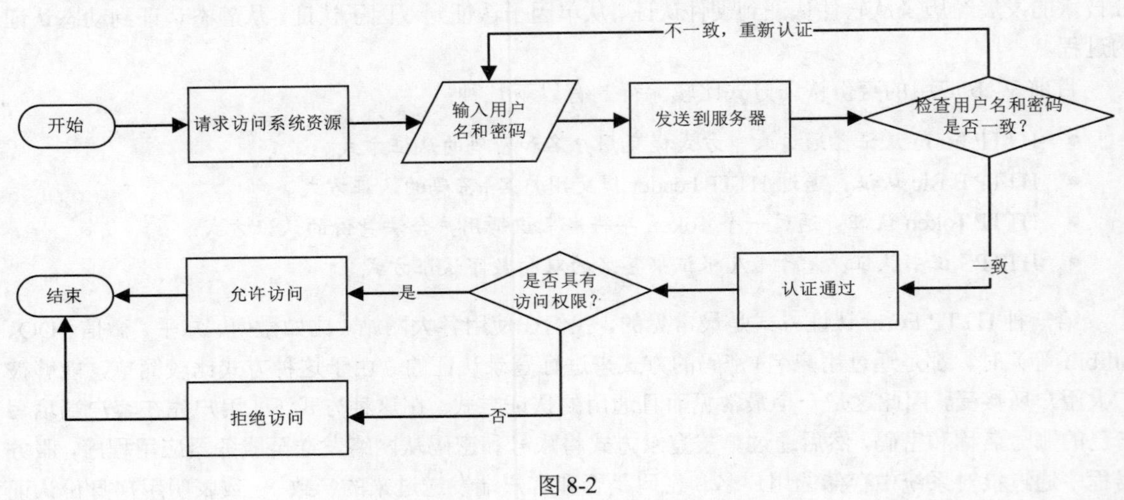

图 8-2

3. 权限控制

只要有用户参与的系统一般都要有权限控制，权限控制实现对用户访问系统的控制，按照安全规则或者安全策略控制用户可以访问而且只能访问自己被授权的资源。很多人常将“用户身份认证”“密码加密”“系统管理”等概念与权限控制概念混淆，这一点需要注意。当说到程序的权限管理时，人们往往会想到角色这一概念。角色是代表一系列可执行的操作或责任的实体，用于限定你在软件系统中能做什么、不能做什么。用户账号往往与角色相关联，因此一个用户在软件系统中能做什么取决于与之关联的各个角色。

例如，一个用户以关联了“管理员”角色的账号登录系统，那么这个用户可以得到管理员的所有操作权限。既然角色代表了可执行的操作这一概念，一个合乎逻辑的做法是在软件开发中使用角色来控制对软件功能和数据的访问。这种权限控制方法就叫基于角色的访问控制（Role-Based Access Control，或简称为 RBAC）。

RBAC 又分为 RBAC0、RBAC1、RBAC2、RBAC3。RBAC0 是其他版本的基础，也是权限控制的核心部分，主要由用户（User）、角色（Role）、许可（Permission）、会话（Session）四个

元素组成。

这里通过一个类图（见图 8-3）来展示一下 RBAC0 模型的结构。

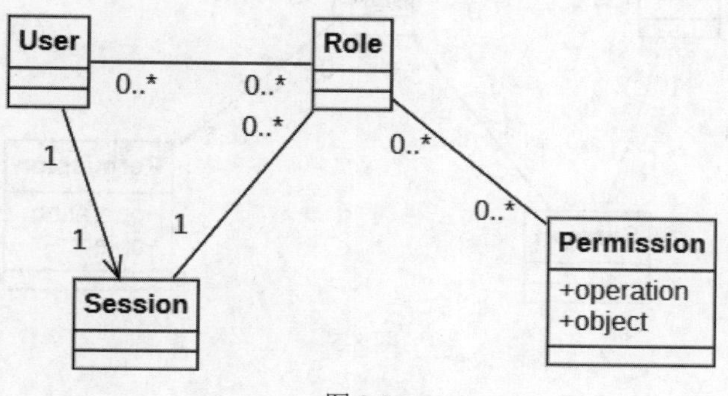

图 8-3

RBAC0 可以解决大部分场景下的权限问题，是最通用的。图 8-3 描述了用户和角色是多对多的关系，表示一个用户可以在不同的场景下拥有不同的角色（例如项目经理同时也是员工）、一个角色可以赋给多个用户（比如项目开发人员这一角色可以赋给多个项目经理、开发组组长、组员等）；角色和许可也是多对多的关系，表示角色可以拥有多个许可操作，同一个许可操作可以授权给多个角色。用户在登录以后会以一种角色建立一次会话，而这次会话会关联到对应的角色，以该角色身份进行被授权许可的操作。

使用 RBAC0 这种权限模型时，我们可以把图 8-3 中的多对多关系分解成两对多对一（使用中间表），建立 5 张数据表来表示这个模型，这 5 张表的 ER 图如图 8-4 所示。

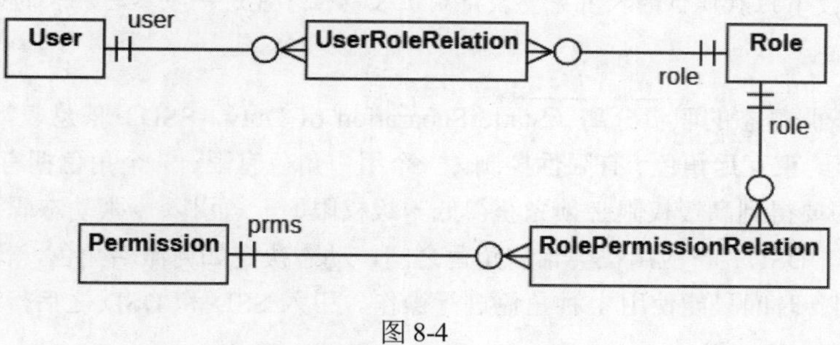

图 8-4

这样在项目中就可以用这 5 张表来进行基本的权限管理，在 User 表中保存用户信息，在角色表中保存角色信息，在许可表中保存能够执行的操作和对象，然后通过 UserRoleRelation 中间表来维护用户和角色多对多的关系，通过 RolePermissionRelation 中间表来维护角色和许可之间多对多的关系。这样就很方便地完成了权限的管理。

RBAC1 模型是在 RBAC0 的基础上的，对角色的划分更加细化，突出角色的层级关系，使得角色具有继承和包含关系，比如在一个图书管理系统中，超级管理员角色可以包含系统管理员角色所有的权限，而系统管理员又可以拥有图书管理员角色的所有权限，这样就形成了一种树形的分层关系，这种模型层次分明，包含关系比较明确。虽然 RBAC0 模型同样能够实现类似的控制，但是会有数据冗余，层次关系不明确，没有 RBAC1 模型更加面向对象。RBAC1 模型的类图如图 8-5 所示。

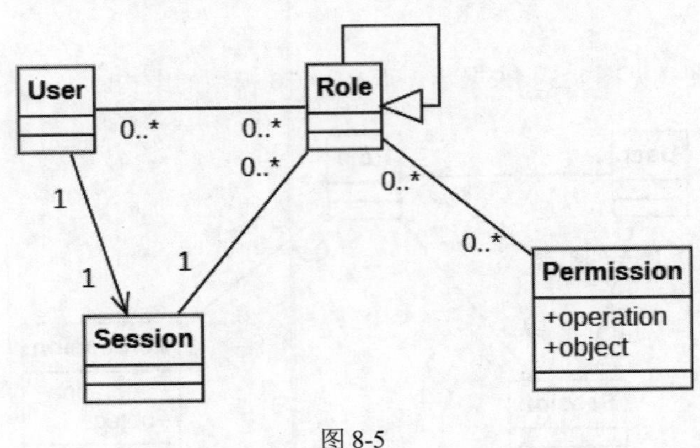

图 8-5

RBAC2 同样也是基于 RBAC0 模型的基础改进而来的，目的是实现对角色权限的限制。其中包含下面三种限制：

- 角色互斥：互斥角色是指各自权限相互制约的两个角色。对于此类角色，一个用户一次活动只能获得其中一种角色，不能同时具有两种角色，比如在一次审计活动中用户不能同时被分配给会计角色和审计角色。
- 有限性控制：角色的权限不能是无限制的，用户可拥有的角色不能是无限制的，用户可能也是有限制的，不能无限制地分配用户，比如公司领导人就是有数量限制的。
- 先决条件约束：想要获得更高级别的权限，首先要拥有低一级的权限。这很像职位晋升，想要获得 L3 的权限，必须从 L1 晋升到 L2 之后才有资格晋升 L3，比如要求班长必须从副班长中选出，没有班级职位的人首先要获得副班长职位才能进一步参与班长的选拔，从而成为班长。

由此，引入一张静态的职责分离（Static Separation of Duty，SSD）信息表，应用在用户和角色之间，主要用于约束互斥角色、有限性控制（一个用户角色有限、一个角色拥有的许可有限等）、先决条件约束（想要得到高级权限必须先获得低一级权限），再引入一张动态职责分离（Dynamic Separation of Duty，DSD），应用在会话和角色之间，动态决定如何激活角色，比如一个用户拥有 5 种角色，但是同一时间只能使用 1 种角色进行操作。引入 SSD 和 DSD 之后，RBAC2 的类图如图 8-6 所示。

RBAC3 是 RBAC1 和 RBAC2 的整合版，既实现了 RBAC1 的角色层次关系，又实现了 RBAC2 对角色和权限的限制，因此 RBAC3 的权限控制是最全面的，但同时又是最复杂的。RBAC3 的类图如图 8-7 所示。

在平常的权限系统中想完全遵循 RBAC 模型是很难的，因为难免在系统业务上有一些差异化的业务考量，所以在设计之初不要太理想、太追求严格的 RBAC 模型设计，因为这样会使得系统处处鸡肋无法拓展，不能因为追求完美导致过度设计，从而带来不必要的复杂化。

RBAC 是一种模型，是一种思想，可以在此基础之上结合实际的业务进行调整，适合的才是最好的。任何系统都会涉及权限管理的模块，无论复杂还是简单，都可以以 RBAC 模型为基础进行相关的灵活运用。

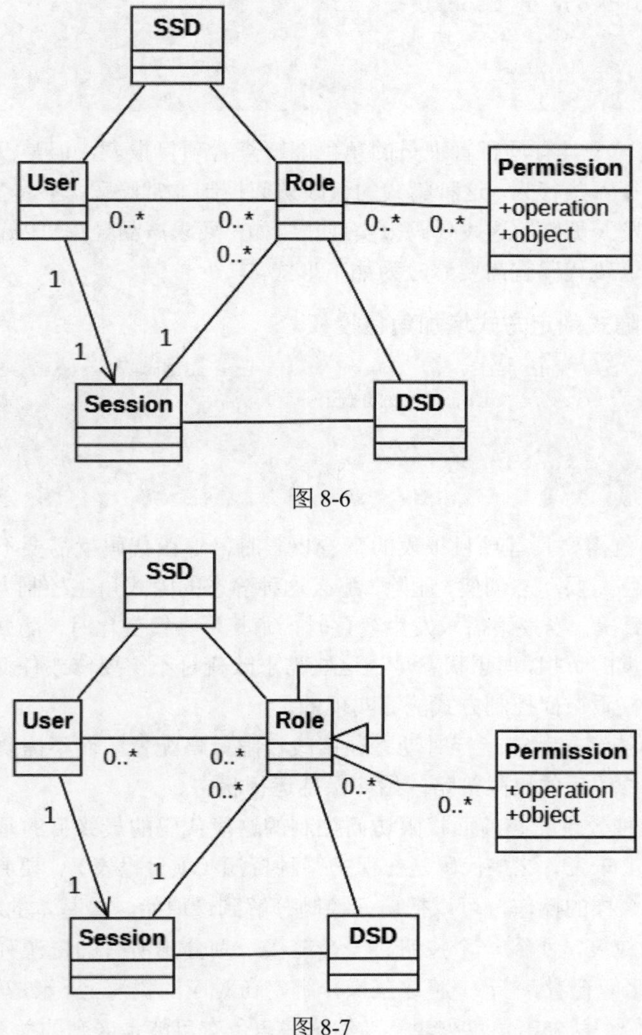

图 8-6

图 8-7

基于角色的权限控制是对系统用户进行分门别类，在一般的 Web 系统中通常会有超级管理员（Super Administrator）、系统管理员（System Administrator）、一般用户（User）这三个基本分类，根据业务或者受众还可以继续对用户进行更细粒度的划分，比如区分是否是 VIP 用户、某个区域用户、某个业务的用户等。基于角色的权限控制在划分粒度上较粗，一般会根据主体（Subject）所拥有的角色来决定主体能否访问一组资源。

在实际使用 RBAC 模型进行编码时有两种方式：一种是隐式（模糊）的，一种是显式（明确）的。

隐式的访问控制是直接通过角色来验证用户有没有操作权限，以角色为单位粒度来判定是否具有访问权。假设有“Project Manager”（项目经理）这样一个角色，只有具有项目经理角色的用户才能访问查看项目报表，控制访问的方式就是根据是否存在这一角色来决定是否显示“查看项目报表”这一按钮，下面通过一段代码来展示如何使用隐式的判定方式。

示例代码 8-1　隐式判定方式角色授权

```
if (user.hasRole("Project Manager") ) {
```

```
    //show the project report button
} else {
    //don't show the button
}
```

这一段代码假设一个具有项目管理员的角色能够查看项目报表，但是没有明确语句来规定这一个角色到底能够执行哪些行为。这种隐式的权限访问控制非常脆弱，不具备可扩展性，需求一旦发生变化，就需要开发人员修改这段代码来实现。例如，要求产品经理“Product Manager”也能够查看项目报表，那么这段代码就需要修改为如下形式。

示例代码 8-2　隐式判定方式增加角色授权

```
if (user.hasRole("Project Manager") || user.hasRole("Product Manager")) {
    //show the project report button
} else {
    //don't show the button
}
```

当越来越多的角色需要具有项目报表的查看权限时，这段代码就需要不断地添加角色判断，而且如果需求要求角色需要动态创建、删除，那么这种静态的隐式角色控制方式就不能够满足需求了。从这个例子能够看出，当权限需求发生变化时，尤其是频繁变化时，隐式的权限访问控制方式会给程序开发带来沉重的负担，理想状态是当这些需求改变时不需要修改任何代码就能动态满足这种需求，这是显式的权限访问控制方式能够实现的。

显式的访问控制能够在程序运行时动态地修改权限策略配置，而不需要重构代码重新部署系统，就能够迅速适应修改后的权限策略，始终保持运行状态。

怎么才能达到这种效果呢？前面权限访问控制的这段代码的最终目的是什么，最终要做什么样的控制呢？从根本上来说，这些代码是在保护某种资源（项目报表），要界定一个用户能不能对其进行操作、能做什么样的操作，可以把访问控制分解到最原始、最基本的粒度上。

首先，以资源为粒度，小到一个按钮、一个链接，划分出所有的资源列表；然后抽象出这些资源所能做的操作，比如查看、修改，将这些操作绑定在对应的资源上，形成一个完整的权限列表；其次，根据用户的身份，划分出角色列表，当然这些角色有可能是嵌套的关系；最后，将一组资源权限列表赋予某个角色，这一个角色就可以拥有很多个对资源进行操作的权限了，这一步操作就是角色与权限的绑定。当一个角色不能操作某个资源的时候，只需要将这个资源的权限和角色的绑定关系解除（从关联表中删除）即可。

以前面提到的项目管理员需要具有查看项目报表的权限为例，使用显式的访问控制方式对其进行修改，相关代码就变得比较简单了。

示例代码 8-3　显式判定方式资源授权

```
if (user.isPermitted("projectReport:view")) {
    //show the project report button
} else {
    //don't show the button
}
```

在修改之后的这段代码中，我们只需要判断用户是不是有被授权查看项目报表的权限，而不必关心用户是什么角色，这样即使用户的角色发生改变，只要其现有的角色还拥有查看项目报表的这个权限，用户就依然能够查看项目报表。采用这种方式进行资源授权的好处是，不同的角色可以被动态地（在项目运行过程中）授予相同的资源访问权限，不必拘泥于具体哪个角色，也不需要修改相关权限判定的代码了。这种基于资源细粒度的授权方式使得代码的可扩展性和健壮性得到了很大的提升，在这种显式的访问控制模型下不必再局限于角色，角色退化为只是用来对资源权限进行分组，用户可以没有任何角色，直接给用户分配相关资源的权限，这样新型的 RBAC 模型，可以称为“Resource-Based Access Control”。

8.1.2 集成 Apache Shiro

1. Shiro 基本概念

当你真正使用 RBAC 来设计一套权限管理系统的时候，就会觉得很复杂，复杂到每一个地方每一个细节都要仔细考虑，会徒增很多开发成本，尤其是在开发众多项目时，每个项目都需要一套权限管理系统，大多数项目对于权限管理的需求都是差不多的，那么有没有被多个系统使用的比较通用的权限管理框架呢？答案是有的。权限管理框架的出现解决了重复造轮子的问题，将绝大多数项目对权限要求的功能进行抽象，尽可能地覆盖更多的需求场景，与业务系统解耦合，是比较现实的需求。Apache Shiro 就是这样一个通用型的开源安全权限管理框架，是基于资源的（Resource-Based），下面是官方给它的定义。

Apache Shiro 是一个功能强大且灵活的开源安全框架，可以干净地处理身份验证、授权、企业会话管理和加密。Apache Shiro 的首要目标是易于使用和理解。安全有时可能非常复杂，甚至会很痛苦，但这不是必需的。框架应尽可能掩盖复杂性，并公开简洁直观的 API，以简化开发人员确保其应用程序安全地工作。

你可以使用 Shiro 进行以下主要操作：

- 身份认证：验证用户身份，通常称为用户“登录”。
- 授权：对用户进行访问控制，比如确定是否为用户分配了特定的安全角色，确定是否允许用户做某个操作等。
- 信息加密：保护或隐藏数据以防窥视。
- 会话管理：即使在没有 Web 或者 EJB 容器的情况下，也可以在任何环境中使用基于会话（Session）的 API。
- 单点登录：实现 SSO 的功能，这一部分将在 8.2 节中介绍。
- 汇总数据源：将多个保存用户信息的安全数据源进行汇总和复合，形成单个复合的用户“视图”。
- 记住我：这项功能用于在勾选后对用户进行关联，从而实现用户登录之后在一定日期内无须再次登录即可访问系统服务。

Shiro 的目标是在所有应用程序环境中实现从简单的命令行应用程序到大型企业应用程序，不必强加对其他第三方框架、容器或应用程序服务器的依赖，而可以在任何环境中直接使用它。这并不是说 Shiro 就抛弃了与框架和环境的集成，相反，Shiro 旨在尽可能地集成到第三方主流框架或者环境中，更好地实现协同工作。

在这些操作中，身份认证和授权控制这两部分在上一节介绍过，这是权限框架最基本的部分。这里涉及密码学部分，因为用户账户密码需要加密，包括一些权限隐私数据都需要加密算法的支持。会话管理是保持一个用户在登录之后整个会话活动与系统之间交互的一个跟踪过程，通常 Web 服务器保持会话状态的目的是在一段时间内保持用户的登录状态，用户在登录成功之后会下发一个登录凭证，用户的客户端持有这个登录凭证可以持续访问应用程序提供的服务，当用户浏览器关闭、会话超时后，这个登录凭证就会失效，然后提示用户需要重新登录。

Shiro 中有以下三个非常重要的概念：

- Subject：主题，可以理解为“当前执行操作的主体”。在 Shiro 中，Subject 不单单是指人（用户），也可以是第三方程序（进程），任何与当前软件系统进行交互的事物都可以，其中包含了当前用户、角色和授权的相关信息。获取到 Subject 之后，可以立即访问当前用户使用 Shiro 进行绝大部分操作，例如登录、注销、访问其会话、执行授权检查等。
- SecurityManager：安全管理器，用来管理 Subject。Subject 代表了当前用户的安全操作，而 SecurityManager 管理所有用户的安全操作，它是 Shiro 体系结构的核心，其内部包含了众多的安全组件，通常一个程序只需要一个单例的 SubjectManager，在对其初始化配置之后，就不需要再对进行设置了，更多的时间将花在操作 Subject 的 API 上。
- Realm：领域，它充当了 Shiro 与数据持久化存储（例如数据库）的“桥梁”或者“连接器”，当对用户执行认证（登录）和授权（访问控制）验证时，Shiro 会从应用配置的一个或多个 Realms 中查找用户及其权限信息。Realms 本质上是安全相关的 DAO，它封装了数据源的连接详细信息，并根据需要使关联数据可用于 Shiro。在配置 Shiro 时，必须至少指定一个（可以配置多个）Realm 用于身份验证和/或授权。

Shiro 的核心体系结构概念除了 Subject、SecurityManager、Realms 之外，还包括 Authenticator（认证）、Authorizer（授权）、SessionManager（会话管理）、CacheManager（缓存管理）、Cryptography（密码学），如图 8-8 所示。

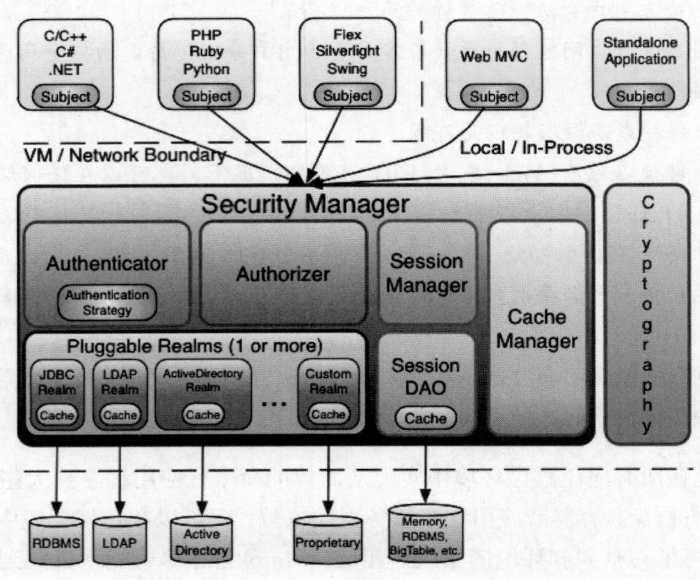

图 8-8

下面对其他几个概念进行简单介绍：

- Authenticator：认证器，负责执行用户的登录验证，当用户提交登录时，这部分的处理逻辑交给 Authenticator 负责，它将从 Realms 中获取用户相关的账户信息，验证其身份是否合法，并反馈验证结果。
- Authorizer：授权和访问控制器，负责确定用户具有的权限，以便决定是否允许用户做某项操作，它将从 Realms 中获取用户的角色和权限（许可）信息，然后确定用户是否能够执行相关操作。
- SessionManager：会话管理器，负责创建和管理用户的会话（Session）生命周期，它能够在任何环境中在本地管理用户会话，即使没有 Web/Servlet/EJB 容器，也一样可以保存会话。默认情况下，Shiro 会检测当前环境中现有的会话机制（比如 Servlet 容器）进行适配，如果没有（比如独立应用程序或者非 Web 环境），它将会使用内置的企业会话管理器来提供相应的会话管理服务，其中还涉及一个名为 SessionDAO 的对象。SessionDAO 负责 Session 的持久化操作（CRUD），允许 Session 数据写入到后端持久化数据库。
- CacheManager：缓存管理器，负责创建和管理缓存，因为 Shiro 支持多个后端数据源同时提供信息以供身份验证、授权和会话管理，因此有必要引入缓存机制来提高性能。
- Cryptography：密码学，引入密码学的相关技术是必要的，因为 Shiro 旨在提供一个高安全性的企业级服务。Shiro 提供了很多精心设计过的加解密类 API（比如 Hash、SHA 等），非常易于使用和理解，使得加密技术能够非常方便地运用到项目中来。

在了解了 Shiro 的基本概念以后，接下来学习如何对 Shiro 进行配置使用。

2. 使用 INI 方式对 Shiro 进行配置

要想使用 Shiro，就必须先进行配置，也就是对 SecurityManager 进行配置，默认的 SecurityManager 是使用 POJO 来进行配置的，也可以使用任何与 POJO 兼容的配置机制，比如 Java 代码、Spring XML、YAML、.properties 和.ini 文件等。Shiro 默认使用基于文本的 INI 文件的方式来对其进行配置，下面的示例配置 8-4 是一个基于 INI 配置 Shiro 的简单示例。

示例代码 8-4 使用 INI 配置 Shiro，文件名为 shiro.ini

```
[main]
cm = org.apache.shiro.authc.credential.HashedCredentialsMatcher
cm.hashAlgorithm = SHA-512
cm.hashIterations = 1024
#是否使用 Base64 编码
cm.storedCredentialsHexEncoded = false
iniRealm.credentialsMatcher = $cm
[users]
jdoe = TWFuIGlzIGRpc3Rpbmd1aXNoZWQsIG5vdCBvbmx5IGJpcyByZWFzb2
asmith = IHNpbmd1bGFyIHBhc3Npb24gZnJvbSBvdGhlciBhbXNoZWQsIG5vdCB
```

在这份示例配置中可以看到用于配置 SecurityManager 实例的 INI 配置示例，其中包含了两个部分：[main]和[users]。

[main]部分是配置 SecurityManager 对象或者 SecurityManager 使用的任何对象（如 Realms）的地方。在此示例中，我们看到两个对象被配置：

- cm 对象：Shiro 的 HashedCredentialsMatcher 类的实例。cm 实例的各种属性是通过以“.”符号嵌套的语法配置的——该语法配置由示例代码 8-5 中第 1 步操作所示的 IniSecurityManagerFactory 对象来装载配置。
- iniRealm 对象：SecurityManager 用来表示以 INI 格式定义的用户账户的组件。

在[users]部分中，可以给出一些用户账户，以“用户名=密码加密串”的形式给出。这种方式一般用于比较简单的程序或者测试。

示例代码 8-5　加载 shiro.ini 配置文件

```
import org.apache.shiro.SecurityUtils;
import org.apache.shiro.config.IniSecurityManagerFactory;
import org.apache.shiro.mgt.SecurityManager;
import org.apache.shiro.util.Factory;
    //省略部分内容
    //1. Load the INI configuration
    Factory<SecurityManager> factory =
        new IniSecurityManagerFactory("classpath:shiro.ini");
    //2. Create the SecurityManager
    SecurityManager securityManager = factory.getInstance();
    //3. Make it accessible
    SecurityUtils.setSecurityManager(securityManager);
```

在示例代码 8-5 中，加载 INI 配置分为了以下 3 个步骤：

（1）加载 INI 配置文件，装载到 IniSecurityManagerFactory 的工厂实例对象中。

（2）通过 IniSecurityManagerFactory 工厂实例创建 SecurityManager 实例，这里使用了工厂设计模式。

（3）将 SecurityManager 实例注册到 SecurityUtils 中，从而使得全局共享同一个 SecurityManager 实例，以单例的模式对外提供服务，它会是一个 VM（虚拟机级别的）静态单例（基于静态内存），但这不是必需的，可以自己根据需要决定是否要将其注册为静态内存单例模式。

这里只是对 INI 配置和使用进行了一些简单的介绍，有关 INI 配置的更多详细信息，请参阅 Shiro 的官方文档：http://shiro.apache.org/documentation.html。

使用 INI 方式配置和使用 Shiro 是比较传统的一种方式，它可以不依赖于任何框架（比如 Spring 家族系列），从而能够快速集成到你的应用程序当中，有这方面需求的读者可以自行阅读官方文档学习使用。接下来重点介绍将 Shiro 和 Spring Boot 框架进行集成，以及在 Spring Boot 中如何使用它来进行安全认证和权限管理，读者可以通过在 Spring Boot 应用中运用 Shiro 来进一步学习和了解 Shiro。

3. 在 Spring Boot 中使用 Shiro

得益于 Spring Boot 的特性，在 Spring Boot 项目中集成 Shiro 变得非常简单。这里将通过案例的方式来讲解如何在 Spring Boot Web 项目中使用 Shiro。

（1）创建工程，引入依赖

首先使用 IntelliJ IDEA 创建一个 Spring Boot Web 项目（参照 3.3.3 小节），项目配置信息如

图 8-9 所示。Spring Boot 版本选择“2.3.6”，项目依赖项勾选 Web 一栏中的“Spring Web”，SQL 一栏中勾选“Spring Data JPA”，取消勾选 Security 一栏中的“Spring Security”，然后依次单击“Next”按钮完成项目的创建。

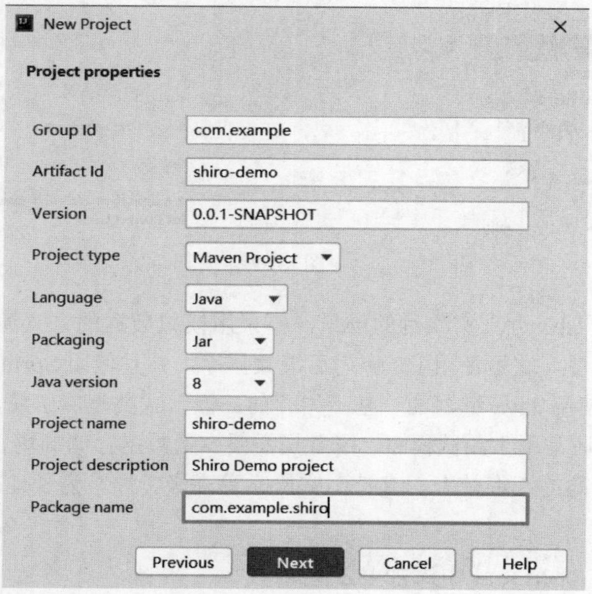

图 8-9

在 pom.xml 的<dependencies></dependencies>节点配置中添加 Shiro 的依赖：

示例代码 8-6　在 pom.xml 中添加 Shiro 依赖

```
<dependency>
    <groupId>org.apache.shiro</groupId>
    <artifactId>shiro-spring-boot-web-starter</artifactId>
    <version>1.7.0</version>
</dependency>
```

添加 Lombok 依赖（Lombok 能通过注解的方式在编译时自动为属性生成构造器、getter/setter、equals、hashcode、toString 方法，简化代码，提高开发效率）：

示例代码 8-7　在 pom.xml 中添加 Lombok 依赖

```
<dependency>
   <groupId>org.projectlombok</groupId>
   <artifactId>lombok</artifactId>
</dependency>
```

添加 Lombok 依赖之后，还需要在 IntelliJ IDEA 中安装 Lombok 插件：依次单击菜单 File→Settings 打开设置界面，找到左侧的 Plugins 一项打开，搜索 Lombok 进行安装，如图 8-10 所示。

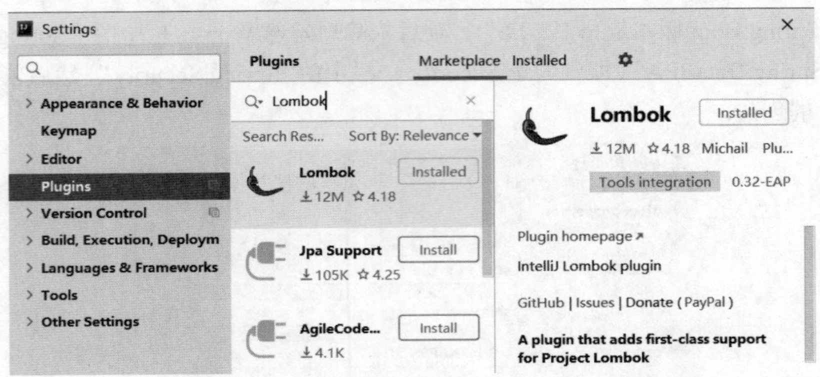

图 8-10

为了使示例项目更加接近真实项目的场景，不采用模拟数据的方式演示，同时避免引入较重量的数据库而带来复杂性，这里使用 H2 作为后端数据库，并且使用 Spring Data JPA 作为 Dao 层的数据源。H2 是一个使用 Java 编写的、基于内存的、嵌入式数据库引擎，H2 文件体积非常小，安装、启动非常简单，通常用于测试程序，或者只需要少量数据存储的场景。当然，把它用作缓存也不错，因为它具有较高的查询效率，支持全文检索等高级特性。为了使用它，同样需要在 POM 中添加 H2 的依赖。

示例代码 8-8 在 pom.xml 中添加 H2 依赖

```
<dependency>
    <groupId>com.h2database</groupId>
    <artifactId>h2</artifactId>
</dependency>
```

添加完上述依赖项后，完整的 pom.xml 文件内容如下所示。

示例代码 8-9 完整 pom.xml 文件内容

```
<?xml version="1.0" encoding="UTF-8"?>
<project xmlns="http://maven.apache.org/POM/4.0.0"
xmlns:xsi="http://www.w3.org/2001/XMLSchema-instance"
    xsi:schemaLocation="http://maven.apache.org/POM/4.0.0
https://maven.apache.org/xsd/maven-4.0.0.xsd">
    <modelVersion>4.0.0</modelVersion>
    <parent>
        <groupId>org.springframework.boot</groupId>
        <artifactId>spring-boot-starter-parent</artifactId>
        <version>2.3.6.RELEASE</version>
        <relativePath/> <!-- lookup parent from repository -->
    </parent>
    <groupId>com.example</groupId>
    <artifactId>shiro-demo</artifactId>
    <version>0.0.1-SNAPSHOT</version>
    <name>shiro-demo</name>
    <description>Shiro Demo project</description>
    <properties>
        <java.version>1.8</java.version>
```

```
    </properties>
    <dependencies>
        <dependency>
            <groupId>org.springframework.boot</groupId>
            <artifactId>spring-boot-starter-data-jpa</artifactId>
        </dependency>
        <dependency>
            <groupId>org.springframework.boot</groupId>
            <artifactId>spring-boot-starter-web</artifactId>
        </dependency>
        <dependency>
            <groupId>org.apache.shiro</groupId>
            <artifactId>shiro-spring-boot-web-starter</artifactId>
            <version>1.7.0</version>
        </dependency>
        <dependency>
            <groupId>org.projectlombok</groupId>
            <artifactId>lombok</artifactId>
            <scope>provided</scope>
        </dependency>
        <dependency>
            <groupId>com.h2database</groupId>
            <artifactId>h2</artifactId>
        </dependency>
        <dependency>
            <groupId>org.springframework.boot</groupId>
            <artifactId>spring-boot-starter-test</artifactId>
            <scope>test</scope>
        </dependency>
    </dependencies>
    <build>
        <plugins>
            <plugin>
                <groupId>org.springframework.boot</groupId>
                <artifactId>spring-boot-maven-plugin</artifactId>
            </plugin>
        </plugins>
    </build>
</project>
```

（2）创建实体类

创建 com.example.shiro.entity 包，我们使用 RBAC0 模型（使用中间表）创建 5 个实体，分别是 User、Role、Permission、UserRole、RolePermission，将其放在 com.example.shiro.entity 包下。为了容易理解，实体类模型没有采用 Spring Data JPA 的一对多、多对一、多对多映射（OneToMany、ManyToOne、ManyToMany）关系，也就是说生成的数据库表是没有物理外键约束的（我们在逻辑层面认为是有外键关联的），改为由我们手动维护。在每一个实体类中都包含两个 Instant 类型的字段（gmtCreate 和 gmtModified），分别用来标记数据记录的创建时间和修改时间，都包含一个 Long 类型的 id 字段，它是每个数据库表的逻辑主键。

下面是这五个实体类的代码。

示例代码 8-10　用户实体类 User.java

```
package com.example.shiro.entity;
import …
@Data
@AllArgsConstructor
@NoArgsConstructor
@Builder
@Entity
@Table
public class User {
    @Id
    @GeneratedValue(strategy = GenerationType.IDENTITY)
    private Long id;
    @Column(unique = true, nullable = false)
    private String username;
    @Column(nullable = false)
    private String password;
    /*加密盐*/
    @Column(nullable = false)
    private String salt;
    @Column
    private Instant gmtModified;
    @Column
    private Instant gmtCreate;
}
```

其中，字段 username 用以保存用户名，字段 password 存储密码加密后的密文（防止泄露），字段 salt 存储对密码加密时用的盐。

示例代码 8-11　角色实体类 Role.java

```
package com.example.shiro.entity;
import …
@Data
@AllArgsConstructor
@NoArgsConstructor
@Entity
@Table
public class Role {
    @Id
    @GeneratedValue(strategy = GenerationType.IDENTITY)
    private Long id;
    @Column
    private String name;
    @Column
    private Instant gmtModified;
    @Column
    private Instant gmtCreate;
}
```

字段 name 存储角色名字。

示例代码 8-12　权限许可实体类 Permission.java

```
package com.example.shiro.entity;
import …
@Data
@AllArgsConstructor
@NoArgsConstructor
@Entity
@Table
public class Permission {
    @Id
    @GeneratedValue(strategy = GenerationType.IDENTITY)
    private Long id;
    @Column
    private String name;
    @Column
    private Instant gmtModified;
    @Column
    private Instant gmtCreate;
}
```

字段 name 用来存储权限许可的名字，以冒号隔开，例如 user:list 表示获取用户列表，user:delete 表示可以删除用户，user:add 表示可以增加用户，user:view 表示可以查看用户信息，冒号之前是资源名称（也可以嵌套子级资源），冒号之后是允许执行的动作。

示例代码 8-13　用户-角色关系实体类 UserRole.java

```
package com.example.shiro.entity;
import …
@Data
@AllArgsConstructor
@NoArgsConstructor
@Entity
@Table
public class UserRole {
    @Id
    @GeneratedValue(strategy = GenerationType.IDENTITY)
    private Long id;
    @Column
    private Long userId;
    @Column
    private Long roleId;
    @Column
    private Instant gmtModified;
    @Column
    private Instant gmtCreate;
}
```

其中，字段 userId 对应 User 实体类的 id，roleId 对应 Role 实体类的 id，用户和角色之间多对多的关联关系可以通过 UserRole 实体类体现出来。

示例代码 8-14　角色-许可关系实体类 RolePermission.java

```
package com.example.shiro.entity;
import …
@Data
@AllArgsConstructor
@NoArgsConstructor
@Entity
@Table
public class RolePermission {
    @Id
    @GeneratedValue(strategy = GenerationType.IDENTITY)
    private Long id;
    @Column
    private Long roleId;
    @Column
    private Long permissionId;
    @Column
    private Instant gmtModified;
    @Column
    private Instant gmtCreate;
}
```

其中，字段 roleId 对应 Role 实体类的 id，permissionId 对应 Permission 实体类的 id，角色和资源权限许可之间的多对多关联关系可以通过 RolePermission 实体类体现出来。

（3）创建实体类对应的 DAO 接口

使用 Spring Data JPA 时，DAO 的名字以 Repository 结尾，遵循 Spring Data JPA 规范，只需要继承自 JpaRepository 接口即可。分别创建 UserRepository、RoleRepository、PermissionRepository 三个接口，代码如下。

示例代码 8-15　UserRepository.java

```
package com.example.shiro.repository;
import …
@Repository
public interface UserRepository extends JpaRepository<User, Long> {
    List<User> getByUsername(@NonNull String username);
}
```

这里扩展了一个方法 getByUsername(String username)，通过 User 实体类的字段 username 来获取用户信息，返回查询到的用户列表。

示例代码 8-16　RoleRepository.java

```
package com.example.shiro.repository;
import …
@Repository
public interface RoleRepository extends JpaRepository<Role, Long> {
    @Query("select r from Role r join UserRole ur on r.id = ur.roleId where
```

```
ur.userId=:userId")
        List<Role> getRoles(@Param("userId") Long userId);
    }
```

其中扩展了一个方法 getRoles(@Param("userId") Long userId)，通过@Query 注解编写了一个自定义 HQL 查询语句，来实现通过用户 id 查询得到关联的角色列表。

示例代码 8-17　PermissionRepository.java

```
    package com.example.shiro.repository;
    import …
    @Repository
    public interface PermissionRepository extends JpaRepository<Permission, Long>
{
        @Query("select p from Permission p join UserRole ur on p.id = ur.roleId where
ur.roleId=:roleId")
        List<Permission> getPermissions(@Param("roleId") Long roleId);
    }
```

和 RoleRepository 类似，扩展了一个方法 getPermissions(@Param("roleId") Long roleId)，通过@Query 注解编写了一个自定义 HQL 查询语句，来实现通过角色 id 查询关联的权限许可列表。

（4）创建对应的服务类

首先创建 com.example.shiro.service、com.example.shiro.service.impl 包，并创建 UserService、RoleService、PermissionService 接口，创建实现类 UserServiceImpl、RoleServiceImpl、PermissionServiceImpl。对应的代码内容分别如示例代码 8-18~示例代码 8-23 所示。

示例代码 8-18　UserService.java

```
    package com.example.shiro.service;
    import …
    public interface UserService {
        /**
         * 通过用户名获取账户
         *
         * @param username
         * @return
         */
        Optional<User> getByUsername(@NonNull String username);
    }
```

示例代码 8-19　UserServiceImpl.java

```
    package com.example.shiro.service.impl;
    import …
    @Service
    public class UserServiceImpl implements UserService {
        @Resource
        private UserRepository userRepository;
        @Override
        public Optional<User> getByUsername(@NonNull String username) {
            List<User> result = userRepository.getByUsername(username);
```

```
        return Optional.ofNullable(result.isEmpty() ? null : result.get(0));
    }
}
```

示例代码 8-20　RoleService.java

```
package com.example.shiro.service;
import …
public interface RoleService {
    List<Role> getRoles(@NonNull Long userId);
}
```

示例代码 8-21　RoleServiceImpl.java

```
package com.example.shiro.service.impl;
import …
@Service
public class RoleServiceImpl implements RoleService {
    @Resource
    private RoleRepository roleRepository;
    @Override
    public List<Role> getRoles(@NonNull Long userId) {
        return roleRepository.getRoles(userId);
    }
}
```

示例代码 8-22　PermissionService.java

```
package com.example.shiro.service;
import …
public interface PermissionService {
    public List<Permission> getPermissions(@NonNull Long roleId);
}
```

示例代码 8-23　PermissionServiceImpl.java

```
package com.example.shiro.service.impl;
import …
@Service
public class PermissionServiceImpl implements PermissionService {
    @Resource
    private PermissionRepository permissionRepository;
    @Override
    public List<Permission> getPermissions(@NonNull Long roleId) {
        return permissionRepository.getPermissions(roleId);
    }
}
```

（5）修改 application.yml，配置数据源、JPA、H2

完整的 application.yml 配置内容如下。

示例代码 8-24　配置文件 application.yml

```
spring:
datasource:
```

```
    # 设置 h2 的链接 URL，以文件方式存储，文件名为 h2，模式采用 MYSQL
    url:
dbc:h2:file:./h2;AUTO_SERVER=TRUE;DB_CLOSE_DELAY=-1;MODE=MYSQL;AUTO_RECONN
CT=TRUE
    # 用户名默认 sa
    username: sa
    # 密码为空即可
    password:
    # 驱动名称
    driver-class: org.h2.Driver
    name: datasource
    platform: h2
    continue-on-error: true
  jpa:
    # 输出生成的 SQL 语句
    show-sql: true
    open-in-view: true
    generate-ddl: true
    hibernate:
# 设置自动创建数据库表
      ddl-auto: create
    properties:
      hibernate:
        enable_lazy_load_no_trans: true
        # 设置 hibernate 方言为 H2Dialect
        dialect: org.hibernate.dialect.H2Dialect
        # 格式化输出 SQL 语句
        format_sql: true
        hbm2ddl:
          # 支持初始化 import.sql 文件的多行 SQL 命令
          import_files_sql_extractor:
org.hibernate.tool.hbm2ddl.MultipleLinesSqlCommandExtractor
  h2:
    console:
      # 开启 h2 的 Web 客户端访问控制台
      enabled: true
      # 设置 h2 的 Web 客户端的地址
      path: /h2-console
```

（5）添加配置类 BeanConfig.java，配置扫描注入

新建 com.example.shiro.config 包，然后创建 BeanConfig.java 配置类，添加包扫描，其代码内容如下。

示例代码 8-25　BeanConfig.java 文件

```
package com.example.shiro.config;
import …
@Configuration
//配置扫描的包
@ComponentScans({
    // 服务类所在的包
```

```
    @ComponentScan("com.example.shiro.service.**"),
    // Controller 类所在的包
    @ComponentScan("com.example.shiro.controller.**")
})
//开启 JPA Repository 接口，自动扫描 com.example.shiro.repository 包下的接口
@EnableJpaRepositories("com.example.shiro.repository")
//扫描 com.example.shiro.entity 包下面的实体类
@EntityScan("com.example.shiro.entity")
public class BeanConfig {}
```

设置好之后，运行项目，访问 h2 的 Web 客户端地址：http://localhost:8080/h2-console，登录界面如图 8-11 所示。单击“Connect”按钮登录，成功登录之后就能看到 Spring Data JPA 已经自动创建好了五个实体类对应的数据库表，如图 8-12 所示。此时的数据库表中是没有任何记录的，因为我们还没有添加数据，后面会预置数据在数据库表中。

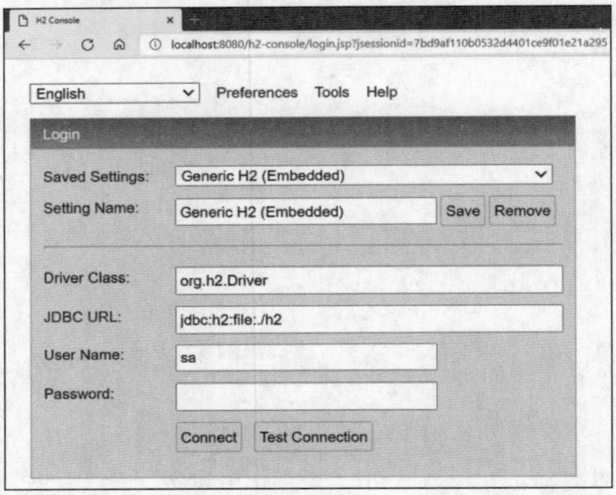

图 8-11

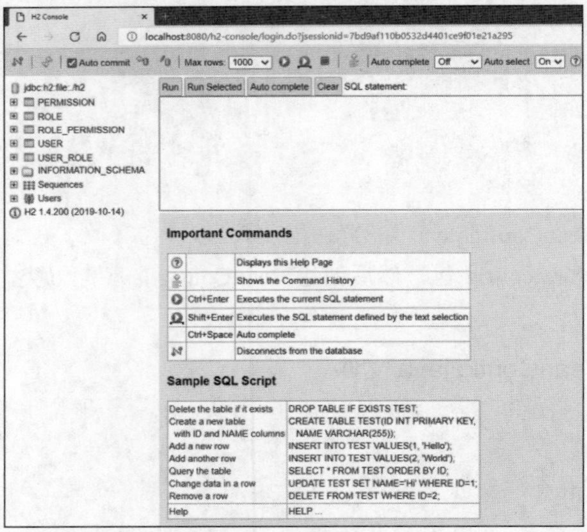

图 8-12

（7）实现自定义的 ShiroRealm，提供登录和鉴权

这一步是给 Shiro 提供一个 Realm，告诉 Shiro 如何进行登录认证和权限鉴定，当然不是我们来告诉 Shiro 怎么做，而是按照 Shiro 约定的方式（继承 AuthorizingRealm 并实现两个方法）提供给 Shiro 登录信息和权限许可信息，剩下的认证和鉴权是交给 Shiro 框架做的事情，我们需要做的仅仅是继承 AuthorizingRealm 这个抽象类，实现其中的 AuthenticationInfo doGetAuthenticationInfo(AuthenticationToken token) 方法和 AuthorizationInfo doGetAuthorizationInfo(PrincipalCollection principals)方法：前者提供登录信息，即根据登录的用户名查询数据库表中的用户信息，将其加密后的密码和盐封装成 AuthenticationInfo 对象提供给 Shiro，Shiro 来比对用户输入的密码是否和查出来的用户密码一致；后者提供授权信息，当登录认证通过之后，访问相关资源时就会进行权限鉴定，而权限的鉴定是通过后面这个方法来提供授权信息的，需要根据用户信息查出相关的角色、资源许可列表，将其封装成 AuthorizationInfo 对象提供给 Shiro，Shiro 根据对应资源需要的许可和用户拥有的资源许可做对比，能够匹配的到时才允许用户访问，否则就会抛出相关异常。首先创建包 com.example.shiro.dao，因为 Realm 的定位是一个 DAO，因此放在 dao 包下再合适不过了。完整的 ShiroRealm 示例实现如示例代码 8-26 所示。

示例代码 8-26　ShiroRealm.java 文件

```
package com.example.shiro.dao;
import …
public class ShiroRealm extends AuthorizingRealm {
    @Resource
    private UserService userService;
    @Resource
    private RoleService roleService;
    @Resource
    private PermissionService permissionService;
    public ShiroRealm() {}
    public ShiroRealm(CredentialsMatcher matcher) {
        super(matcher);
    }
    @Override
    protected AuthorizationInfo doGetAuthorizationInfo(PrincipalCollection
principals) {
        SimpleAuthorizationInfo authorizationInfo = new
impleAuthorizationInfo();
        /*在 doGetAuthenticationInfo 方法的返回对象中，塞进了根据用户名查到的 User 对
象，将其取出*/
        User user = (User) principals.getPrimaryPrincipal();
        /*查询用户的角色信息*/
        List<Role> roles = roleService.getRoles(user.getId());
        roles.forEach(r -> {
            /*将角色名添加到 authorzationInfo 对象中*/
            authorizationInfo.addRole(r.getName());
            /*查询该角色拥有的资源许可权限列表*/
            List<String> permissions = permissionService
                .getPermissions(r.getId())
                .stream()
```

```
                .map(Permission::getName)
                .collect(Collectors.toList());
            /*将资源许可列表添加到 authorizationInfo 对象中*/
            authorizationInfo.addStringPermissions(permissions);
        });
        return authorizationInfo;
    }
    @Override
    protected AuthenticationInfo doGetAuthenticationInfo(AuthenticationToken
token) throws AuthenticationException {
        /*获取用户名，去除用户名前后的空白字符*/
        String username = StringUtils.trimWhitespace(
            Optional.ofNullable(token)
                .map(AuthenticationToken::getPrincipal)
                .map(Object::toString).orElse(null));
        /*用户名不为空*/
        if (!StringUtils.isEmpty(username)) {
            /*根据用户名获取用户信息（主要是密码和盐）*/
            Optional<User> user = userService.getByUsername(username);
            if (user.isPresent()) {
                /*返回查询出来的用户信息,封装成 SimpleAuthenticationInfo 提供给 Shiro
                  判定用户名密码是否正确*/
                return new SimpleAuthenticationInfo(
                            user.get(),
                            user.get().getPassword(),
                            ByteSource.Util.bytes(user.get().getSalt()),
                            getName());
            }
        }
        return null;
    }
}
```

在“return new SimpleAuthenticationInfo(user.get()，user.get().getPassword()，ByteSource.Util.bytes(user.get().getSalt())，getName())”中，SimpleAuthenticationInfo 这个类构造方法的第一个参数是提供 user 对象，以供在 doGetAuthorizationInfo 方法中获取，第二个参数是传递对应用户的正确密码（加密后的），第三个参数是提供盐的 ByteSource 类型的对象。由于我们在数据库中存储的 salt 是原始字符串，因此需要通过 ByteSource.Util.bytes(salt)工具类方法进行转换。创建用户时 salt 的处理方式与这里保持一致，最后一个参数是提供当前的 realmName，通过 getName()方法获取即可（父类已经实现）。

（8）提供 ShiroConfig 配置类，对 Shiro 进行配置

前面有介绍过，Shiro 默认是通过在 Resources 目录下提供 shiro.ini 文件的方式来对 Shiro 进行配置，但是在 Spring Boot 中更推荐使用配置类的方式来做。ShiroConfig 类的完整示例代码如下。

示例代码 8-27　ShiroConfig.java

```
package com.example.shiro.config;
import …
```

```
@Configuration
public class ShiroConfig {
    /*md5 加密算法*/
    public static String HASH_ALGORITHM_NAME = "md5";
    /*2 次加密*/
    public static int HASH_ITERATIONS = 1;
    @Bean
    public HashedCredentialsMatcher md5HashedCredentialsMatcher() {
        HashedCredentialsMatcher hashedCredentialsMatcher =
            new HashedCredentialsMatcher();
        /*设置加密算法*/
        hashedCredentialsMatcher.setHashAlgorithmName(HASH_ALGORITHM_NAME);
        /*设置加密次数*/
        hashedCredentialsMatcher.setHashIterations(HASH_ITERATIONS);
        return hashedCredentialsMatcher;
    }
    /**
     * 提供自定义的 Realm，并将其注册名设置为 authorizer
     * @return
     */
    @Bean
    public ShiroRealm authorizer() {
        return new ShiroRealm(md5HashedCredentialsMatcher());
    }
    /**
     * filter 工厂，用来设置过滤条件和跳转条件
     * @param securityManager
     * @return
     */
    @Bean
    public ShiroFilterFactoryBean shiroFilterFactoryBean(
                        SecurityManager securityManager) {
        ShiroFilterFactoryBean shiroFilterFactoryBean =
            new ShiroFilterFactoryBean();
        shiroFilterFactoryBean.setSecurityManager(securityManager);
        Map<String, String> map = new HashMap<>();
        //登出
        map.put("/user/logout", "logout");
        //登录
        shiroFilterFactoryBean.setLoginUrl("/user/login");
        //首页
        shiroFilterFactoryBean.setSuccessUrl("/index");
        //错误页面，认证不通过跳转
        shiroFilterFactoryBean.setUnauthorizedUrl("/error");
        shiroFilterFactoryBean.setFilterChainDefinitionMap(map);
        return shiroFilterFactoryBean;
    }
    /**
     * 开启 Shiro 注解(如@RequiresRoles,@RequiresPermissions),
     * 需借助 SpringAOP 扫描使用 Shiro 注解的类,并在必要时进行安全逻辑验证
```

```
     * 配置以下两个 bean(DefaultAdvisorAutoProxyCreator 和
authorizationAttributeSourceAdvisor)
     */
    @Bean
    public DefaultAdvisorAutoProxyCreator advisorAutoProxyCreator() {
        DefaultAdvisorAutoProxyCreator advisorAutoProxyCreator =
            new DefaultAdvisorAutoProxyCreator();
        advisorAutoProxyCreator.setProxyTargetClass(true);
        return advisorAutoProxyCreator;
    }
    /**
     * 开启 aop 注解支持
     */
    @Bean
    public AuthorizationAttributeSourceAdvisor
authorizationAttributeSourceAdvisor(SecurityManager securityManager) {
        AuthorizationAttributeSourceAdvisor advisor =
            new  AuthorizationAttributeSourceAdvisor();
        advisor.setSecurityManager(securityManager);
        return advisor;
    }
}
```

在 shiroFilterFactoryBean 的配置中，我们设置了“/user/logout”为登出路径、“/user/login”为登录路径、“/index”为登录成功之后的跳转路径、“/error”为登录失败后的错误提示跳转路径。在 md5HashedCredentialsMatcher 的配置中，配置了使用 md5 为加密算法、加密次数为 1 次。

（9）预置用户数据

数据库中没有任何数据，我们要进行登录和授权测试就需要预先在数据库表中插入一些数据。一般情况下，会在程序启动成功创建数据库表之后向数据库表中添加一些预置的用户（比如超级管理员用户）、角色（提供一些基本的角色，比如管理员角色）、权限许可（系统中存在的资源和允许的操作），我们可以利用 Hibernate 的预置数据脚本 import.sql（当 hibernate.ddl-auto 配置项为 create 时会默认执行此脚本导入数据）来实现。这里提供了一些简单的数据供测试使用，import.sql 脚本内容如下。

示例代码 8-28　import.sql 文件

```
insert into user(id, username, password, salt, gmt_create, gmt_modified)
select
   NULL,'admin','e5d17c24dfd73845c44818fabef23d19', 'abcd1234', now(), now()
from dual
where not exists(
   select 1 from user where username = 'admin'
);
insert into permission(id, name, gmt_create, gmt_modified)
select NULL, 'user:list', now(), now() from dual
where not exists(
   select 1 from permission where name='user:list'
);
```

```
insert into role(id, name, gmt_create, gmt_modified)
select NULL, 'system', now(), now() from dual
where not exists(
    select 1 from role where name='system'
);
with t as (select
    u.id as user_id, r.id as role_id
    from user u, role r
    where u.username='admin' and r.name='system')
insert into user_role(id, user_id, role_id, gmt_create, gmt_modified)
select
    NULL, user_id, role_id, now(), now()
from t
where not exists(
    select 1 from user_role ur
    join t    where ur.user_id = t.user_id and ur.role_id = t.role_id
);
with t as (select
    r.id as role_id, p.id as permission_id
    from permission p, role r
    where p.name='user:list' and r.name='system')
insert into role_permission(id, role_id, permission_id, gmt_create,
gmt_modified)
select
    NULL, role_id, permission_id, now(), now()
from t
where not exists(
    select 1 from role_permission rp
    join t
    where rp.role_id = t.role_id and rp.permission_id = t.permission_id
);
commit;
```

创建了一个管理员用户“admin”、一个系统角色“system”，并将“system”角色赋予“admin”用户，在“permission”表中创建了一个资源许可“user:list”（查看所有用户列表），并把这个许可授予“system”角色。在创建用户时，需要给出加密后的用户密码，按照前面的配置，我们使用md5 加密算法对原始密码“123”加盐“abcd1234”后进行一次哈希，这可以通过创建一个测试类PasswordHelperTest 来实现。

示例代码 8-29　加密测试类 PasswordHelperTest.java

```
package com.example.shiro;
import …
public class PasswordHelperTest {
    @Test
    public void pwdHash() {
        String salt = "abcd1234";
        String password = "123";
        System.out.println(ByteSource.Util.bytes("abcd1234"));
        SimpleHash simpleHash = new SimpleHash(
```

```
            ShiroConfig.HASH_ALGORITHM_NAME, password,
            ByteSource.Util.bytes(salt),
            ShiroConfig.HASH_ITERATIONS);
        System.out.println(simpleHash.toHex());
    }
}
```

通过执行 pwdHash()测试方法，我们能够得到加密后的密码，然后将其通过预置脚本语句插入到 user 表中。

（10）登录认证测试

为了能够更容易理解，避免引入页面带来复杂性，我们将通过实现纯 REST 风格的接口来进行测试，而不是通过传统的登录页面来实现，这在前后端分离的项目中很常见。首先创建一个用来和前端交互的领域模型 UserVO，然后创建一个用来返回给前端结果数据的模型 JsonResult。

示例代码 8-30　UserVO.java

```
package com.example.shiro.vo;
import …
@Data
@AllArgsConstructor
@NoArgsConstructor
public class UserVO {
    private Long id;
    private String username;
    @JsonInclude(JsonInclude.Include.NON_NULL)
    private String password;
    private Long timestamp;
}
```

示例代码 8-31　JsonResult.java

```
package com.example.shiro.vo;
import …
@Data
@NoArgsConstructor
@AllArgsConstructor
@Builder
public class JsonResult {
    private Integer code;
    private String msg;
    private Object data;
    public interface JsonResultCode{
        Integer ERROR = -1;
        Integer SUCCESS = 1;
    }
}
```

JsonResult 实体有三个字段：code 用来标记执行成功与否，通过 1（SUCCESS）和-1（ERROR）来标识执行成功和失败；msg 用来存储一些提示信息；data 用来存储相应的结果数据。

然后创建 com.example.shiro.controller 包和 UserRestController 控制器类，提供一个 login 的方法。

示例代码 8-32　UserRestController.java

```
package com.example.shiro.controller;
import …
@RestController
@RequestMapping("/user")
public class UserRestController {
    @Resource
    private UserService userService;
    @PostMapping("/login")
    public JsonResult login(@RequestBody UserVO userVO) {
        if (Objects.isNull(userVO)) {
            return new JsonResult(
                        JsonResult.JsonResultCode.ERROR,
                        "数据无效！", null);
        }
        String username = StringUtils.trimWhitespace(userVO.getUsername());
        String pwd = StringUtils.trimWhitespace(userVO.getPassword());
        if (StringUtils.isEmpty(username) || StringUtils.isEmpty(pwd)) {
            return new JsonResult(
                        JsonResult.JsonResultCode.ERROR,
                        "用户名或密码不能为空！", null);
        }
        UsernamePasswordToken token = new UsernamePasswordToken(username, pwd);
        try {
            SecurityUtils.getSubject().login(token);
            return new JsonResult(
                        JsonResult.JsonResultCode.SUCCESS,
                        "登录成功！", userVO);
        } catch (UnknownAccountException e) {
            return new JsonResult(
                        JsonResult.JsonResultCode.ERROR,
                        "用户账户不存在！", userVO);
        } catch (AuthenticationException e) {
            return new JsonResult(
                        JsonResult.JsonResultCode.ERROR,
                        "用户名或密码错误！", userVO);
        }
    }
}
```

其中规定了“/user/login”路径只能通过 Post 请求方式来访问。接下来启动项目，没有出现错误就可以进行登录测试了。我们编写的是 RESTful 登录接口，需要借助工具来进行测试，这里使用 Postman 来进行测试（也可以选择其他类似的工具），创建一个 post 请求，在 body 中给出 raw（原始）格式的数据，以提供登录的用户名和密码：

```
{
    "username": "admin",
    "password": "123"
}
```

在 Header 中设置“Content-Type”为“application/json”，然后就可以发送请求来测试登录了，不出意外的话将会返回登录成功的信息：

```
{
    "code": 1,
    "msg": "登录成功！",
    "data": {
        "id": null,
        "username": "admin",
        "password": "123",
        "timestamp": null
    }
}
```

在 Postman 中的返回结果如图 8-13 所示。

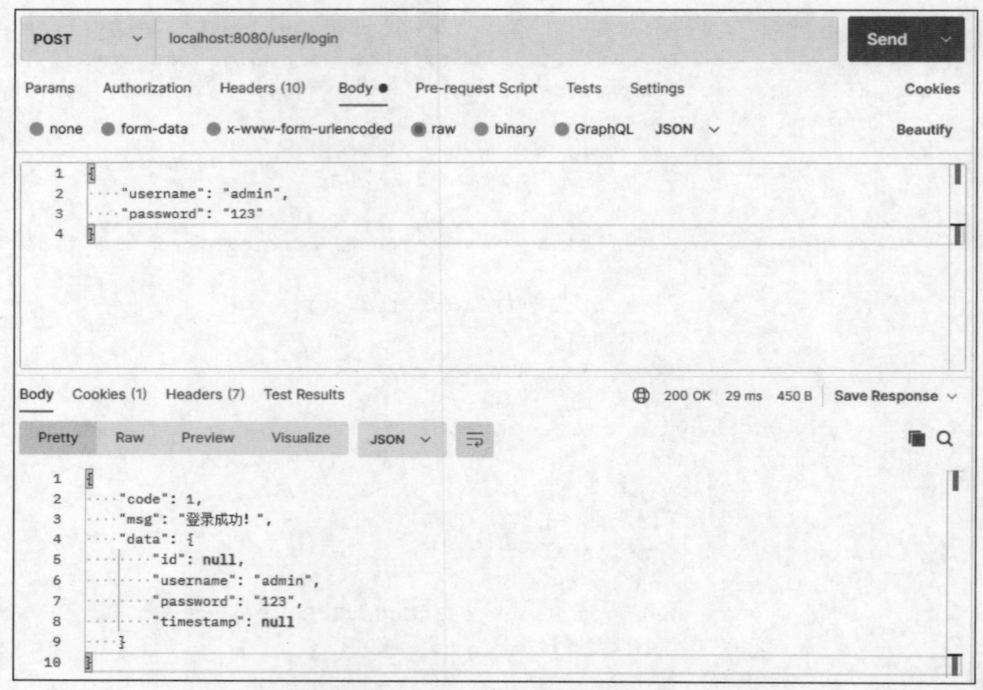

图 8-13

返回结果可能是如下错误信息：

```
{
    "timestamp": "2021-02-14T17:40:32.969+00:00",
    "status": 405,
    "error": "Method Not Allowed",
    "message": "",
    "path": "/user/login"
}
```

这是因为我们编写的 login 方法限制了只能使用 POST 请求来访问，而你使用的是 GET 请求，这是不被允许的，将请求方法改为 POST 方式提交即可。

（11）权限授权测试

在上面的 login 测试中，由于我们在 ShiroConfig 中设置了“/user/login”路径为登录路径，因此 Shiro 不会对这个路径进行任何拦截。我们现在测试一个设定了权限要求的接口，即查看所有用户的列表（只能是拥有“user:list”授权许可的用户才可以）。在前面的预置数据中，我们已经将“user:list”权限授予了“system”角色，而“system”角色又授予了“admin”管理员用户，因此“admin”用户是能够进行访问的。我们来实现这部分逻辑验证推理。

首先在 UserService 接口中增加 List<User> findAll()方法，并在 UserServiceImpl 中实现：

示例代码 8-33　UserServiceImpl 类实现 List<UserVO> findAll()方法

```
@Override
public List<UserVO> findAll() {
    List<UserVO> userList = userRepository.findAll().stream().map(user -> {
        UserVO userVO = new UserVO();
        userVO.setId(user.getId());
        userVO.setUsername(user.getUsername());
        return userVO;
    }).collect(Collectors.toList());
    return userList;
}
```

然后在 UserRestController 类中实现方法 JsonResult getUserList()：

示例代码 8-34　UserRestController 类实现 JsonResult getUserList()方法

```
@RequiresPermissions("user:list")
@GetMapping("/list")
public JsonResult getUserList() {
    try {
        List<UserVO> userList = userService.findAll();
        return new JsonResult(
                JsonResult.JsonResultCode.SUCCESS, "获取数据成功！", userList);
    } catch (Exception e) {
        e.printStackTrace();
        return new JsonResult(
                    JsonResult.JsonResultCode.ERROR, "获取数据失败！", null);
    }
}
```

@RequiresPermissions 注解的作用是：要求当前的用户必须具有指定的资源许可才能够执行操作，这里要求用户必须拥有“user:list”许可。在 Shiro 中同时还有另外四个常用注解：@RequiresGuest、@RequiresUser、@RequiresRoles({""})、@RequiresAuthentication。其中，@RequiresGuest 表示允许匿名访问，无须登录；@RequiresUser 表示要求用户登录后才可以访问；@RequiresRoles 表示要求用户必须同时具有相应的角色才可以访问、执行操作；@RequiresAuthentication 表示当前 Subject 必须在 session 中已经过认证。

在前一步以 admin 用户登录测试成功之后，可以通过 GET 请求访问 localhost:8080/user/list 来验证能否获得用户列表，同样使用 Postman 发起请求，返回结果如图 8-14 所示。返回的 JsonResult

结果 code 为 1，msg 提示获取数据成功：

```
{
    "code": 1,
    "msg": "获取数据成功！",
    "data": [
        {
            "id": 1,
            "username": "admin",
            "timestamp": null
        }
    ]
}
```

这表示我们使用具有相关权限许可的 admin 用户获取用户列表成功了。shiro 在验证权限之后通过了相关访问操作。读者可以自行测试创建其他用户，在不具备“user:list”权限的情况下来访问“/user/list”看看能否得到用户列表、是不是会被拒绝。在本小节项目的代码附件中实现了注册用户的功能，通过“/user/register”就可以创建用户，然后通过新建的用户来测试。

图 8-14

总　结

Shiro 是一个小巧而功能非常强大的安全权限管理框架，要想完整地利用 Shiro 来实现权限系统的管理，还需要做很多事情，比如统一的异常拦截等，本书无法一一讲解此类内容，读者可以查阅 Shiro 官方文档或者相关图书资料来进一步学习 Shiro 这个强大的框架。

8.1.3 集成 Spring Security

1. Spring Security 基本背景

Spring Security 是一个提供身份验证、授权和保护以防止常见攻击的框架，凭借对命令式和响应式应用程序的一流支持，成为用于保护基于 Spring 应用程序的事实标准。

Spring Security 是 Spring 家族中的一个安全管理框架。实际上，在 Spring Boot 出现之前，Spring Security 就已经发展了多年，但是使用的并不多（配置复杂、过于繁重），安全权限管理这个领域曾经一直是 Shiro 的天下。

相对于 Shiro，在 SSM/SSH 中整合 Spring Security 都是比较麻烦的操作，所以 Spring Security 虽然功能比 Shiro 强大，但是使用反而没有 Shiro 多（Shiro 虽然功能没有 Spring Security 多，但是对于大部分项目而言够用了）。

自从有了 Spring Boot 之后，Spring Boot 对于 Spring Security 提供了自动化配置方案，可以零配置使用 Spring Security。

2. Spring Security 核心概念

AuthenticationManager：用户认证的管理类，所有的认证请求（比如 login）都会通过提交一个 AuthenticationToken 给 AuthenticationManager 的 authenticate()方法来实现。当然事情肯定不是它来做，具体校验动作会由 AuthenticationManager 将请求转发给具体的实现类来做。根据实现反馈的结果再调用具体的 Handler 来给用户以反馈。这个类基本等同于 Shiro 的 SecurityManager。

AuthenticationProvider：认证的具体实现类，一个 provider 是一种认证方式的实现，比如提交的用户名、密码是通过和数据库中查出的 user 记录做比对实现的，就有一个 DaoAuthenticationProvider；如果是通过 CAS 请求单点登录系统实现的，就有一个 CASAuthenticationProvider；如果是通过 JWT 请求单点登录系统实现的，就有一个 JwtAuthenticationProvider。AuthenticationProvider 类似于 Shiro 的 Realm 定义，Spring Security 已经将主流的认证方式提供了默认实现，比如 DAO、LDAP、CAS、OAuth2 等。

AuthenticationManager 只是一个代理接口，真正的认证是由 AuthenticationProvider 来做的。一个 AuthenticationManager 可以包含多个 Provider，每个 Provider 通过实现一个 support 方法来表示自己支持哪一种 AuthenticationToken 的认证。AuthenticationManager 默认的实现类是 ProviderManager。

UserDetailService：用户认证通过 Provider 来做，所以 Provider 需要拿到系统已经保存的认证信息，获取用户信息的接口 spring-security 抽象成 UserDetailService。虽然叫 Service，但是更合适的是把它认为是我们系统里经常有的 UserDao。

AuthenticationToken：所有提交给 AuthenticationManager 的认证请求都会被封装成一个 Token 的实现，比如最容易理解的 UsernamePasswordAuthenticationToken。这个就不再多说了，连名字都跟 Shiro 中一样。

SecurityContext：当用户通过认证之后就会为这个用户生成一个唯一的 SecurityContext，里面包含用户的认证信息 Authentication。通过 SecurityContext 我们可以获取到用户的标识 Principle 和授权信息 GrantedAuthrity。在系统的任何地方只要通过 SecurityHolder.getSecruityContext()就可以获取到 SecurityContext。在 Shiro 中通过 SecurityUtils.getSubject()到达同样的目的。

3. 对 Web 系统的支持

对于 Spring 框架使用最多的还是 Web 系统。对于 Web 系统来说，进入认证的最佳入口就是过滤器（Filter）。Spring Security 不仅实现了认证的逻辑，还通过 Filter 实现了常见的 Web 攻击的防护。

下面按照 Request 进入的顺序列举一下常用的 Filter：

- SecurityContextPersistenceFilter：用于将 SecurityContext 放入 Session 的 Filter。
- UsernamePasswordAuthenticationFilter：使用用户名和密码进行登录认证的 Filter，类似的还有 CasAuthenticationFilter、BasicAuthenticationFilter 等。在这些 Filter 中，生成用于认证的 Token，提交到 AuthenticationManager，如果认证失败就会直接返回。
- RememberMeAuthenticationFilter，通过 cookie 来实现 Remember Me 功能的 Filter。
- AnonymousAuthenticationFilter：如果一个请求在到达这个 Filter 之前 SecurityContext 没有初始化，那么这个 Filter 会默认生成一个匿名 SecurityContext。这在支持匿名用户的系统中非常有用。
- ExceptionTranslationFilter：捕获所有 Spring Security 抛出的异常，并决定处理方式。
- FilterSecurityInterceptor：权限校验的拦截器，访问的 URL 权限不足时会抛出异常。

既然用了上面那么多 filter，它们在 FilterChain 中的先后顺序就显得非常重要了。对于每一个系统或者用户自定义的 Filter，Spring Security 都要求必须指定一个 Order，用来做排序。对于系统的 Filter 默认顺序，是在一个 FilterComparator 类中定义的，核心实现如下：

```
FilterComparator() {
    int order = 100;
    put(ChannelProcessingFilter.class, order);
    order += STEP;
    put(ConcurrentSessionFilter.class, order);
    order += STEP;
    put(WebAsyncManagerIntegrationFilter.class, order);
    order += STEP;
    put(SecurityContextPersistenceFilter.class, order);
    order += STEP;
    put(HeaderWriterFilter.class, order);
    order += STEP;
    put(CorsFilter.class, order);
    order += STEP;
    put(CsrfFilter.class, order);
    order += STEP;
    put(LogoutFilter.class, order);
    order += STEP;
    filterToOrder.put(
        "org.springframework.security.oauth2.client.web" +
        ".OAuth2AuthorizationRequestRedirectFilter", order);
    order += STEP;
    put(X509AuthenticationFilter.class, order);
```

```
        order += STEP;
        put(AbstractPreAuthenticatedProcessingFilter.class, order);
        order += STEP;
        filterToOrder.put(

"org.springframework.security.cas.web.CasAuthenticationFilter",order);
        order += STEP;
    filterToOrder.put(
                "org.springframework.security.oauth2.client.web" +
                .OAuth2LoginAuthenticationFilter", order);
        order += STEP;
        put(UsernamePasswordAuthenticationFilter.class, order);
        order += STEP;
        put(ConcurrentSessionFilter.class, order);
        order += STEP;
        filterToOrder.put(
                org.springframework.security.openid.OpenIDAuthenticationFilter",
                order);
        order += STEP;
        put(DefaultLoginPageGeneratingFilter.class, order);
        order += STEP;
        put(ConcurrentSessionFilter.class, order);
        order += STEP;
        put(DigestAuthenticationFilter.class, order);
        order += STEP;
        put(BasicAuthenticationFilter.class, order);
        order += STEP;
        put(RequestCacheAwareFilter.class, order);
        order += STEP;
        put(SecurityContextHolderAwareRequestFilter.class, order);
        order += STEP;
        put(JaasApiIntegrationFilter.class, order);
        order += STEP;
        put(RememberMeAuthenticationFilter.class, order);
        order += STEP;
        put(AnonymousAuthenticationFilter.class, order);
        order += STEP;
        put(SessionManagementFilter.class, order);
        order += STEP;
        put(ExceptionTranslationFilter.class, order);
        order += STEP;
        put(FilterSecurityInterceptor.class, order);
        order += STEP;
        put(SwitchUserFilter.class, order);
    }
```

对于自定义的 Filter，如果要加入 Spring Security 的 FilterChain 中，就必须指定加到已有的哪个 Filter 之前或者之后。

4. 在 Spring Boot 中集成 SpringSecurity

（1）创建工程项目

和上一节 Shiro 示例项目一样，这一节中仍然会通过一个示例项目来介绍 Spring Security。与 Shiro 示例项目类似，首先需要创建一个名为“spring-security-demo”的项目，基础包名为“com.example.spring.security”，在创建时勾选添加“Spring Security”“H2 Database”“Lombok”“Spring Web”“Spring Data JPA”。需要注意的是，Spring Security 需要 Java 8 或更高版本的运行时环境。创建完成后，完整的 pom.xml 文件内容如下。

示例代码 8-35 pom.xml 文件

```
<?xml version="1.0" encoding="UTF-8"?>
<project xmlns="http://maven.apache.org/POM/4.0.0"
xmlns:xsi=http://www.w3.org/2001/XMLSchema-instance
    xsi:schemaLocation="http://maven.apache.org/POM/4.0.0
https://maven.apache.org/xsd/maven-4.0.0.xsd">
    <modelVersion>4.0.0</modelVersion>
    <parent>
        <groupId>org.springframework.boot</groupId>
        <artifactId>spring-boot-starter-parent</artifactId>
        <version>2.3.6.RELEASE</version>
        <relativePath/> <!-- lookup parent from repository -->
    </parent>
    <groupId>com.example</groupId>
    <artifactId>spring-security-demo</artifactId>
    <version>0.0.1-SNAPSHOT</version>
    <name>spring-security-demo</name>
    <description>Spring Security Demo project for Spring Boot</description>
    <properties>
        <java.version>1.8</java.version>
    </properties>
    <dependencies>
        <dependency>
            <groupId>org.springframework.boot</groupId>
            <artifactId>spring-boot-starter-data-jpa</artifactId>
        </dependency>
        <dependency>
            <groupId>org.springframework.boot</groupId>
            <artifactId>spring-boot-starter-security</artifactId>
        </dependency>
        <dependency>
            <groupId>org.springframework.boot</groupId>
            <artifactId>spring-boot-starter-web</artifactId>
        </dependency>
        <dependency>
            <groupId>com.h2database</groupId>
            <artifactId>h2</artifactId>
            <scope>runtime</scope>
        </dependency>
        <dependency>
```

```
            <groupId>org.projectlombok</groupId>
            <artifactId>lombok</artifactId>
            <optional>true</optional>
        </dependency>
        <dependency>
            <groupId>org.springframework.boot</groupId>
            <artifactId>spring-boot-starter-test</artifactId>
            <scope>test</scope>
            <exclusions>
                <exclusion>
                    <groupId>org.junit.vintage</groupId>
                    <artifactId>junit-vintage-engine</artifactId>
                </exclusion>
            </exclusions>
        </dependency>
        <dependency>
            <groupId>org.springframework.security</groupId>
            <artifactId>spring-security-test</artifactId>
            <scope>test</scope>
        </dependency>
    </dependencies>
    <build>
        <plugins>
            <plugin>
                <groupId>org.springframework.boot</groupId>
               <artifactId>spring-boot-maven-plugin</artifactId>
               <configuration>
                   <excludes>
                       <exclude>
                           <groupId>org.projectlombok</groupId>
                           <artifactId>lombok</artifactId>
                       </exclude>
                   </excludes>
               </configuration>
            </plugin>
        </plugins>
    </build>
</project>
```

（2）Spring Security 初体验

在项目依赖添加之后，Spring Security 相关的接口都已经添加进来了。相比于 Shiro，Spring Security 的零配置使得项目变得非常简单，甚至会让你感到惊讶。

创建 controller 包，并创建 HelloController 类，仅仅提供一个输出“Hello Spring Security”的方法。

示例代码 8-36　HelloController.java

```
package com.example.spring.security.controller;
import org.springframework.web.bind.annotation.RestController;
@RestController
```

```
public class HelloController {
    @GetMapping("/hello")
    public String hello(){
        return "Hello Spring Security!";
    }
}
```

编译并运行项目，能够看到部分输出结果如下：

```
  .   ____          _            __ _ _
 /\\ / ___'_ __ _ _(_)_ __  __ _ \ \ \ \
( ( )\___ | '_ | '_| | '_ \/ _` | \ \ \ \
 \\/  ___)| |_)| | | | | || (_| |  ) ) ) )
  '  |____| .__|_| |_|_| |_\__, | / / / /
 =========|_|==============|___/=/_/_/_/
 :: Spring Boot ::        (v2.3.8.RELEASE)
(部分内容省略)
2021-02-15 02:42:38.643  INFO 24328 ---
[           main] .s.d.r.c.RepositoryConfigurationDelegate : Bootstrapping Spring
Data JPA repositories in DEFAULT mode.
(部分内容省略)
2021-02-15 02:42:40.081  INFO 24328 --- [           main]
org.hibernate.Version                    : HHH000412: Hibernate ORM core version
5.4.27.Final
(部分内容省略)
2021-02-15 02:42:40.337  INFO 24328 --- [           main]
org.hibernate.dialect.Dialect            : HHH000400: Using dialect:
org.hibernate.dialect.H2Dialect
(部分内容省略)
2021-02-15 02:42:40.902  INFO 24328 ---
[           main] .s.s.UserDetailsServiceAutoConfiguration :

Using generated security password: 5fdbb2d6-87f8-4456-8157-fbf014f69706

2021-02-15 02:42:40.985  INFO 24328 --- [           main]
o.s.s.web.DefaultSecurityFilterChain     : Creating filter chain: any request,
[org.springframework.security.web.context.request.async.WebAsyncManagerIntegra
tionFilter@4b41587d,
(部分内容省略)
2021-02-15 02:42:41.020  INFO 24328 --- [           main]
o.s.b.w.embedded.tomcat.TomcatWebServer  : Tomcat started on port(s): 8080 (http)
with context path ''
(部分内容省略)
2021-02-15 02:43:03.381  INFO 24328 --- [nio-8080-exec-1]
o.s.web.servlet.DispatcherServlet        : Completed initialization in 5 ms
```

其中，加黑代码标出的内容是 Spring Security 提供的默认用户名为“user”的随机生成的密码，使用它就可以用来登录。打开浏览器访问 http://localhost:8080/hello，将会跳转到登录页面，如图 8-15 所示。

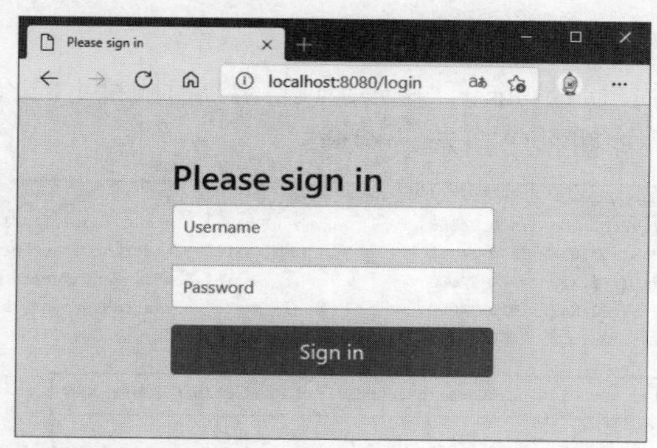

图 8-15

输入用户名“user”和控制台输出的密码，单击“Sign in”按钮就能够成功登录，可以看到输出的“Hello Spring Security”内容，如图 8-16 所示。

当用户从浏览器发送请求访问/hello 接口时，服务端会返回 302 响应码，让客户端重定向到/login 页面。用户在 /login 页面登录，登录成功之后就会自动跳转到/hello 接口。另外，也可以使用 Postman 来发送请求。使用 Postman 发送请求时，可以将用户信息放在请求头中（可以避免重定向到登录页面），如图 8-17 所示。

图 8-16

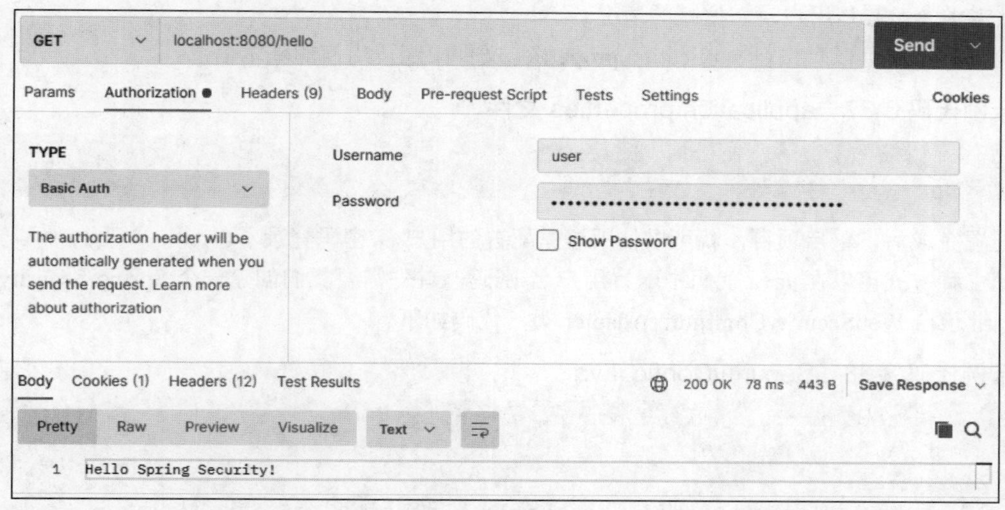

图 8-17

通过以上两种不同的登录方式，可以看出 Spring Security 支持两种不同的认证方式：

- 可以通过 Form 表单来认证。
- 可以通过 HttpBasic 来认证。

（3）用户名密码配置

默认情况下，登录的用户名是 user，密码是项目启动时随机生成的字符串，可以从启动的控制台日志中看到默认密码，如图 8-18 所示。

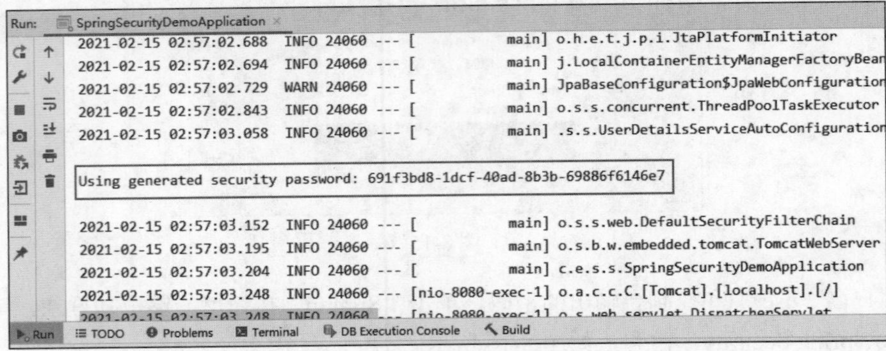

图 8-18

这个随机生成的密码每次启动时都会变。对登录的用户名/密码进行配置，有以下三种不同的方式：

- 在 application.properties 中进行配置。
- 通过 Java 代码配置在内存中。
- 通过 Java 从数据库中加载。

前两种比较简单，第三种方式使用数据库，会有较大的代码量。对于一些简单的项目来说，使用前两种方式非常方便；对于企业级项目来说，前两种方式一是不安全，二是不能满足大量用户的存储需求，尤其是用户动态增减的情况。

第一种方式可以直接在 application.properties 文件中配置用户的基本信息。

示例代码 8-37　application.properties 文件

```
spring.security.user.name=admin
spring.security.user.password=123
```

配置完成后，重启项目，就可以使用这里配置的用户名/密码登录了。

第二种方式可以在 Java 代码中配置用户名/密码，首先需要我们创建一个 Spring Security 的配置类，继承自 WebSecurityConfigurerAdapter 类，代码如下。

示例代码 8-38　SecurityConfig.java

```
package com.example.spring.security.config;
import …
@Configuration
public class SecurityConfig extends WebSecurityConfigurerAdapter {
    @Override
    protected void configure(AuthenticationManagerBuilder auth)
                            throws Exception {
        String encryptedPwd = passwordEncoder().encode("123");
        //下面这两行配置表示在内存中配置了两个用户
        auth.inMemoryAuthentication()
```

```
                .withUser("admin")
                .roles("admin")
                .password(encryptedPwd)
                .and()
                .withUser("test")
                .roles("user")
                .password(encryptedPwd);
        }
        @Bean
        PasswordEncoder passwordEncoder() {
            return new BCryptPasswordEncoder();
        }
    }
```

这里我们在 configure 方法中配置了两个用户 admin 和 test，用户的密码都是加密之后的字符串（明文是 123）。需要注意的是，从 Spring5 开始强制要求密码要加密，如果不想加密，可以提供一个过期的 PasswordEncoder 的实例 NoOpPasswordEncoder，但是不建议这么做，毕竟不安全。

Spring Security 中提供了 BCryptPasswordEncoder 密码编码工具，可以非常方便地实现密码的加密加盐，相同明文加密出来的结果总是不同，这样就不需要用户去额外保存盐的字段了，这一点比 Shiro 要方便得多。关于密码加密以及存储的更多介绍可以查阅官方文档：https://docs.spring.io/spring-security/site/docs/5.4.2/reference/html5/#authentication-password-storage。

第三种方式使用数据库来存储用户密码信息以及权限、角色等信息，会比前两者复杂得多。这里我们先把 Shiro 示例项目 shiro-demo 中的 entity 包、service 包、repository 包、vo 包及包下面的所有类文件整体复制到本项目 com.example.spring.security 包下，修改其中报错误引用（import）的地方，依次修改，重新导入包（在错误引入的类名上按快捷键 Ctrl+Shift+Space），然后按快捷键 Ctrl+Shift+Alt+L，选择“Whole file”，勾选“Optimize imports”“Rearrange code”“Code cleanup”，单击“Run”按钮执行自动优化和冗余代码清理，如图 8-19 所示。

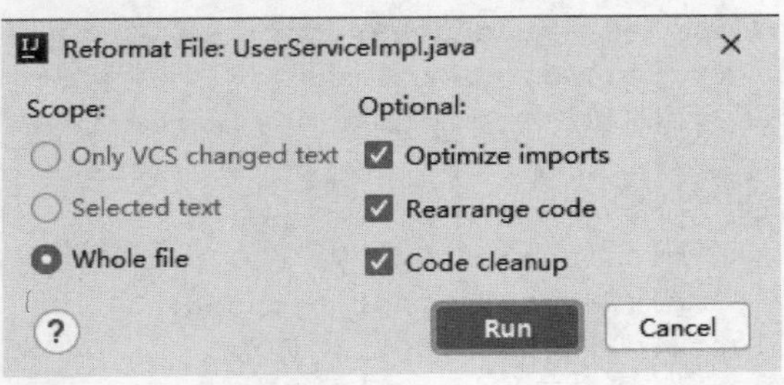

图 8-19

然后将 shiro-demo 项目 resources 目录下的 application.yml 和 import.sql 文件一并复制到 spring-security-demo 项目的 resources 目录下，删除原来的 application.properties 文件。将 shiro-demo 项目中 com.example.shiro.config 包下的 BeanConfig.java 文件复制到当前项目的 com.example.spring.security.config 包下面。

上面这些代码和配置都是可以复用的，因为没有牵扯到与 shiro 相关的内容，无论是使用 Shiro 框架还是 Spring Security 框架都是通用的。

最关键的一部分是实现 Spring Security 提供的 UserDetailsService 接口，它和 Shiro 框架的 Realm

抽象类似，整个 Spring Security 连接数据库的核心就是实现 UserDetaisService 接口。我们要做的就是编写 UserDetailsServiceImpl 类，实现该接口。

示例代码 8-39 UserDetailsServiceImpl.java

```
    package com.example.spring.security.service.impl;
    import …
    @Service
    public class UserDetailsServiceImpl implements UserDetailsService {
        @Resource
        private UserService userService;
        @Resource
        private RoleService roleService;
        @Resource
        private PermissionService permissionService;
        @Override
        public UserDetails loadUserByUsername(String username) throws
UsernameNotFoundException {
            Optional<User> user = userService.getByUsername(username);
            return user.map(u -> {
                List<Role> roles = roleService.getRoles(u.getId());
                List<Permission> permissions = permissionService
    .getPermissions(u.getId());
                List<GrantedAuthority> grantedAuthorities = new
ArrayList<>(roles.size() + permissions.size());
                /*在 Spring Security 中，角色和权限统称为 GrantAuthority，角色和权限都交给
GrantAuthenty 管理，而区分角色和权限的方式，就是在角色名称前加前缀 ROLE_以表示角色*/
                // 将角色信息转换为 SimpleGrantedAuthority 对象类型
                List<SimpleGrantedAuthority> roleAuthorities = roles.stream()
                    //给角色名称增加前缀 ROLE_
                    .map(role -> "ROLE_" + role.getName())
                    .map(SimpleGrantedAuthority::new)
                    .collect(Collectors.toList());
                // 将授权许可信息转换为 SimpleGrantedAuthority 对象类型
                List<SimpleGrantedAuthority> permissionAuthorities =
    permissions.stream()
                    .map(Permission::getName)
                    .map(SimpleGrantedAuthority::new)
                    .collect(Collectors.toList());
                /*将角色和授权许可合并到 grantedAuthorities 列表*/
                grantedAuthorities.addAll(roleAuthorities);
                grantedAuthorities.addAll(permissionAuthorities);
                UserDetails userDetails = org.springframework.security.core
    .userdetails.User.builder()
                    //设置用户名和密码
                    .username(u.getUsername())
                    .password(u.getPassword())
                    //设置权限列表
                    .authorities(grantedAuthorities)
                    .build();
                return userDetails;
```

```
        }).orElse(null);
    }
}
```

这样就提供了一个 UserDetailsService 实现，有了它 Spring Security 就能够从中获取用户的认证信息和权限信息了。只实现这个 UserDetailsServiceImpl 还不能够直接运行，还需要在 SecurityConfig 类中修改配置。

示例代码 8-40　修改配置 SecurityConfig.java

```
package com.example.spring.security.config;
import …
@EnableWebSecurity
public class SecurityConfig extends WebSecurityConfigurerAdapter {
    @Resource
    private UserDetailsService userDetailsService;
    @Bean
    PasswordEncoder passwordEncoder() {
        return new BCryptPasswordEncoder();
    }
    @Override
    protected void configure(AuthenticationManagerBuilder auth) throws
Exception {
        // 使用内存方式存储用户名和密码
        /*String encryptedPwd = passwordEncoder().encode("123");
        //下面这两行配置表示在内存中配置了两个用户
        auth.inMemoryAuthentication()
            .withUser("admin")
            .roles("admin")
            .password(encryptedPwd)
            .and()
            .withUser("test")
            .roles("user")
            .password(encryptedPwd);*/
        // 使用自定义实现的 userDetailsServiceImpl 来加载数据库中的用户和权限信息
        auth.userDetailsService(userDetailsService)
            // 设置加密方式，这里要和注册、创建用户时的密码加密方式使用同一个
// PasswordEncoder
            .passwordEncoder(passwordEncoder());
    }
}
```

至此，Spring Security 对接数据库的逻辑已经全部实现，接下来进行登录测试。在复制过来的 import.sql 中已经预置了一个用户名为“admin”的账户，但是密码仍然是原先 shiro-demo 项目中使用的加密方式，在这里已经不再适用，需要使用新的密码加密工具 BCryptPasswordEncoder 重新获取密码。我们通过创建一个 PasswordEncoderTest 测试类来实现，在 test 源码包中新建测试类 PasswordEncoderTest，其内容如下。

示例代码 8-41　PasswordEncoderTest.java

```
package com.example.spring.security;
```

```
import …
public class PasswordEncoderTest {
    @Test
    public void encodePwd() {
        String passowrd = "123";
        BCryptPasswordEncoder passwordEncoder = new BCryptPasswordEncoder();
        String encryptedPassword = passwordEncoder.encode(passowrd);
        System.out.println(encryptedPassword);
    }
}
```

运行测试方法 encodePwd，在控制台输出中获得加密后的密码“$2a$10$SljMzzv/fAxZEBjv3ckLoOgTfZYp82Br/ZHZRt4E6tM465HJR9wG.”。需要注意的是，每次运行生成的加密串并不一样，因此运行时出现和本书给出的密码不一样是正常的。将输出的加密串复制到 import.sql 中的 user 表插入语句中替换原来的密码，并把加密盐 salt 置为空串（不需要盐了，废弃字段），这部分内容的完整代码如下。

示例代码 8-42 向 user 表中插入 admin 用户的 SQL 语句

```
insert into user(id, username, password, salt, gmt_create,gmt_modified)
select
NULL,'admin','$2a$10$SljMzzv/fAxZEBjv3ckLoOgTfZYp82Br/ZHZRt4E6tM465HJR9wG.
','',now(),now()
from dual
where not exists(
    select 1 from user where username = 'admin'
);
```

准备工作已经完成，运行项目看一下能否正常启动。结果发现运行报错，提示找不到 UserRepository（以及其他 Repository）。这是因为我们从 shiro-demo 项目中复制过来的 BeanConfig 文件中配置的扫描包仍然是原先 shiro 项目的包，而不是本项目的包，为了让扫描配置能够更加通用，可以使用“**”通配符来匹配，修改后的 BeanConfig.java 内容如下。

示例代码 8-43 修改后的 BeanConfig.java 文件

```
package com.example.spring.security.config;
import …
@Configuration
//配置扫描的包
@ComponentScans({
    // 服务类所在的包
    @ComponentScan("com.example.**.service.**"),
    // Controller 类所在的包
    @ComponentScan("com.example.**.controller.**")
})
//开启 JPA Repository 接口，自动扫描 com.example.shiro.repository 包下的接口
@EnableJpaRepositories("com.example.**.repository")
//扫描 com.example.shiro.entity 包下面的实体类
@EntityScan("com.example.**.entity")
public class BeanConfig {}
```

再次启动项目编译运行，看到输出结果如下所示即表示启动成功了。

示例代码 8-44　spring-security 启动编译运行结果输出

```
(部分内容省略)
  .   ____          _            __ _ _
 /\\ / ___'_ __ _ _(_)_ __  __ _ \ \ \ \
( ( )\___ | '_ | '_| | '_ \/ _` | \ \ \ \
 \\/  ___)| |_)| | | | | || (_| |  ) ) ) )
  '  |____| .__|_| |_|_| |_\__, | / / / /
 =========|_|==============|___/=/_/_/_/
 :: Spring Boot ::        (v2.3.8.RELEASE)
(部分内容省略)
2021-02-15 13:24:14.214  INFO 16220 --- [           main]
org.hibernate.dialect.Dialect            : HHH000400: Using dialect:
org.hibernate.dialect.H2Dialect
(部分内容省略)
Hibernate:
    create table permission (
       id bigint generated by default as identity,
        gmt_create timestamp,
        gmt_modified timestamp,
        name varchar(255),
        primary key (id)
    )
Hibernate:
    create table role (
       id bigint generated by default as identity,
        gmt_create timestamp,
        gmt_modified timestamp,
        name varchar(255),
        primary key (id)
    )
Hibernate:
    create table role_permission (
       id bigint generated by default as identity,
        gmt_create timestamp,
        gmt_modified timestamp,
        permission_id bigint,
        role_id bigint,
        primary key (id)
    )
Hibernate:
    create table user (
       id bigint generated by default as identity,
        gmt_create timestamp,
        gmt_modified timestamp,
        password varchar(255) not null,
        salt varchar(255) not null,
        username varchar(255) not null,
        primary key (id)
```

```
        )
    Hibernate:
        create table user_role (
           id bigint generated by default as identity,
            gmt_create timestamp,
            gmt_modified timestamp,
            role_id bigint,
            user_id bigint,
            primary key (id)
        )
    Hibernate:
        alter table user
           add constraint UK_sb8bbouer5wak8vyiiy4pf2bx unique (username)
    2021-02-15 13:24:14.719  INFO 16220 --- [           main]
o.h.t.schema.internal.SchemaCreatorImpl  : HHH000476: Executing import script
'file:/D:/code/idea/spring-security-demo/target/classes/import.sql'
    Hibernate:
        insert into user(id,username,password,salt,gmt_create,gmt_modified) select
NULL,'admin','$2a$10$SljMzzv/fAxZEBjv3ckLoOgTfZYp82Br/ZHZRt4E6tM465HJR9wG.','',
now(), now() from dual where not exists(    select 1 from user where username =
'admin' )
    Hibernate:
        insert into permission(id, name, gmt_create, gmt_modified) select    NULL,
'user:list', now(), now() from dual where not exists(    select 1 from permission
where name='user:list' )
    Hibernate:
        insert into role(id, name, gmt_create, gmt_modified) select    NULL,
'system', now(), now() from dual where not exists(    select 1 from role where
name='system' )
    Hibernate:
        with t as (select    u.id as user_id, r.id as role_id    from user u, role
r    where u.username='admin' and r.name='system') insert into user_role(id,
user_id, role_id, gmt_create, gmt_modified) select    NULL, user_id, role_id, now(),
now() from t where not exists(    select 1 from user_role ur    join t    where
ur.user_id = t.user_id and ur.role_id = t.role_id )
    Hibernate:
        with t as (select    r.id as role_id, p.id as permission_id    from
permission p, role r    where p.name='user:list' and r.name='system') insert into
role_permission(id, role_id, permission_id, gmt_create, gmt_modified) select
NULL, role_id, permission_id, now(), now() from t where not exists(    select 1 from
role_permission rp    join t    where rp.role_id = t.role_id and rp.permission_id
= t.permission_id )
    Hibernate:
        commit
    （部分内容省略）
    2021-02-15 13:24:15.714  INFO 16220 --- [           main]
o.s.b.w.embedded.tomcat.TomcatWebServer  : Tomcat started on port(s): 8080 (http)
with context path ''
    2021-02-15 13:24:15.720  INFO 16220 --- [           main]
c.e.s.s.SpringSecurityDemoApplication    : Started SpringSecurityDemoApplication
```

```
in 4.103 seconds (JVM running for 4.443)
```

可以看到相关数据库表已经建立，预置用户数据已经成功插入数据库表中。

（4）安全配置

对于登录接口，登录成功后的响应、登录失败后的响应都可以在 WebSecurityConfigurerAdapter 的实现类中配置。

示例代码 8-45　安全配置实现 SecurityConfig.java

```
    @Override
    protected void configure(HttpSecurity http) throws Exception {
        http
            //开启登录配置
            .authorizeRequests()
            //设置不需要认证的路径，一律允许通过，这种方式会走 Spring Security 的过滤器链，
当遇到这些访问路径时直接允许通过
            .antMatchers("/index", "/home").permitAll()
            //设置需要角色的路径，表示访问 /user/view 接口，需要具备 system 角色
            .antMatchers("/user/view/**").hasRole("system")
            //设置需要权限许可的路径，表示访问 /user/list 接口，需要具有“user:list”权限许可
            .antMatchers("/user/list").hasAuthority("user:list")
            //设置其他路径都需要登录后才可访问
            .anyRequest().authenticated()
            //开启一个新的设置项
            .and()
            //设置登录表单
            .formLogin()
            /*//定义登录页面，未登录时会自动跳转到该页面
            .loginPage("/login")
            //登录处理接口
            .loginProcessingUrl("/doLogin")
            //定义登录时，用户名的 key，默认为 username
            .usernameParameter("username")
            //定义登录时，用户密码的 key，默认为 password
            .passwordParameter("password")
            //登录成功的处理器
            .successHandler(new AuthenticationSuccessHandler() {
                @Override
                public void onAuthenticationSuccess(HttpServletRequest req,
HttpServletResponse resp, Authentication authentication) throws IOException,
ServletException {
                    resp.setContentType("application/json;charset=utf-8");
                    PrintWriter out = resp.getWriter();
                    out.write("success");
                    out.flush();
                }
            })
            //登录失败后的处理器
            .failureHandler(new AuthenticationFailureHandler() {
                @Override
```

```
            public void onAuthenticationFailure(HttpServletRequest req,
HttpServletResponse resp, AuthenticationException exception) throws IOException,
ServletException {
                resp.setContentType("application/json;charset=utf-8");
                PrintWriter out = resp.getWriter();
                out.write("fail");
                out.flush();
            }
        })*/
        //和表单登录相关的接口直接允许通过
        .permitAll()
        //开启一个新的设置项
        .and()
        //设置登出（注销登录）
        .logout()
        //设置登出路径
        .logoutUrl("/logout")
        //设置登出成功后的处理器
        .logoutSuccessHandler(new LogoutSuccessHandler() {
            @Override
            public void onLogoutSuccess(HttpServletRequest req,
HttpServletResponse resp, Authentication authentication) throws IOException,
ServletException {
                resp.setContentType("application/json;charset=utf-8");
                PrintWriter out = resp.getWriter();
                out.write("logout success");
                out.flush();
            }
        })
        //和登出相关的接口允许通过
        .permitAll()
        .and()
        //启用 Http Basic 认证
        .httpBasic()
        .and()
        //关闭 csrf 跨域支持
        .csrf().disable();
    }
    @Override
    public void configure(WebSecurity web) throws Exception {
        //设置不需要认证的路径，这种方式不走 Spring Security 过滤器链，推荐使用
        web.ignoring().antMatchers("/verifycode");
    }
```

在上面的配置中，设定“/user/view/**”路径需要具有 system 角色，“/user/list”路径需要拥有“user:list”权限许可，“/index”“/home”这两个角色允许直接访问，“/verifycode”路径不需要验证，无须走过滤器。在前面我们定义了“/hello”，按照设定规则需要登录才能够访问。

（5）测试验证

为了更好地验证角色和权限许可策略是否生效，我们同样创建一个 UserRestController 控制器

类，编写两个 Rest 接口，分别是“/user/view/{id}”“/user/list”，其内容如下。

示例代码 8-46　UserResetController.java

```
package com.example.spring.security.controller;
import …
@RestController
@RequestMapping("/user")
public class UserRestController {
    @Resource
    private UserService userService;
    @GetMapping("/view/{id}")
    public JsonResult getUserInfo(@PathVariable("id") Long id) {
        User user = userService.getById(id);
        return new JsonResult(JsonResult.JsonResultCode.SUCCESS, "获取数据成功！
", user);
    }
    @GetMapping("/list")
    public JsonResult getUserList() {
        try {
            List<UserVO> userList = userService.findAll();
            return new JsonResult(JsonResult.JsonResultCode.SUCCESS, "获取数据
成功！", userList);
        } catch (Exception e) {
            e.printStackTrace();
            return new JsonResult(JsonResult.JsonResultCode.ERROR, "获取数据失
败！", null);
        }
    }
}
```

在 UserService 接口中添加一个方法 User getById(Long id)，然后在 UserServiceImpl 实现类中实现这个方法。

示例代码 8-47　UserServiceImpl 类实现 User getById（Long id）方法

```
@Override
public User getById(Long id) {
    return userRepository.getOne(id);
}
```

然后重新编译运行项目，待启动完成之后打开浏览器访问 http://localhost:8080/hello，就会跳转到登录页面，使用预置的用户“admin”和密码“123”进行登录之后成功输出“Hello Spring Security!”，说明“/hello”按照规则是要登录后才能访问的设定生效了。访问 http://localhost:8080/logout 退出登录，然后访问 http://localhost:8080/user/list，同样会跳转到登录页面，再登录之后会成功看到输出的用户列表结果：

```
{
    "code":1,
    "msg":"获取数据成功",
    "data":[{
```

```
        "id":1,
        "username":"admin",
        "timestamp":null
    }]
}
```

在预置用户“admin”的角色数据中，给定它拥有“system”角色和“user:list”权限，因此访问“/user/list”是可以获取到的，接下来测试一下能否正常获取“/user/view/1”接口的数据，结果出现一个错误，如图 8-20 所示。

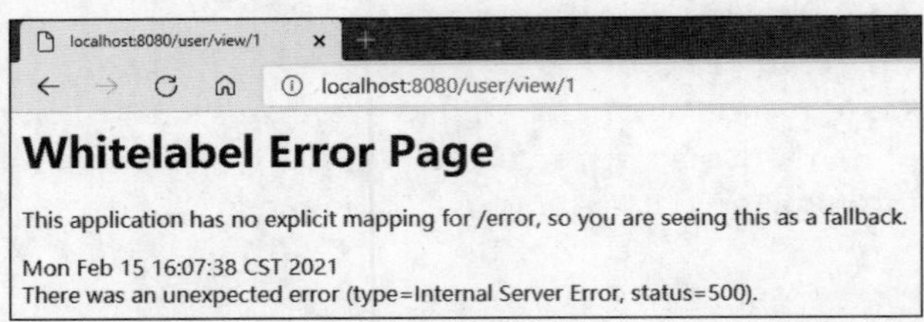

图 8-20

查看控制台打印的错误日志，发现一行关键性错误提示：

```
    com.fasterxml.jackson.databind.exc.InvalidDefinitionException: No
serializer found for class
org.hibernate.proxy.pojo.bytebuddy.ByteBuddyInterceptor and no properties
discovered to create BeanSerializer (to avoid exception, disable
SerializationFeature.FAIL_ON_EMPTY_BEANS) (through reference chain:
com.example.spring.security.vo.JsonResult["data"]->com.example.spring.security
.entity.User$HibernateProxy$yFgBCH3h["hibernateLazyInitializer"])
```

这个错误出现的原因是：jsonplugin 用的是 Java 的内审机制，Hibernate 会给被管理的 pojo 加入一个 hibernateLazyInitializer 属性，jsonplugin 会把 hibernateLazyInitializer 也拿出来操作，并读取里面一个不能被反射操作的属性。解决这个错误的方法是在 User 实体类上打上注解 @JsonIgnoreProperties(value = { "hibernateLazyInitializer", "handler" })，或者不要把 Entity 领域模型的对象直接返回给前端，而是在 Service 层返回数据时将其重新封装成 VO 再返回给前端，或者在 UserRestController#getUserInfo(@PathVariable("id") Long id)方法中对数据重新进行封装：

示例代码 8-48 重新将 User 对象封装成 UserVO 对象

```
    @GetMapping("/view/{id}")
    public JsonResult getUserInfo(@PathVariable("id") Long id) {
        User user = userService.getById(id);
        UserVO userVO = new UserVO();
        userVO.setId(user.getId());
        userVO.setUsername(user.getUsername());
        userVO.setPassword(user.getPassword());
        userVO.setTimestamp(user.getGmtModified().toEpochMilli());
        return new JsonResult(JsonResult.JsonResultCode.SUCCESS, "获取数据成功！",
userVO);
    }
```

再次运行项目并访问 http://localhost:8080/user/view/1，登录之后就能够成功获取用户信息：

```
{
    "code": 1,
    "msg": "获取数据成功！",
    "data": {
        "id": 1,
        "username": "admin",
        "password":
"$2a$10$SljMzzv/fAxZEBjv3ckLoOgTfZYp82Br/ZHZRt4E6tM465HJR9wG.",
        "timestamp": 1613377682282
    }
}
```

（6）使用注解进行权限控制

通过 SecurityConfig 配置权限控制不利于扩展和管理，当权限控制规则非常多时，维护起来费时费力，同时会导致整个 SecurityConfig 文件代码量过多。Spring Security 也支持像 Shiro 那样使用注解的方式来控制角色、权限等要求，Spring Security 默认是禁用注解的，要想开启注解，需要在继承 WebSecurityConfigurerAdapter 的类上加@EnableGlobalMethodSecurity 注解，判断用户对某个控制层的方法是否具有访问权限。我们只需要在 SecurityConfig 配置类上打上注解@EnableGlobalMethodSecurity(prePostEnabled = true)，并将其中的.antMatchers("/user/view/**").hasRole("system")、.antMatchers("/user/list").hasAuthority("user:list")注释，然后修改 UserRestController 的两个方法，加上注解@PreAuthorize。@PreAuthorize 注解使用切面来控制方法是否被调用，它的值是一个表达式，比如 hasRole("system")表示是否具有 system 角色，hasAuthrity("usr:list")表示是否具有 user:list 权限。它的值支持 SpEL 表达式（关于 SpEL 表达式的更多介绍请查阅官方文档：https://docs.spring.io/spring-framework/docs/4.2.x/spring- framework-reference/html/expressions.html）。权限相关支持的表达式在类 SecurityExpressionRoot 中已经给出，其中包括以下几种：

- hasAuthority(String authority)：是否具有某个权限。
- hasAnyAuthority(String... authorities)：是否具有给出的权限列表中的任意一个。
- hasRole(String role)：是否具有角色，即判断含有前缀 ROLE_*相关的权限。
- hasAnyRole(String... roles)：是否具有给出的角色列表中的任意一个。
- hasAnyAuthorityName(String prefix，String... roles)：是否具有给出的权限列表中的任意一个，前缀可以自行指定，而不必须是 ROLE_。
- permitAll()：允许所有。
- denyAll()：拒绝所有。
- isAnonymous()：判断是否匿名用户。
- isAuthenticated()：判断用户身份是否非匿名用户，即通过登录认证或记住我来访问的都算 authenticated。
- isRememberMe()：判断用户是否是通过“记住我”的功能访问。
- isFullyAuthenticated()：判断用户是否是非匿名用户并且非“记住我”访问，即必须完整登录认证才被认为是 authenticated。

这里只需要用到 hasRole 和 hasAuthority 两个表达式，修改后的 UserRestController 内容如下代码所示。

示例代码 8-49 UserRestController.java（使用注解方式限制方法的执行）

```
package com.example.spring.security.controller;
import …
@RestController
@RequestMapping("/user")
public class UserRestController {
    @Resource
    private UserService userService;
    //限制只有 system 角色的用户才可以访问
    @PreAuthorize("hasRole('system')")
    @GetMapping("/view/{id}")
    public JsonResult getUserInfo(@PathVariable("id") Long id) {
        User user = userService.getById(id);
        UserVO userVO = new UserVO();
        userVO.setId(user.getId());
        userVO.setUsername(user.getUsername());
        userVO.setPassword(user.getPassword());
        userVO.setTimestamp(user.getGmtModified().toEpochMilli());
        return new JsonResult(JsonResult.JsonResultCode.SUCCESS, "获取数据成功！
", userVO);
    }
    //限制必须拥有 user:list 权限的用户才可访问
    @PreAuthorize("hasAuthority('user:list')")
    @GetMapping("/list")
    public JsonResult getUserList() {
        try {
            List<UserVO> userList = userService.findAll();
            return new JsonResult(JsonResult.JsonResultCode.SUCCESS, "获取数据
成功！", userList);
        } catch (Exception e) {
            e.printStackTrace();
            return new JsonResult(JsonResult.JsonResultCode.ERROR, "获取数据失
败！", null);
        }
    }
}
```

修改完成后，再次运行，分别访问 http://localhost:8080/user/view/1 和 http://localhost:8080/user/list 两个路径，在登录之后都成功返回了结果。

总 结

本小节带领大家对 Spring Security 有了一个初步的认识，并从一个实际的项目示例来学习它是如何配置和使用的，相信读者能从中看出和 Shiro 的不同之处。无论是 Shiro 还是 Spring Security 框架，其中涉及到的内容依然还有很多，限于篇幅，本书仅能带领大家入门，无法更深入地讲解。想要学习更多知识，读者可以参考相关图书和文档，进一步探究其中的奥秘。

8.2 实现单点登录

单点登录（Single Sign On，SSO）就是通过用户的一次性鉴别登录。用户在身份认证服务器上登录一次以后即可获得访问单点登录系统中其他关联系统和应用软件的权限，同时这种实现是不需要管理员对用户的登录状态或其他信息进行修改的，这意味着在多个应用系统中用户只需一次登录就可以访问所有相互信任的应用系统。这种方式减少了由登录产生的时间消耗，辅助了用户管理，是目前比较流行的。单点登录在目前微服务架构中非常流行，当项目模块众多、单体应用过于庞大无法提供所有服务时就会涉及系统单元的拆分，应用不再是由一个项目独立完成，而是由众多项目和模块同时运行提供服务，从而形成一个应用服务集群，同时对外提供服务。在集群内部，项目和模块之间是松耦合的，但是对于用户端来说是无感知的，在用户看来就好像是一个应用在为其提供服务。单点登录的目的就是当用户访问一个模块登录之后（不用再登录即可访问其他模块功能），即在集群内部多个项目模块之间能够同时共享用户的身份认证信息和授权信息。

单点登录是一种帮助用户快捷访问网络中多个站点的安全通信技术。单点登录系统基于一种安全的通信协议，该协议通过多个系统之间的用户身份信息的交换来实现单点登录。使用单点登录系统时，用户只需要登录一次就可以访问多个系统，不需要记忆多个口令密码。单点登录使用户可以快速访问网络，从而提高工作效率，同时也能提高系统的安全性。

8.2.1 Redis+Session 认证

实现单点登录的一种常见方式是把传统的基于 Web 容器的 Session 会话机制与 Web 容器解耦，将 Session 单独保存在一个能够共享访问的系统当中，多个服务都能够通过这个共享 Session 数据的系统来获取用户认证信息。Redis 就是这样一个能够实现 Session 共享的好方式。

Redis 是完全开源的，遵守 BSD 协议，是一个高性能的 key-value 数据库。Redis 与其他 key - value 缓存产品有以下三个特点：

- Redis 支持数据的持久化，可以将内存中的数据保存在磁盘中，重启的时候可以再次加载进行使用。
- Redis 不仅仅支持简单的 key-value 类型的数据，还提供 list、set、zset、hash 等数据结构的存储。
- Redis 支持数据的备份，即 master-slave 模式的数据备份。

Redis 具有以下优势：

- 性能极高：Redis 能读的速度是 110000 次/s，写的速度是 81000 次/s。
- 丰富的数据类型：Redis 支持二进制类型的 Strings、Lists、Hashes、Sets 及 Ordered Sets 数据类型操作。
- 原子：Redis 的所有操作都是原子性的，意思就是要么成功执行要么失败全不执行。单个操作是原子性的。多个操作也支持事务，即原子性，通过 MULTI 和 EXEC 指令包起来。

- 丰富的特性：Redis 还支持 publish/subscribe、通知、key 过期等特性。

利用 Redis 的这些优点，把 session 数据存放在 Redis 里，统一管理，向外提供服务接口。Redis 可以设置过期时间，对应 session 的失效时间，具有存取速度快、效率高、无单点故障、可以部署集群等优点。

1. 下载并安装 Redis

Redis 下载地址为 https://github.com/tporadowski/redis/releases，可自行选择安装路径（默认安装在 C:\Program Files\Redis 目录下，这里选择安装在 D:\Program Files\Redis 目录下）。进入安装目录下，执行"redis-server.exe redis.windows.conf"命令运行 Redis。在 Windows 下首次安装完成后可能会遇到无法启动的情况，需要执行"redis-cli.exe"连接到 Redis 服务，输入"shutdown"指令关闭 Redis 服务，然后才可以手动输入命令启动，如图 8-21 所示。

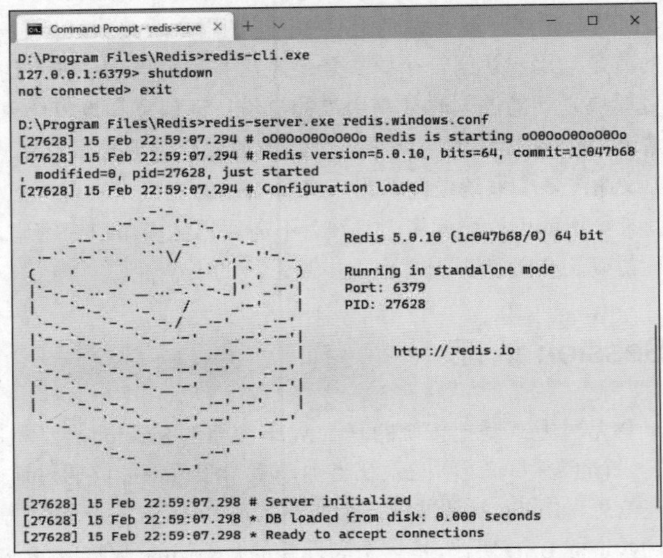

图 8-21

2. 引入 Redis 相关依赖包

以 shiro-demo 项目为基础，对其进行改造，加入 Redis 以实现 session 共享。首先复制 shiro-demo 项目目录，并重命名为 shiro-redis-sso-demo。然后在 IDEA 中打开 shiro-redis-sso-demo 项目目录，在项目模块名称上右键单击鼠标，依次选择菜单按钮"Refactor->Rename"（或直接按快捷键 Shift+F6）来修改项目名称，修改为：shiro-redis-sso-demo，然后打开 pom.xml 修改 artifactId、name 和 Description，并加入 Redis 和 common-pool2 依赖包。

示例代码 8-50 引入 Redis 相关依赖

```
<dependency>
    <groupId>org.springframework.boot</groupId>
    <artifactId>spring-boot-starter-data-redis</artifactId>
</dependency>
<dependency>
    <groupId>org.apache.commons</groupId>
```

```
    <artifactId>commons-pool2</artifactId>
</dependency>
```

其中加粗部分为修改后的内容。特别注意这里额外引入了 commons-pool2 依赖，因为在 Spring Boot 在 2.x 版本中默认使用 Lettuce、在 1.x 版本默认使用 Jedis。Jedis 和 Lettuce 都是 Redis 客户端，两者的区别如下：

- Jedis 是直连模式，在多个线程间共享一个 Jedis 实例时是线程不安全的。
- 如果想要在多线程环境下使用 Jedis，就需要使用连接池。
- 每个线程都去拿 Jedis 实例，当连接数量增多时，物理连接成本就较高了。
- Lettuce 是基于 Netty 的。Netty 是一个多线程、事件驱动的 I/O 框架，连接实例可以在多个线程间共享，通过异步的方式可以让我们更好地利用系统资源，而不用浪费线程等待网络或磁盘 I/O。

Spring Boot 2.x 版本使用 Lettuce，需要依赖于 commons-pool2，如果不引入它就会报一个 java.lang.ClassNotFoundException: org.apache.commons.pool2.impl.GenericObjectPoolConfig 异常。

3. 修改 application.yml 文件并加入 Redis 相关配置

这里提供了一个简单的配置，内容如下所示。

示例代码 8-51 在 application.yml 中加入 Redis 配置内容

```
spring:
  redis:
    database: 0
    host: 127.0.0.1
    port: 6379
    password:
    lettuce:
      pool:
        min-idle: 5
        max-idle: 8
        max-active: 8
        max-wait: 1ms
        shutdown-timeout: 100ms
...
```

4. 创建 RedisSessionDAO、修改 ShiroConfig 配置

Shiro 使用 Redis 来管理 Session，需要实现 SessionDAO，默认使用基于内存的 SessionDAO，这里我们需要实现一个 RedisSessionDAO，完整代码如下。

示例代码 8-52 RedisSessionDAO.java

```
package com.example.shiro.dao;
import …
public class RedisSessionDAO extends AbstractSessionDAO {
    // Session 超时时间，单位为毫秒
    private long expireTime = 120000;
    //注入 redis 操作客户端
```

```
    @Resource
    private RedisTemplate redisTemplate;
    public RedisSessionDAO() {
        super();
    }
    public RedisSessionDAO(long expireTime) {
        this.expireTime = expireTime;
    }
    @Override
    protected Serializable doCreate(Session session) {
        System.out.println("===============doCreate================");
        Serializable sessionId = this.generateSessionId(session);
        this.assignSessionId(session, sessionId);
        redisTemplate.opsForValue()
        .set(session.getId(), session, expireTime, TimeUnit.MILLISECONDS);
        return sessionId;
    }
    @Override
    protected Session doReadSession(Serializable sessionId) {
        System.out.println("==============doReadSession=================");
        if (sessionId == null) {
            return null;
        }
        return (Session) redisTemplate.opsForValue().get(sessionId);
    }
    @Override
    public void update(Session session) throws UnknownSessionException {
        System.out.println("===============update================");
        if (session == null || session.getId() == null) {
            return;
        }
        session.setTimeout(expireTime);
        redisTemplate.opsForValue()
        .set(session.getId(), session, expireTime, TimeUnit.MILLISECONDS);
    }
    @Override
    public void delete(Session session) {
        System.out.println("===============delete================");
        if (null == session) {
            return;
        }
        redisTemplate.opsForValue().getOperations().delete(session.getId());
    }
    /**
     * 获取当前活跃的会话数，可以用来统计在线人数。要实现这个功能，可以在 session 保存到
     * redis 时加入一个前缀，统计的时候则使用 keys("session-prefix*")的方式来模糊查找
     * redis 中所有的 session 集合
     * @return
     */
    @Override
```

```
    public Collection<Session> getActiveSessions() {
        System.out.println("==============getActiveSessions============
====");
        return redisTemplate.keys("*");
    }
}
```

其中涉及 Redis 操作的部分，使用到了 RedisTemplate。RedisTemplate 是 Spring Data Redis 提供给用户的最高级抽象客户端，用户可直接通过 RedisTemplate 进行多种操作。我们实现的 RedisSessionDAO 主要是实现 Session 的创建、获取、更新和删除以及计数操作。

实现了 RedisSessionDAO 之后，还需要将其注册为 Bean，提供给 SessionManager。在 ShiroConfig 配置类中加入如下配置即可。

示例代码 8-53　为 RedisSessionDAO 注册 Bean，并注册 SessionManager 关键代码

```
/**
 * 注册自定义的{@link RedisSessionDAO}实例
 * @return
 */
@Bean
public RedisSessionDAO redisSessionDAO() {
    return new RedisSessionDAO();
}
/**
 * 提供会话管理器，使用默认的 Web 会话管理器，并将自定义的{@link #redisSessionDAO()}注
册到 sessionManager 中
 *
 * @return
 */
@Bean
public SessionManager sessionManager() {
    DefaultWebSessionManager defaultWebSessionManager =
    new DefaultWebSessionManager();
    defaultWebSessionManager.setSessionDAO(redisSessionDAO());
    return defaultWebSessionManager;
}
```

5. 启动测试

完成前面的配置之后就可以编译运行项目了，不出什么问题的话项目能够正常启动，如果启动报错，就需要查看一下项目是不是遗漏了前面的配置。运行成功之后，打开 Postman 创建一个 POST 请求执行登录，在 Body 中给出用户名和密码，单击“Send”按钮执行登录请求，这部分操作在 8.1.2 小节最后的测试部分介绍过，这里不再说明。在我们执行登录请求之后服务器返回了 500 的错误。

打开 IDEA 的控制台输出，查看异常情况，可以看到如下错误信息：

```
（部分内容省略）
==============doReadSession=================
Hibernate:
    select
```

```
        user0_.id as id1_3_,
        user0_.gmt_create as gmt_crea2_3_,
        user0_.gmt_modified as gmt_modi3_3_,
        user0_.password as password4_3_,
        user0_.salt as salt5_3_,
        user0_.username as username6_3_
    from
        user user0_
    where
        user0_.username=?
===============doCreate================
===============update================
==============doReadSession=================
===============update================
2021-02-16 01:17:35.569 ERROR 16936 --- [nio-8080-exec-1]
o.a.c.c.C.[.[.[/].[dispatcherServlet]    : Servlet.service() for servlet
[dispatcherServlet] in context with path [] threw exception [Request processing
failed; nested exception is
org.springframework.data.redis.serializer.SerializationException: Cannot
serialize; nested exception is
org.springframework.core.serializer.support.SerializationFailedException:
Failed to serialize object using DefaultSerializer; nested exception is
java.io.NotSerializableException: com.example.shiro.entity.User] with root cause

java.io.NotSerializableException: com.example.shiro.entity.User
    at java.io.ObjectOutputStream.writeObject0(ObjectOutputStream.java:1184)
~[na:1.8.0_265]
     at java.io.ObjectOutputStream.writeObject(ObjectOutputStream.java:348)
~[na:1.8.0_265]
     at java.util.HashSet.writeObject(HashSet.java:288) ~[na:1.8.0_265]
（部分内容省略）
     at
org.apache.shiro.session.mgt.DefaultSessionManager.onChange(DefaultSessionMana
ger.java:212) ~[shiro-core-1.7.0.jar:1.7.0]
     at
org.apache.shiro.session.mgt.AbstractNativeSessionManager.setAttribute(Abstrac
tNativeSessionManager.java:258) ~[shiro-core-1.7.0.jar:1.7.0]
（部分内容省略）
==============doReadSession=================
```

分析这段日志输出，不难看出在登录请求到达服务器后端之后显示读取 Session 信息，但是没有读到（因为是第一次登录），于是查询数据库表获取用户信息执行了登录验证逻辑，完成之后将生成的 Session 信息保存到 Redis（从 doCreate、doUpdate 两行的输出可以看出），但是之后发生了一个异常 NotSerializableException: com.example.shiro.entity.User。这就很明显了，是因为将 User 对象序列化保存时出现了不能序列化的错误，解决办法是修改 User 实体类，实现 Serializable 接口：

```
public class User implements Serializable{
    //…
}
```

再次运行项目，使用 Postman 执行登录请求就可以看到登录成功的返回结果了，如图 8-22 所示。

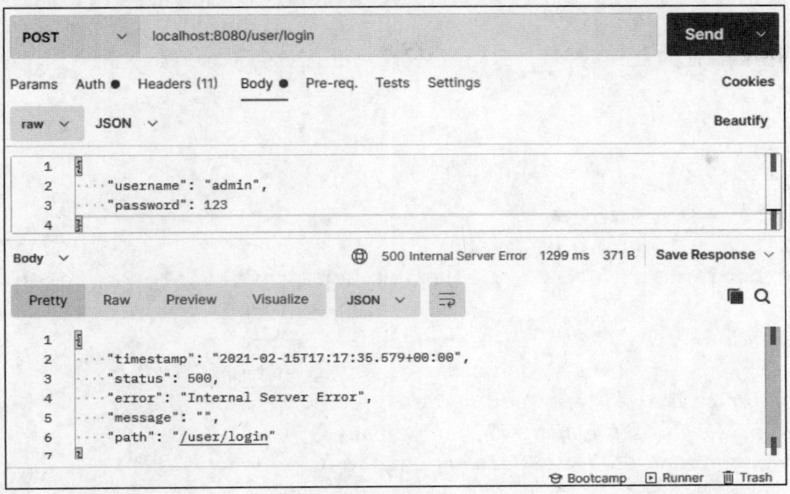

图 8-22

至此，我们已经完成了 shiro-redis-sso-demo 项目的改造，引入 Redis 来保存 Session 信息。虽然能够正常登录、获取数据等操作，但是还是无法体现出多个项目共享 Session 会话实现单点登录的特点。

6. 改造 shiro-redis-sso-demo 项目加密算法使用 BCrypt

BCrypt 这种加密方式比 MD5 更加安全，Spring Security 也一直极力在推荐它，并且已经在最新版将 MD5 剔除。BCrypt 算法将随机生成的 salt 保存在加密后的字符串中，而无须单独保存 salt，每次执行加密得到的加密串都不一样。Shiro 本身并不支持 BCrypt 算法，要想让 Shiro 也使用 BCrypt 这种加密算法，就需要自行实现。

首先，在 pom.xml 中引入 jBCrypt 依赖包。

示例代码 8-54　引入 jBCrypt 依赖包

```
<!--引入 BCrypt 加密算法支持-->
<dependency>
    <groupId>de.svenkubiak</groupId>
    <artifactId>jBCrypt</artifactId>
    <version>0.4.3</version>
</dependency>
```

然后，新建 util 工具包，在 util 包下新建自定义密码凭证匹配器，继承自 SimpleCredentialsMatcher。

示例代码 8-55　BCryptCredentialsMatcher.java

```
package com.example.shiro.util;
import org.apache.shiro.authc.AuthenticationInfo;
import org.apache.shiro.authc.AuthenticationToken;
import org.apache.shiro.authc.credential.SimpleCredentialsMatcher;
import org.mindrot.jbcrypt.BCrypt;
/**
 * 自定义实现 BCrypt 加密算法的凭证匹配器
```

```
 */
public class BCryptCredentialsMatcher extends SimpleCredentialsMatcher {
    /**
     * 重写密码验证的方法，验证一致返回 true，不一致返回 false
     *
     * @param token
     * @param info
     * @return
     */
    @Override
    public boolean doCredentialsMatch(AuthenticationToken token,
                                      AuthenticationInfo info) {
        //获得前台传过来的密码
        char[] pwd = (char[]) token.getCredentials();
        String originalPassword = new String(pwd);
        //这是数据库里查出来的密码
        String hashedPw = (String) info.getCredentials();
        //进行比对
        return BCrypt.checkpw(originalPassword, hashedPw);
    }
}
```

接着修改 ShiroConfig 配置类中针对 ShiroRealm 的定义，为其注入 BCryptCredentialsMatcher，不再使用 HashedCredentialsMatcher 实现。

示例代码 8-56　修改 ShiroConfig 类中 ShiroRealm 注册 Bean 的定义（部分关键代码）

```
@Configuration
public class ShiroConfig {
    /**
     * 提供自定义的 Realm，并将其注册名设置为 authorizer
     *
     * @return
     */
    @Bean
    public ShiroRealm authorizer() {
        return new ShiroRealm(bCryptCredentialsMatcher());
    }
    //省略其他配置内容
}
```

ShiroRealm 中 doGetAuthenticationInfo 方法部分也需要修改，在返回的 SimpleAuthenticationInfo 对象的 salt 参数设置为 null：

```
@Override
protected AuthenticationInfo doGetAuthenticationInfo(AuthenticationToken
token) throws AuthenticationException {
    /*获取用户名，去除用户名前后的空白字符*/
    String username = StringUtils.trimWhitespace(
    Optional.ofNullable(token)
        .map(AuthenticationToken::getPrincipal)
        .map(Object::toString).orElse(null));
```

```
    /*用户名不为空*/
    if (!StringUtils.isEmpty(username)) {
        /*根据用户名获取用户信息（主要是密码和盐）*/
        Optional<User> user = userService.getByUsername(username);
        if (user.isPresent()) {
            /*返回查询出来的用户信息，封装成 SimpleAuthenticationInfo 提供给 Shiro 去判
            /定用户名、密码是否正确*/
            return new SimpleAuthenticationInfo(user.get(),
                        user.get().getPassword(), null, getName());
        }
    }
    return null;
}
```

这里将其中的 salt 参数设置为 null，其实改不改无所谓，因为 salt 已经用不到了。

最后将 import.sql 预置数据脚本中插入 admin 用户语句的密码修改成和 spring-security-demo 中的 admin 一样的用户密码，salt 字段置空：

```
insert into user(id, username, password, salt, gmt_create, gmt_modified)
select
    NULL,'admin',
    '$2a$10$SljMzzv/fAxZEBjv3ckLoOgTfZYp82Br/ZHZRt4E6tM465HJR9wG.', '',
    now(), now()
from dual
where not exists(
    select 1 from user where username = 'admin'
    );
```

修改完成后，重新启动项目，使用 Postman 进行登录测试，可以成功登录。

7. 引入 Spring Session

Spring Session 是 Spring 的项目之一。Spring Session 提供了一套创建和管理 Servlet HttpSession 的方案，默认采用外置的 Redis 来存储 Session 数据，以此来解决 Session 共享的问题。我们可以直接使用 Spring Session 来管理，而无须自行实现 RedisSessionDAO。要想使用 Spring Session，只需要以下三步操作：

步骤 01　在 pom.xml 中引入 Spring Session 依赖：

```
<dependency>
    <groupId>org.springframework.session</groupId>
    <artifactId>spring-session-data-redis</artifactId>
</dependency>
```

步骤 02　新建 SessionConfig 配置类，开启注解@EnableRedisHttpSession。

示例代码 8-57　SessionConfig.java

```
package com.example.shiro.config;
import …
@Configuration
@EnableRedisHttpSession
```

```
public class SessionConfig {}
```

步骤03 将 ShiroConfig 配置类中 redisSessionDAO 和 sessionManager 这两个 Bean 的注册注释掉（或删除）。

8. 多应用 SSO 测试

接下来进行多个应用的单点登录测试。

首先使用命令“mvnw clean package”对本项目进行打包，打包完成后将 target 目录下生成的 shiro-redis-sso-demo-0.0.1-SNAPSHOT.jar 文件复制到其他地方，比如 D:\code\目录下，并通过命令行运行此 jar 包：

```
java -jar shiro-redis-sso-demo-0.0.1-SNAPSHOT.jar
```

然后修改当前 IDEA 中的项目 application.yml 文件，添加服务器配置，修改端口号为 8081：

```
server:
   port: 8081
# 其他配置省略
```

在 IDEA 中启动它，将运行在 8081 端口。命令行执行的 jar 包会运行在 8080 端口。这就相当于是同一个应用多个 Tomcat 实例部署的模式了。接下来使用 Postman 请求 http://localhost:8080/user/login 访问服务器进行登录，再请求 http://localhost:8080/user/list 接口获取用户列表数据，以此来验证单点登录是否成功。

执行结果如图 8-23、图 8-24 所示，可以看到通过 http://localhost:8080/user/login 成功登录之后 Postman 获取到了 Session 信息“SESSION= NmQwNzQxNjktMzVlZi00Mzc1LTllYm-QtY2Q4YTNiMWVhMjM5; Path=/; HttpOnly;”，之后携带这个 cookie 信息成功获取到了 http://localhost:8080/user/list 接口的数据。结果证明单点登录配置成功。需要注意的一点是，共享 Session 绝不仅仅是为了同一个应用多个 Tomcat 实例运行解决登录问题，更多的意义在于多个子系统之间共享会话信息，免除登录的问题。

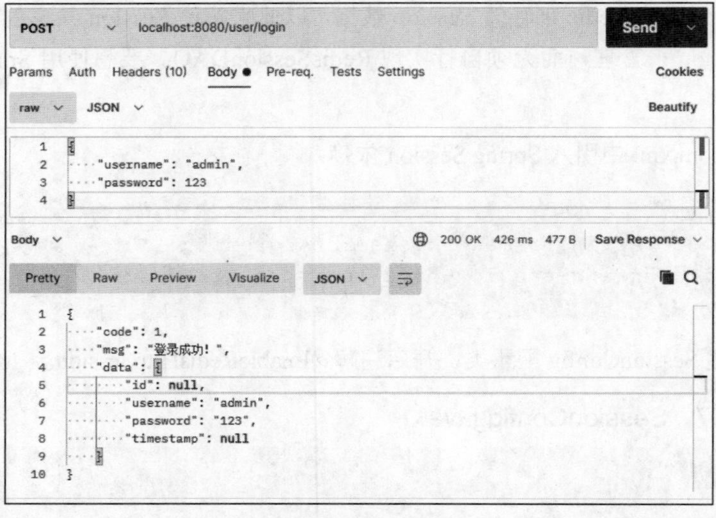

图 8-23

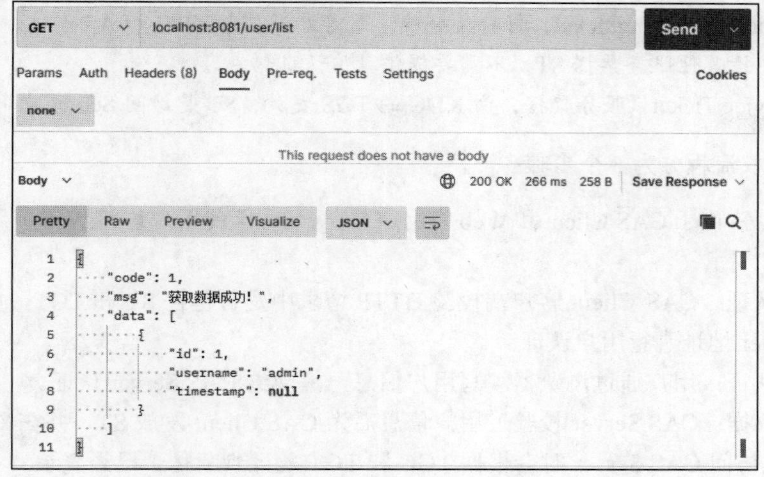

图 8-24

8.2.2　CAS 认证

实现单点登录除了共享 Session 的方式之外，还有一种常见的方式——使用中央认证服务（Central Authentication Service，CAS）。它的认证流程如图 8-25 所示。

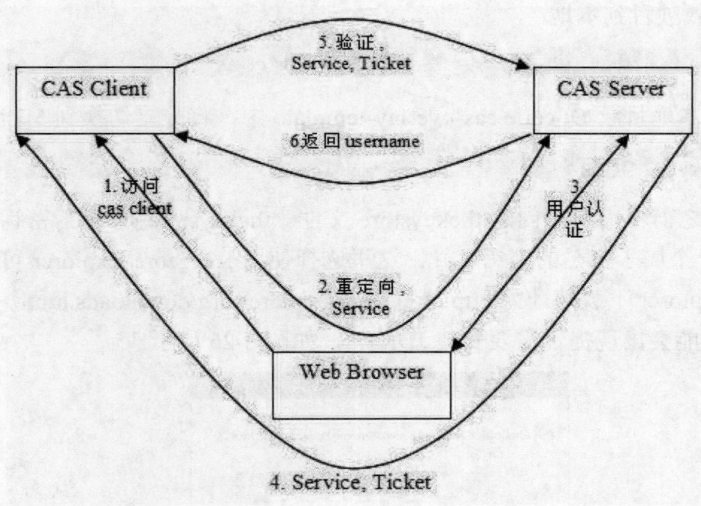

图 8-25

其中涉及几个名词需要了解一下：

- AS：Authentication Service，认证服务，发放 TGT。
- KDC：Key Distribution Center，密钥发放中心。
- TGS：Ticket-Granting Service，票据授权服务，索取 TGT，发放 ST。
- TGC：Ticket-Granting Cookie，授权的票据证明，由 CAS Server 通过 SSL 方式发送给终端用户。该值存在 Cookie 中，根据 TGC 可以找到 TGT。

- TGT：Ticket Granting ticket，俗称大令牌，或者票根，由 KDC 和 AS 发放。获取该票据后，可直接申请其他服务票据 ST，不需要提供身份认证信息。
- ST：Service Ticket，服务票据，由 KDC 的 TGS 发放，ST 是访问 Server 内部的令牌。

其认证和授权流程分为 6 个步骤：

（1）访问服务：由于 CAS Client 和 Web 应用部署在一起，因此用户访问 Web 应用时，CAS Client 就会处理请求。

（2）定向认证：CAS Client 客户端校验 HTTP 请求中是否包含 ST 和 TGT，如果没有就会重定向到 CAS Server 地址进行用户认证。

（3）用户认证：用户通过浏览器填写用户信息，提交给 CAS Server 认证。

（4）发放票据：CAS Server 校验过用户信息后为 CAS Client 发放 ST，并在浏览器 Cookie 中设置 TGC，下次访问 CAS Server 时会根据 TGC 和 TGT 验证判断是否已经登录。

（5）验证票据：CAS Client 拿到 ST 后，再次请求 CAS Server 验证 ST 合法性，验证通过后允许客户端访问。

（6）传输用户信息：CAS Server 校验过 ST 后传输用户信息给 CAS Client。

接下来我们以 8.1.3 小节中的 spring-security-demo 项目为基础，将其改造引入 CAS 认证服务。

1. 下载并安装运行 CAS Server

使用 Git 克隆下载项目到本地：

```
git clone https://github.com/apereo/cas-overlay-template.git
```

这里将其克隆到本地目录 D:\code\cas-overlay-template 中，将分支切换到 5.3 版本：

```
git checkout 5.3
```

在执行构件编译之前，需要先创建 thekeystore 文件。thekeystore 是 SSL 密钥文件，我们需要为 CAS 服务器创建一个属于自己的密钥文件。这里选择使用 KeyStore Explorer 可视化工具进行密钥创建。KeyStore Explorer 下载地址为 http://keystore-explorer.org/downloads.html。安装完成之后，运行 kse.exe，此时可能会出现提示需要安装 JDK1.8，如图 8-26 所示。

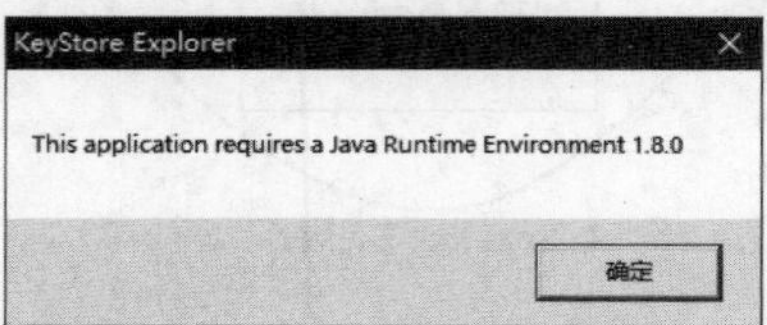

图 8-26

系统中已经安装过 JDK1.8，为什么还会提示安装 JDK1.8 呢？这是因为我们安装的 JDK1.8 是采用免安装版的压缩包解压后配置环境变量来使用的，并没有添加到系统注册表，KeyStore Explorer 无法识别到 JDK。

这里有以下两种解决办法：

一种是重装 JDK1.8，比如重新安装 AdoptOpenJDK msi 版本，勾选“JavaSoft(Oracle) registry keys”，如图 8-27 所示。

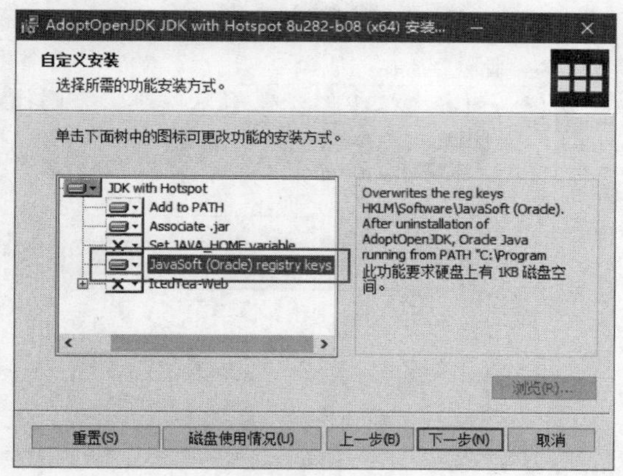

图 8-27

另一种解决办法是不运行 kse.exe，而在命令行中使用 java -jar kse.jar 来运行：

```
C:\Program Files (x86)\KeyStore Explorer>java -jar kse.jar
```

KeyStore Explorer 的运行界面如图 8-28 所示。单击“Create a new KeyStore”，在弹出的对话框中选择“PKCS #12”类型，单击“OK”按钮，然后单击“Generate Key Pair”按钮，选择“RSA”加密算法、再单击“OK”按钮进入下一步证书信息设置界面，这一部分操作如图 8-29、图 8-30 所示。

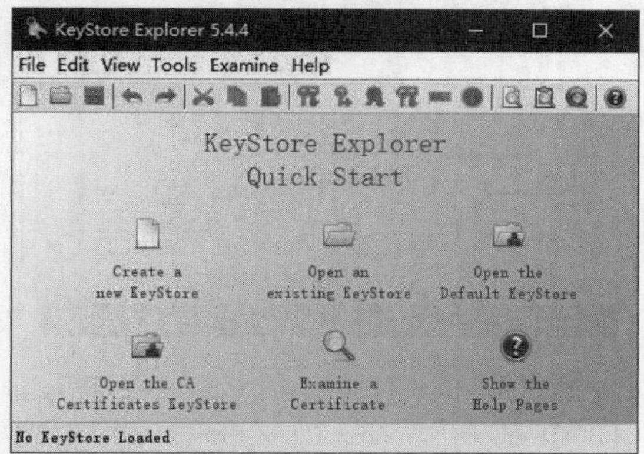

图 8-28

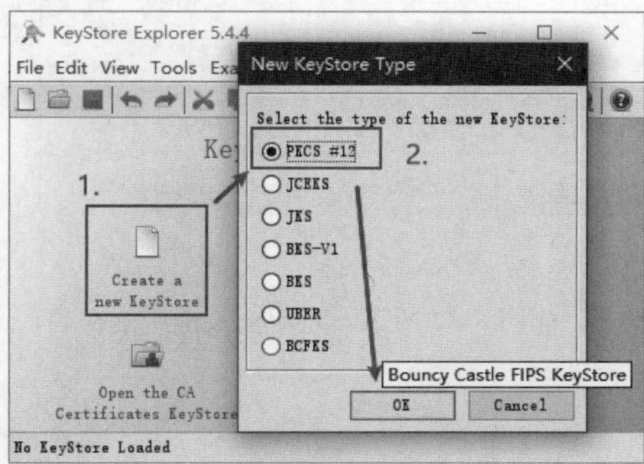

图 8-29

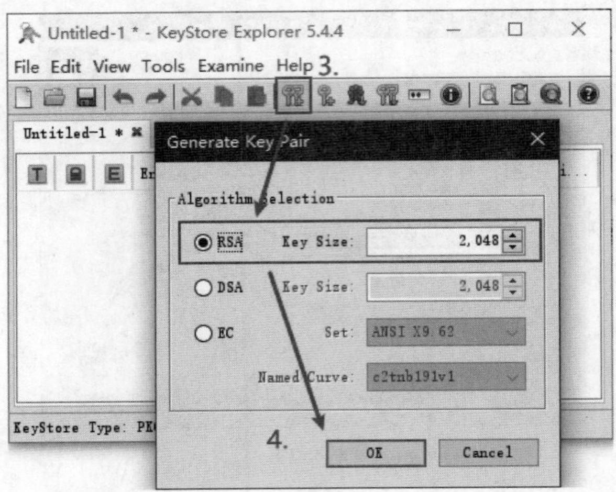

图 8-30

在证书详情设置中，单击 Name 后面的小图标，填写各种名称信息，Common Name 填写域名信息（domain），这里使用“localhost”，Country 填写“cn”（中国），如图 8-31 所示。

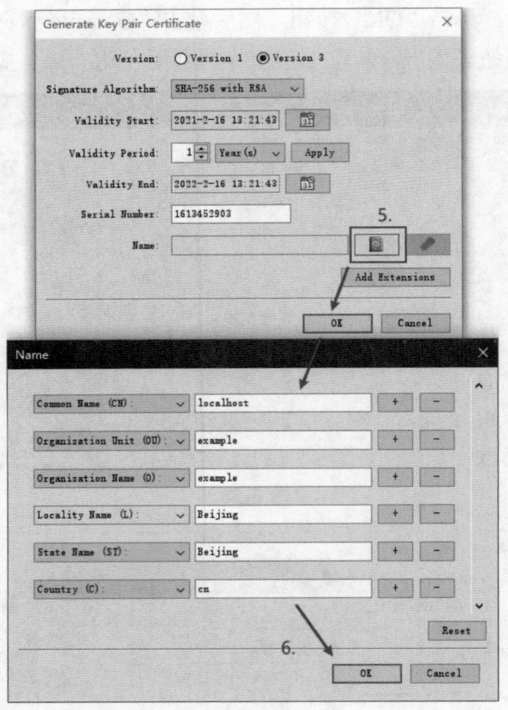

图 8-31

之后会提示输入证书别名，如图 8-32 所示。之后输入密钥的密码，这里输入“123”确认即可成功生成密钥，如图 8-33、图 8-34 所示。

图 8-32

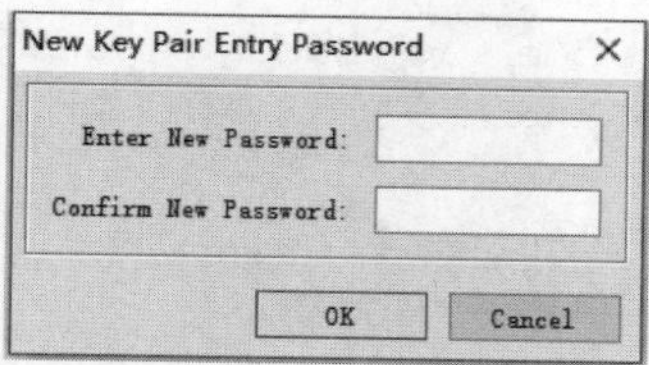

图 8-33

图 8-34

选择文件保存位置：在 cas-overlay-template 项目目录下新建目录 src/main/resources，然后在创建好的密钥上方保存按钮，输入密码“123”，将其保存在之前克隆的“cas-overlay-template”项目目录中的“src/main/resources”目录，并修改名称为“thekeystore”。这里保存的路径为“D:\code\cas-overlay-template\src\main\resources\thekeystore”并将 target/cas/WEB-INF/classes/目录下的 application.properties 文件和 log4j2.xml 文件复制到 src/main/resources 目录下，cmd 命令如下：

```
> mkdir src\main\resources
> copy target\cas\WEB-INF\classes\application.properties src\main\resources
> copy target\cas\WEB-INF\classes\log4j2.xml src\main\resources
```

然后修改 src/main/resources 目录下的 application.properties，将其中的 ssl 配置进行修改：

```
server.ssl.key-store=classpath:thekeystore
server.ssl.key-store-password=123
server.ssl.key-password=123
```

以上操作完成之后，打开 cmd 控制台，进入 cas-overlay-template 目录下，执行“build.cmd bootrun”命令：

```
Microsoft Windows [版本 10.0.19042.630]
(c) 2020 Microsoft Corporation. 保留所有权利。
C:\Users\Administrator>D:
D:\>cd D:\code\cas-overlay-template
D:\code\cas-overlay-template>build.cmd bootrun
[INFO] Scanning for projects...
（部分内容省略）
    _    ____  _____ ____  _____ ___     ____    _    ____
   / \  |  _ \| ____|  _ \| ____/ _ \   / ___|  / \  / ___|
  / _ \ | |_) |  _| | |_) |  _|| | | | | |     / _ \ \___ \
 / ___ \|  __/| |___|  _ <| |__| |_| | | |___ / ___ \ ___) |
/_/   \_\_|   |_____|_| \_\_____\___/   \____/_/   \_\____/
CAS Version: 5.3.16
CAS Commit Id: 1808d979b784500d4b4dd846003a0230f326b57d
CAS Build Date/Time: 2021-02-16T04:11:12.886Z
Spring Boot Version: 1.5.18.RELEASE
Spring Version: 4.3.25.RELEASE
```

```
    Java Home: D:\Develop\Java\jdk8u265\jre
    Java Vendor: AdoptOpenJDK
    Java Version: 1.8.0_265
    JVM Free Memory: 726 MB
    JVM Maximum Memory: 7 GB
    JVM Total Memory: 939 MB
    JCE Installed: Yes
    Node Version: v14.8.0
    NPM Version: N/A
    OS Architecture: amd64
    OS Name: Windows 10
    OS Version: 10.0
    OS Date/Time: 2021-02-16T14:23:19.804
    OS Temp Directory: C:\Users\ADMINI~1\AppData\Local\Temp\
    ------------------------------------------------------------
    Apache Tomcat Version: Apache Tomcat/8.5.47
    ------------------------------------------------------------
    2021-02-16 14:23:19,846 INFO
[org.apereo.cas.configuration.DefaultCasConfigurationPropertiesSourceLocator] -
<Configuration directory [\etc\cas\config] is not a directory or cannot be found
at the specific path>
    （部分内容省略）
      ____ ______ ___  ____  _
     / ___|_   _/ _ \|  _ \| |
     \___ \ | || | | | |_) | |
      ___) || || |_| |  __/|_|
     |____/ |_| \___/|_|   (_)
    CAS is configured to accept a static list of credentials for authentication.
While this is generally useful for demo purposes, it is STRONGLY recommended that
you DISABLE this authentication method (by setting 'cas.authn.accept.users' to a
blank value) and switch to a mode that is more suitable for production.>
    （部分内容省略）
      ____  _____    _    ______   __
     |  _ \| ____|  / \  |  _ \ \ / /
     | |_) |  _|   / _ \ | | | \ V /
     |  _ <| |___ / ___ \| |_| || |
     |_| \_\_____/_/   \_\____/ |_|
    2021-02-16 14:23:44,380 INFO
[org.apereo.cas.support.events.listener.DefaultCasEventListener] - <>
    2021-02-16 14:23:44,380 INFO
[org.apereo.cas.support.events.listener.DefaultCasEventListener] - <Ready to
process requests @ [2021-02-16T06:23:44.380Z]>
```

最后输出 READY 的信息，表示 CAS Server 已经成功启动！打开浏览器访问 https://localhost:8443/cas/login 登录，输入用户名“casuser”、密码“Mellon”即可登录中央认证服务器，登录页面如图 8-35 所示。

图 8-35

2. Spring Security 中 CAS 认证流程

（1）当用户访问需要认证的路径时，由于缺乏 Authentication 对象，会引发 AuthenticationException 异常，或者是 AccessDeniedException 异常，然后交由 ExceptionTranslationFilter 处理，之后调用对应的 AuthenticationEntry 实现类来处理该异常。使用 CAS 时，此类是 CasAuthenticationEntryPoint。

（2）CasAuthenticationEntryPoint 会将用户的浏览器重定向到 CAS 服务器地址，并传递一个 service 参数，这个参数是 Spring Security 项目（开发的项目应用）的登录回调地址（URL）。

（3）浏览器被重定向到 CAS 服务器之后，将会提示输入用户名和密码进行登录，如果具有 cookie 就不再提示登录（这是一个例外的情况）。然后 CAS 将使用 AuthenticationHandler 来判定用户名和密码是否有效。

（4）成功登录之后，CAS 服务器会将用户浏览器重定向到之前 service 指定的回调地址，并传递一个 ticket 参数，这个参数代表了“服务票证”，它是一个加密的不可读的字符串。

（5）回到开发的 Web 应用服务程序中，CasAuthenticationFilter 就会拦截到访问 service 指定路径（可以配置）的请求，之后该过滤器会构造一个 UsernamePasswordAuthenticationToken 对象，将其传递给 AuthenticationManager 来进行验证。

（6）AuthenticationManager 的实现 ProviderManager 会调用 CasAuthenticationProvider 来进行处理，该 Provider 只会处理以下两种 Token：

- UsernamePasswordAuthenticationToken 实例的 principal 是 CasAuthenticationFilter.CAS_STATEFUL_IDENTIFIER 或者 CasAuthenticationFilter.CAS_STATELESS_IDENTIFIERlia 这两种类型的 Token。

- CasAuthenticationToken 实例的 Token。

（7）之后 CasAuthenticationProvider 会调用 TicketValidator 来验证服务票据 ticket，例如 Cas20ServiceTicketValidator。TicketValidator 会向 CAS 服务器发出 HTTPS 请求来验证这个 ticket 的有效性，并传递之前的回调地址 service。

（8）CAS 服务器在收到验证请求之后，将会对比提供的 ticket 和签发的 service url 是否匹配，若匹配则将返回包含用户名的 TicketResponse。

（9）Cas20TicketValidator 收到 TicketResponse，解析用户名等信息。

（10）CasAuthenticationProvider 进行验证，请求 AuthenticationUserDetailsService 实例来加载用户信息（UserDetails），该实例的实现类比如是 GrantedAuthorityFromAssertionAttributesUserDetailsService。

（11）如果验证没有问题，CasAuthenticationProvider 会构造一个 CasAuthenticationToken，其中包含了 TicketResponse 和具体的权限信息 GrantedAuthority。

（12）CasAuthenticationFilter 将 CasAuthenticationToken 放置在 SecurityContext（安全上下文）中。

（13）最终用户的浏览器会被重定向到引起 AuthenticationException 异常的原始访问路径上。

3. 改造 spring-security-demo 项目，接入 CAS 认证

复制 spring-security-demo，将项目重命名为 spring-security-cas-demo，并修改 pom.xml 中的 artifactId、name 为 spring-security-cas-demo，修改 description：

```
<groupId>com.example</groupId>
<artifactId>spring-security-cas-demo</artifactId>
<version>0.0.1-SNAPSHOT</version>
<name>spring-security-cas-demo</name>
<description>Spring Security CAS Demo project for Spring Boot</description>
```

然后引入依赖包：

```
<!--引入 Spring Security CAS 支持-->
<dependency>
    <groupId>org.springframework.security</groupId>
    <artifactId>spring-security-cas</artifactId>
</dependency>
```

配置一个 ServiceProperties 类型的 Bean，指定 service 参数地址：

```
/**
 * 指定 service 相关信息
 */
@Bean
public ServiceProperties serviceProperties() {
    ServiceProperties serviceProperties = new ServiceProperties();
    // 本机服务，访问该路径时进行 CAS 校验登录
    serviceProperties.setService("http://localhost:8080/login/cas");
    serviceProperties.setAuthenticateAllArtifacts(true);
    return serviceProperties;
}
```

然后配置 CasAuthenticationFilter 和 CasAuthenticationEntryPoint 两个 Bean：

```
/**
 * CAS 认证过滤器
 */
@Bean
public CasAuthenticationFilter casAuthenticationFilter() throws Exception {
    CasAuthenticationFilter casAuthenticationFilter =
    new CasAuthenticationFilter();
    casAuthenticationFilter.setAuthenticationManager
(authenticationManager());
    return casAuthenticationFilter;
}
/**
 * 设置认证入口
 */
@Bean
public CasAuthenticationEntryPoint casAuthenticationEntryPoint() {
    CasAuthenticationEntryPoint casAuthenticationEntryPoint =
    new CasAuthenticationEntryPoint();
    casAuthenticationEntryPoint
    .setLoginUrl("https://localhost:8443/cas/login");
    casAuthenticationEntryPoint.setServiceProperties(serviceProperties());
    return casAuthenticationEntryPoint;
}
```

为了启用 CAS 身份验证流程，需要在 WebSecurityConfigurerAdapter 实现类 CasSecurityConfig 中的 configure 方法中对这两个 Bean 进行配置：

```
http.exceptionHandling()
    .authenticationEntryPoint(casAuthenticationEntryPoint())
    .and()
    .addFilter(casAuthenticationFilter())
```

接下来配置 CasAuthenticationProvider Bean，用于进行 Ticket 的验证，因为 CasAuthenticationProvider 还会用到 TicketValidator，因此也需要配置此 Bean。另外，该 Provider 还需要使用 AuthenticationUserDetailsService 实例，因此总共需要配置 3 个 Bean：

```
@Bean
public AuthenticationUserDetailsService authenticationUserDetailsService() {
    return new UserDetailsByNameServiceWrapper(userDetailsService);
}
@Bean
public Cas20ServiceTicketValidator ticketValidator() {
    // 指定 CAS 校验器，进行 Ticket 校验
    return new Cas20ServiceTicketValidator("https://localhost:8443/cas");
}
/**
 * CAS 认证 Provider
 */
@Bean
```

```
public CasAuthenticationProvider casAuthenticationProvider() {
    CasAuthenticationProvider casAuthenticationProvider =
    new CasAuthenticationProvider();
    casAuthenticationProvider.setAuthenticationUserDetailsService(
    authenticationUserDetailsService());
    casAuthenticationProvider.setServiceProperties(serviceProperties());
    casAuthenticationProvider.setTicketValidator(ticketValidator());
    casAuthenticationProvider.setKey("casAuthenticationProviderKey");
    return casAuthenticationProvider;
}
```

其中还用到了一个 Bean：UserDetailsService。该 Bean 就是之前我们自定义实现的 UserDetailsServiceImpl 实例，用以提供用户信息（UserDetails）。这样就完成了 CAS 的基本配置。

完整的 SecurityCASConfig 配置类如下。

示例代码 8-58　SecurityCASConfig.java

```
package com.example.spring.security.config;
import …
@Configuration
@EnableWebSecurity
@EnableGlobalMethodSecurity(prePostEnabled = true)
//启用判断，只有当 spring.cas.enabled=true 时才会启用
@ConditionalOnProperty(name = {"spring.cas.enabled"}, havingValue = "true")
public class SecurityCASConfig extends WebSecurityConfigurerAdapter {
    @Resource
    private UserDetailsService userDetailsService;
    @Override
    protected void configure(AuthenticationManagerBuilder auth)
    throws Exception {
        super.configure(auth);
        auth.userDetailsService(userDetailsService)
        .passwordEncoder(passwordEncoder())
        .and()
        .authenticationProvider(casAuthenticationProvider());
    }
    @Override
    protected void configure(HttpSecurity http) throws Exception {
        http.exceptionHandling()
        .authenticationEntryPoint(casAuthenticationEntryPoint())
        .and()
        .addFilter(casAuthenticationFilter());
    }
    @Bean
    PasswordEncoder passwordEncoder() {
        return new BCryptPasswordEncoder();
    }
    /**
     * CAS 认证过滤器
     */
    @Bean
```

```
    public CasAuthenticationFilter casAuthenticationFilter()
                throws Exception {
        CasAuthenticationFilter casAuthenticationFilter =
        new CasAuthenticationFilter();
        casAuthenticationFilter
        .setAuthenticationManager(authenticationManager());
        return casAuthenticationFilter;
    }
    /**
     * 设置认证入口
     */
    @Bean
    public CasAuthenticationEntryPoint casAuthenticationEntryPoint() {
        CasAuthenticationEntryPoint casAuthenticationEntryPoint =
        new CasAuthenticationEntryPoint();
        casAuthenticationEntryPoint
        .setLoginUrl("https://localhost:8443/cas/login");
        casAuthenticationEntryPoint.setServiceProperties
(serviceProperties());
        return casAuthenticationEntryPoint;
    }
    /**
     * 指定 service 相关信息
     */
    @Bean
    public ServiceProperties serviceProperties() {
        ServiceProperties serviceProperties = new ServiceProperties();
        // 本机服务，访问该路径时进行 Cas 校验登录
        serviceProperties.setService("http://localhost:8080/login/cas");
        serviceProperties.setAuthenticateAllArtifacts(true);
        return serviceProperties;
    }
    @Bean
    public AuthenticationUserDetailsService
authenticationUserDetailsService()
  {
        return new UserDetailsByNameServiceWrapper(userDetailsService);
    }
    @Bean
    public Cas20ServiceTicketValidator ticketValidator() {
        // 指定 CAS 校验器，进行 Ticket 校验
        return new Cas20ServiceTicketValidator("https://localhost:8443/cas");
    }
    /**
     * CAS 认证 Provider
     */
    @Bean
    public CasAuthenticationProvider casAuthenticationProvider() {
        CasAuthenticationProvider casAuthenticationProvider =
        new CasAuthenticationProvider();
```

```
        casAuthenticationProvider.setAuthenticationUserDetailsService(
        authenticationUserDetailsService());
        casAuthenticationProvider.setServiceProperties(serviceProperties());
        casAuthenticationProvider.setTicketValidator(ticketValidator());
        casAuthenticationProvider.setKey("casAuthenticationProviderKey");
        return casAuthenticationProvider;
    }
}
```

启动 CAS 服务器，然后运行开发好的示例项目，不出意外就能启动成功。我们通过浏览器访问 http://localhost:8080/user/list 来获取用户列表测试 CAS 认证是否正常，然后浏览器被重定向到了 https://localhost:8443/cas/login?service=http%3A%2F%2Flocalhost%3A 8080%2 Flogin%2Fcas，但是出现了一个提示，如图 8-36 所示。CAS 服务器提示我们“未认证授权的服务”，告诉我们没有定义该服务。这是因为从 CAS 4.*开始，CAS 默认只支持 HTTPS 的方式。现在这个应用程序是 HTTP 协议的，CAS 服务器版本是 5.3，所以 CAS 不认。

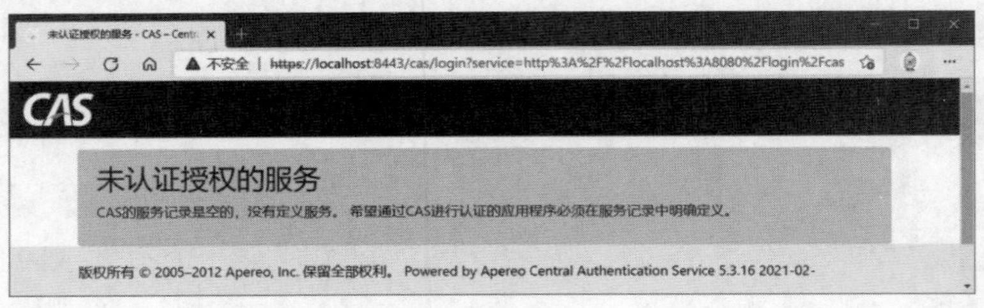

图 8-36

解决这个问题有两种方案：一种是将开发的项目改为 HTTPS 协议，需要创建 SSL 证书（这里不讲）；另一种是修改 CAS 服务器的配置，使其支持 HTTP 协议。我们采用第二种方案来暂时解决这个问题。

在 cas-overlay-template 目录下，找到 target/cas/WEB-INF/classes/services 目录下的 HTTPSandIMAPS-10000001.json 文件，将其复制到 src/main/resources/services 目录下，修改其内容：

```
{
  "@class": "org.apereo.cas.services.RegexRegisteredService",
  "serviceId": "^(https|http|imaps)://.*",
  "name": "HTTPS and IMAPS",
  "id": 10000001,
  "description": "This service definition authorizes all application urls that support HTTPS and IMAPS protocols.",
  "evaluationOrder": 10000
}
```

其中的“serviceId”配置为其添加了“http”，这样就能够允许 HTTP 服务使用 CAS。然后还需要修改 src/main/resources 目录下的 application.properties，在最后添加加粗字体的两项配置：

```
##
# CAS Authentication Credentials
#
cas.authn.accept.users=casuser::Mellon
cas.tgc.secure=false
```

```
cas.serviceRegistry.initFromJson=true
```

这样就能够使 CAS 服务器从 HTTPSandIMAPS-10000001.json 中读取相关配置了。配置完成之后，重新运行 CAS 服务器：build.cmd bootrun。然后使用浏览器访问 http://localhost:8080/user/list，正常跳转到 CAS 服务器登录页面，如图 8-37 所示。

输入用户名“admin”和密码“123”进行登录，但是提示我们认证信息无效，这是为什么呢？查看 CAS 运行服务的控制台输出日志，可以看到如下信息：

```
    2021-02-25 00:22:52,675 ERROR
[org.apereo.cas.authentication.PolicyBasedAuthenticationManager] -
<Authentication has failed. Credentials may be incorrect or CAS cannot find
authentication handler that supports [UsernamePasswordCredential(username=admin)]
of type [UsernamePasswordCredential]. Examine the configuration to ensure a method
of authentication is defined and analyze CAS logs at DEBUG level to trace the
authentication event.>
    2021-02-25 00:22:52,683 INFO
[org.apereo.inspektr.audit.support.SLF4jLoggingAuditTrailManager] - <Audit trail
record BEGIN
    =============================================================
    WHO: admin
    WHAT: Supplied credentials: [UsernamePasswordCredential(username=admin)]
    ACTION: AUTHENTICATION_FAILED
    APPLICATION: CAS
    WHEN: Thu Feb 25 00:22:52 CST 2021
    CLIENT IP ADDRESS: 0:0:0:0:0:0:0:1
    SERVER IP ADDRESS: 0:0:0:0:0:0:0:1
    =============================================================
    >
    2021-02-25 00:22:58,128 INFO
[org.apereo.cas.services.AbstractServicesManager] - <Loaded [1] service(s) from
[InMemoryServiceRegistry].>
```

认证失败了，最后一行提示我们 CAS 服务器当前使用的是“InMemoryServiceRegistry”，也就是说使用的是基于内存的服务注册方式，而用户名密码是由 CAS 服务器来提供的。细心的你可能会发现，我们在前面修改 application.properties 添加两项配置时，有一行配置是“cas.authn.accept.users=casuser::Mellon”，这不就是提供的用户名和密码信息吗？我们只需要修改这个配置，加入和项目连接的数据库中预置的用户名和密码就可以了（事实上，可以自己修改配置实现连接和项目使用同一个数据库来获取用户信息，这里不展开讲解，有兴趣的读者可以自行阅读 CAS 官方文档来修改）。将接受的用户名和密码配置如下：

```
cas.authn.accept.users=casuser::Mellon,admin::123
```

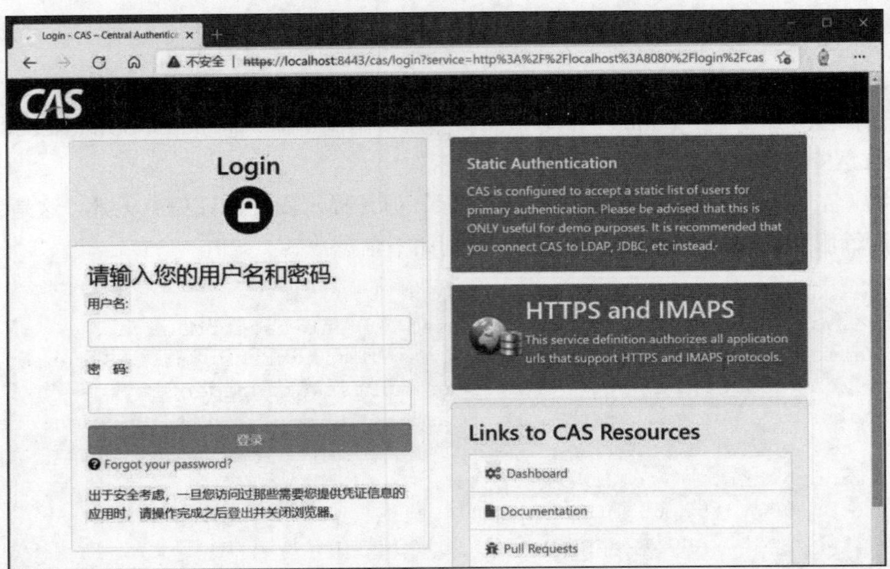

图 8-37

重启 CAS 服务器，再次尝试登录，登录成功后跳转回服务地址，出现如图 8-38 所示的页面。查看后台控制台，发现如下报错信息：

```
    SSL error getting response from host: localhost : Error Message:
sun.security.validator.ValidatorException: PKIX path building failed:
sun.security.provider.certpath.SunCertPathBuilderException: unable to find valid
certification path to requested target
```

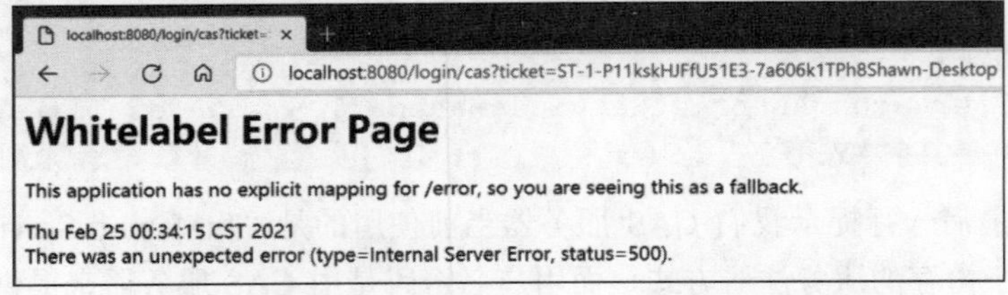

图 8-38

通过错误信息发现依然是 HTTPS 的问题，提示无法找到合法的 SSL 证书。这是因为应用程序会自动查找 JAVA_HOME 目录下的 cacerts 证书，可以通过如下命令来查看当前的证书列表：

```
    > keytool -list -keystore $JAVA_HOME/jre/lib/security/cacerts
    keytool 错误: java.lang.Exception: 密钥库文件不存在:
$JAVA_HOME/jre/lib/security/cacerts
```

我们需要把为 CAS 服务器创建的 SSL 证书导入 cacerts 列表中。首先运行 KeyStore Explorer：

```
    c:\Program Files (x86)\KeyStore Explorer>java -jar kse.jar
```

然后在图形界面中单击“Open an existing keystore”按钮，找到 D:\code\cas-overlay-template\src\main\resources\thekeystore 文件，输入保存 thekeystore 文件时设置的密码“123”，然后在该 keystore 上右击，选择“Export”→“Export Certificate Chain”，如图 8-39 所示。

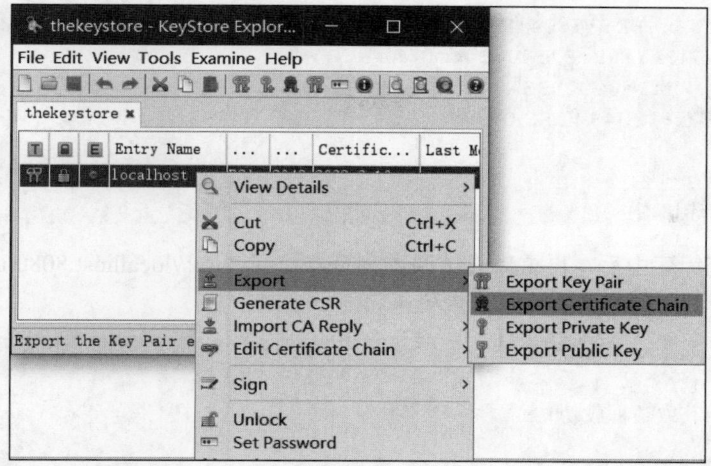

图 8-39

然后设置导出格式为“X.509”，选择导出路径，这里设置为“D:\code\cas-overlay-template\src\main\resources\localhost.cer”，如图 8-40 所示。单击“Export”按钮导出.cer 文件。

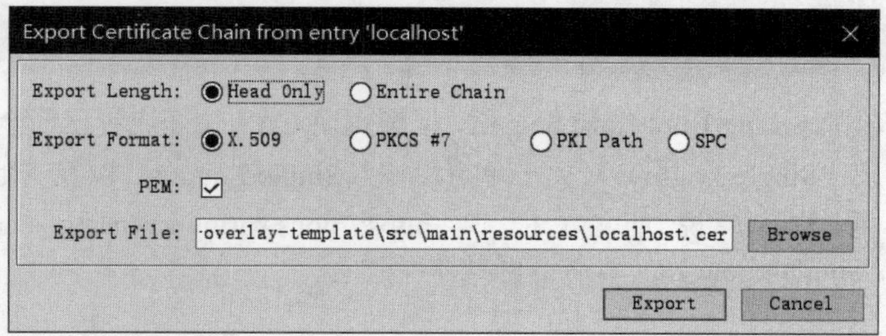

图 8-40

然后将 localhost.cer 文件导入 cacerts 证书列表中，命令如下：

```
    > keytool -import -file
D:\code\cas-overlay-template\src\main\resources\localhost.cer -keystore
$JAVA_HOME/jre/lib/security/cacerts
```

提示输入密钥库口令，该口令为“changeit”，输入之后会提示是否导入，输入“是”就能成功导入该证书了。这一步操作的输出信息如下：

```
    > keytool -import -file
D:\code\cas-overlay-template\src\main\resources\localhost.cer
-keystore %JAVA_HOME%/jre/lib/security/cacerts
    输入密钥库口令:
    所有者: CN=localhost, OU=example, O=example, L=Beijing, ST=Beijing, C=cn
    发布者: CN=localhost, OU=example, O=example, L=Beijing, ST=Beijing, C=cn
    序列号: 602b5667
    有效期为 Tue Feb 16 13:21:43 CST 2021 至 Wed Feb 16 13:21:43 CST 2022
    证书指纹:
            MD5:  9D:D9:40:D8:9C:A4:FA:F8:A4:27:52:EF:6B:DE:B7:03
            SHA1: 98:3E:10:9D:42:1A:A1:E6:8D:03:F6:25:CF:36:A4:AD:1B:9E:AF:25
```

```
        SHA256: BE:34:E4:EE:C2:84:0A:60:59:DE:35:8E:E5:D8:1B:CD:1B:FA:94:
67:F3:4A:DE:7A:B4:A7:08:F7:40:B4:A0:A4
    签名算法名称: SHA256withRSA
    主体公共密钥算法: 2048 位 RSA 密钥
    版本: 3
    是否信任此证书? [否]:  是
    证书已添加到密钥库中
```

证书添加到密钥库中后，再重新尝试通过浏览器访问 http://localhost:8080/user/list。由于之前我们已经登录成功了，因此无须登录，直接返回用户列表信息：

```
{
    "code": 1,
    "msg": "获取数据成功！",
    "data": [
        {
            "id": 1,
            "username": "admin",
            "timestamp": null
        }
    ]
}
```

至此，一个简单的 Spring Boot 结合 Security 框架实现 CAS 认证流程已经全部完成。上面仅仅配置实现了单点登录（Single Login），关于单点注销（Single Logout）以及其他高级配置的实现，读者可以查看官方说明文档（https://docs.spring.io/spring-security/site/docs/5.4.1/reference/html5/#cas-singlelogout），本书不再具体展开。

8.2.3 JWT 认证

1. 什么是 JWT

Json Web Token（JWT）是为了在网络应用环境间传递声明而执行的一种基于 JSON 的开放标准（RFC 7519）。该 Token 被设计为紧凑且安全的，特别适用于分布式站点的单点登录（SSO）场景。JWT 的声明一般被用来在身份提供者和服务提供者间传递被认证的用户身份信息，以便于从资源服务器获取资源，也可以增加一些额外的其他业务逻辑所必需的声明信息。该 Token 可以直接被用于认证，也可以被加密。

2. 传统 Session 认证所显露的问题

HTTP 本身是一种无状态的协议，这就意味着如果用户向我们的应用提供了用户名和密码来进行用户认证，那么下一次请求时用户还要再一次进行用户认证，因为根据 HTTP 协议，我们并不能知道是哪个用户发出的请求，所以为了让我们的应用能识别是哪个用户发出的请求，我们只能在服务器存储一份用户登录的信息，这份登录信息会在响应时传递给浏览器，告诉其保存为 cookie，以便下次请求时发送给应用，这样我们的应用就能识别请求来自哪个用户了。这就是传统的基于 session 认证。这种基于 session 的认证使应用本身很难得到扩展，随着不同客户端用户的增加，独立的服务器已无法承载更多的用户，这时基于 session 认证应用的问题就会暴露出来：

- 内存：每个用户经过应用认证之后都要在服务端做一次记录，以方便用户下次请求的鉴别，通常 session 都是保存在内存中的，随着认证用户的增多，服务端的开销会明显增大。
- 扩展性：用户认证之后，服务端做认证记录，如果认证的记录被保存在内存中，就意味着用户下次请求还必须在这台服务器上，这样才能拿到授权的资源，在分布式的应用上相应地限制了负载均衡器的能力。这也意味着限制了应用的扩展能力。
- CSRF：因为是基于 cookie 来进行用户识别的，所以如果 cookie 被截获，那么用户就很容易受到跨站请求伪造的攻击。

3. 基于 Token 的鉴权机制

基于 Token 的鉴权机制类似于 HTTP 协议也是无状态的，它不需要在服务端保留用户的认证信息或者会话信息。这就意味着基于 Token 认证机制的应用不需要去考虑用户在哪一台服务器登录了，为应用的扩展提供了便利。

流程上是这样的：

①用户使用用户名、密码来请求服务器进行登录。

②服务器验证用户的登录信息。

③服务器通过验证发送给用户一个 Token。

④客户端存储 Token，并在每次请求时附送上 Token 值。

⑤服务端验证 Token 值，并返回数据。

这个 Token 必须在每次请求时传递给服务端，它应该保存在请求头里。另外，服务端要支持 CORS（跨来源资源共享）策略，一般我们在服务端这么做就可以了。

```
Access-Control-Allow-Origin: *
```

4. JWS 以及其构成

JWS（JSON Web Signature）是 JWT 的一种实现，除了 JWS 外，JWE(JSON Web Encryption) 也是 JWT 的一种实现。它们的关系如图 8-41 所示。

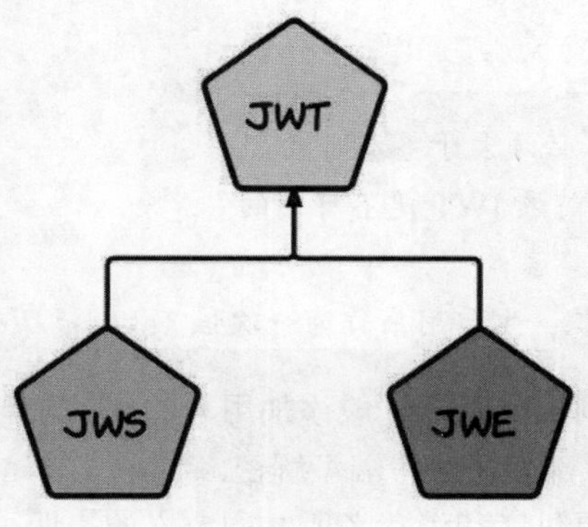

图 8-41

签了名（Signed）的 JWT 称为 JWS（JSON Web Signature）。事实上，JWT 本身并不存在——它必须是 JWS 或 JWE（JSON Web 加密）二者之一。JWT 它就像一个抽象类——而 JWS 和 JWE 是它的具体实现。JWS 是目前最常用的一种实现，下面主要介绍一下 JWS 的构成。

JWS 它规定了一个简单的具有统一的（三个部分）表达形式的字符串。第一部分称为头部（header），第二部分称为有效载荷（payload），第三部分是签证或称签名（signature）。

（1）header

JWT 的头部承载两部分信息：

- 声明类型，这里是 JWT。
- 声明加密的算法，通常直接使用 HMAC、SHA256。

完整的头部就像下面这样的 JSON：

```
{
   "typ": "JWT",
   "alg": "HS256"
}
```

然后将头部进行 base64 加密（该加密是可以对称解密的），构成第一部分。可以在 Linux Shell 或者 Git bash 中对 header 内容 base64 加密测试，得到的字符串就是 header 的内容：

```
$ echo -n '{"alg":"HS256","typ":"JWT"}' | base64
eyJhbGciOiJIUzI1NiIsInR5cCI6IkpXVCJ9
```

（2）payload

有效载荷就是存放有效信息的地方。载荷就像是飞机、火车上承载的货品，这些有效信息包含三个部分的声明（Claim）：

- 注册的声明（Registered Claims）。
- 公开的声明（Public Claims）。
- 私有的声明（Private Claims）。

JWT 规定了 7 个注册的声明（建议但不强制使用）：

- iss：JWT 签发者。
- sub：JWT 所面向的用户。
- aud：接收 JWT 的一方。
- exp：JWT 的过期时间，必须大于签发时间。
- nbf：定义在什么时间之前该 JWT 是不可用的。
- iat：JWT 的签发时间。
- jti：JWT 的唯一身份标识，主要用来作为一次性 Token，从而回避重放攻击。

公开的声明：可以添加任何的信息，一般添加用户的相关信息或其他业务需要的必要信息，但不建议添加敏感信息，因为该部分在客户端可解密。

私有的声明：私有声明是提供者和消费者所共同定义的声明，一般不建议存放敏感信息，因为 base64 是对称解密的，意味着该部分信息可以归类为明文信息。

定义一个 payload：

```
{
   "sub": "1234567890",
```

```
    "name": "John Doe",
    "admin": true
}
```

然后将其进行 base64 加密，得到 JWT 的第二部分，即：

```
echo -n '{
> "sub": "1234567890",
> "name": "John Doe",
> "admin": true
> }
> ' | base64
ewoic3ViIjogIjEyMzQ1Njc4OTAiLAoibmFtZSI6ICJKb2huIERvZSIsCiJhZG1pbiI6IHRydWUK
fQo=
```

（3）signature

JWT 的第三部分是一个签证信息，由三部分组成：

- header（base64 加密后的）。
- payload（base64 加密后的）。
- secret。

这部分需要 base64 加密后的 header 和 base64 加密后的 payload 使用“.”连接组成的字符串，然后通过 header 中声明的加密方式进行加盐 secret 组合加密，构成 JWT 的第三部分。

将 header、payload、signature 这三部分 base64 加密后的字符串用字符“.”连接成一个完整的字符串，构成了最终的 JWT，如图 8-42 所示。

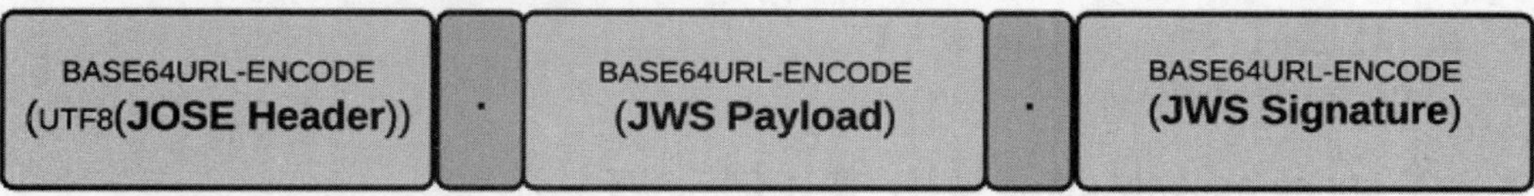

图 8-42

注　意

secret 是保存在服务器端的，JWT 的签发生成也是在服务器端的，secret 就是用来进行 JWT 的签发和 JWT 验证的，所以它就是服务端的私钥，在任何场景都不应该流露出去。一旦客户端得知这个 secret，就意味着客户端可以自我签发 JWT 了。

5. 如何使用 JWT

一般是在请求头里加入 Authorization，并加上 Bearer 标注：

```
fetch('api/user/1', {
  headers: {
    'Authorization': 'Bearer ' + token
  }
})
```

服务端会验证 Token，如果验证通过就会返回相应的资源。整个流程如图 8-43 所示。

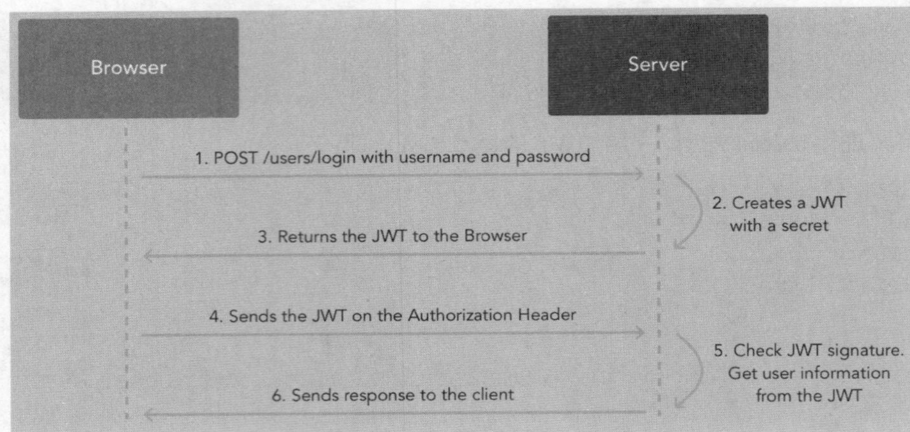

图 8-43

6. JWT 的优点

- 因为 JSON 的通用性，所以 JWT 是可以进行跨语言支持的，比如 Java、JavaScript、Node.js、PHP 等很多语言都可以使用。
- 因为有了 payload 部分，所以 JWT 可以在自身存储一些其他业务逻辑所必要的非敏感信息。
- JWT 的构成非常简单，字节占用很小，所以是非常便于传输的。
- 它不需要在服务端保存会话信息，易于应用的扩展。

7. 使用 JWT 的注意事项

- 不应该在 JWT 的 payload 部分存放敏感信息，因为该部分是客户端可解密的部分。
- 保护好 secret 私钥，该私钥非常重要。
- 尽量使用[HTTPS]协议。

8. 改造 spring-security-demo 项目，使其支持 JWT 认证

我们通过一个项目案例接入 JWT，主要实现如下目标：

- 支持用户通过 REST 方式使用用户名和密码登录。
- 登录后通过 Http Header 返回 Bearer Token，每次请求，客户端需通过 Header 将 Bearer Token 带回，用于权限校验。
- 服务端负责 Token 的定期刷新。

下面进行项目的搭建。首先复制 spring-security-demo 项目，将其重命名为 spring-security-jwt-demo，修改 pom.xml 中的以下内容：

```
<groupId>com.example</groupId>
<artifactId>spring-security-jwt-demo</artifactId>
<version>0.0.1-SNAPSHOT</version>
<name>spring-security-jwt-demo</name>
<description>Spring Security JWT Demo project for Spring Boot</description>
```

并引入 JWT 依赖包：

```
<dependency>
   <groupId>io.jsonwebtoken</groupId>
   <artifactId>jjwt</artifactId>
   <version>0.9.1</version>
</dependency>
```

创建使用 REST 方式来进行登录认证的拦截器，对于用户登录行为，Security 通过定义一个 Filter 来拦截“/login”（默认路径）。Spring Security 默认支持 Form 方式登录，所以对于使用 JSON 发送登录信息的情况，需要我们自定义一个 Filter，这个 Filter 可以直接从 AbstractAuthenticationProcessingFilter 继承，也可以通过 OncePerRequestFilter 实现。只需要实现两部分：一部分是 RequestMatcher，指定拦截的 Request 类型；另一部分是从 Request Body 中提取出 username 和 password 提交给 AuthenticationManager。该实现代码示例如下。

示例代码 8-59　JsonUsernamePasswordAuthenticationFilter.java

```
package com.example.spring.security.jwt.filter;
import …
/**
 * 拦截登录请求，解析 JSON 获取用户名和密码
 */
@Getter
@Setter
public class JsonUsernamePasswordAuthenticationFilter extends
AbstractAuthenticationProcessingFilter {
    public static final String SPRING_SECURITY_FORM_USERNAME_KEY = "username";
    public static final String SPRING_SECURITY_FORM_PASSWORD_KEY = "password";
    private String usernameParameter = SPRING_SECURITY_FORM_USERNAME_KEY;
    private String passwordParameter = SPRING_SECURITY_FORM_PASSWORD_KEY;
    public JsonUsernamePasswordAuthenticationFilter() {
        //拦截 URL 为“/usr/login”的 POST 请求
        super(new AntPathRequestMatcher("/user/login", "POST"));
    }
    public JsonUsernamePasswordAuthenticationFilter(
            AuthenticationManager authenticationManager) {
        //拦截 URL 为“/usr/login”的 POST 请求
        super(new AntPathRequestMatcher("/user/login", "POST"));
        Assert.notNull(authenticationManager,
            "authenticationManager cannot be null");
        this.setAuthenticationManager(authenticationManager);
    }
    @Override
    public Authentication attemptAuthentication(HttpServletRequest request,
            HttpServletResponse response)
        throws AuthenticationException, IOException, ServletException {
        if (!request.getMethod().equals("POST")) {
            throw new AuthenticationServiceException(
                "Authentication method not supported: " + request.getMethod());
        }
        //获取 request body
        String body = StreamUtils.copyToString(request.getInputStream(),
```

```
                Charset.forName("UTF-8"));
        String username = "", password = "";
        //解析 body json，获取用户名和密码
        if (StringUtils.hasText(body)) {
            ObjectMapper mapper = new ObjectMapper();
            HashMap<String, String> jsonObj = mapper.readValue(body,
                HashMap.class);
            username = StringUtils.trimWhitespace(
                jsonObj.get(usernameParameter));
            password = StringUtils.trimWhitespace(
                jsonObj.get(passwordParameter) + "");
        }
        //封装到 token
        UsernamePasswordAuthenticationToken authToken =
            new UsernamePasswordAuthenticationToken(username, password);
        //将 token 交给 authenticationManager 验证
        return this.getAuthenticationManager().authenticate(authToken);
    }
}
```

指定拦截“/user/login”路径的 POST 请求，从 Request Body 中取出用户名和密码，封装成 UsernamePasswordAuthenticationToken 类型的 Token 对象交给 AuthenticationManager 进行登录认证。封装后的 Token 最终是交给 AuthenticationProvider 来处理的。对于 UsernamePassword AuthenticationToken 类型处理的 provider，Spring Security 已经提供了一个默认实现 DaoAuthenticationProvider。DaoAuthenticationProvider 使用 UserDetailsService 来获取用户的登录信息，进行比对，因此我们需要实现一个 UserDetailsService 来从数据库中查询用户的信息，示例代码如下。

示例代码 8-60 UserDetailsServiceImpl.java

```
package com.example.spring.security.service.impl;
import …
@Service
public class UserDetailsServiceImpl implements UserDetailsService {
    @Resource
    private UserService userService;
    @Resource
    private RoleService roleService;
    @Resource
    private PermissionService permissionService;
    @Override
    public UserDetails loadUserByUsername(String username) throws
UsernameNotFoundException {
        try {
            Optional<User> user = userService.getByUsername(username);
            return user.map(u -> {
                List<Role> roles = roleService.getRoles(u.getId());
                List<Permission> permissions = permissionService
                    .getPermissions(u.getId());
                List<GrantedAuthority> grantedAuthorities = new
```

```
ArrayList<>(roles.size() + permissions.size());
                /*在 Spring Security 中，角色和权限统称为 GrantAuthority，角色和权限都
                  交给 GrantAuthenty 管理，而区分角色和权限的方式就是在角色名称前加前缀
                  ROLE_以表示角色*/
                // 将角色信息转换为 SimpleGrantedAuthority 对象类型
                List<SimpleGrantedAuthority> roleAuthorities = roles.stream()
                    //给角色名称增加前缀 ROLE_
                    .map(role -> "ROLE_" + role.getName())
                    .map(SimpleGrantedAuthority::new)
                    .collect(Collectors.toList());
                // 将授权许可信息转换为 SimpleGrantedAuthority 对象类型
                List<SimpleGrantedAuthority> permissionAuthorities =
                    permissions.stream()
                    .map(Permission::getName)
                    .map(SimpleGrantedAuthority::new)
                    .collect(Collectors.toList());
                /*将角色和授权许可合并到 grantedAuthorities 列表*/
                grantedAuthorities.addAll(roleAuthorities);
                grantedAuthorities.addAll(permissionAuthorities);
                UserDetails userDetails = org.springframework.security.core
                    .userdetails.User.builder()
                    //设置用户名和密码
                    .username(u.getUsername())
                    .password(u.getPassword())
                    //设置权限列表
                    .authorities(grantedAuthorities)
                    .build();
                return userDetails;
            }).orElse(null);
        } catch (Exception e) {
            throw new UsernameNotFoundException("查找用户出现错误！", e);
        }
    }
}
```

我们数据库中用户的密码加密方式采用的是 BCrypt 算法，Spring Security 默认使用 BCryptPasswordEncoder 类来实现密码加密后比较。AuthenticationProvider 校验的结果无非成功或者失败两种，对于这两种结果，我们只需要实现两个 Handler 接口（AuthenticationSuccessHandler 和 AuthenticationFailureHandler），然后 set 到 Filter 里，Filter 在收到 Provider 的处理结果后会回调这两个 Handler 的方法。

登录成功后，需要返回给客户端一个 JWT Token，因此需要一个 JwtUtil 工具类来生成 Token。JwtUtil 示例代码如下。

示例代码 8-61　JwtUtil.java

```
public class JwtUtil {
    // 携带 token 的请求头名字
    public final static String TOKEN_HEADER = "Authorization";
    //token 的前缀
    public final static String TOKEN_PREFIX = "Bearer ";
```

```
    // 默认密钥
    public final static String DEFAULT_SECRET = "123123...";
    // token 有效期
    private static final long TOKEN_EXPIRATION = 24 * 60 * 60 * 1000;
    private static final SignatureAlgorithm ALGORITHM =
SignatureAlgorithm.HS256;
    private static String secret;
    /**
     * 创建 token
     * @param username 账户主体
     * @return
     */
    public static String createToken(String username) {
        Instant instant = Instant.now().plusMillis(TOKEN_EXPIRATION);
        String token = Jwts.builder()
            //设置账户主体：sub
            .setSubject(username)
            //设置签发时间：iat
            .setIssuedAt(Date.from(instant))
            //设置过期时间：exp，必须大于签发时间
            .setExpiration(Date.from(instant))
            //签名信息，采用 secret 作为私钥
            .signWith(ALGORITHM, getSecret())
            .compressWith(CompressionCodecs.GZIP).compact();
        return token;
    }
    /**
     * 解析 token 串为 jws
     * @param token
     * @return
     */
    public static Jws<Claims> decode(String token) throws
        ExpiredJwtException, SignatureException, UnsupportedJwtException {
        return
Jwts.parser().setSigningKey(getSecret()).parseClaimsJws(token);
    }
    /**
     * 从 token 中获取用户名
     * @param token
     * @return
     */
    public static String getUsernameFromToken(String token) {
        if (StringUtils.isEmpty(token))
            return null;
        String username = Jwts.parser()
            .setSigningKey(getSecret())
            .parseClaimsJws(token)
            .getBody().getSubject();
        return username;
    }
```

```
    public static String getSecret() {
        return Optional.ofNullable(secret).orElse(DEFAULT_SECRET);
    }
    public static void setSecret(String secret) {
        JwtUtil.secret = secret;
    }
}
```

其中包含了一个创建 Token 的方法 createToken、一个解析 Token 的方法 decode（解析之后获取 Jws Claims）和一个从 Token 中获取用户名的方法 getUsernameFromToken。

然后创建 JwtLoginSuccessHandler 来处理登录成功后生成 Token 返回给客户端的逻辑，示例代码如下。

示例代码 8-62　JwtLoginSuccessHandler.java

```
package com.example.spring.security.jwt.handler;
import …
/**
 * 调用 {@link org.springframework.security.authentication.
AuthenticationManager#authenticate(Authentication)}
 * 方法对登录认证通过之后的回调处理器，用于给客户端派发 JWT Token
 */
public class JwtLoginSuccessHandler implements AuthenticationSuccessHandler
{
    @Override
  public void onAuthenticationSuccess(HttpServletRequest request,
        HttpServletResponse response, Authentication authentication)
           throws IOException, ServletException {
        UsernamePasswordAuthenticationToken authToken =
            (UsernamePasswordAuthenticationToken) authentication;
        // 创建 Token
        String token = JwtUtil.createToken(authToken.getName());
        response.setContentType("application/json;charset=UTF-8");
        //在响应中将 Token 放入 header
        response.setHeader(JwtUtil.TOKEN_HEADER, JwtUtil.TOKEN_PREFIX +
token);
        response.setStatus(HttpStatus.OK.value());
        response.getWriter().println("登录成功！");
        response.getWriter().flush();
    }
}
```

创建 SimpleJsonLoginFailureHandler 类来实现登录失败后的处理逻辑，示例代码如下。

示例代码 8-63　SimpleJsonLoginFailureHandler.java

```
package com.example.spring.security.jwt.handler;
import …
/**
 * 一个简单的认证失败后的回调处理器，返回 403 错误
 */
public class SimpleJsonLoginFailureHandler implements
```

```
AuthenticationFailureHandler {
    @Override
    public void onAuthenticationFailure(HttpServletRequest request,
                                        HttpServletResponse response,
                                        AuthenticationException exception)
        throws IOException, ServletException {
        // 返回 403 拒绝错误
        response.addHeader("Authorization", "Bearer No Token");
        response.setContentType("application/json;charset=UTF-8");
        PrintWriter writer = response.getWriter();
        writer.write("服务器已拒绝，认证失败！这可能是由于提供的用户名或密码不正确导致！
");
        writer.flush();
        response.sendError(HttpStatus.FORBIDDEN.value(),
            HttpStatus.FORBIDDEN.getReasonPhrase());
    }
}
```

这里实现比较简单，只是简单返回一个 403 拒绝访问的错误，并提示认证失败的原因。

上面登录认证的逻辑已经处理完了，使用一个 JsonAuthenticationConfigurer 来将它们组合在一起，代码的关键部分在 configure(B http)方法中，如示例代码 8-64 所示，完整代码请下载项目源码查看。

示例代码 8-64 JsonAuthenticationConfigurer.java

```
package com.example.spring.security.jwt.config;
import …
@Setter
@Getter
@Accessors(chain = true, fluent = true)
public class JsonAuthenticationConfigurer<T extends
                            JsonAuthenticationConfigurer<T, B>,
                            B extends HttpSecurityBuilder<B>>
                            extends AbstractHttpConfigurer<T, B> {
    //其他部分代码省略
    @Override
    public void configure(B http) throws Exception {
        AuthenticationManager authenticationManager = http
            .getSharedObject(AuthenticationManager.class);
        JsonUsernamePasswordAuthenticationFilter jsonAuthFilter = new
            JsonUsernamePasswordAuthenticationFilter(authenticationManager);
        // 设置登录成功后的处理器，用来颁发 jwt token
        jsonAuthFilter.setAuthenticationSuccessHandler(
            new JwtLoginSuccessHandler());
        // 设置登录失败后的处理器
        jsonAuthFilter.setAuthenticationFailureHandler(
            new SimpleJsonLoginFailureHandler());
        jsonAuthFilter.setUsernameParameter("username");
        jsonAuthFilter.setPasswordParameter("password");
        RememberMeServices rememberMeServices =
            http.getSharedObject(RememberMeServices.class);
```

```
        if (rememberMeServices != null) {
            jsonAuthFilter.setRememberMeServices(rememberMeServices);
        }
        jsonAuthFilter = postProcess(jsonAuthFilter);
        //指定登录认证 Filter 的位置
        http.addFilterAfter(jsonAuthFilter, LogoutFilter.class);
    }
}
```

其中主要是对 JsonUsernamePasswordAuthenticationFilter 做了一些配置，设置了其成功和失败的处理器，设置了要过滤的请求，以及注销成功的回调 Handler。当调用 configure 方法时，这个 filter 就会加入 security FilterChain 的指定位置。

同时还设置了默认的 AuthenticationEntryPoint 的实现 JwtAuthenticationEntryPoint，它的作用是匿名用户访问需要登录认证权限的路径时返回一个 401 错误，提示需要进行登录认证，示例代码如下。

示例代码 8-65　JwtAuthenticationEntryPoint.java

```
package com.example.spring.security.jwt;
import …
/**
 * 未进行认证，即匿名用户访问无权限时返回 401 未认证的错误
 */
public class JwtAuthenticationEntryPoint implements AuthenticationEntryPoint
{
    @Override
    public void commence(HttpServletRequest request,
                         HttpServletResponse response,
                         AuthenticationException authException)
                         throws IOException, ServletException {
        // 返回 401 认证错误
        response.addHeader("Authorization", "Bearer No Token");
        response.sendError(HttpStatus.UNAUTHORIZED.value(),
            HttpStatus.UNAUTHORIZED.getReasonPhrase());
    }
}
```

客户端拿到 JWT Token 之后，需要携带这个 Token 来访问服务，因此需要另外一个 Filter 来拦截，识别 JWT Token，进行校验，获取权限信息，起名为 JwtAuthenticationFilter，示例代码如下。

示例代码 8-66　JwtAuthenticationFilter.java

```
package com.example.spring.security.jwt.filter;
import …
/**
 * 授权过滤器，用于验证 Token 正确性
 */
@SLF4j
public class JwtAuthenticationFilter extends OncePerRequestFilter {
    private final RequestHeaderRequestMatcher requiresRequestHeaderMatcher;
    private AuthenticationManager authenticationManager;
```

```
    private AuthenticationSuccessHandler successHandler =
        new SavedRequestAwareAuthenticationSuccessHandler();
    private AuthenticationFailureHandler failureHandler =
        new SimpleUrlAuthenticationFailureHandler();
    //非强制认证的请求匹配器集合，用于匹配不需要进行强制认证的 URL
    private List<RequestMatcher> permissiveRequestMatchers;
    public JwtAuthenticationFilter(
        AuthenticationManager authenticationManager){
        //拦截 header 中带 Authorization 的请求
        this.requiresRequestHeaderMatcher = new
RequestHeaderRequestMatcher(JwtUtil.TOKEN_HEADER);
        this.authenticationManager = authenticationManager;
    }
    @Override
    protected void doFilterInternal(HttpServletRequest request,
                                    HttpServletResponse response,
                                    FilterChain filterChain)
        throws ServletException, IOException {
        /*
         * header 中不包含 Authorization 请求头，直接放行，因为部分 URL 匿名用户也可以访问
         * 即使有些路径需要鉴权（非匿名用户不允许访问），但是没有携带 Token，这里放行也是
         * 没问题的，因为 SecurityContext 中没有认证信息，后面会被权限控制模块拦截
         */
        if (!requiresRequestHeader(request, response)) {
            filterChain.doFilter(request, response);
            return;
        }
        request.setCharacterEncoding("UTF-8");
        response.setContentType("application/json;charset=UTF-8");
        AuthenticationException failed = null;
        Authentication authResult = null;
        try {
            String token = getJwtToken(request);
            JwtAuthenticationToken jwtAuthToken =
                new JwtAuthenticationToken(token);
            authResult = authenticationManager.authenticate(jwtAuthToken);
        } catch (InternalAuthenticationServiceException e) {
            logger.error("An internal error occurred while trying to
authenticate the user.");
            failed = e;
        } catch (AuthenticationException e) {
            failed = e;
        }
        if (authResult != null) {
            successfulAuthentication(request, response, filterChain,
authResult);
        } else if (!permissiveRequest(request)) {
            //若请求不在非强制认证列表中，则进行失败认证处理
            //若请求在非强制认证列表中，则不会进行失败处理，即不返回错误
            failedAuthentication(request, response, failed);
```

```
        }
        filterChain.doFilter(request, response);
    }
    /**
     * 处理非强制的请求
     * @param request
     * @return
     */
    private boolean permissiveRequest(HttpServletRequest request) {
        if (permissiveRequestMatchers == null)
            return false;
        for (RequestMatcher permissiveMatcher : permissiveRequestMatchers) {
            if (permissiveMatcher.matches(request))
                return true;
        }
        return false;
    }
    /**
     * 验证失败处理
     * @param request
     * @param response
     * @param failed
     * @throws IOException
     * @throws ServletException
     */
    private void failedAuthentication(HttpServletRequest request,
                                      HttpServletResponse response,
                                      AuthenticationException failed)
        throws IOException, ServletException {
        //清理上下文
        SecurityContextHolder.clearContext();
        failureHandler.onAuthenticationFailure(request, response, failed);
    }
    /**
     * 验证成功处理
     * @param request
     * @param response
     * @param filterChain
     * @param authResult
     * @throws IOException
     * @throws ServletException
     */
    private void successfulAuthentication(HttpServletRequest request,
                                          HttpServletResponse response,
                                          FilterChain filterChain,
                                          Authentication authResult)
        throws IOException, ServletException {
        //设置上下文
        SecurityContextHolder.getContext().setAuthentication(authResult);
        successHandler.onAuthenticationSuccess(request,response,authResult);
```

```
}
/**
 * 请求头匹配，只匹配包含 Authentication 的请求
 * @param request
 * @param response
 * @return
 */
protected boolean requiresRequestHeader(HttpServletRequest request,
                                        HttpServletResponse response) {
    return requiresRequestHeaderMatcher.matches(request);
}
/**
 * 从请求头中获取 Token
 * @param request
 * @return
 */
private String getJwtToken(HttpServletRequest request) {
    String header = request.getHeader(JwtUtil.TOKEN_HEADER);
    if (header == null || !header.startsWith(JwtUtil.TOKEN_PREFIX)) {
        log.info("请求头不含 JWT token，调用下一个过滤器");
        return null;
    }
    //去掉 token prefix
    String token = header.split(" ")[1].trim();
    return token;
}
/**
 * 设置不需要验证的 URL
 * @param urls
 */
public void setPermissiveUrl(String... urls) {
    if (permissiveRequestMatchers == null)
        permissiveRequestMatchers = new ArrayList<>();
    if (urls == null)
        return;
    for (String url : urls)
        permissiveRequestMatchers.add(new AntPathRequestMatcher(url));
}
public void setAuthenticationManager(AuthenticationManager
                                        authenticationManager) {
    Assert.notNull(authenticationManager,
        "authenticationManager could not be null.");
    this.authenticationManager = authenticationManager;
}
public void setSuccessHandler(AuthenticationSuccessHandler
                                successHandler) {
    Assert.notNull(successHandler,
        "authenticationSuccessHandler could be null.");
    this.successHandler = successHandler;
}
```

```
    public void setFailureHandler(AuthenticationFailureHandler
                                        failureHandler) {
        Assert.notNull(failureHandler,
            "authenticationFailureHandler could be null.");
        this.failureHandler = failureHandler;
    }
    @Override
    public void afterPropertiesSet() throws ServletException {
        Assert.notNull(authenticationManager,
            "authenticationManager could not be null.");
        Assert.notNull(successHandler,
            "authenticationSuccessHandler could be null.");
        Assert.notNull(failureHandler,
            "authenticationFailureHandler could be null.");
    }
}
```

这个 Filter 的实现跟登录的 Filter 有点区别：

- 经过 Filter 的请求会继续经过 FilterChain 中的其他 Filter。跟登录请求不一样，Token 只是为了识别用户。
- 如果 header 中没有认证信息或者认证失败，就会判断请求的 URL 是否为强制认证的（通过 permissiveRequestUrl 方法判断）。如果请求不是强制认证，也会放过，比如博客类应用匿名用户访问查看页面、未登录用户进行登出操作。

其他逻辑跟登录一样，组装一个 Token 提交给 AuthenticationManager。然后需要一个 provider 来接收 JWT 的 Token，在收到 Token 请求后对 Token 做验证（有效性），示例代码如下。

示例代码 8-67　JwtAuthenticationProvider.java

```
package com.example.spring.security.jwt.provider;
import …
/**
 * JWT Token 认证器，用于在{@link JwtAuthenticationFilter}
 * 过滤器取得 Token，交予 authenticationManager，调用本认证提供器来认证
 */
public class JwtAuthenticationProvider implements AuthenticationProvider {
    private UserDetailsService userDetailsService;
    public JwtAuthenticationProvider(UserDetailsService userDetailsService) {
        this.userDetailsService = userDetailsService;
    }
    @Override
    public Authentication authenticate(Authentication authentication)
                                    throws AuthenticationException {
        //获取 JWT Token
        JwtAuthenticationToken jwtAuthenticationToken =
            (JwtAuthenticationToken) authentication;
        String token = jwtAuthenticationToken.getToken();
        //解析 JWS
        Jws<Claims> jws;
```

```
        try {
            jws = JwtUtil.decode(token);
        } catch (ExpiredJwtException e) {
            throw new NonceExpiredException("Jwt token expires.", e);
        } catch (JwtException e) {
            throw new BadCredentialsException("JWT token is not correct.", e);
        }
        // 查询用户的权限信息，此处应该从缓存中取出用户权限信息
        // （因为用户登录时获取了权限信息，那时应该将 userdetails 信息放入缓存）
        String username = jws.getBody().getSubject();
        UserDetails userDetails = userDetailsService
            .loadUserByUsername(username);
        JwtAuthenticationToken jwtToken = new JwtAuthenticationToken(
            username, token, userDetails, userDetails.getAuthorities());
        return jwtToken;
    }
    @Override
    public boolean supports(Class<?> authentication) {
        return
authentication.isAssignableFrom(JwtAuthenticationToken.class);
    }
}
```

从 Header 中取出 JWT Token 进行解析，然后取得用户名，从数据库中查询获得权限信息之后，将 UserDetails、Token、权限信息封装成一个 JwtAuthenticationToken 对象。JwtAuthenticationToken 类实现如下。

示例代码 8-68 JwtAuthenticationToken.java

```
package com.example.spring.security.jwt.config;
import …
/**
 * JWT Token 凭证
 */
@Setter
public class JwtAuthenticationToken extends AbstractAuthenticationToken {
    private String principal;
    private String credentials;
    private String token;
    public JwtAuthenticationToken(String token) {
        super(Collections.emptyList());
        this.token = token;
    }
    public JwtAuthenticationToken(String principal, String token,
                    UserDetails userDetails,
                    Collection<? extends GrantedAuthority> authorities) {
        super(authorities);
        this.principal = principal;
        this.token = token;
        this.setDetails(userDetails);
    }
```

```
        @Override
        public void setDetails(Object details) {
            super.setDetails(details);
        }
        @Override
        public Object getCredentials() {
            return credentials;
        }
        @Override
        public Object getPrincipal() {
            return principal;
        }
        public String getToken() {
            return this.token;
        }
        @Override
        public Collection<GrantedAuthority> getAuthorities() {
            return (Collection<GrantedAuthority>) ((UserDetails)
getDetails()).getAuthorities();
        }
    }
```

如果 Token 认证失败，并且不在 permissiveRequestUrls 列表中，就会调用 FailureHandler，这个 Handler 和登录行为一致，返回 403 错误，并提示失败原因，示例代码如下。

示例代码 8-69　JwtAuthenticationFailureHandler.java

```
    package com.example.spring.security.jwt.handler;
    import …
    /**
     * JWT 鉴定 Token 失效之后的错误处理器，返回 403 错误
     */
    public class JwtAuthenticationFailureHandler implements
AuthenticationFailureHandler {
        @Override
        public void onAuthenticationFailure(HttpServletRequest request,
                HttpServletResponse response,
                AuthenticationException exception)
                throws IOException, ServletException {
            response.setContentType("application/json;charset=UTF-8");
            PrintWriter writer = response.getWriter();
            writer.write("服务器已拒绝，认证失败！这可能是由于 Token 失效的导致的！");
            writer.flush();
            response.sendError(HttpStatus.FORBIDDEN.value(),
                HttpStatus.FORBIDDEN.getReasonPhrase());
        }
    }
```

Token 校验成功之后，在继续执行 FilterChain 中的其他 Filter 之前检查一下 Token 是否需要更新，以防止 Token 过期需要用户再次登录，刷新成功之后，将 Token 重新放入 header 中返回给客户端。所以新增一个 JwtRefreshSuccessHandler 来处理 Token 认证成功之后的刷新操作，示例代码

如下。

示例代码 8-70　JwtRefreshSuccessHandler.java

```
package com.example.spring.security.jwt.handler;
import …
/**
 * 定时刷新 Token，默认时间间隔 5 分钟
 */
public class JwtRefreshSuccessHandler implements AuthenticationSuccessHandler
{
    //Token 刷新间隔时间，默认 5 分钟
    private static final Integer TOKEN_REFRESH_INTERVAL = 5 * 60;
    private Integer tokenRefreshInterval;
    public JwtRefreshSuccessHandler(int tokenRefreshInterval) {
        this.tokenRefreshInterval = tokenRefreshInterval;
    }
    public JwtRefreshSuccessHandler() {
        this.tokenRefreshInterval = TOKEN_REFRESH_INTERVAL;
    }
    @Override
    public void onAuthenticationSuccess(HttpServletRequest request,
       HttpServletResponse response,
       Authentication authentication)
       throws IOException, ServletException {
        //获取 Token
        JwtAuthenticationToken jwtAuthToken =
                (JwtAuthenticationToken) authentication;
        String token = jwtAuthToken.getToken();
        Jws<Claims> jws = JwtUtil.decode(token);
        //取出签发时间
        Date issuedAt = jws.getBody().getIssuedAt();
        //判断是否需要刷新（在有效期内，签发时间加上刷新时间间隔后早于当前时间，则需要刷新
token）
        boolean needRefresh = issuedAt.toInstant()
            .plusSeconds(getTokenRefreshInterval())
            .isBefore(Instant.now());
        if (needRefresh) {
            //重新生成 token
            String newToken = JwtUtil.createToken(jwtAuthToken.getName());
            response.setHeader(JwtUtil.TOKEN_HEADER, newToken);
        }
    }
    public int getTokenRefreshInterval() {
        if (tokenRefreshInterval == null)
            return TOKEN_REFRESH_INTERVAL;
        return tokenRefreshInterval;
    }
    public void setTokenRefreshInterval(int tokenRefreshInterval) {
        this.tokenRefreshInterval = tokenRefreshInterval;
    }
```

```
}
```

这样 Token 认证的处理逻辑就完成了，同样新建一个 JwtAuthenticationConfigurer 类来初始化和配置 JwtAuthenticationFilter，示例代码如下。

示例代码 8-71　JwtAuthenticationFilter.java

```
package com.example.spring.security.jwt.config;
import …
@Setter
@Getter
@Accessors(chain = true, fluent = true)
public class JwtAuthenticationConfigurer<T extends
JwtAuthenticationConfigurer<T, B>,
    B extends HttpSecurityBuilder<B>>
    extends AbstractHttpConfigurer<T, B> {
    private AuthenticationSuccessHandler successHandler =
        new JwtRefreshSuccessHandler();
    private AuthenticationFailureHandler failureHandler =
        new JwtAuthenticationFailureHandler();
    @Setter(AccessLevel.NONE)
    private String[] permissiveRequestUrls;
    public JwtAuthenticationConfigurer() {}
    @Override
    public void configure(B http) throws Exception {
        JwtAuthenticationFilter jwtAuthFilter =
            new JwtAuthenticationFilter(
                http.getSharedObject(AuthenticationManager.class));
        jwtAuthFilter
            .setAuthenticationManager(
                http.getSharedObject(AuthenticationManager.class));
        jwtAuthFilter.setSuccessHandler(this.successHandler);
        jwtAuthFilter.setFailureHandler(this.failureHandler);
        jwtAuthFilter.setPermissiveUrl(permissiveRequestUrls);
        //将 filter 放在 LogoutFilter 之前
        JwtAuthenticationFilter filter = postProcess(jwtAuthFilter);
        http.addFilterBefore(filter, LogoutFilter.class);
    }
    public JwtAuthenticationConfigurer<T, B> permissiveRequestUrls(
                                    String... permissiveRequestUrls) {
        this.permissiveRequestUrls = permissiveRequestUrls;
        return this;
    }
}
```

使用 JSON 和 JWT Token 进行登录认证的整个流程已经全部实现。我们需要将这些组件集成到 Spring Security 中。Spring Security 提供了一个 WebSecurityConfigurerAdapter 配置适配器来做配置集成，因此我们需要实现一个 SecurityConfig 类来组织配置，示例代码如下。

示例代码 8-72　SecurityConfig.java

```
package com.example.spring.security.config;
```

```
import …
//启用 WebSecurity 配置
@EnableWebSecurity
//开启注解支持
@EnableGlobalMethodSecurity(prePostEnabled = true)
public class SecurityConfig extends WebSecurityConfigurerAdapter {
    @Resource
    private UserDetailsService userDetailsService;
    private AuthenticationManagerBuilder auth;
    @Bean
    PasswordEncoder passwordEncoder() {
        return new BCryptPasswordEncoder();
    }
    @Override
    protected void configure(AuthenticationManagerBuilder auth)
                                throws Exception {
        auth.userDetailsService(userDetailsService)
            .passwordEncoder(passwordEncoder())
            .and()
            .authenticationProvider(jwtAuthenticationProvider());
    }
    @Override
    protected void configure(HttpSecurity http) throws Exception {
        http.authorizeRequests()
            .anyRequest().authenticated()
            //禁用 csrf，因为不使用 session
            .and().csrf().disable()
            //前后端分离采用 JWT，禁用 session
            .sessionManagement()
            .sessionCreationPolicy(SessionCreationPolicy.STATELESS)
            .disable()
            //支持跨域
            .cors().and()
            //添加 header 设置，支持跨域和 Ajax 请求
            .headers()
            .addHeaderWriter(new StaticHeadersWriter(Arrays.asList(
                new Header("Access-control-Allow-Origin", "*"),
                new Header("Access-Control-Expose-Headers",
                        JwtUtil.TOKEN_HEADER)
            ))).and()
            .addFilterAfter(new OptionsRequestFilter(), CorsFilter.class)
            // 添加登录 Filter
            .apply(new JsonAuthenticationConfigurer<>()).and()
            //添加 Token 认证 Filter
            .apply(new JwtAuthenticationConfigurer<>())
            .permissiveRequestUrls("/user/logout").and()
            //使用默认的 logoutFilter
            .logout().addLogoutHandler(logoutHandler())
            .logoutSuccessHandler(
                new HttpStatusReturningLogoutSuccessHandler());
```

```
    }
    //…
    /**
     * JWT 校验认证器
     * @return
     */
    @Bean
    public AuthenticationProvider jwtAuthenticationProvider() {
        return new JwtAuthenticationProvider(userDetailsService());
    }
    @Bean
    public TokenClearLogoutHandler logoutHandler() {
        return new TokenClearLogoutHandler(userDetailsService);
    }
    @Bean
    protected CorsConfigurationSource corsConfigurationSource() {
        CorsConfiguration configuration = new CorsConfiguration();
        configuration.setAllowedOrigins(Arrays.asList("*"));
        configuration.setAllowedMethods(
            Arrays.asList("GET", "POST", "HEAD", "OPTION"));
        configuration.setAllowedHeaders(Arrays.asList("*"));
        configuration.addExposedHeader(JwtUtil.TOKEN_HEADER);
        UrlBasedCorsConfigurationSource source =
            new UrlBasedCorsConfigurationSource();
        source.registerCorsConfiguration("/**", configuration);
        return source;
    }
}
```

上面的配置开启了跨域支持 http.cors()，这样 Security 就会从 CorsConfigurationSource 中读取跨域配置，因此我们需要提供该 Bean 来进行配置：

```
@Bean
protected CorsConfigurationSource corsConfigurationSource() {
    CorsConfiguration configuration = new CorsConfiguration();
    configuration.setAllowedOrigins(Arrays.asList("*"));
    configuration.setAllowedMethods(
        Arrays.asList("GET", "POST", "HEAD", "OPTION"));
    configuration.setAllowedHeaders(Arrays.asList("*"));
    configuration.addExposedHeader(JwtUtil.TOKEN_HEADER);
    UrlBasedCorsConfigurationSource source =
        new UrlBasedCorsConfigurationSource();
    source.registerCorsConfiguration("/**", configuration);
    return source;
}
```

CSRF 攻击是针对使用 session 的情况，这里是不需要的，所以禁用了 CSRF。关于 CSRF 的内容可参考 https://docs.spring.io/spring-security/site/docs/5.0.7.RELEASE/reference/htmlsingle/#csrf。另外，还禁用了默认的 form 表单登录方式，禁用了 session 管理（因为 JWT 方式是无状态的，不需

要 session）。启用了 logout 支持，Spring Security 已经默认支持 logout filter，会拦截/logout 请求，交给 LogoutHandler 处理，同时在 logout 成功后调用 LogoutSuccessHandler。对于 logout，我们可以在 LogoutHandler 中清除保存的 Token 信息（同样要在登录成功之后将 Token 信息保存到 Redis 缓存或数据库），这样在 logout 之后再使用 Token 访问就会失败。本示例项目并未实现保存 Token 和清除 Token 的逻辑，读者可以自己尝试实现。

示例代码 8-73　TokenClearLogoutHandler.java

```
package com.example.spring.security.jwt.handler;
import …
public class TokenClearLogoutHandler implements LogoutHandler {
    private UserDetailsService userDetailsService;
    public TokenClearLogoutHandler(UserDetailsService userDetailsService) {
        this.userDetailsService = userDetailsService;
    }
    @Override
    public void logout(HttpServletRequest request, HttpServletResponse
response, Authentication authentication) {
        clearToken(authentication);
    }
    protected void clearToken(Authentication authentication) {
        if (authentication == null)
            return;
        UserDetails user = (UserDetails) authentication.getDetails();
        if (user != null && user.getUsername() != null)
            // 这里可以执行清理缓存中的 Token
            ;
    }
}
```

对于 Ajax 的跨域请求，浏览器在发送真实请求之前，会向服务端发送 OPTIONS 请求，看服务端是否支持。对于 options 请求，我们只需要返回 header，不需要再进入其他的 filter，所以我们加了一个 OptionsRequestFilter，填充 header 后就直接返回，示例代码如下。

示例代码 8-74　OptionsRequestFilter.java

```
package com.example.spring.security.jwt.filter;
import …
public class OptionsRequestFilter extends OncePerRequestFilter {
    @Override
    protected void doFilterInternal(HttpServletRequest request,
                HttpServletResponse response, FilterChain filterChain)
                throws ServletException, IOException {
        if(request.getMethod().equals("OPTIONS")) {
            response.setHeader("Access-Control-Allow-Methods",
                "GET,POST,OPTIONS,HEAD");
            response.setHeader("Access-Control-Allow-Headers",
                response.getHeader("Access-Control-Request-Headers"));
            return;
```

```
        }
        filterChain.doFilter(request, response);
    }
}
```

这样整个 Spring Boot 结合 Security 实现 JWT 登录认证的流程就结束了。接下来运行该项目进行测试。

首先使用 Postman 发送一个 Post 请求到/user/login 进行登录，如图 8-44 所示。

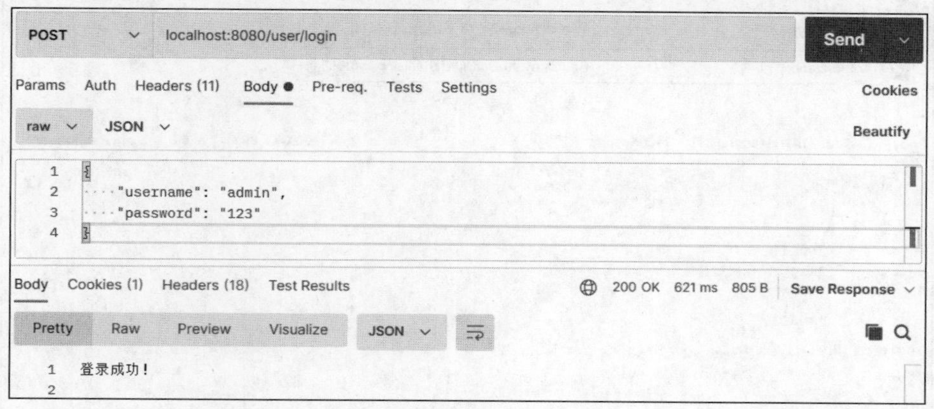

图 8-44

可以看到提示登录成功的消息，响应头中的 Header 信息如图 8-45 所示，正确返回了 JWT Token 信息。

Body　Cookies (1)　Headers (18)　Test Results　200 OK　621 ms　805 B　Save Response

KEY	VALUE
Vary	Origin
Vary	Access-Control-Request-Method
Vary	Access-Control-Request-Headers
Authorization	Bearer eyJhbGciOiJIUzI1NiIsInppcCI6IkdaSVAifQ.H4sIAAAA...
X-Content-Type-Options	nosniff
X-XSS-Protection	1; mode=block
Cache-Control	no-cache, no-store, max-age=0, must-revalidate

图 8-45

复制 Token 信息，并使用这个 Token 来请求/user/list 路径获取用户列表信息。在 Postman 中创建一个 Get 请求，设置请求头“Authorization”值为复制的 Token 信息，然后发送该请求，可以看到返回结果成功获取到了用户列表信息，如图 8-46 所示。

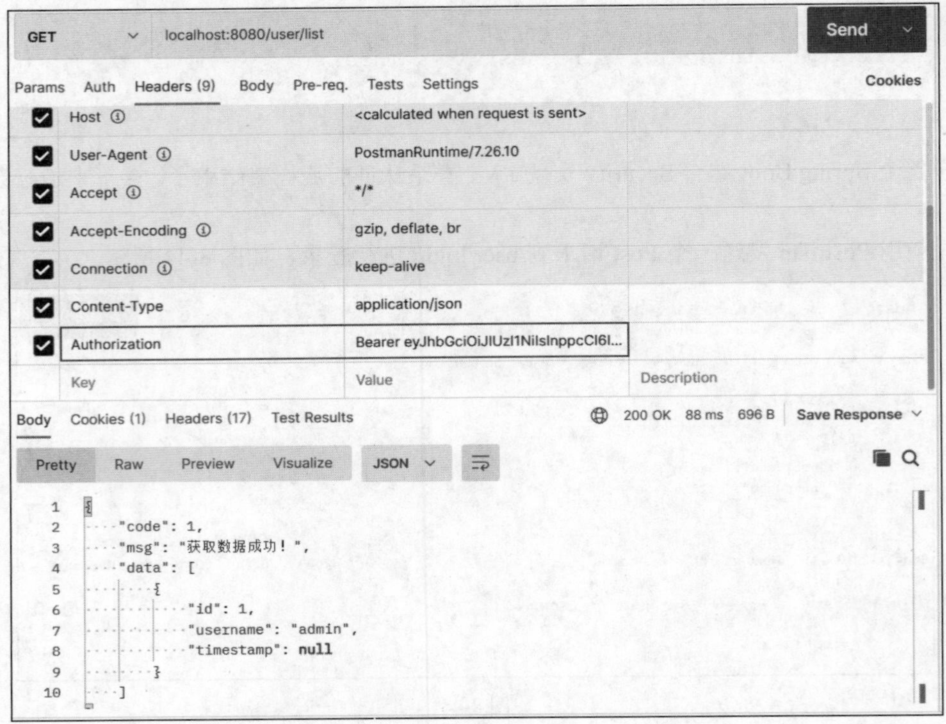

图 8-46

至此，我们在 Spring Boot 项目中使用 Security 框架实现了一个简单的 JWT 登录认证逻辑。本示例中还有不完善的地方，比如各种错误异常处理、Token 信息缓存、以及 Token 缓存清理等都还未实现，读者可以根据项目需求来具体实现。

8.3 第三方登录（OAuth 2.0）

8.3.1 什么是 OAuth 2.0

OAuth 2.0 通常被设计为访问第三方服务，提供一套统一的方案实现本地认证和第三方认证，对于系统能力开放是非常有意义的。

OAuth 2.0 是一个应用之间彼此访问数据的开源授权框架。通俗地讲，它是用来授权第三方应用程序的，通过提供 OAuth 2.0 协议的 API 使得第三方应用程序能够使用自己应用程序中的用户数据进行登录和授权，这样用户就能够直接利用自己的应用程序账户来登录和使用第三方程序了。2.0 是目前比较流行的做法，率先被 Google、Yahoo、Microsoft、Facebook 等使用。之所以标注为 2.0，是因为最初有一个 1.0 框架，但是这个 1.0 框架太复杂，易用性差，所以没有得到普及。2.0 是一个新的设计，协议简单清晰，但是并不兼容 1.0，可以说与 1.0 没什么关系。

举个大家熟知的例子，我们经常会使用 QQ 或者微信账号来登录其他网站应用，比如 Bilibili、知乎、腾讯视频等。其中利用的技术就是 OAuth 2.0。有的第三方应用可以直接利用 QQ 或者微信账号登录，而无须注册成为该网站用户，之后就能够合理合法地正常使用该网站的功能了；有的第三方应用需要先注册成为该网站用户，然后通过绑定 QQ、微信账户之后才可以直接利用 QQ 或者微信账户来进行登录和使用第三方应用。这两种方式无论是哪一种，其原理都是一样的，都是利用 OAuth 来进行授权的，只是实现的方式有些不一样。

简单来说，OAuth 就是一种授权机制。账户所有者（用户本身）告诉系统（进行授权），“我”同意授权第三方应用进入系统，获取“我”在系统中存储的一些账户数据。系统得到所有者授权之后，会向第三方应用下发一个短期的令牌（Token），用来取代用户密码，供第三方应用使用，第三方应用就可以通过该令牌来进行登录、获取用户信息了。

令牌（Token）与密码（password）的作用是一样的，都可以进入系统，但是有三点差异。

（1）令牌是短期的，到期会自动失效，用户自己无法修改。密码一般长期有效，用户不修改就不会发生变化。

（2）令牌可以被数据所有者撤销，会立即失效。比如微信、QQ 可以随时撤销第三方应用持有的令牌，令牌立即失效，第三方应用就无法再使用。但是密码一般不允许被他人撤销。

（3）令牌有权限范围（scope），比如一个人拿了令牌之后只能进入小区 2 号楼。对于网络服务来说，只读令牌就比读写令牌更安全，但密码一般是完整权限。

上面这些设计保证了令牌既可以让第三方应用获得权限，同时又随时可控，不会危及系统安全。这就是 OAuth 2.0 的优点。

需要注意的是，只要获得了令牌，持有令牌就能进入系统。系统一般不会再次确认身份，所以令牌必须保密，泄漏令牌与泄漏密码的后果是一样的。这也是令牌的有效期一般都设置得很短的原因。这就跟古装电视剧里的情节一样，只要偷拿到令牌，向守卫亮出令牌，就可以进入守卫大门，只要获得了调兵令牌（无论是通过什么方式获得）就能够调兵遣将，无人会再质疑身份的真假性（因为令牌代表了授权），即使是假冒身份，只要令牌是真的即可，士兵将士见令牌后一律服从令牌持有者在令牌所具有的权限范围内的所有命令和指示。所以，保护令牌的重要性不言而喻。

OAuth 2.0 对于如何颁发令牌的细节都规定得非常详细。具体来说，一共分成四种授权类型（Authorization Grant），对应四种颁发令牌的方式，适用于不同的互联网场景。在 8.3.6 小节中将会详细介绍这四种授权方式。

8.3.2 角色定义

OAuth 2.0 为用户和应用定义了如下角色：

- 资源拥有者（Resource Owner）：拥有共享数据的人或应用，比如 QQ 或者微信的用户。他们拥有的资源就是他们的数据，通常来说是指人，但也可以是一个应用。
- 资源服务器（Resource Server）：托管资源的服务器，比如 QQ、微信或 Facebook。
- 客户端应用（Client Application）：请求访问存储在资源服务器上的资源的应用，也可以称之为第三方客户端应用。资源被资源拥有者所拥有。客户端应用可以是一个请求访问用户 QQ 账号的第三方游戏或者第三方视频网站等。

- 授权服务器（Authorization Server）：授权客户端应用能够访问资源拥有者所拥有的资源。授权服务器和资源服务器可以是同一个服务器，也可以是不同的服务器。如果这两个服务器是分开的，OAuth 2.0 没有讨论这两个服务器应该如何通信，由资源服务器和授权服务器开发者自己设计决定。

8.3.3 客户端角色

OAuth 2.0 客户端角色被细分为一系列类型和配置，本节将阐述这些类型和配置。OAuth 2.0 规范定义了以下两种客户端类型：

- 保密的：保密的客户端能够对外部保持客户端密码保密。该客户端密码是由授权服务器分配给客户端应用的。为了避免欺骗，该密码是授权服务器用来识别客户端的。例如，一个保密的客户端可以是 Web 应用，除了管理员之外，没有任何人能够访问服务器且无法看到该密码。
- 公共的：公共的客户端不能使客户端密码保密，比如手机应用或桌面应用会将密码嵌入在应用内部。这样的应用可能被破解，并且泄漏密码。类似这样公共的客户端还包括在用户的浏览器上运行的 JavaScript 应用。用户可以使用一个 JavaScript 调试器来寻找到应用程序，并查看、获取客户端密码。

OAuth 2.0 规范也提到了一系列客户端应用配置文件。这些配置文件是具体类型的应用程序，可以是保密或公开的。这些配置文件有：

- Web 应用（Web Apps）：运行在 Web 服务器内的应用。实际上，Web 应用是由浏览器部分和服务端部分组成的。比如 Web 应用需要访问资源服务器（如 QQ 账号），而客户端密码被保存在服务器端上，因此密码是保密的。示例如图 8-47 所示。

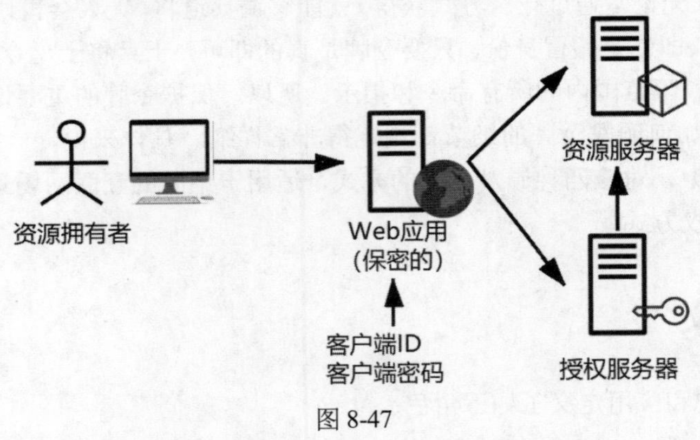

图 8-47

- 用户代理应用（User Agent Apps）：典型的例子是运行在浏览器上的 JavaScript 应用，其中浏览器是用户代理。用户代理应用可以托管保存在 Web 服务器上，然后通过下载到本地浏览器上运行，比如一个 JavaScript 游戏只能运行在浏览器里。这类应用的客户端密码被保存在浏览器的 JavaScript 程序中，可以被用户破解查看，因此是公有的。示例如图 8-48 所示。

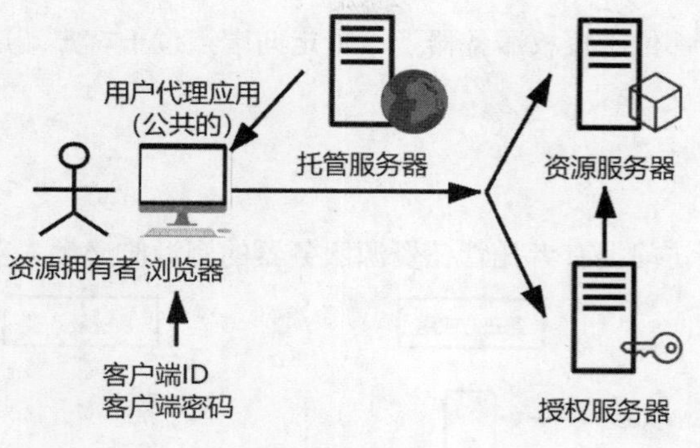

图 8-48

- 原生应用（Native Apps）：原生应用（比如桌面应用或手机应用）被安装在用户计算机或设备（手机、平板等）上，因此客户端密码相应地也被存储在用户计算机或设备上，这类客户端密码是公有的。示例如图 8-49 所示。

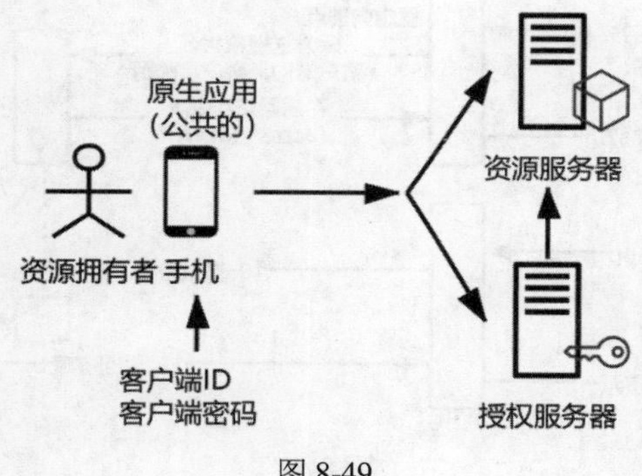

图 8-49

虽然 OAuth 2.0 划分了这三种客户端应用配置模型，但是它们亦可以是混合应用，比如原生应用也可以有服务器端部分，将客户端密码保存在服务端部分，同样可以使得密码是保密的。这种混合型的应用同样能够适配 OAuth 2.0 提供的这些配置文件的认证模型。

8.3.4　端点

OAuth 2.0 定义了一系列端点（Endpoint），比如一个 Java Servlet、JSP 网页、PHP 网页、ASP.NET 网页、Spring Boot 的 REST Endpoint 等。端点最典型的就是 Web 服务器上的 URI。

这些端点定义划分为以下三种：

- 授权端点：资源拥有者所登录的授权服务器，并授权给客户端应用访问的端点（Authorization Endpoint），在这个端点登录成功之后会发放授权凭证（有四种方式，比如授权码），通过这个凭证来进一步获取访问令牌。
- 重定向端点：在授权端点授权以后，资源拥有者被重定向到客户端应用的端点（Redirect URI Endpoint）。
- 令牌端点：在授权服务器上生成一个访问令牌的端点（Token Endpoint）。客户端应用访问这个端点需要提交授权凭证（比如授权码）、客户端 ID 和客户端密钥。

授权端点和令牌端点都位于授权服务器上，重定向端点位于客户端应用上。

8.3.5 授权过程

一个客户端应用想要访问拥有者托管在资源服务器的资源时必须先获得授权，如图 8-50 所示。

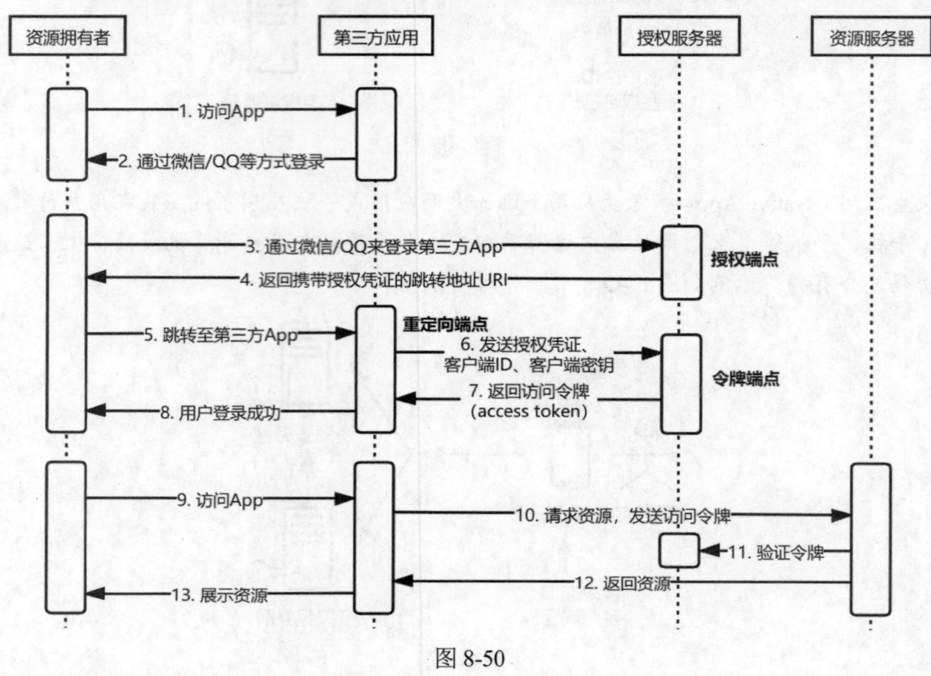

图 8-50

（1）第一步，用户访问第三方客户端 Web 应用。应用登录页有按钮“通过 QQ 登录”（或者其他的系统，如微信、Facebook、Google 或 Twitter）。

（2）第二步，当用户单击了该按钮后，会被重定向到授权的应用（如 QQ）。用户登录并确认授权应用中的数据给第三方客户端应用。

（3）第三步，授权应用服务器将用户重定向到第三方客户端应用提供的 URI，提供这种重定向的 URI 通常是实现客户端应用程序注册账户与授权应用程序的用户账户绑定。在注册中，第三方客户端应用的系统注册该重定向 URI，在注册过程中授权应用也会给客户端应用发放客户端标识 ID 和密码，并在 URI 后追加一个授权凭证。该授权凭证代表了授权。

（4）第四步，用户在客户端应用访问网页被定位到重定向的 URI。在这个过程中，客户端应用首先会连接授权应用，并且发送在重定向请求参数中接收到的客户端标识 ID、客户端密码和授权凭证。然后授权应用校验成功之后将返回一个访问令牌（access Token）。

一旦客户端有了访问令牌，该令牌便可以被发送到 QQ、微信、Facebook、Google、Twitter 等来访问登录用户的资源。

8.3.6 OAuth 2.0 的四种授权方式

作用域（Scope）是 OAuth 2.0 中的一种机制，用于限制应用程序对用户账户的访问。第三方应用程序可以请求一个或多个作用域，然后将请求的作用域展示给用户，在用户同意授权之后，系统颁发给该应用程序的访问令牌将被限制于所授予的作用域范围内。

在 RFC6749 标准中解释了 OAuth 是什么：OAuth 通过引入授权层，将客户的角色与资源的角色分开。在 OAuth 中，客户端请求访问受控资源，资源所有者同意以后，资源服务器向客户端颁发令牌，客户端通过持有该令牌去请求数据。

OAuth 的核心就是向第三方应用颁发令牌，OAuth 2.0 规定了四种获得令牌的流程，你可以选择最适合自己的那一种向第三方应用颁发令牌。最新的四种授权方式是授权码（Authorization Code）、客户端凭证（Client Credentials）、设备码（Device Code）、密码式（Refresh Token）。此外，还有一种授权方式：隐式授权（Implicit Code），在之前比较推荐用于原生应用和用户代理应用中请求访问授权，但是因为涉及安全风险（包括密码式授权方式），现在已经过时，不再推荐使用，而是推荐基于授权码的扩展 PKCE 的授权方式。

不管哪一种授权方式，在第三方应用申请令牌之前都必须先到系统备案，说明自己的身份，然后会拿到两个身份识别码：客户端 ID（Client ID）和客户端密钥（Client Secret）。这是为了防止令牌被滥用，没有备案过的第三方应用，是不会拿到令牌的。

接下来依次解释一下这四种授权方式。

1. 授权码

授权码方式指的是第三方应用先申请一个授权码，然后用该码获取访问令牌。这是目前流行的推荐方式，这种方式总共是 2 个请求加 2 个响应，即（一个授权请求+响应）和一个令牌请求+响应。这种方式是最常用的流程，安全性也最高，适用于那些有后端的 Web 应用。这种方式适用于图 8-47 所示的情形。

首先，客户端 Web 应用构建一个授权请求，跳转到授权端点。该请求地址路径如下所示。

```
https://authorization-server.com/auth?
   response_type=code&
   client_id=29352735982374239857&
   redirect_uri=https://example-app.com/callback&
   scope=read&
   state=xcoivjuywkdkhvusuye3kch
```

该授权请求的参数说明如表 8-1 所示。

当用户访问此授权请求时，会向用户询问是否允许授权，在用户允许后将返回授权响应。授权响应包含了需要用来获取访问令牌的授权码，响应参数如表 8-2 所示。

表 8-1　授权请求的参数及其说明

请求参数（值）	释义	要求
response_type=code	这个参数告诉授权服务器要求返回授权码	必需
client_id	客户端应用标识，授权服务器根据这个参数来明确知道是谁在请求。这个参数值是开发人员通过向授权服务注册来获得的	必需
redirect_uri	跳转地址，告诉授权服务器在批准请求后将用户发送回何处	可选
scope	一个或多个用空格分隔的字符串，指示客户端应用请求的作用域	可选
state	客户端状态，客户端应用通过生成一个随机字符串，将其包含在该请求中。然后，在用户授权应用后检查是否返回了相同的值，用于防止 CSRF 攻击	可选，推荐使用

表 8-2　响应参数及其说明

响应参数（值）	释义	要求
code	返回的授权码	必需
state	客户端状态，和请求时发送的 state 值一致	如果请求中包含 state，那么响应中就必须包含

当用户批准请求后，授权服务器会将用户浏览器重定向到 redirect_uri，并添加相应参数，如下所示。

```
https://example-app.com/callback?
    code=g0ZGZmNjVmOWIjNTk2NTk4ZTYyZGI3&
    state=xcoiv98y2kd22vusuye3kch
```

code 有效期非常短，通常只有 1~10 分钟的有效期，这取决于 OAuth 服务。

一旦获取到授权码，就立即向令牌端点发起令牌请求，例如：

```
https://authentication-server.com/oauth/token?
    grant_type=authorization_code&
    code=g0ZGZmNjVmOWIjNTk2NTk4ZTYyZGI3&
    redirect_uri=https://example-app.com/redirect&
    client_id=xxxxxxxxxx&
    client_secret=xxxxxxxxxx
```

该令牌请求参数释义如表 8-3 所示。

表 8-3　令牌请求的参数及其说明

请求参数（值）	释义	要求
grant_type=authorization_code	告诉令牌端点客户端应用正在使用授权码类型	必需
code	客户端应用在授权响应重定向 URI 中获得的授权代码	必需
redirect_uri	与授权请求中使用相同的重定向 URI。有些 API 不需要此参数，需要根据授权服务器提供的 API 文档来确定	可选
client_id	客户端应用的标识（ID）	必需
client_secret	客户端应用的密钥。这样可以确保仅从客户端应用发出获取访问令牌的请求，而不是从可能已经拦截了授权代码的潜在攻击者发出请求	必需

令牌服务器端点收到该令牌请求时会颁发访问令牌（Access Token），具体做法是向令牌请求中给出的 redirect_uri 指定的网址发送一段 JSON 数据：

```
{
    "access_token": "MTQ0NjJkZmQ5OTM2NDE1ZTZjNGZmZjI",
    "token_type": "bearer",
    "expires_in": 3600,
    "refresh_token": "IwOGYzYTlmM2YxOTQ5MGE3YmNmMDFkNTVk",
    "scope": "read",
    "uid": 100101,
    "info": {...}
}
```

该令牌响应参数释义如表 8-4 所示。

表 8-4　令牌响应的参数及其说明

响应参数	释义	要求
access_token	授权服务器分配的访问令牌	必需
token_type	被授权服务器分配的令牌类型	必需
expires_in	失效时间，访问令牌过多少秒后就不再有效	可选
refresh_token	令牌过期后的刷新令牌。一旦 access_token 失效，就可以通过访问令牌刷新端点，将 refresh_token 传递过去获取一个新的 access_token	可选

除了这四项响应参数以外，还可能返回其他的一些信息，具体需要参考对应的授权服务器提供商给出的 API 文档。以上就是使用授权码方式来获取访问令牌的详细请求和响应说明。读者可以从官方提供的模拟环境进行学习：https://www.oauth.com/playground/authorization-code.html。

此外，授权码类型还有一个扩展类型：PKCE（RFC 7636）。它可以防止多种攻击并能够安全地从公共客户端执行 OAuth 交换，具体请访问官方文档（https://www.oauth.com/pkce/）。这种方式是最安全、最完善的，适用于保密的和公关的两种客户端类型。

PKCE 模拟环境学习地址为 https://www.oauth.com/playground/authorization-code-with-pkce.html。

2. 客户端凭证

当应用程序请求访问令牌而不是代表用户访问自己的资源时，将使用“客户端凭证”的方式获取访问令牌，这种方式是单个请求+响应。请求获取访问令牌的方式如下所示。

```
https://authentication-server.com/oauth/token?
    grant_type=client_credentials&
    client_id=xxxxxxxxxx&
    client_secret=xxxxxxxxxx
```

与授权码方式不同的是，grant_type 的值为 client_credentials，指定为客户端凭证方式，client_id 和 client_secret 两个参数与授权码方式令牌请求中的这两个参数是一样的，这两个参数也可以通过 HTTP Basic auth 方式放置在 Header 中提供。

发送令牌请求成功后，授权服务器颁发令牌，做出响应，响应内容与授权码方式的令牌响应内容一致。

3. 设备代码

无浏览器或受输入限制的设备可以使用设备代码授权类型来将先前获得的设备代码交换为访问令牌。设备代码授权类型值为 urn:ietf:params:oauth:grant-type:device_code。

首先设备向授权服务器注册获得设备代码（device_code），在用户允许授权之前，设备开始以固定时间间隔轮询请求访问令牌，直到用户允许授权或者拒绝。

设备代码方式发送的令牌请求如下：

```
https://authentication-server.com/oauth/token?
    grant_type=urn:ietf:params:oauth:grant-type:device_code&
    client_id=a17c21ed&
    device_code=NGU5OWFiNjQ5YmQwNGY3YTdmZTEyNzQ3YzQ1YSA
```

其中的参数 device_code 就是设备代码。关于响应以及错误码的说明请参考官方说明（https://www.oauth.com/oauth2-servers/device-flow/token-request/）。

4. 密码式

密码式授权类型是一种将用户凭据交换为访问令牌的方式。因为客户端应用程序必须收集用户的密码并将其发送到授权服务器，所以不建议使用此授权。密码式授权只包含单个请求+响应。

密码式申请令牌请求的示例如下：

```
https://authentication-server.com/oauth/token?
    grant_type=password&
    username=user@example.com&
    password=1234luggage&
    client_id=xxxxxxxxxx&
 client_secret=xxxxxxxxxx
```

各项请求参数说明如表 8-5 所示。

表 8-5　各项请求参数及其说明

请求参数（值）	释义	要求
grant_type="password"	指定授权类型为密码式	必需
username	用户名	必需
password	用户密码	必需
scope	客户端应用请求的作用域	可选
client_id	客户端应用的标识（ID）	必需
client_secret	客户端应用的密钥。这样可以确保仅从客户端应用发出获取访问令牌的请求，而不是从可能已经拦截了授权代码的潜在攻击者发出请求	必需

响应说明可参考授权码中的。

8.3.7 OpenID Connect

1. 概述

OpenID Connect（OIDC） 1.0 是在 OAuth 2.0 框架之上的简单身份认证层，允许客户端应用基于授权服务器执行的身份验证来验证最终用户的身份，并以类似于 REST 方式的可互操作来获取有关最终用户的基本配置文件信息。关于可互操作性（Interoperability）：又称互用性，是指不同的计算机系统、网络、操作系统和应用程序一起工作并共享信息的能力。

OpenID Connect 允许所有类型的客户端（包括基于 Web 的客户端、移动客户端和 JavaScript 客户端）请求并接收有关经过身份验证的会话和最终用户的信息。规范套件是可扩展的，允许使用可选功能，例如身份数据加密、OpenID 提供器（Providers）的发现以及会话管理。

使用 OIDC，客户端应用可以请求 Identity Token，会和 Access Token 一同返回客户端应用。这个 Identity Token 可以被用来登录客户端应用程序，客户端应用还可以使用 Access Token 来访问 API 资源。

OpenID Connect 允许开发者验证跨网站和应用的用户，而无须拥有和管理密码文件（数据表）。OIDC 已经有很多的企业在使用，比如 Google 的账号认证授权体系、Microsoft 的账号体系等。当然，这些企业有的也是 OIDC 背后的推动者。

2. 与 OAuth 2.0 的区别

OAuth 2.0 授权框架[RFC6749]和 OAuth 2.0 Bearer Token[RFC6750]使用规范为第三方应用程序提供了获取和使用对 HTTP 资源的有限访问权限的通用框架。它们定义了获取和使用访问令牌访问资源的机制，但是没有定义提供身份信息的标准方法。

OAuth 2.0 不是身份认证（Authentication）协议，而是一个规范性的框架。这个规范没有强制性的约束力，不像 HTTP，它没有确切固定的格式，无法提供身份认证服务。OpenID Connect 在 OAuth 2.0 框架基础之上添加了一些组件来提供身份认证的能力，可以进行身份认证（Authentication），所以它兼容 OAuth 2.0，但同样没有非常固定的格式，没有人规定一定要怎么做，因为它们都只是一个规范性的约定。

OAuth 2.0 提供了 Access Token 来解决授权第三方客户端访问受保护资源的问题，OIDC 在这个基础上提供了 ID Token 来解决第三方客户端标识用户身份认证的问题。二者都是基于 JWT 技术的。OIDC 的核心在于在 OAuth 2.0 的授权流程中一并提供用户的身份认证信息（ID Token）到第三方客户端，ID Token 使用 JWT 格式来包装，得益于 JWT（JSON Web Token）的自包含性、紧凑性以及防篡改机制，使得 ID Token 可以安全地传递给第三方客户端程序并且容易被验证。此外，还提供了 UserInfo 的接口，用于获取用户更完整的信息。

3. 主要术语

表 8-6 提供了 OIDC 使用的主要术语，并展示了术语与 OAuth 2.0 的对应关系。

表 8-6　OIDC 的主要术语及其与 OAuth 2.0 的对应关系

术语	释义	与 OAuth 2.0 对应术语
End User	简写为 EU，表示一个人类用户（非系统程序）	对应于资源拥有者（Resource Owner）
Relying Party	依赖部分，简写为 RP，用来代指 OAuth 2.0 中的受信任的客户端、身份认证和授权信息的消费方	对应于客户端应用（Client Application）
OpenID Provider	简写为 OP，有能力提供 EU 认证的服务，用来为 RP 提供 EU 的身份认证信息	对应于授权服务器和资源服务器（Authorization Server、Resource Server）
ID Token	JWT 格式的数据，包含 EU 身份认证的信息	无
UserInfo Endpoint	用户信息接口（受 OAuth 2.0 保护），当 RP 使用 Access Token 访问时返回授权用户的信息，此接口必须使用 HTTPS	无

4. 工作流程

从抽象的角度来看，OIDC 工作的流程主要有以下 5 个步骤：

①RP 发送一个认证请求给 OP。
②OP 对 EU 进行身份认证，然后提供授权。
③OP 将生成 ID Token 和 Access Token，返回给 RP。
④RP 使用 Access Token 发送请求到 UserInfo Endpoint，获取用户身份信息。
⑤UserInfo Endpoint 返回给 RP 对应 EU 的用户信息（使用 Claims 承载）。

5. ID Token

ID Token 是一个安全令牌，是一个授权服务器提供的包含用户信息（由一组 Cliams 构成以及其他辅助的 Cliams）的 JWT 格式的数据结构，主要构成部分如表 8-7 所示。

表 8-7　ID Token 的主要构成部分及其说明

字段	全称	释义	要求
iss	Issuer Identifier	提供认证信息者的唯一标识，一般是一个 HTTPS 的 URL（不包含 querystring 和 fragment 部分）	必需
sub	Subject Identifier	iss 提供的 EU 标识，在 iss 范围内唯一。它会被 RP 用来标识唯一的用户，最长为 255 个 ASCII 字符	必需
aud	Audience(s)	标识 ID Token 的受众，必须包含 OAuth 2.0 的 client_id	必需
exp	Expiration time	过期时间，超过此时间的 ID Token 会作废不再被验证通过	必需
iat	Issued At Time	JWT 的构建时间	必需
auth_time	AuthenticationTime	EU 完成认证的时间。如果 RP 发送 AuthN 请求的时候携带 max_age 的参数，则此 Claim 是必需的	可选
Nonce	Nonce	RP 发送请求的时候提供的随机字符串，用来减缓重放攻击，也可以用来关联 ID Token 和 RP 本身的 Session 信息	可选

（续表）

字段	全称	释义	要求
acr	Authentication Context Class Reference	表示一个认证上下文引用值，可以用来标识认证上下文类	可选
amr	Authentication Methods References	表示一组认证方法	可选
azp	Authorized party	结合 aud 使用。只有在被认证的一方和受众（aud）不一致时才使用此值，一般情况下很少使用	可选

ID Token 通常情况下还会包含其他的 Claims（毕竟上述 claim 中只有 sub 是和 EU 相关的，这在一般情况下是不够的，还必须有 EU 的用户名、头像等其他资料，OIDC 提供了一组公共的 cliams，可从 http://openid.net/specs/openid-connect-core-1_0.html #StandardClaims 查看详细说明。另外，ID Token 必须使用 JWS（JWT 的一种实现）进行签名、使用 JWE（JWT 的另一种实现，保证了安全性）加密，从而提供认证的完整性、不可否认性以及可选的保密性。

一个简单的 ID Token 示例如下：

```
{
   "iss": "https://server.example.com",
   "sub": "24400320",
   "aud": "s6BhdRkqt3",
   "nonce": "n-0S6_WzA2Mj",
   "exp": 1311281970,
   "iat": 1311280970,
   "auth_time": 1311280969,
   "acr": "urn:mace:incommon:iap:silver"
}
```

6. 认证授权方式

OIDC 的授权认证方式主要有三种：

- 基于授权码（Authorization Code），及其扩展方式 PKCE。
- 隐式（Implicit Code），已经不再推荐使用。
- 混合模式（Implicit Code + Authorization Code）。

具体的认证授权流程说明可查看官方文档（https://openid.net/specs/openid-connect-core -1_0.html）。

8.4 优雅地生成接口文档

“接口文档”又称为“API 文档”，一般是由开发人员所编写的，用来描述系统所提供的接口信息的文档。大家都根据这个接口文档进行开发，并需要一直维护和遵守。编写接口文档的目的是：

- 能够让前端开发与后台开发人员更好地配合，在项目开发过程中需要有一个统一的文件进行沟通交流开发，提高工作效率。

- 项目迭代或者项目人员更迭时，方便后期人员查看和维护。
- 方便测试人员进行接口测试。

良好的接口文档有助于提高沟通和工作效率，因此规范编写接口文档是一个良好的习惯。大量的 API 文档存在于源码中时，对于开发人员来说，可以通过查看源码的 javadoc 来了解接口信息；对于非开发人员来说，查看源码可能会产生一些障碍。因此，有必要将 API 文档从源码中抽取出来，形成一个统一的文档，方便随时查阅。借助一些接口文档生成工具可以快速完成这一项工作，apidoc 和 swagger 就是两款非常流行的接口文档生成工具。

8.4.1 apidoc

apidoc 是一个简单的 RESTful API 文档生成工具，从代码注释中提取特定格式的内容生成文档。apidoc 支持多种主流的编码语言，包括 Java、C、C#、PHP 和 JavaScript。一般情况下，语言会有多种注释方法，例如在 Java 中就有普通风格的多行注释和 Javadoc 风格的注释。apidoc 并不支持所有的注释，比如 Java 中仅支持 Javadoc 风格的注释。需要说明的是，apidoc 并不具备语义识别能力，不会发现代码中是否有 BUG，仅仅通过文件后缀来判断语言类型。

它对代码没有侵入性，只需要写好相关的注释即可，并且它仅通过写简单的配置就可以生成高颜值的 api 接口页面。它是基于 Node.js 的，所以需要安装 Node.js 环境。

检查是否安装 Node.JS 命令是 npm-version，如果已安装就会输出版本号，如果没有安装则会输出错误信息“不是内部或外部命令”，需要安装 Node.js。Node.js 的下载地址为 https://node.js.org/zh-cn/，安装步骤不再赘述。

1. 安装 apidoc

使用 npm 命令进行安装：

```
npm install apidoc -g
```

检查 apidoc 安装是否成功：

```
apidoc -h
```

出现帮助提示则说明安装成功。

2. 在 Spring Boot 项目中使用 apidoc 生成接口文档

首先创建一个空的 Spring Boot 项目，项目名为 apidoc-demo。在项目根路径下创建 apidoc.json 文件，内容如下所示。

示例代码 8-75 package.json

```
{
    "name": "测试 api 文档",
    "version": "0.1.0",
    "description": "这只是一个测试的页面",
    "title": "APIDOC 测试",
    "url" : "http://localhost:8080",
    "sampleUrl": "http://localhost:8080"
```

```
}
```

创建 IndexController 类，并创建 helloApidoc 方法、添加注释信息（使用官方注解），示例代码如下。

示例代码 8-76　IndexController.java

```
package com.example.demo.controller;
import …
/**
 * 接口文档测试首页
 */
@RestController
public class IndexController {
    /**
     * @api {get} /hello-apidoc 测试
     * @apiName hello-apidoc
     * @apiDescription api 接口文档测试
     * @apiGroup index
     * @apiSampleRequest /hello-apidoc
     */
    @GetMapping("/hello-apidoc")
    public String helloApidoc() {
        return "Hello Apidoc";
    }
    /**
     * @api {get} /memberlist 会员列表
     * @apiDescription 会员用户列表展示
     * @apiName memberlist
     * @apiParam {String} type 会员类型
     * @apiParamExample {json} Request-Example:
     * {
     * "type":"svip"
     * }
     * @apiGroup index
     * @apiSampleRequest /memberlist
     */
    @PostMapping("/memberlist")
    public List<String> memberList( String type) {
        List<String> list = new ArrayList<>();
        list.add("小明");
        list.add("小李");
        list.add("小红");
        return list;
    }
```

}最后生成接口文档，需要在控制台进入项目的外面一层目录，例如本示例项目路径为 D:\code\idea\apidoc-demo，就需要在 D:\code\idea 目录下执行以下命令生成接口文档：

```
apidoc -i apidoc-demo/ -o apidoc-demo-doc/
```

其中，-i 参数表示输入的路径，-o 参数表示输出路径。执行完成后，生成的接口文档在

apidoc-demo-doc 目录下，在浏览器中打开 index.html 页面，就可以看到如图 8-51 所示的接口文档信息。

图 8-51

3. apidoc.json 配置详解

apidoc.json 放置在项目根目录中，可选内容包括有关项目的常用信息，例如标题、简短描述、版本和配置选项（例如页眉/页脚设置或模板特定的选项）。下面是一个配置示例：

示例代码 8-77　apidoc.json

```
{
    "name": "example",
    "version": "0.1.0",
    "description": "apiDoc basic example",
    "title": "Custom apiDoc browser title",
    "url" : "https://api.github.com/v1"
}
```

apidoc.json 所有的配置项如表 8-8 所示。

表 8-8　apidoc.json 所有的配置项及其说明

配置项	说明
name	项目名称
version	项目版本号
description	项目简介

（续表）

配置项		说明
title		apidoc 在浏览器中显示的标题
url		接口地址的前缀，比如 https://api.github.com/v1
useHostUrlAsSampleUrl		如果设置为 true，那么主机 URL 地址将使用 sampleUrl 配置的地址，默认为 false
sampleUrl		如果设置此地址，那么将会展示一个测试请求该 api 的表单，详细介绍请看@apiSampleRequest 注解
header	title	设置页眉标题
	filename	页眉文件的文件名（markdown 文件）
footer	title	页脚文本
	filename	页脚文件的文件名（markdown 文件）
order		用于排序输出的 api 名称/组名称的列表，未定义的名称将自动显示在最后。 "order": ["Error", "Define", "PostTitleAndError", "PostError"]

4. apidoc 命令详解

使用命令“apidoc -h”可以查看所有支持的命令和参数。重要的参数说明如表 8-9 所示。

表 8-9　apidoc 命令的重要参数及其说明

参数	描述
-f，--file-filters	RegEx-Filter 选择要分析的文件（可以使用许多-f），默认为 .cs .dart .erl .go .java .js .php .py .rb .ts。 示例（仅解析.js 和.ts 文件）：apidoc -f ".*\\.js$" -f ".*\\.ts$"
-i，--input	输入/源目录名、项目文件的位置。 示例：apidoc -i myapp/
-o，--output	输出目录名、放置生成的文档的位置 示例：apidoc -o apidoc/
-t，--template	使用模板输出文件，可以创建和使用自己的模板 示例：apidoc -t mytemplate/
-e，--exclude-filters	RegEx-Filter 选择不应该解析的文件/目录（可以使用许多-e），默认值为[] 示例：apidoc -e node_modules

5. apidoc 注释详解

一个比较完整的 apidoc 注释示例如下面的代码所示。

示例代码 8-77　apidoc 注释示例

```
/**
 * @api {get} /user/:id Request User information
 * @apiName GetUser
 * @apiGroup User
 *
 * @apiParam {Number} id Users unique ID.
 *
 * @apiSuccess {String} firstname Firstname of the User.
 * @apiSuccess {String} lastname  Lastname of the User.
 *
 * @apiSuccessExample Success-Response:
 *     HTTP/1.1 200 OK
 *     {
 *       "firstname": "John",
 *       "lastname": "Doe"
 *     }
 *
 * @apiError UserNotFound The id of the User was not found.
 *
 * @apiErrorExample Error-Response:
 *     HTTP/1.1 404 Not Found
 *     {
 *       "error": "UserNotFound"
 *     }
 */
```

apidoc 文档块以/**开头、以*/结尾。本示例描述了一种通过获取请求方式来获取来用户信息的方法。在该示例 apidoc 注释中：

- @api {get} /user/:id Request User information 是强制性的，没有@apiapiDoc 会忽略文档块。
- @apiName 必须是唯一名称，并且应始终使用，格式为“方法+路径（例如，获取+用户）”
- @apiGroup 应该始终使用，并用于将相关的 API 组合在一起。
- 其他字段都是可选的，具体说明可以参考官方文档（https://apidocjs.com/#params）。

8.4.2　Swagger

1. Swagger 是什么

Swagger 是一系列用于 RESTful API 开发的工具。开源的部分包括：

- OpenAPI Specification：API 规范，规定了如何描述一个系统的 API。
- Swagger Codegen：用于通过 API 规范生成服务端和客户端代码。
- Swagger Editor：用来编写 API 规范。

- Swagger UI：用于展示 API 规范。

非开源的部分包括：

- Swagger Hub：云服务，相当于 Editor + Codegen + UI。
- Swagger Inspector：手动测试 API 的工具。
- SoapUI Pro：功能测试和安全测试的自动化工具。
- LoadUI Pro：压力测试和性能测试的自动化工具。

2. 为什么要用 Swagger

在微服务的盛行下，成千上万的接口文档编写不可能靠人力来编写，于是 Swagger 就产生了，它采用自动化实现并解决了人力编写接口文档的问题。它通过在接口及实体上添加几个注解的方式能够在项目启动后自动化生成接口文档。

Swagger 提供了一个全新的维护 API 文档的方式，有四大优点：

- 只需要少量的注解，Swagger 就可以根据代码自动生成 API 文档，很好地保证了文档的时效性。
- 跨语言性，支持 40 多种语言。
- Swagger UI 呈现出来的是一份可交互式的 API 文档，我们可以直接在文档页面尝试 API 的调用，省去了准备复杂的调用参数的过程。
- 可以将文档规范导入相关的工具（例如 SoapUI），这些工具将会为我们自动创建自动化测试。

3. Swagger 的发展

Swagger 工具的强大功能始于 OpenAPI 规范——RESTful API 设计的行业标准，通过这套规范，你只需要按照它的规范去定义接口及接口相关的信息，再通过 Swagger 衍生出来的一系列项目和工具就可以做到生成各种格式的接口文档，生成多种语言的客户端/服务端的代码以及在线接口调试页面等。如果按照新的开发模式，在开发新版本或者迭代版本的时候只需要更新 Swagger 描述文件就可以自动生成接口文档和客户端服务端代码，做到调用端代码、服务端代码以及接口文档的一致性。

即便如此，对于许多开发来说编写 yml 或 JSON 格式的描述文件本身也是有一定负担的，特别是在后面持续迭代开发的时候，往往会忽略更新描述文件而直接更改代码。久而久之，这个描述文件就和实际项目渐行渐远了，基于该描述文件生成的接口文档也失去了参考意义。Spring 迅速将 Swagger 规范纳入自身的标准，建立了 Spring-swagger 项目，改成了现在的 Springfox。通过在项目中引入 Springfox，可以扫描相关的代码，生成该描述文件，进而生成与代码一致的接口文档和客户端代码。这种通过代码生成接口文档的形式在后面需求持续迭代的项目中显得尤为重要和高效。

Springfox 其实是一个通过扫描代码提取代码中的信息、生成 API 文档的工具。API 文档的格式不止 Swagger 的 OpenAPI Specification，还有 RAML、jsonapi。Springfox 的目标同样包括支持这些格式。

在 Swagger 的教程中都会提到@Api、@ApiModel、@ApiOperation 注解，这些注解其实不是 Springfox 的，而是 Swagger 的（springfox-swagger2 包依赖了 swagger-core 包，这些注解正是这个包里的）。swagger-core 包只支持 JAX-RS2，并不支持常用的 Spring MVC，而是由 springfox-swagger

将用于 JAX-RS2 的注解适配到 Spring MVC 上。

除了 Spring MVC 外，Springfox 还支持如下库：

- Spring Data REST。
- JSR 303（这项标准的参考实现是 Hibernate Validator）。

4. 案例实战：将 Spring Boot 接口自动生成接口文档

我们以 8.1.2 小节中实现的 shiro-demo 为基础，在 pom.xml 中添加 Springfox 依赖包。

示例代码 8-78　在 pom.xml 中添加 springfox-swagger 依赖

```
<!--引入 Springfox（swagger）支持包-->
<dependency>
   <groupId>io.springfox</groupId>
   <artifactId>springfox-swagger2</artifactId>
   <version>2.9.2</version>
</dependency>
<!--swagger-ui-->
<dependency>
   <groupId>io.springfox</groupId>
   <artifactId>springfox-swagger-ui</artifactId>
   <version>2.9.2</version>
</dependency>
```

在 application.yml 配置文件中添加 swagger 配置：

```
spring:
  # 启用 swagger 支持
  swagger2:
    enabled: true
  #...
```

新建 SwaggerConfig.java 配置类：

示例代码 8-79　SwaggerConfig.java 配置类

```
@Configuration
@EnableSwagger2
@EnableWebMvc
public class SwaggerConfig implements WebMvcConfigurer {
   private static final String BASE_PACKAGE = "com.example.shiro";
   @Value(value = "${spring.swagger2.enabled}")
   private Boolean swaggerEnabled;
   @Override
   public void configureDefaultServletHandling(
            DefaultServletHandlerConfigurer configurer) {
      configurer.enable();
   }
   @Bean
   public Docket createRestApi() {
      return new Docket(DocumentationType.SWAGGER_2)
      .apiInfo(apiInfo())
```

```
            //设置是否启用
            .enable(swaggerEnabled)
            //选择为哪些目标接口生成接口文档
            .select()
            //设置需要生成接口文档的目录包
            .apis(RequestHandlerSelectors.basePackage(BASE_PACKAGE))
            //任意路径
            .paths(PathSelectors.any())
            .build();
    }
    /**
     * 设置ApiInfo,
     * @return
     */
    private ApiInfo apiInfo() {
        return new ApiInfoBuilder()
        .title("Shiro-Demo接口文档")
        .description("Spring Boot Shiro Demo")
        //配置服务网站
        // .termsOfServiceUrl("")
        .version("v0.0.1")
        .build();
    }
    /**
     * 解决swagger被拦截的问题
     * @param registry
     */
    @Override
    public void addResourceHandlers(ResourceHandlerRegistry registry) {
    // 解决静态资源无法访问的问题
        registry.addResourceHandler("/**")
        .addResourceLocations("classpath:/static/");
        // 解决swagger无法访问的问题
        registry.addResourceHandler("/swagger-ui.html")
        .addResourceLocations("classpath:/META-INF/resources/");
        registry.addResourceHandler("/doc.html")
        .addResourceLocations("classpath:/META-INF/resources/");
        // 解决swagger的js文件无法访问的问题
        registry.addResourceHandler("/webjars/**")
        .addResourceLocations("classpath:/META-INF/resources/webjars/");
    }
}
```

修改ShiroConfig.java配置类，将Shiro对Swagger的权限放开：

示例代码8-80 放开Swagger相关路径的权限

```
   @Bean
   public ShiroFilterFactoryBean shiroFilterFactoryBean(SecurityManager
securityManager) {
      ShiroFilterFactoryBean shiroFilterFactoryBean =
```

```
        new ShiroFilterFactoryBean();
    shiroFilterFactoryBean.setSecurityManager(securityManager);
    Map<String, String> map = new HashMap<>();
    //登出
    map.put("/user/logout", "logout");
    map.put("/swagger-ui.html", "anon");
    map.put("/webjars/springfox-swagger-ui/**", "anon");
    map.put("/swagger-resources/**", "anon");
    map.put("/v2/api-docs", "anon");
    …
    shiroFilterFactoryBean.setFilterChainDefinitionMap(map);
    return shiroFilterFactoryBean;
}
```

接下来修改 UserRestController，编写符合 Swagger 规范的接口文档。

示例代码 8-81　修改 UserRestController.java

```
@RestController
@RequestMapping("/user")
@Api(tags = "用户管理")
public class UserRestController {
    @Resource
    private UserService userService;
    @ApiOperation(value = "用户注册", notes = "使用用户名和密码注册用户",
        httpMethod = "POST")
    @ApiImplicitParams({
        @ApiImplicitParam(name = "username", value = "用户名",
            required = true, dataTypeClass = String.class),
        @ApiImplicitParam(name = "password", value = "密码", required = true,
            dataTypeClass = String.class),
    })
    @ApiResponses({
        @ApiResponse(code = 1, message = "success", response = JsonResult.class),
        @ApiResponse(code = -1, message = "error")
    })
    @RequiresGuest
    @PostMapping("/register")
    public JsonResult create(@RequestBody UserVO userVO) {
        …
    }
}
```

编写完成后，编译运行项目，访问 http://localhost:8080/swagger-ui.html 就可以看到接口文档已经生成成功，如图 8-52 所示。

关于更多 Swagger 注解的说明，请查阅官方文档：https://springfox.github.io/springfox/docs/current/ #overriding-descriptions-via-properties。

关于更多 Springfox 的知识和介绍，请查阅文档：https://springfox.github.io/springfox/docs/current/#introduction。

图 8-52

8.5　集成日志框架打印日志

在计算机领域，日志文件（logfile）是一个记录了发生在运行中的操作系统或其他软件中的事件的文件，或者记录了在网络聊天软件的用户之间发送的消息。日志记录（Logging）是指保存日志的行为。最简单的做法是将日志写入单个存放日志的文件。

日志通常有以下几个级别：

- FATAL：表示需要立即被处理的系统级错误。当该错误发生时，表示服务已经出现了某种程度的不可用，系统管理员需要立即介入。这属于最严重的日志级别，因此该日志级别必须慎用，如果这种级别的日志经常出现，则该日志也失去了意义。通常情况下，一个进程的生命周期中应该只记录一次 FATAL 级别的日志，即该进程遇到无法恢复的错误而退出时。当然，如果某个系统的子系统遇到了不可恢复的错误，那么该子系统的调用方也可以记入 FATAL 级别日志，以便通过日志报警提醒系统管理员修复。
- ERROR：该级别的错误也需要马上被处理，但是紧急程度要低于 FATAL 级别。当 ERROR 错误发生时，已经影响了用户的正常访问。从该意义上来说，实际上 ERROR 错误和 FATAL 错误对用户的影响是相当的。FATAL 相当于服务已经挂了，而 ERROR 相当于好死不如赖活着，然而活着却无法提供正常的服务，只能不断地打印 ERROR 日志。特别需要注意的是，ERROR 和 FATAL 都属于服务器自己的异常，是需要马上得到人工介入并处理的。用户自己操作不当（如请求参数错误等），是不应该记为 ERROR 日志的。
- WARN：表示系统可能出现问题，也可能没有，比如网络的波动等。对于那些目前还不是错误，然而不及时处理也会变为错误的情况，也可以记为 WARN 日志，比如一个存储系统的磁盘使用量超过阀值或者系统中某个用户的存储配额快用完等。对于 WARN 级别的日志，虽然不需要系统管理员马上处理，也是需要及时查看并处理的。因此，此种级别的日志不应太多，能不打 WARN 级别的日志，就尽量不要打。

- INFO：该种日志记录系统的正常运行状态，例如某个子系统的初始化、某个请求的成功执行等。通过查看 INFO 级别的日志，可以很快地对系统中出现的 WARN、ERROR、FATAL 错误进行定位。INFO 日志不宜过多，通常情况下，INFO 级别的日志应该不大于 TRACE 日志的 10%。
- DEBUG 或 TRACE：这两种日志具体的规范应该由项目组自己定义，该级别日志的主要作用是对系统每一步的运行状态进行精确的记录。通过该种日志，可以查看某一个操作每一步的执行过程，可以准确定位是何种操作、何种参数、何种顺序导致了某种错误的发生。可以保证在不重现错误的情况下也可以通过 DEBUG（或 TRACE）级别的日志对问题进行诊断。需要注意的是，DEBUG 日志也需要规范日志格式，应该保证除了记录日志的开发人员外，其他的人员也可以通过 DEBUG（或 TRACE）日志来定位问题。

这些日志级别的优先级从低到高排列为：

TRACE < DEBUG < INFO < WARN < ERROR < FATAL

日志记录了系统行为的时间、地点、状态等相关信息，能够帮助我们了解并监控系统状态，在发生错误或者接近某种危险状态时能够及时提醒我们处理，同时在系统产生问题时能够帮助我们快速定位、诊断并解决问题。因此妥善记录系统日志是十分有必要的。

在 Java 程序中常用日志框架可以分为两类：

- 无具体实现的抽象门面框架，如 Commons Logging、SLF4J。
- 具体实现的框架，如 Log4j、Log4j 2、Logback、Jul。

8.5.1 Java 程序日志框架发展史

Java 程序日志的发展已有很多年历史，如今的日志框架已经发展得非常成熟，并形成了体系。

最早可以追溯到 1996 年早期，欧洲安全电子市场项目组决定编写它自己的程序跟踪 API（Tracing API）。经过不断的完善，这个 API 终于成为一个十分受欢迎的 Java 日志软件包，即 Log4j。Apache 发现了这个优秀的框架之后，说服了 Log4j 作者 Ceki Gülcü将 Log4j 贡献给 Apache 并加入 Apache 参与 Log4j 的维护。其间 Log4j 近乎成了 Java 社区的日志标准。

2002 年 Java 1.4 发布，Sun 推出了自己的日志库 JUL（Java Util Logging），其实现基本模仿了 Log4j 的实现。在 JUL 出来以前，Log4j 已经成为一项成熟的技术，使得 Log4j 在选择上占据了一定的优势。

接着，Apache 推出了 Jakarta Commons Logging（JCL），只是定义了一套门面日志接口（其内部也提供一个 Simple Log 的简单实现），支持运行时动态加载日志组件的实现。也就是说，在应用代码里，只需调用 Commons Logging 的接口，底层实现可以是 Log4j 也可以是 Java Util Logging，JCL 实现了各种日志系统的兼容，避免了各个组件日志系统不一致引起的混乱问题。

2006 年，Log4j 的作者 Ceki Gülcü觉得 JCL 并不好用，而且不适应 Apache 的工作方式，便离开了 Apache，Apache 之后，Ceki 自己创建了门面日志接口组件 SLF4J 来替代 Commons Logging，之后 Ceki 觉得 Log4j 存在一些性能等问题，于是又重新写了一套日志系统叫 Logback，并回瑞典创建了 QOS 公司。QOS 官网上是这样描述 Logback 的：The Generic，Reliable Fast&Flexible Logging Framework（一个通用、可靠、快速且灵活的日志框架）。因为 Logback 比 Log4j 大约快 10 倍、消耗更少的内存，迁移成本也很低，自动压缩日志、支持多样化配置、不需要重启就可以恢复 I/O 异常等优势，又名噪一时。

Apache 并不甘示弱，眼看有被 Logback 反超的势头，于 2012-07 重写了 Log4j 1.x，成立了新的项目 Log4j 2，Log4j 2 具有 Logback 的所有特性。

现今，Java 日志领域被划分为两大体系阵营：Commons Logging 阵营和 SLF4J 阵营。

Commons Logging 在 Apache 大树的笼罩下有很大的用户基数，不过有证据表明形势正在发生变化。2013 年年底有人分析了 GitHub 上的 30000 个项目，统计出最流行的 100 个 Libraries，结果发现 SLF4J 的发展趋势更好。

8.5.2 第一代日志框架 Log4j

Log4j 可以很方便地集成到 Spring Boot 应用程序中。下面以一个简单的项目示例来介绍如何将 Log4j 运用到 Spring Boot 项目中。

1. 为项目添加 Log4j 依赖

首先创建名为 log4j-demo 的项目，在 pom.xml 中引入 Log4j 依赖包：

```
<!--Log4j 依赖包-->
<dependency>
    <groupId>log4j</groupId>
    <artifactId>log4j</artifactId>
    <version>1.2.17</version>
</dependency>
```

2. Log4j 配置

Log4j 日志的配置可以用 log4j.xml 和 log4j.properties 这两种形式，推荐使用 log4j.properties。Log4j 的配置文件一般放在 resources 文件夹下。下面给出一个简单的配置示例：

示例代码 8-82 一个简单的 log4j.properties 配置

```
log4j.rootLogger=INFO,Console,File
log4j.appender.Console=org.apache.log4j.ConsoleAppender
log4j.appender.Console.Target=System.out
log4j.appender.Console.layout = org.apache.log4j.PatternLayout
log4j.appender.Console.layout.ConversionPattern=[%p] [%d{yyyy-MM-dd
HH\:mm\:ss}][%c - %L]%m%n
log4j.appender.File = org.apache.log4j.RollingFileAppender
log4j.appender.File.File = logs/info/info.log
log4j.appender.File.MaxFileSize = 10MB
log4j.appender.File.Threshold = ALL
```

```
    log4j.appender.File.layout = org.apache.log4j.PatternLayout
    log4j.appender.File.layout.ConversionPattern =[%p] [%d{yyyy-MM-dd
HH\:mm\:ss}][%c - %L]%m%n
```

3. 使用 Log4j 打印日志

Log4j 提供了一个 org.apache.log4j.Logger 对象来进行日志的打印，可以通过 Logger.getLogger(*.class)来获取这个 Logger 对象。下面演示一下如何使用 Log4j 在 Spring Boot 项目中打印日志。在项目中新建一个控制器类 IndexController，代码如下。

示例代码 8-83　IndexController.java

```
@RestController
public class IndexController {
    private static Logger LOGGER = Logger.getLogger(IndexController.class);
    @GetMapping("/")
    public String index() {
        LOGGER.info("访问了首页！");
        return "This is index page!";
    }
    @GetMapping("/hello")
    public String hello() {
        LOGGER.info("访问了欢迎页！");
        return "Welcome to log4j!";
    }
}
```

启动并运行项目，通过浏览器分别访问 http://localhost:8080/hello 和 http://localhost:8080/两个接口，查看控制台输出内容：

```
    …
    2021-02-17 10:49:15.100  INFO 3804 --- [  restartedMain]
o.s.b.w.embedded.tomcat.TomcatWebServer  : Tomcat started on port(s): 8080 (http)
with context path ''
    2021-02-17 10:49:15.107  INFO 3804 --- [  restartedMain]
com.example.demo.DemoApplication         : Started DemoApplication in 1.875 seconds
(JVM running for 2.258)
    2021-02-17 10:49:23.463  INFO 3804 --- [nio-8080-exec-1]
o.a.c.c.C.[Tomcat].[localhost].[/]       : Initializing Spring DispatcherServlet
'dispatcherServlet'
    2021-02-17 10:49:23.464  INFO 3804 --- [nio-8080-exec-1]
o.s.web.servlet.DispatcherServlet        : Initializing Servlet
'dispatcherServlet'
    2021-02-17 10:49:23.467  INFO 3804 --- [nio-8080-exec-1]
o.s.web.servlet.DispatcherServlet        : Completed initialization in 3 ms
    [INFO] [2021-02-17 10:49:23][com.example.demo.controller.IndexController - 20]
访问了欢迎页！
    [INFO] [2021-02-17 10:49:37][com.example.demo.controller.IndexController - 14]
访问了首页！
```

最后两行打印了两个 INFO 级别的日志，内容是我们自己定义的。这说明 Log4j 打印日志成功。

上述日志仅限于我们自己编写的程序，在使用了 Log4j 1.x 版本的 Logger 打印日志的情况下才可以打印出来。Spring Boot 应用程序框架本身的日志并不会使用我们配置的 Log4j 1.x 打印日志。在 Spring Boot 框架中，其实已经提供了 artifactId 为 spring-boot-starter-log4j 的包，可以很方便地打印框架本身的日志，但是目前这个包已经过时不再维护，Spring Boot 只有 1.3.x 和 1.3.x 以下版本才支持 Log4j 1.x 的日志配置，1.3.x 以上版本只支持 Log4j 2。

Log4j 是一个组件化设计的日志系统，架构大致如下：

```
log.info("User signed in.");
 ├──>| Appender |───>|  Filter  |───>|  Layout  |───>| Console|
 ├──>| Appender |───>|  Filter  |───>|  Layout  |───>|   File|
 └──>| Appender |───>|  Filter  |───>| Layout  |───>|  Socket |
```

当我们使用 Log4j 输出一条日志时，Log4j 自动通过不同的 Appender 把同一条日志输出到不同的目的地，比如 console（输出到屏幕）、file（输出到文件）、socket（通过网络输出到远程计算机）、jdbc（输出到数据库）。

在输出日志的过程中，通过 Filter 来过滤哪些 log 需要被输出、哪些 log 不需要被输出，比如仅输出 ERROR 级别的日志。

最后，通过 Layout 来格式化日志信息，比如自动添加日期、时间、方法名称等信息。

上述结构虽然复杂，但是在实际使用的时候并不需要关心 Log4j 的 API，而是通过配置文件来配置。

8.5.3　简单日志门面框架 SLF4J

SLF4J 是一个日志门面框架，采用门面设计模式，定义了一套日志接口规范，与底层日志框架的实现无关，SLF4J 可作为各种日志记录框架（例如 java.util.logging，logback，Log4j）的简单外观或抽象，允许最终用户在部署时插入所需的日志记录框架。SLF4J 改进了 Commons Logging 框架的一些缺陷。

下面是一个简单的 SLF4J 示例：

示例代码 8-84　简单的 SLF4J 日志打印示例

```
import org.SLF4j.Logger;
import org.SLF4j.LoggerFactory;
public HelloWorld {
   public static void main (String [] args) {
       Logger logger = LoggerFactory.getLogger (HelloWorld.class) ;
       logger.info (“Hello World”) ;
   }
}
```

与直接使用 Log4j 不同的是，它不再直接依赖 Logger.getLogger 来获取日志对象，而是使用 LoggerFactory 来获取，得到的 Logger 对象也不再是 Log4j 的 Logger，是 SLF4J 自己提供的 Logger 对象。SLF4J 内部实现了与各种日志框架实现的对接工作，因此使用 SLF4J 无须关心底层日志框架使用什么，而只需关注日志的打印，具体由谁打印可以在运行部署时决定，开发阶段无须关心。

SLF4J 只是一个门面接口，这意味着它不提供完整的日志记录解决方案。使用 SLF4J 无法执行配置 appender 或设置日志记录级别等操作。因此，在某个时间点，任何非平凡的应用程序都需要直接调用底层日志记录系统。换句话说，独立应用程序无法完全独立于 API 底层日志记录系统。然而，SLF4J 减少了这种依赖对近乎无痛水平的影响。

假设 CRM 应用程序使用 Log4j 进行日志记录，但是一个重要客户端要求通过 JDK 1.4 日志记录执行日志记录。如果应用程序充斥着成千上万的直接 Log4j 调用，那么迁移到 JDK 1.4 将是一个相对冗长且容易出错的过程。更糟糕的是，可能需要维护两个版本的 CRM 软件。如果一直在调用 SLF4J API 而不是 Log4j，则可以在几分钟内通过将一个 jar 文件替换为另一个 jar 文件来完成迁移。

SLF4J 的 Maven 依赖包如下：

```
<dependency>
    <groupId>org.SLF4j</groupId>
    <artifactId>SLF4j-api</artifactId>
    <version>1.7.30</version>
</dependency>
```

8.5.4 使用 Logback

Logback 与 Log4j 是同一个作者创建的，对 Log4j 做了很多实质性的改进和优化，旨在作为流行的 Log4j 项目的后继者。Logback 在概念上与 Log4j 非常相似，因此如果你对 Log4j 十分熟悉，那么对 Logback 的切换会是无感知的。相较于 Log4j，Logback 组件不但速度快，而且内存占用空间较小。

Logback 的体系结构可以应用在不同的情况下。目前，Logback 分为三个模块：logback-core、logback-classic 和 logback-access。logback-core 模块是核心，是其他两个模块的基础。logback-classic 模块看作是 Log4j 的改进版本。此外，logback-classic 原生实现了 SLF4J API，因此结合使用 SLF4J API 可以轻松地在 Logback 和其他日志框架（例如 Log4j 或 java.util.logging（JUL））之间来回切换。Logback-access 模块与 Servlet 容器（例如 Tomcat 和 Jetty）集成，以提供 HTTP 访问日志功能。有需要的话，也可以轻松地在 logback-core 模块之上构建自己的模块。

接下来我们展示一个项目案例，实现在 Spring Boot 应用中结合使用 SLF4j 和 Logback。

1. 为项目添加 Logback 和 SLF4J 依赖包

在 Spring Boot 中集成 Logback 和 SLF4J 非常简单，因为 Spring Boot 官方已经提供了名为 spring-boot-starter-logging 的包，该包包含了 logback-classic、logback-core，而 logback-classic 又默认依赖了 SLF4j-api，因此无须多余的配置，只需在 Spring Boot 项目中引入 spring-boot-starter-logging 依赖包即可。创建名为 logback-demo 的 Spring Boot 项目，引入该依赖包：

```
<!--spring-boot-starter-logging-->
<dependency>
    <groupId>org.springframework.boot</groupId>
    <artifactId>spring-boot-starter-logging</artifactId>
</dependency>
```

事实上，我们无须引入此包，Spring Boot 默认选择的是 SLF4J + Logback 的组合，如果不需要更改为其他日志系统（如 Log4j2 等），则无须多余的配置，Logback 默认会将日志打印到控制台上。

由于新建的 Spring Boot 项目一般都会引用 spring-boot-starter 或者 spring-boot-starter-web，而这两个起步依赖中都已经包含了对于 spring-boot-starter-logging 的依赖，如图 8-53 所示，所以我们无须额外添加依赖。

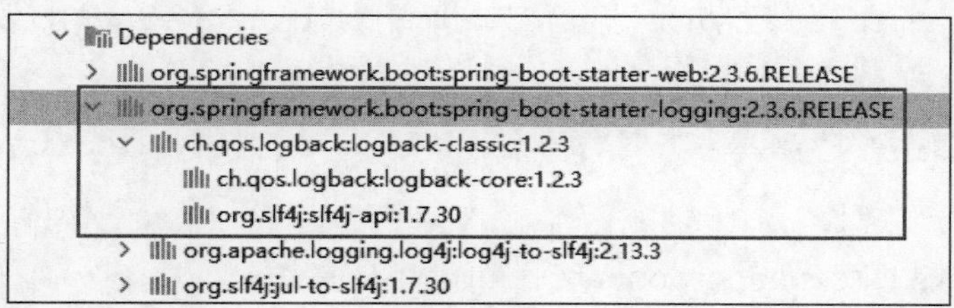

图 8-53

2. 配置 Logback

在 resources 目录下创建 logback-spring.xml 配置文件，填写 Logback 的配置。下面提供一个示范的配置。

示例代码 8-85　logback-spring.xml

```
<?xml version="1.0" encoding="UTF-8"?>
    <!-- 日志级别从低到高分为 TRACE < DEBUG < INFO < WARN < ERROR < FATAL，如果设置为
WARN，则低于 WARN 的信息都不会输出 -->
    <!-- scan:当此属性设置为 true 时，配置文件如果发生改变，将会被重新加载，默认值为 true -->
    <!-- scanPeriod:设置监测配置文件是否有修改的时间间隔，如果没有给出时间单位，默认单位是毫
秒。当 scan 为 true 时，此属性生效。默认的时间间隔为 1 分钟。 -->
    <!-- debug:当此属性设置为 true 时，将打印出 logback 内部日志信息，实时查看 logback 运行
状态，默认值为 false。 -->
    <configuration scan="true" scanPeriod="10 seconds">
    <!--<include resource="org/springframework/boot/logging/logback/base.xml"
/>-->
    <contextName>logback</contextName>
    <!-- name 值是变量的名称，value 值是变量定义的值。通过定义的值会被插入 logger 上下文中。
定义变量后，可以用“${}”来使用变量。 -->
    <property name="log.path" value="/logs/"/>
    <!-- 彩色日志 -->
    <!-- 彩色日志依赖的渲染类 -->
    <conversionRule conversionWord="clr"
converterClass="org.springframework.boot.logging.logback.ColorConverter"/>
    <conversionRule conversionWord="wex" converterClass=
"org.springframework.boot.logging.logback.WhitespaceThrowableProxyConverter"/>
    <conversionRule conversionWord="wEx" converterClass=
     "org.springframework.boot.logging.logback.ExtendedWhitespaceThrowableProxy
Converter"/>
    <!-- 彩色日志格式 -->
    <property name="CONSOLE_LOG_PATTERN"
            value="${CONSOLE_LOG_PATTERN:
                -%clr(%d{yyyy-MM-dd HH:mm:ss.SSS}){faint}
            %clr(${LOG_LEVEL_PATTERN:-%5p})
                %clr(${PID:- }){magenta} %clr(---){faint}
            %clr([%15.15t]){faint} %clr(%-40.40logger{39}){cyan}
```

```
            %clr(:){faint}
          %m%n${LOG_EXCEPTION_CONVERSION_WORD:-%wEx}}"/>
  <!--输出到控制台-->
  <appender name="CONSOLE" class="ch.qos.logback.core.ConsoleAppender">
     <!--此日志 appender 是为开发使用的，只配置最低级别，控制台输出的日志级别是大于或等于此
级别的日志信息-->
     <filter class="ch.qos.logback.classic.filter.ThresholdFilter">
        <level>info</level>
     </filter>
     <encoder>
        <Pattern>${CONSOLE_LOG_PATTERN}</Pattern>
        <!-- 设置字符集 -->
        <charset>UTF-8</charset>
     </encoder>
  </appender>
  <!--输出到文件-->
  <!-- 时间滚动输出 level 为 DEBUG 日志 -->
  <appender name="DEBUG_FILE"
          class="ch.qos.logback.core.rolling.RollingFileAppender">
     <!-- 正在记录的日志文件的路径及文件名 -->
     <file>${log.path}/log_debug.log</file>
     <!--日志文件输出格式-->
     <encoder>
        <pattern>
         %d{yyyy-MM-dd HH:mm:ss.SSS} [%thread] %-5level %logger{50} - %msg%n
        </pattern>
        <charset>UTF-8</charset> <!-- 设置字符集 -->
     </encoder>
     <!-- 日志记录器的滚动策略，按日期、大小记录 -->
     <rollingPolicy
             class="ch.qos.logback.core.rolling.TimeBasedRollingPolicy">
        <!-- 日志归档 -->
        <fileNamePattern>
              ${log.path}/debug/log-debug-%d{yyyy-MM-dd}.%i.log
           </fileNamePattern>
        <timeBasedFileNamingAndTriggeringPolicy
           class="ch.qos.logback.core.rolling.SizeAndTimeBasedFNATP">
           <maxFileSize>100MB</maxFileSize>
        </timeBasedFileNamingAndTriggeringPolicy>
        <!--日志文件保留天数-->
        <maxHistory>15</maxHistory>
     </rollingPolicy>
     <!-- 此日志文件只记录 debug 级别的 -->
     <filter class="ch.qos.logback.classic.filter.LevelFilter">
        <level>debug</level>
        <onMatch>ACCEPT</onMatch>
        <onMismatch>DENY</onMismatch>
     </filter>
  </appender>
  <!-- 时间滚动输出 level 为 INFO 日志 -->
```

```
    <appender name="INFO_FILE"
                 class="ch.qos.logback.core.rolling.RollingFileAppender">
        <!-- 正在记录的日志文件的路径及文件名 -->
        <file>${log.path}/log_info.log</file>
        <!--日志文件输出格式-->
        <encoder>
            <pattern>
             %d{yyyy-MM-dd HH:mm:ss.SSS} [%thread] %-5level %logger{50} - %msg%n
            </pattern>
            <charset>UTF-8</charset>
        </encoder>
        <!-- 日志记录器的滚动策略，按日期、大小记录 -->
        <rollingPolicy
                class="ch.qos.logback.core.rolling.TimeBasedRollingPolicy">
            <!-- 每天日志归档路径以及格式 -->
            <fileNamePattern>
                ${log.path}/info/log-info-%d{yyyy-MM-dd}.%i.log
            </fileNamePattern>
            <timeBasedFileNamingAndTriggeringPolicy
                class="ch.qos.logback.core.rolling.SizeAndTimeBasedFNATP">
                <maxFileSize>100MB</maxFileSize>
            </timeBasedFileNamingAndTriggeringPolicy>
            <!--日志文件保留天数-->
            <maxHistory>15</maxHistory>
        </rollingPolicy>
        <!-- 此日志文件只记录 info 级别的 -->
        <filter class="ch.qos.logback.classic.filter.LevelFilter">
            <level>info</level>
            <onMatch>ACCEPT</onMatch>
            <onMismatch>DENY</onMismatch>
        </filter>
    </appender>
    <!-- 时间滚动输出 level 为 WARN 日志 -->
    <appender name="WARN_FILE"
                 class="ch.qos.logback.core.rolling.RollingFileAppender">
        <!-- 正在记录的日志文件的路径及文件名 -->
        <file>${log.path}/log_warn.log</file>
        <!--日志文件输出格式-->
        <encoder>
            <pattern>
             %d{yyyy-MM-dd HH:mm:ss.SSS} [%thread] %-5level %logger{50} - %msg%n
            </pattern>
            <!-- 此处设置字符集 -->
            <charset>UTF-8</charset>
        </encoder>
        <!-- 日志记录器的滚动策略，按日期、大小记录 -->
        <rollingPolicy
class="ch.qos.logback.core.rolling.TimeBasedRollingPolicy">
            <fileNamePattern>
                ${log.path}/warn/log-warn-%d{yyyy-MM-dd}.%i.log
```

```
        </fileNamePattern>
        <timeBasedFileNamingAndTriggeringPolicy
            class="ch.qos.logback.core.rolling.SizeAndTimeBasedFNATP">
            <maxFileSize>100MB</maxFileSize>
        </timeBasedFileNamingAndTriggeringPolicy>
        <!--日志文件保留天数-->
        <maxHistory>15</maxHistory>
    </rollingPolicy>
    <!-- 此日志文件只记录 WARN 级别的 -->
    <filter class="ch.qos.logback.classic.filter.LevelFilter">
        <level>warn</level>
        <onMatch>ACCEPT</onMatch>
        <onMismatch>DENY</onMismatch>
    </filter>
</appender>
<!-- 时间滚动输出 level 为 ERROR 日志 -->
<appender name="ERROR_FILE"
            class="ch.qos.logback.core.rolling.RollingFileAppender">
    <!-- 正在记录的日志文件的路径及文件名 -->
    <file>${log.path}/log_error.log</file>
    <!--日志文件输出格式-->
    <encoder>
        <pattern>
            %d{yyyy-MM-dd HH:mm:ss.SSS} [%thread] %-5level %logger{50} - %msg%n
        </pattern>
        <charset>UTF-8</charset> <!-- 此处设置字符集 -->
    </encoder>
    <!-- 日志记录器的滚动策略，按日期、大小记录 -->
    <rollingPolicy
            class="ch.qos.logback.core.rolling.TimeBasedRollingPolicy">
        <fileNamePattern>
            ${log.path}/error/log-error-%d{yyyy-MM-dd}.%i.log
        </fileNamePattern>
        <timeBasedFileNamingAndTriggeringPolicy
            class="ch.qos.logback.core.rolling.SizeAndTimeBasedFNATP">
            <maxFileSize>100MB</maxFileSize>
        </timeBasedFileNamingAndTriggeringPolicy>
        <!--日志文件保留天数-->
        <maxHistory>15</maxHistory>
    </rollingPolicy>
    <!-- 此日志文件只记录 ERROR 级别的 -->
    <filter class="ch.qos.logback.classic.filter.LevelFilter">
        <level>ERROR</level>
        <onMatch>ACCEPT</onMatch>
        <onMismatch>DENY</onMismatch>
    </filter>
</appender>
<!--部分内容省略-->
<!--开发环境:打印控制台-->
<springProfile name="dev">
```

```
        <logger name="com.nmys.view" level="debug"/>
    </springProfile>
    <root level="info">
        <appender-ref ref="CONSOLE"/>
        <appender-ref ref="DEBUG_FILE"/>
        <appender-ref ref="INFO_FILE"/>
        <appender-ref ref="WARN_FILE"/>
        <appender-ref ref="ERROR_FILE"/>
    </root>
    <!--生产环境:输出到文件-->
    <!--<springProfile name="pro">-->
    <!--<root level="info">-->
    <!--<appender-ref ref="CONSOLE" />-->
    <!--<appender-ref ref="DEBUG_FILE" />-->
    <!--<appender-ref ref="INFO_FILE" />-->
    <!--<appender-ref ref="ERROR_FILE" />-->
    <!--<appender-ref ref="WARN_FILE" />-->
    <!--</root>-->
    <!--</springProfile>-->
</configuration>
```

在 application.properties 中添加 Logback 的配置：

```
logging.config=classpath:logback-spring.xml
```

3. 测试 Logback

创建 LogController，内容如下。

示例代码 8-86　LogController.java

```
package com.example.demo.controller;
import lombok.extern.SLF4j.SLF4j;
import org.springframework.web.bind.annotation.GetMapping;
import org.springframework.web.bind.annotation.RestController;
@RestController
@SLF4j
public class LogController {
    @GetMapping("/test-logback")
    public void testLogback() {
        log.trace("Trace 日志...");
        log.debug("Debug 日志...");
        log.info("Info 日志...");
        log.warn("Warn 日志...");
        log.error("Error 日志...");
    }
}
```

其中我们使用了@SLF4j 注解，这个注解是 Lombok 提供的，在编译时会自动插入如下代码：

```
public class LogController {
    private static final org.SLF4j.Logger log =
           org.SLF4j.LoggerFactory.getLogger(LogController.class);
```

```
    //…
}
```

这样就省去了我们手写创建 Logger 对象的语句，非常方便。当然，在不使用 Lombok 时，上述代码需要我们手动编写。

启动项目，通过浏览器访问 http://localhost:8080/，查看控制台日志，可以看到如下信息：

```
…
2021-02-17 13:13:32.459  INFO 25620 --- [nio-8080-exec-1]
c.example.demo.controller.LogController  : Info 日志...
2021-02-17 13:13:32.460  WARN 25620 --- [nio-8080-exec-1]
c.example.demo.controller.LogController  : Warn 日志...
2021-02-17 13:13:32.460 ERROR 25620 --- [nio-8080-exec-1]
c.example.demo.controller.LogController  : Error 日志...
```

我们在代码中写了 TRACE、DEBUG、INFO、WARN、ERROR 五个级别的日志，而输出只有 3 个，分别是 INFO、WARN、ERROR，这是因为我们在 logback 配置文件里限定了日志的输出级别：

```
<!--输出到控制台-->
<appender name="CONSOLE" class="ch.qos.logback.core.ConsoleAppender">
    <!--此日志 appender 是为开发使用的，只配置最低级别，控制台输出的日志级别是大于或等于此级别的日志信息-->
    <filter class="ch.qos.logback.classic.filter.ThresholdFilter">
        <level>info</level>
    </filter>
    <encoder>
        <Pattern>${CONSOLE_LOG_PATTERN}</Pattern>
        <!-- 设置字符集 -->
        <charset>UTF-8</charset>
    </encoder>
</appender>
```

8.5.5 升级版 Log4j2

Apache Log4j 2 是 Log4j 的升级版，对 Log4j 的前身 Log4j 1.x 进行了重大改进，并提供了 Logback 中可用的许多改进，同时解决了 Logback 体系结构中的一些固有问题。

1. 主要改进

主要的变化有如下几个方面。

（1）API 分离

Log4j 2 的 API 与实现是分开的，从而使应用程序开发人员可以清楚地知道他们可以使用哪些类和方法，同时确保向前的兼容性。这使 Log4j 团队可以安全且兼容的方式改进实施。

Log4j API 是一个日志门面，可以与 Log4j 实现一起使用，也可以与其他日志实现（例如 Logback）一起使用。与 SLF4J 相比，Log4j API 具有多个优点：

- Log4j API 支持记录消息而不只是字符串。

- Log4j API 支持 Lambda 表达式。
- 与 SLF4J 相比，Log4j API 提供了更多的日志记录方法。
- 除了 SLF4J 支持的"参数化日志记录"格式外，Log4j API 还支持使用 java.text.MessageFormat 语法以及 printf 样式消息的事件。
- Log4j API 提供了 LogManager.shutdown()方法。基础的日志记录必须实现 Terminable 接口，该方法才能生效。
- 完全支持其他结构，例如标记、日志级别和 ThreadContext（aka MDC）。

（2）性能提升

Log4j 2 包含基于 LMAX Disruptor 库的下一代异步记录器。在多线程方案中，与 Log4j 1.x 和 Logback 相比，异步 Logger 的吞吐量高 18 倍，延迟降低了几个数量级。Log4j 2 明显优于 Log4j 1.x、Logback 和 java.util.logging，尤其是在多线程应用程序中。

（3）支持多种 API

Log4j 2 API 在提供更高的性能同时提供对 Log4j 1.2、SLF4J、Commons Logging 和 java.util.logging（JUL）API 的支持。

（4）避免锁定

使用 Log4j 2 API 的应用程序始终可以选择使用任何 SLF4J 的兼容库，这个功能通过 Log4j-to-SLF4j 适配器来实现与 SLF4J 兼容库的对接。

（5）自动重新加载配置

与 Logback 一样，Log4j 2 可以在修改后自动重新加载其配置。比 Logback 更好的是，它不会在进行重新配置时丢失日志事件。

（6）Java 8 Lambda 支持

以前如果构建日志消息的成本很高，那么通常会在构建消息之前显式检查是否启用了请求的日志级别。例如：

```
if(log.isDebugEnabled()){
   log.debug("Debug 日志...");
}
```

在 Java 8 上运行的客户端代码可以受益于 Log4j 的 Lambda 支持。如果未启用请求的日志级别，由于 Log4j 不会评估 Lambda 表达式，因此可以用更少的代码获得相同的效果。

（7）可以自定义日志级别

在 Log4j 2 中，可以通过代码或配置轻松定义自定义日志级别。

总之，Log4j 2 做了大量的改进和优化工作，是目前最先进、性能十分优良的一代日志框架。关于 Log4j2 的更多介绍，请查阅官方说明文档：https://logging.apache.org/log4j/2.x/index.html。

2. 在 Spring Boot 中集成 Log4j 2

以前面的 logback-demo 项目为基础，对其进行修改，以支持 Log4j 2。首先复制 logback-demo 项目，重命名为 log4j2-demo，然后修改依赖项为：

```
    <dependency>
        <groupId>org.springframework.boot</groupId>
        <artifactId>spring-boot-starter-web</artifactId>
        <!-- Spring Boot 默认是用 logback 日志框架的，所以需要排除 Logback，不然会出现 jar
依赖冲突的报错 -->
        <exclusions>
        <!-- 将 spring-boot 的默认 logging 包排除-->
            <exclusion>
                <groupId>org.springframework.boot</groupId>
                <artifactId>spring-boot-starter-logging</artifactId>
            </exclusion>
        </exclusions>
    </dependency>
    <!--引入 spring-boot-starter-log4j2-->
    <dependency>
        <groupId>org.springframework.boot</groupId>
        <artifactId>spring-boot-starter-log4j2</artifactId>
    </dependency>
```

由于 Spring Boot 默认集成 Logback 框架，因此切换为使用 Log4j 2 时需要先将 spring-boot-starter-logging 包排除，否则会引起 jar 包依赖冲突，然后将 Spring Boot 官方对 Log4j 2 的封装包 spring-boot-starter-log4j2 引入即可。

3. 配置 Log4j 2

在 resources 目录下新建 log4j2.xml 配置文件，示例内容如下。

示例代码 8-87　log4j2.xml

```
<?xml version="1.0" encoding="UTF-8"?>
    <configuration status="info">
        <appenders>
            <!--输出控制台的配置-->
            <Console name="STDOUT">
                <!--控制台只输出 level 及以上级别的信息（onMatch)，其他的直接拒绝
(onMismatch) -->
                <ThresholdFilter level="info" onMatch="ACCEPT" onMismatch="DENY"/>
                <!-- 输出日志的格式-->
                <PatternLayout pattern="%d{yyyy-MM-dd HH:mm:ss.SSS} [%t] %-5level
[%logger{50}:%L] - %msg%n"  charset="UTF-8"/>
            </Console>
            <!--sql 专用输出-->
            <Console name="STDOUT2">
                <ThresholdFilter level="info" onMatch="ACCEPT" onMismatch="DENY"/>
                <PatternLayout pattern=
              "%d{yyyy-MM-dd HH:mm:ss.SSS} [%t] %-5level [%logger{50}:%L] - %msg%n"
charset="UTF-8"/>
            </Console>
            <!--下面的 XXX 可以修改为自定义的文件名-->
            <RollingRandomAccessFile name="FILE-INFO"
fileName="logs/XXX-info.log" filePattern="logs/XXX-info.%d{yyyy-MM-dd-HH}.log">
```

```
            <ThresholdFilter level="info" onMatch="ACCEPT" onMismatch="DENY"/>
            <PatternLayout pattern="%d{yyyy-MM-dd HH:mm:ss.SSS} [%t] %-5level
[%logger{50}:%L] - %msg%n" charset="UTF-8"/>
            <TimeBasedTriggeringPolicy interval="1"/>
            <DefaultRolloverStrategy max="1">
                <Delete basePath="logs" maxDepth="2">
                    <IfFileName glob="*XXX-info.*.log"/>
                    <IfLastModified age="1d"/>
                </Delete>
            </DefaultRolloverStrategy>
        </RollingRandomAccessFile>
        <RollingRandomAccessFile name="FILE-DEBUG"
                 fileName="logs/XXX-debug.log"
                filePattern="logs/XXX-debug.%d{yyyy-MM-dd-HH}.log">
            <ThresholdFilter level="debug" onMatch="ACCEPT" onMismatch="DENY"/>
            <PatternLayout pattern="%d{yyyy-MM-dd HH:mm:ss.SSS} [%t] %-5level
[%logger{50}:%L] - %msg%n" charset="UTF-8"/>
            <TimeBasedTriggeringPolicy interval="1"/>
            <DefaultRolloverStrategy max="1">
                <Delete basePath="logs" maxDepth="2">
                    <IfFileName glob="*XXX-debug.*.log"/>
                    <IfLastModified age="1d"/>
                </Delete>
            </DefaultRolloverStrategy>
        </RollingRandomAccessFile>
        <RollingRandomAccessFile name="FILE-WARN"
fileName="logs/XXX-warn.log" filePattern="logs/XXX-warn.%d{yyyy-MM-dd-HH}.log">
            <ThresholdFilter level="warn" onMatch="ACCEPT" onMismatch="DENY"/>
            <PatternLayout pattern="%d{yyyy-MM-dd HH:mm:ss.SSS} [%t] %-5level
[%logger{50}:%L] - %msg%n" charset="UTF-8"/>
            <TimeBasedTriggeringPolicy interval="1"/>
            <DefaultRolloverStrategy max="1">
                <Delete basePath="logs" maxDepth="2">
                    <IfFileName glob="*XXX-warn.*.log"/>
                    <IfLastModified age="1d"/>
                </Delete>
            </DefaultRolloverStrategy>
        </RollingRandomAccessFile>
        <RollingRandomAccessFile name="FILE-ERROR"
             fileName="logs/XXX-error.log"
            filePattern="logs/XXX-error.%d{yyyy-MM-dd-HH}.log">
            <ThresholdFilter level="error" onMatch="ACCEPT" onMismatch="DENY"/>
            <PatternLayout pattern="%d{yyyy-MM-dd HH:mm:ss.SSS} [%t] %-5level
[%logger{50}:%L] - %msg%n" charset="UTF-8"/>
            <TimeBasedTriggeringPolicy interval="1"/>
            <DefaultRolloverStrategy max="1">
                <Delete basePath="logs" maxDepth="2">
                    <IfFileName glob="*XXX-error.*.log"/>
                    <IfLastModified age="1d"/>
                </Delete>
```

```
            </DefaultRolloverStrategy>
        </RollingRandomAccessFile>
    </appenders>
    <loggers>
        <!-- 将业务 dao 接口填写进去，并用控制台输出 -->
        <AsyncLogger name="com.example.demo.dao" level="info"
                     additivity="false">
            <AppenderRef ref="STDOUT2"/>
        </AsyncLogger>
        <!--includeLocation="true" :打印出行号-->
        <AsyncRoot level="info" includeLocation="true">
            <!--<AppenderRef ref="FILE-INFO" />
            <AppenderRef ref="FILE-WARN" />
            <AppenderRef ref="FILE-ERROR" />-->
            <AppenderRef ref="STDOUT"/>
        </AsyncRoot>
    </loggers>
</configuration>
```

在 application.properties 中添加 Log4j 2 文件的配置项：

```
logging.config=classpath:log4j2.xml
```

4. 测试 Log4j 2

修改 LogController 类，将@SLF4j 注解删除，添加@Log4j2 注解：

```
package com.example.demo.controller;
import lombok.extern.log4j.Log4j2;
import …
@RestController
@Log4j2
public class LogController {
    @GetMapping("/test-logback")
    public void testLogback() {
        log.trace("Trace 日志...");
        log.debug("Debug 日志...");
        log.info("Info 日志...");
        log.warn("Warn 日志...");
        log.error("Error 日志...");
    }
}
```

最后运行项目进行测试，报异常错误，提示无法找到类 com.lmax.disruptor.EventFactory，这是因为 Log4j 2 使用 LMAX Disputor 异步记录器，而这里正好缺少这个包。修复这个问题很简单，在 pom.xml 中引入依赖即可：

```
<!--引入 lmax disruptor 异步记录器，Log4j2 需要-->
<dependency>
    <groupId>com.lmax</groupId>
    <artifactId>disruptor</artifactId>
    <version>3.3.0</version>
```

```
</dependency>
```

重新运行项目，通过浏览器访问 http://localhost:8080/，在控制台输出中我们将会看到如下内容：

```
    2021-02-17 14:25:25.236 [http-nio-8080-exec-1] INFO
[com.example.demo.controller.LogController:17] - Info 日志...
    2021-02-17 14:25:25.236 [http-nio-8080-exec-1] WARN
[com.example.demo.controller.LogController:18] - Warn 日志...
    2021-02-17 14:25:25.236 [http-nio-8080-exec-1] ERROR
[com.example.demo.controller.LogController:19] - Error 日志...
```

我们设定的日志级别为 INFO，因此控制台只打印 INFO 级别及更高级别 WARN 和 ERROR 的日志。

第 9 章

Spring Boot 打包、部署、监控

本章将讲解 Spring Boot 应用程序的打包、部署和监控，介绍 jar 包和 war 包两种不同的打包方式，以及部署到云服务器、Docker 容器，配置热部署来提高开发效率。最后通过两种监控方式简要介绍如何对 Spring Boot 应用进行监控：一种是基于 Actuator 监控性能指标，另一种是使用 APM 工具进行全链路的监控和追踪。

9.1 构建可执行 jar 包部署到云服务器

在 3.3.4 小节中，我们学习了如何使用 spring-boot-maven-plugin 插件，运用命令“mvn clean package”或者“mvnw clean package”来对 Spring Boot 应用程序构建成可执行 jar 包，并在本地环境中运行测试，在 cmd 中执行“java -jar xxx.jar”来运行 Spring Boot 程序，但是一旦关闭 cmd 窗口，程序便被关闭而无法访问了，如果将其运行在服务器上肯定是不行的，因为程序是要一直运行并对外提供服务的，那么在服务器上怎么做呢？

一般我们开发的应用程序都是部署在 Linux 服务器上的，通常使用 xshell 等 ssh 工具登录到服务器。同样地，在 shell 终端中用 “java -jar xxx.jar”命令也能启动，但是问题同样存在，一旦与服务器的 ssh 连接断开、关闭 xshell 等工具，那么通过这个命令运行的程序也会相应地被终止。

想要让程序一直运行在服务器上，我们需要使用“nohup COMMAND &”命令让程序一直保持在后台运行。

接下来以第 8 章中的 shiro-demo 项目为例，介绍如何将其运行在阿里云 ECS 服务器上。

9.1.1 环境准备

将 shiro-demo 项目使用“mvn clean package”命令打包，将在 target 目录下生成可执行 jar 包 shiro-demo-0.0.1-SNAPSHOT.jar，准备好下一步要将其上传到服务器上。

准备好阿里云ECS服务器（也可以是其他云厂商提供的服务器），公网IP地址为121.199.22.25，操作系统为Debian 9.1，并安装配置好Java环境，参考3.2.1小节。

9.1.2 使用 XShell 连接到云服务器

下载并安装 XShell 客户端，下载地址为 https://www.netsarang.com/zh/xshell/。新建会话，填写名称、主机地址，如图 9-1 所示。单击左侧导航栏中的用户身份验证，输入用户名和密码，单击“连接”按钮登录到云服务器，如图 9-2 所示。出现未知主机秘钥的警告提示框，单击“接受并保存”按钮，稍等片刻就可以成功登入服务器了。

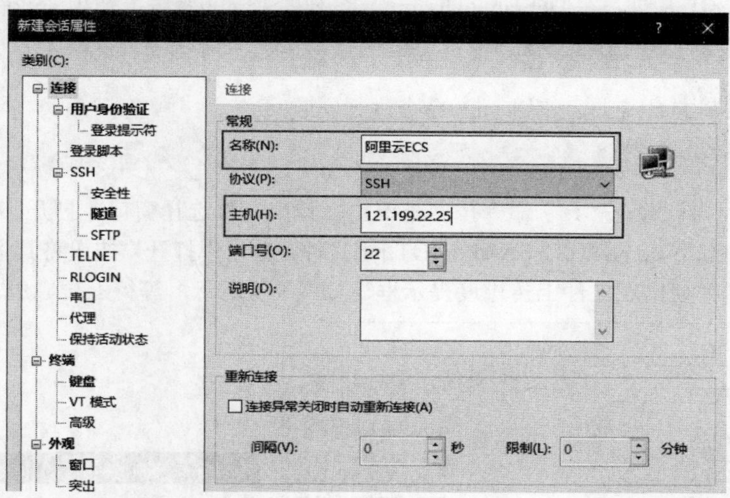

图 9-1

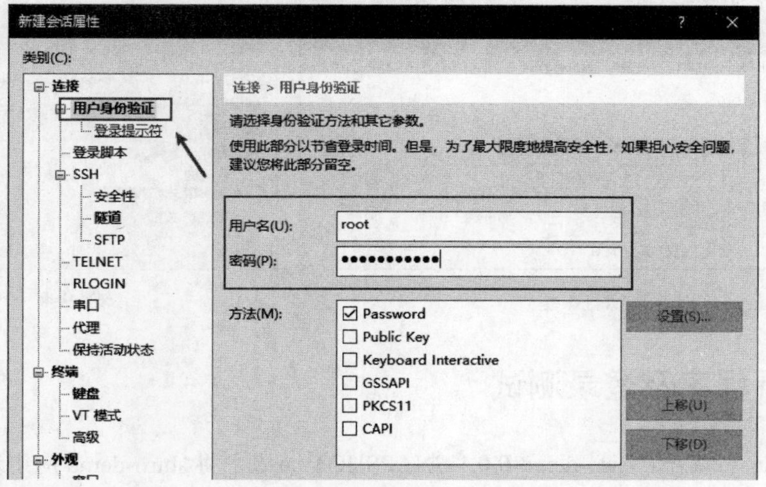

图 9-2

ssh 连接到云服务器后会有如下提示：

```
Connecting to 121.199.22.25:22...
Connection established.
```

```
To escape to local shell, press 'Ctrl+Alt+]'.
WARNING! The remote SSH server rejected X11 forwarding request.
Linux debian 4.9.0-14-amd64 #1 SMP Debian 4.9.246-2 (2020-12-17) x86_64
Welcome to Alibaba Cloud Elastic Compute Service !
Last login: Wed Feb 17 15:39:27 2021 from 49.52.10.216
root@debian:~#
```

9.1.3 上传 jar 包

在 XShell 连接终端窗口，“rz”命令上传 jar 包，此时可能会提示 sz 命令不存在，因为服务器还没有安装 lrzsz 工具包，在 Debian（Ubuntu）系列的 Linux 系统上使用 apt 命令进行安装：

```
sudo apt install -y lrzsz
```

在 CentOS 系列的系统上，使用 yum 命令进行安装：

```
Sudo yum install -y lrzsz
```

安装完成后，再次输入“rz”命令按回车键，会弹出文件选择窗口，打开 shiro-demo 项目的 target 目录，选中 shiro-demo-0.0.1-SNAPSHOT.jar 文件，单击“打开”按钮将其上传到服务器，如图 9-3 所示。接下来会出现文件上传进度提示框，如图 9-4 所示，等待上传完成。

图 9-3

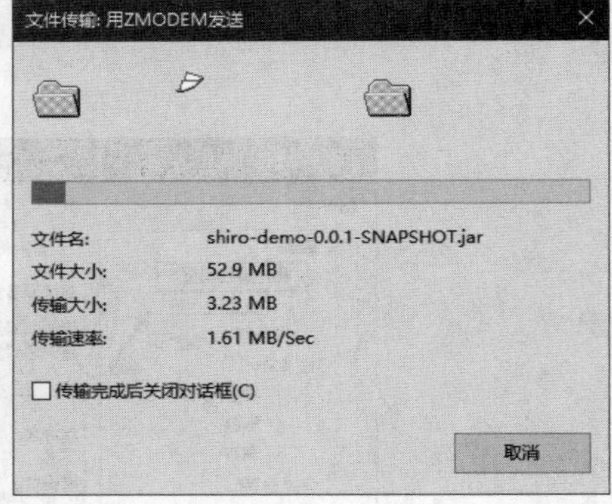

图 9-4

9.1.4 运行程序及登录测试

执行“nohup java -jar shiro-demo-0.0.1-SNAPSHOT.jar &”将 shiro-demo 应用运行在服务器后台，如图 9-5 所示。执行后会输出一个类似“[1] 5103”的内容，这是 shiro-demo 应用在后台运行的进程号。可以通过“ps -ef | grep shiro-demo*”来查看程序在后台运行的状态信息：

```
root@debian:~# ps -ef | grep shiro-demo*
root      5592  3920  0 16:01 pts/0    00:00:00 grep
shiro-demo-0.0.1-SNAPSHOT.jar
```

```
[1]+  Exit 1                  nohup java -jar shiro-demo-0.0.1-SNAPSHOT.jar
root@debian:~#
```

```
阿里云ECS - Xshell 7 (Free for Home/School)
1 阿里云ECS
root@debian:~# ls shiro-demo-0.0.1-SNAPSHOT.jar
shiro-demo-0.0.1-SNAPSHOT.jar
root@debian:~# java -version
java version "1.8.0_121"
Java(TM) SE Runtime Environment (build 1.8.0_121-b13)
Java HotSpot(TM) 64-Bit Server VM (build 25.121-b13, mixed mode)
root@debian:~# nohup java -jar shiro-demo-0.0.1-SNAPSHOT.jar &
[1] 5103
root@debian:~# nohup: ignoring input and appending output to 'nohup.out'
root@debian:~#
```

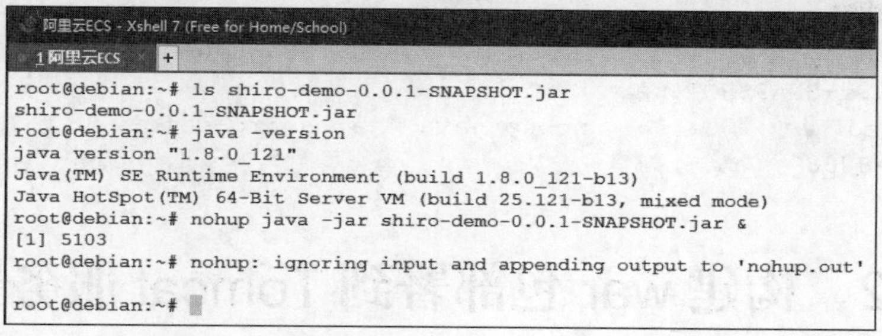

图 9-5

结果意外发现，刚刚执行的 nohup 命令状态是“Exit 1”，退出运行了。此时我们直接通过“java -jar shiro-demo-0.0.1-SNAPSHOT.jar”来查看运行日志输出，发现 8080 端口已经被占用：

```
***************************
APPLICATION FAILED TO START
***************************
Description:
Web server failed to start. Port 8080 was already in use.
Action:
Identify and stop the process that's listening on port 8080 or configure this
application to listen on another port.
```

解决端口占用问题，可以在执行命令时加-server.port=xxx 来修改端口号：

```
nohup java -jar shiro-demo-0.0.1-SNAPSHOT.jar -server.port=8088 &
```

再来用 ps 命令查看一下后台进程信息：

```
root@debian:~# ps -ef | grep shiro-demo*
root      6288  3920 95 16:11 pts/0    00:00:19 java -jar
shiro-demo-0.0.1-SNAPSHOT.jar -server.port=8088
root      6335  3920  0 16:12 pts/0    00:00:00 grep
shiro-demo-0.0.1-SNAPSHOT.jar
```

可以看到 shiro-demo-0.0.1-SNAPSHOT.jar 运行在 pid=6288 的进程上。

使用 curl 命令来测试一下登录：

```
root@debian:~# curl -H "Content-Type: application/json" -X POST -d
'{"username": "admin", "password": "123"}' http://121.199.22.25:8088/user/login
{"code":1,"msg":"登录成功！
","data":{"id":null,"username":"admin","password":"123","timestamp":null}}
```

可以看到返回了登录成功的消息，说明我们的应用程序已经正常运行在服务器的 8088 端口上。

如果需要关闭它，就使用“ps -ef | grep shiro-demo*”找到运行的 pid，然后用“kill -p <pid>”命令执行杀死：

```
root@debian:~# ps -ef | grep shiro-demo*
root      6654  3920  9 16:16 pts/0    00:00:28 java -jar
shiro-demo-0.0.1-SNAPSHOT.jar --server.port=8088
root      6991  3920  0 16:21 pts/0    00:00:00 grep
```

```
shiro-demo-0.0.1-SNAPSHOT.jar
    root@debian:~# kill -9 6654
    root@debian:~# ps -ef | grep shiro-demo*
    root      7007  3920  0 16:21 pts/0    00:00:00 grep
shiro-demo-0.0.1-SNAPSHOT.jar
    [1]+  Killed                  nohup java -jar shiro-demo-0.0.1-SNAPSHOT.jar
--server.port=8088
```

9.2 构建 war 包部署到 Tomcat 服务器

9.2.1 改造 Spring Boot 项目

仍然以 shiro-demo 项目为例，修改 pom.xml 中的 packaging 目标为 war，并将内置 Tomcat 排除，引入外部 Tomcat 依赖 spring-boot-starter-tomcat：

```
…
<groupId>com.example</groupId>
<artifactId>shiro-demo</artifactId>
<version>0.0.1-SNAPSHOT</version>
<packaging>war</packaging>
…
<dependencies>
    …
    <!--排除内置 Tomcat 依赖-->
    <dependency>
        <groupId>org.springframework.boot</groupId>
        <artifactId>spring-boot-starter-web</artifactId>
        <exclusions>
            <exclusion>
                <groupId>org.springframework.boot</groupId>
                <artifactId>spring-boot-starter-tomcat</artifactId>
            </exclusion>
        </exclusions>
    </dependency>
    <!-- 添加外部 Tomcat 包 -->
    <dependency>
        <groupId>org.springframework.boot</groupId>
        <artifactId>spring-boot-starter-tomcat</artifactId>
        <!-- 这里一定要设置为 provided -->
        <scope>provided</scope>
    </dependency>
</dependencies>
```

修改程序入口类 ShiroDemoApplication.java，继承自 SpringBootServletInitializer，并实现 configurer 方法：

示例代码 9-1 改造 ShiroDemoApplication 引入 Servlet

```
package com.example.shiro;
import …
@SpringBootApplication
public class ShiroDemoApplication extends SpringBootServletInitializer {
    public static void main(String[] args) {
        SpringApplication.run(ShiroDemoApplication.class, args);
    }
    /**
     * 重写 configure 方法
     */
    @Override
    protected SpringApplicationBuilder configure(SpringApplicationBuilder builder)
{
        return builder.sources(ShiroDemoApplication.class);
    }
}
```

然后使用“mvnw clean package”执行构建，将在 target 目录下生成“shiro-demo-0.0.1-SNAPSHOT.war”文件。接下来介绍如何在云服务器上部署到 Tomcat 容器。

9.2.2 下载安装 Tomcat

使用 XShell 连接到云服务器，下载 Tomcat 9.0 压缩包：

```
root@debian:~# wget https://downloads.apache.org/tomcat/tomcat-9/v9.0.43/
bin/apache-tomcat-9.0.43.tar.gz
```

将 Tomcat 压缩包解压，并移动到/opt/tomcat 目录：

```
root@debian:~# tar -zxvf apache-tomcat-9.0.43.tar.gz
root@debian:~# mv apache-tomcat-9.0.43 /opt/tomcat
```

9.2.3 上传 war 包

使用 rz 命令将 shiro-demo-0.0.1-SNAPSHOT.war 上传，然后将其移动到/opt/tomcat/webapps/目录下：

```
root@debian:~# mv shiro-demo-0.0.1-SNAPSHOT.war /opt/tomcat/webapps/
```

启动 Tomcat：

```
root@debian:~# sh /opt/tomcat/bin/startup.sh
Using CATALINA_BASE:   /opt/tomcat
Using CATALINA_HOME:   /opt/tomcat
Using CATALINA_TMPDIR: /opt/tomcat/temp
Using JRE_HOME:        /usr/local/java/jdk1.8.0_121/jre
Using CLASSPATH:
/opt/tomcat/bin/bootstrap.jar:/opt/tomcat/bin/tomcat-juli.jar
Using CATALINA_OPTS:
```

```
Tomcat started.
```

通过浏览器访问错误，超链接引用无效。查看 Tomcat 运行状态，如图 9-6 所示。

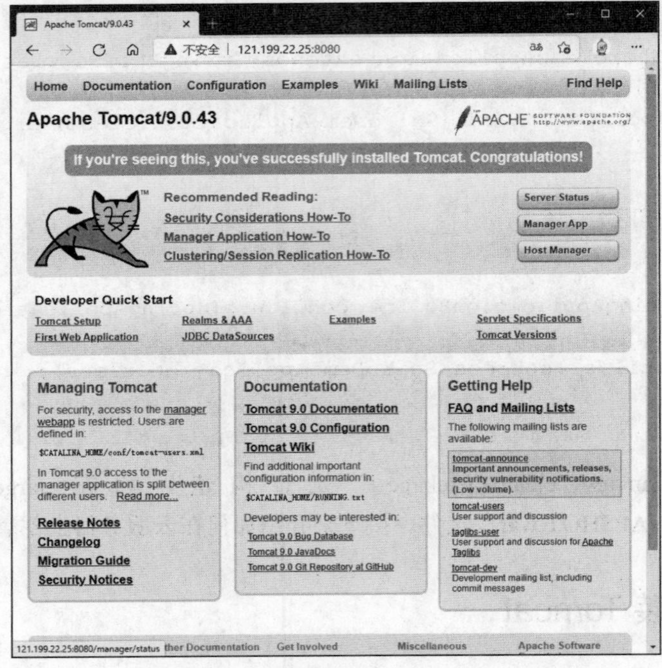

图 9-6

出现 Tomcat 介绍页时说明 Tomcat 启动成功了，我们的 shiro-demo-0.0.1-SNAPSHOT.war 包会被 Tomcat 自动解包放置在 webapps 目录下。

9.2.4 配置 Tomcat

修改/opt/tomcat/conf/server.xml 文件，在<Host>节点中添加<Context>上下文配置：

```
root@debian:~# sudo vim /opt/tomcat/conf/server.xml
…
<Server>
 …
<Service>
 …
<Engine>
 …
    <Host name="localhost"  appBase="webapps"
          unpackWARs="true" autoDeploy="true">
        <!-- SingleSignOn valve, share authentication between Web applications
             Documentation at: /docs/config/valve.html -->
        <!--
        <Valve className="org.apache.catalina.authenticator.SingleSignOn" />
         -->
        <!-- Access log processes all example.
```

```
            Documentation at: /docs/config/valve.html
            Note: The pattern used is equivalent to using pattern="common" -->
        <Valve className="org.apache.catalina.valves.AccessLogValve"
directory="logs"
              prefix="localhost_access_log" suffix=".txt"
              pattern="%h %l %u %t &quot;%r&quot; %s %b" />
        <Context path="/shiro-demo" docBase="shiro-demo-0.0.1-SNAPSHOT"
reloadable="reloadable"/>
        </Host>
    </Engine>
    </Services>
    </Server>
```

在<Context>配置中，path 填写需要映射的上下文路径，docBase 填写项目的名字（war 包的名字去除.war 后缀）。修改完成后保存并重新启动 Tomcat。

通过上面的配置，shiro-demo 项目将会运行在 http://localhost:8080/shiro-demo 路径上。

9.2.5 测试登录

使用 curl 命令测试一下登录：

```
    root@debian:~# curl -H "Content-Type: application/json" -X POST -d
'{"username": "admin", "password": "123"}'
http://121.199.22.25:8080/shiro-demo/user/login
    {"code":1,"msg":"登录成功!
","data":{"id":null,"username":"admin","password":"123","timestamp":null}}
```

至此，构建 war 包并部署在云服务器的 Tomcat 工作已经完成。

9.3 使用 Docker 容器部署

9.3.1 什么是 Docker 容器

Docker 容器是一个开源的应用容器引擎，可以让开发者以统一的方式打包他们的应用以及依赖包到一个可移植的容器中，然后发布到任何安装了 Docker 引擎的服务器上（包括流行的 Linux 机器、Windows 机器），也可以实现虚拟化。容器是完全使用沙箱机制，相互之间不会有任何接口（类似 iPhone 的 app）。几乎没有性能开销，可以很容易地在机器和数据中心运行。最重要的是它们不依赖于任何语言、框架（包括系统）。

Docker 依托于虚拟化技术，它的每一个容器都可以看作一个精简的虚拟机操作系统。它使用镜像（映像）文件作为模板，创建并启动一个虚拟化的精简操作系统，这个系统运行在一个被隔离的沙箱（Sandbox）之中，与宿主操作系统互不影响，不会对宿主机器造成污染。

借助 Docker，你可将容器当作轻巧、模块化的虚拟机使用。同时，你还将获得高度的灵活性，从而实现对容器的高效创建、部署及复制，并能将其从一个环境顺利迁移至另一个环境。

近年来，随着云计算时代的到来，容器化技术变得十分火热，基于云环境的容器化部署成为目前应用程序部署的主流方向。因此，学习使用 Docker 这类容器来部署我们的 Spring Boot 应用程序是十分有必要的。

本节将着重介绍如何使用 Docker 来部署 Spring Boot 应用，而非重点介绍 Docker。关于 Docker 的概念、使用，请读者查看官方网站。

9.3.2 下载并安装 Docker

Docker 的安装步骤在官方文档中已经详细说明，本书不再做介绍。

不同操作系统安装 Docker 的地址如下：

- Debian：https://docs.docker.com/engine/install/debian/。
- CentOS：https://docs.docker.com/engine/install/centos/。
- Ubuntu：https://docs.docker.com/engine/install/ubuntu/。
- Windows：https://docs.docker.com/docker-for-windows/install/。

安装完成后，使用 docker –version 查看安装版本：

```
$ docker --version
Docker version 20.10.2, build 2291f61
```

9.3.3 编写 Dockerfile

在 shiro-demo 项目根目录下新建名为“Dockerfile”的文件，内容如下：

示例代码 9-2 Dockerfile

```
FROM openjdk:8-jre
EXPOSE 8080
ARG JAR_FILE
ADD target/${JAR_FILE} /app.jar
ENTRYPOINT ["java", "-jar","app.jar"]
```

参数说明：

- FROM：基于 openjdk:8-jre 镜像构建。
- EXPOSE：监听 8080 端口。
- ARG：引用 plugin 中配置的 JAR_FILE 文件。
- ADD：将当前 target 目录下的 jar 放置在根目录下，命名为 app.jar，推荐使用绝对路径。
- ENTRYPOINT：执行命令 java -jar /app.jar。

9.3.4 引入 dockerfile-maven-plugin 插件

修改 pom.xml，在<plugins>节点配置中引入 dockerfile-maven-plugin 插件。

```
<?xml version="1.0" encoding="UTF-8"?>
```

```
<project xmlns:xsi="http://www.w3.org/2001/XMLSchema-instance"
xmlns="http://maven.apache.org/POM/4.0.0"
        xsi:schemaLocation="http://maven.apache.org/POM/4.0.0
https://maven.apache.org/xsd/maven-4.0.0.xsd">
    …
    <properties>
        <java.version>1.8</java.version>
        <docker.image.prefix>example</docker.image.prefix>
    </properties>
    …
    <build>
        <plugins>
            …
            <plugin>
                <groupId>com.spotify</groupId>
                <artifactId>dockerfile-maven-plugin</artifactId>
                <version>1.4.13</version>
                <executions>
                    <execution>
                        <id>default</id>
                        <goals>
                            <goal>build</goal>
                            <goal>push</goal>
                        </goals>
                    </execution>
                </executions>
                <configuration>
                    <!--禁用谷歌容器仓库-->
                    <googleContainerRegistryEnabled>
                        false
                        </googleContainerRegistryEnabled>
                    <repository>
                        ${docker.image.prefix}/${project.artifactId}
                    </repository>
                    <tag>${project.version}</tag>
                    <buildArgs>
                        <JAR_FILE>${project.build.finalName}.jar</JAR_FILE>
                    </buildArgs>
                </configuration>
            </plugin>
        </plugins>
    </build>
</project>
```

参数说明：

- repository：指定 Docker 镜像的 repo 名字，要展示在 docker images 中。
- tag：指定 Docker 镜像的 tag，不指定 tag 则默认为 latest。
- buildArgs：指定一个或多个变量，传递给 Dockerfile，在 Dockerfile 中通过 ARG 指令进行引用。JAR_FILE 指定 jar 文件名。

另外，可以在 execution 中同时指定 build 和 push 目标。当运行 mvn package 时，会自动执行 build 目标，构建 Docker 镜像。${docker.image.prefix}变量是用来控制 docker image 构建的目标仓库名的，image 的名字设置为${project.artifactId}变量值，即项目的名字（artifactId），image 的 tag 标记设置为${project.version}（项目的版本号），构建的参数设置了一个“JAR_FILE”，其值为项目构建后的名字，例如 shiro-demo-0.0.1-SNAPSHOT.jar，这个值会提供给上一步中的 Dockerfile 使用，“ARG JAR_FILE”一行表示使用外部参数“JAR_FILE”。这样配置就会变得十分通用，在任何项目里都可以使用这个配置而无须改动。

9.3.5 执行项目构建

使用 mvn clean package 命令执行构建，这时会先执行 spring-boot-maven-plugin 插件进行编译和打包（生成可执行 jar），之后再执行 dockerfile-maven-plugin 插件进行 Docker 镜像的构建，它的作用等同于手动执行构建命令“docker build. --build-arg [ARG_LIST]--tag <repository:tagName>”并传入所需的参数，这个插件代替手工完全自动化工作。

执行构建完成后会输出如下内容：

```
…
[INFO] Results:
[INFO]
[INFO] Tests run: 2, Failures: 0, Errors: 0, Skipped: 0
[INFO]
[INFO]
[INFO] --- maven-jar-plugin:3.2.0:jar (default-jar) @ shiro-demo ---
[INFO] Building jar: D:\code\shiro-demo\target\shiro-demo-0.0.1-SNAPSHOT.jar
[INFO]
[INFO] --- spring-boot-maven-plugin:2.3.6.RELEASE:repackage (repackage) @
shiro-demo ---
[INFO] Replacing main artifact with repackaged archive
[INFO]
[INFO] --- dockerfile-maven-plugin:1.4.13:build (default) @ shiro-demo ---
[INFO] Google Container Registry support is disabled
[INFO] dockerfile: null
[INFO] contextDirectory: D:\code\shiro-demo
[INFO] Building Docker context D:\code\shiro-demo
[INFO] Path(dockerfile): null
[INFO] Path(contextDirectory): D:\code\shiro-demo
[INFO]
[INFO] Image will be built as example/shiro-demo:0.0.1-SNAPSHOT
[INFO]
[INFO] Step 1/5 : FROM java:8
[INFO]
[INFO] Pulling from library/java
[INFO] Digest:
sha256:c1ff613e8ba25833d2e1940da0940c3824f03f802c449f3d1815a66b7f8c0e9d
[INFO] Status: Image is up to date for java:8
[INFO]  ---> d23bdf5b1b1b
[INFO] Step 2/5 : EXPOSE 8080
```

```
    [INFO]
    [INFO]  ---> Using cache
    [INFO]  ---> bcbdf948b376
    [INFO] Step 3/5 : ARG JAR_FILE
    [INFO]
    [INFO]  ---> Using cache
    [INFO]  ---> 9cc0eba0aba1
    [INFO] Step 4/5 : ADD target/${JAR_FILE} /${JAR_FILE}
    [INFO]
    [INFO]  ---> c3d227e9ab0b
    [INFO] Step 5/5 : ENTRYPOINT ["java", "-jar","${JAR_FILE}"]
    [INFO]
    [INFO]  ---> Running in 844659f17c21
    [INFO] Removing intermediate container 844659f17c21
    [INFO]  ---> 84206175550a
    [INFO] Successfully built 84206175550a
    [INFO] Successfully tagged example/shiro-demo:0.0.1-SNAPSHOT
    [INFO]
    [INFO] Detected build of image with id 84206175550a
    [INFO] Building jar:
D:\code\shiro-demo\target\shiro-demo-0.0.1-SNAPSHOT-docker-info.jar
    [INFO] Successfully built example/shiro-demo:0.0.1-SNAPSHOT
    ------------------------------------------------------------------------
    [INFO] BUILD SUCCESS
    ---------------------------------------------------------
    [INFO] Total time:  02:35 min
    [INFO] Finished at: 2021-02-17T19:03:28+08:00
    ------------------------------------------------------------------------
```

这意味着项目的编译、构建、打包、镜像制作都已经成功。从上面的输出中可以看到这么一行内容：

```
    [INFO] Image will be built as example/shiro-demo:0.0.1-SNAPSHOT
```

也就是说，镜像的仓库名为 example，镜像名为 shiro-demo，tag 为 0.0.1-SNAPSHOT。使用 docker images 命令查看生成的镜像信息能够得到验证：

```
$ docker images
REPOSITORY           TAG              IMAGE ID       CREATED        SIZE
example/shiro-demo   0.0.1-SNAPSHOT   84206175550a   2 days ago     699MB
```

9.3.6　启动容器和访问

完成前面的步骤，创建镜像后，我们就可以直接创建 shiro-demo 容器来运行了。

```
docker run --name shiro-demo -d -p 8080:8080 example/shiro-demo:0.0.1-SNAPSHOT
```

其中的参数说明如下：

- --name：指定容器的名字。
- -d：表示在后台运行。

- -p：指定端口号，第一个 8080 为容器内部的端口号，第二个 8080 为外界访问的端口号，将容器内的 8080 端口号映射到外部的 8080 端口号。
- example/shiro-demo:0.0.1-SNAPSHOT：镜像名+版本号。

运行之后会输出一个字符串，该字符串是容器的 id 标识。使用 docker ps 命令查看运行的容器情况：

```
    $ docker ps
    CONTAINER  ID  IMAGE  COMMAND    CREATED    STATUS     PORTS
    NAMES
    19a1a797541f   example/shiro-demo:0.0.1-SNAPSHOT   "java -jar app.jar"   2
days ago   Up About a minute   0.0.0.0:8080->8080/tcp   shiro-demo
```

此时 shiro-demo 容器已经正常运行，并暴露 8080 端口到外部。使用 curl 进行登录测试：

```
    $ curl -H "Content-Type: application/json" -X POST -d '{"username": "admin",
"password": "123"}' http://121.199.22.25:8080/shiro-demo/user/login
    {"code":1,"msg":"登录成功!
","data":{"id":null,"username":"admin","password":"123","timestamp":null}}
```

登录成功，至此使用容器化部署 Spring Boot 应用程序就完成了。

9.4 配置热部署

在开发过程中，可能随时要进行启动测试，修改代码后免不了重新编译、启动 Spring Boot 应用，频繁的修改和重启势必会降低开发效率，热部署（加载）随之出现。热部署就是正在运行状态的应用，修改了源码之后，在不重新启动的情况下能够自动把增量内容编译并部署到服务器上，使得修改立即生效。热部署解决的问题有两个：

- 一是在开发的时候修改代码后不需要重启应用就能看到效果，大大提升开发效率。
- 二是生产上运行的程序可以在不停止运行的情况下进行升级，不影响用户使用。

9.4.1 Spring Boot 开启热部署

Spring 为开发者提供了一个名为 spring-boot-devtools 的模块来使 Spring Boot 应用支持热部署，提高开发者的开发效率，无须手动重启 Spring Boot 应用。devtools 可以实现页面热部署，即页面修改后立即生效（可以直接在 application.properties 文件中配置 spring.thymeleaf.cache=false 实现）；实现类文件热部署，但是类文件修改后不会立即生效；实现对属性文件的热部署。

devtools 会监听 classpath 下的文件变动，并且会立即重启应用（发生在保存时机）。注意：因为其采用的是虚拟机机制，所以该项重启是很快的。

配置了 devtools 之后再修改 java 文件也就支持了热启动，不过这种方式是属于项目重启（速度比较快的项目重启），会清空 session 中的值，也就是说如果有用户登录，项目重启后需要重新登录。

默认情况下，/META-INF/maven、/META-INF/resources、/resources、/static、/templates、/public这些文件夹下的文件修改不会使应用重启，但是会重新加载（devtools 内嵌了一个 LiveReload server，当资源发生改变时，浏览器刷新）。

若要开启热部署，则需要在 pom.xml 中引入依赖：

```
<dependency>
    <groupId>org.springframework.boot</groupId>
    <artifactId>spring-boot-devtools</artifactId>
</dependency>
```

同时要配置 spring-boot-maven-plugin 插件：

```
<plugin>
    <groupId>org.springframework.boot</groupId>
    <artifactId>spring-boot-maven-plugin</artifactId>
    <configuration>
        <!--必须添加这个配置,如果没有该配置，devtools 不会生效 -->
        <fork>true</fork>
    </configuration>
</plugin>
```

配置 fork 为 true，Maven 在编译时会创建一个虚拟机来执行，它的速度会慢一些，fork 需要消耗更多的资源，以及花费更多的时间进行编译，所以如果开发机器性能不高，就应该尽量避免使用这种方式。

在 application.properties 中配置 spring.devtools.restart.enabled=false，此时 restart 类加载器还会初始化，但不会监视文件更新。

在 SprintApplication.run 之前调用 System.setProperty("spring.devtools.restart.enabled"，"false")，可以完全关闭重启支持，配置如下：

```
#热部署生效
spring.devtools.restart.enabled: true
#设置重启的目录
spring.devtools.restart.additional-paths: src/main/java
#classpath 目录下的 WEB-INF 文件夹内容修改不重启
spring.devtools.restart.exclude: WEB-INF/**
```

9.4.2　IntelliJ IDEA 开启热部署

当我们修改了 Java 类后，IDEA 默认是不自动编译的，而 spring-boot-devtools 又是监测 classpath 下的文件发生变化才会重启应用，所以需要设置 IDEA 的自动编译。

首先，在 File→Settings→Build,Execution,Deployment→Compiler 中开启“Build project automatically”，单击“Apply”按钮，如图 9-7 所示。

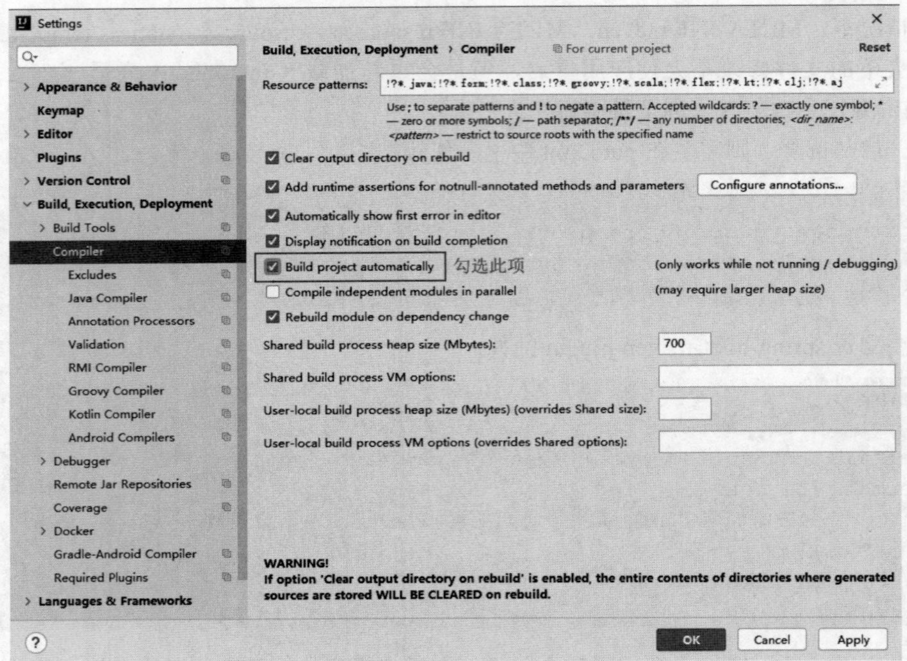

图 9-7

其次，按快捷键“Ctrl+Shift+Alt+/”，选择第一项“Registry...”，勾选其中的“compiler.automake.allow.when.app.running”，如图 9-8 所示。

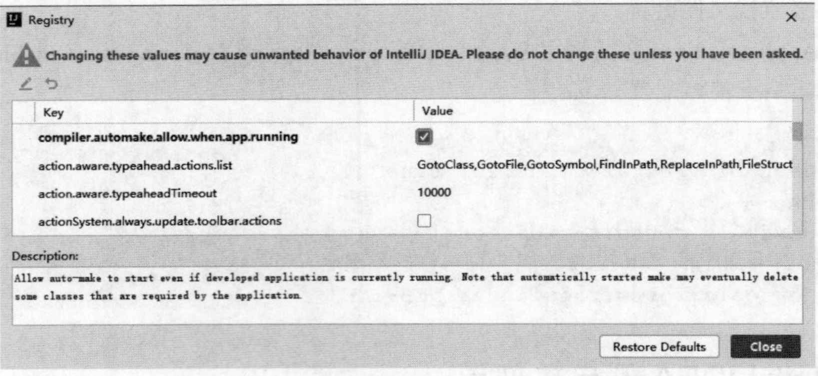

图 9-8

9.4.3 热部署测试

- 修改类→保存：应用会重启。
- 修改配置文件→保存：应用会重启。
- 修改页面→保存：应用不会重启，但会重新加载，页面会刷新（原理是将 spring.thymeleaf.cache 设为 false，参考 Spring Boot 配置模板引擎）。

9.5 应用性能监控

应用性能管理（Application Performance Management，APM）强调的是应用节点的性能管理和监控。它是对企业系统即时监控以实现对应用程序性能管理和故障管理的系统化解决方案。

应用性能管理是一个比较新的网络管理方向，主要指对企业的关键业务应用进行监测、优化，提高企业应用的可靠性和质量，保证用户得到良好的服务，降低 IT 总拥有成本（TCO）。一个企业的关键业务应用的性能强大，可以提高竞争力，并取得商业成功，因此加强应用性能管理（APM）可以产生巨大商业利益。

APM 的技术特点是以代理程序的方式来嵌入自有代码的应用节点中，获得相关业务指标，这类方式即业内所说的 Agent 方式。由于需要嵌入到自有代码中，因此在一定程度上是要消耗性能的，同时要考虑系统的兼容性、稳定性以及系统的支持程度。主流 APM 类产品以支持 Java 平台为主，C/C++平台需要定制开发，其他语言平台也存在不同程度定制的情况。APM 除了能看到应用性能指标之外，还能看到代码之间的调用关系和调用时延，从开发角度来看具有更大的价值。

应用性能监控对于开发人员来说提供了一种故障排查的能力，当出现网络拥堵、请求相应缓慢、出现偶发性错误或异常问题时，追踪数据流，能够通过分析定位到具体出问题的接口、类和方法，快速找到问题产生的原因所在。

对于测试人员和产品人员来说，能够通过链路追踪、服务拓扑来了解整个服务的地图，根据指标数据来分析需求，从而能够在整体上对业务模块、接口进行把控。

对于运维人员来说，有了运维数据可视化可以更快发现、更准确定位、更精准地做出应急决策。

对于 IT 决策者来说，用户体验可量化，并与同业的性能体验比较可以发现自身差距并进行优化，提升客户满意度。

这些都有非常高的企业价值，因此对开发的应用程序的运行情况进行监控是非常有必要的一件事情。APM 的价值就突显了出来。

对于分布式系统的监控，最重要的有三个部分：Logging、Tracing 和 Metrics。

- Logging：用于记录离散的事件。例如，应用程序的调试信息或错误信息。它是我们诊断问题的依据。
- Metrics：用于记录可聚合的数据。例如，队列的当前深度可被定义为一个度量值，在元素入队或出队时被更新；HTTP 请求个数可被定义为一个计数器，新请求到来时进行累加。
- Tracing：用于记录请求范围内的信息。例如，一次远程方法调用的执行过程和耗时。它是我们排查系统性能问题的利器。

这三个部分并非完全分离和独立的，虽有相互重叠的部分，但各自有所偏重。

9.5.1 Spring Boot Actuator

执行器（Actuator）是一个制造业术语，指的是用于移动或控制东西的一个机械装置，一个很小的改变就能让执行器产生大量的运动。

在 Spring Boot 中，它变身成一个性能监控器，Spring Boot Actuator 可以帮助我们全方面监控应用节点，比如健康检查、审计、统计、HTTP 追踪等。Actuator 一词本身的含义用在这里非常形象，意味着监控可能是一个很小的东西，却能带来高价值的回报。

得益于 Spring Boot 的种种优点，Spring Boot 成为 Java 语言开发项目非常流行的框架，使得微服务架构几乎成为业界不二之选，随着 Spring Boot 的发展，它逐步走向了更追逐技术前沿的方向：Spring Cloud 和 Service Mesh（服务网格），尤其是 Service Mesh 微服务架构，将服务开发和服务治理分开来，微服务数量和节点呈现几何级的增长，每一个节点都是系统组成部分，如何保持如此多节点的可用性是一件非常有挑战性的工作。全方位的监控变得越来越重要，当我们遇到 bug 时，总是希望可以看到更多信息，因此一般我们选用的服务开发框架都需要有方便又强大的监控功能支持。

在 Spring Boot 应用中，要实现监控的功能，只需要依赖组件 spring-boot-starter-actuator。它提供了很多监控和管理你的 Spring Boot 应用的 HTTP 或者 JMX 端点，并且可以有选择地开启和关闭部分功能。当 Spring Boot 应用中引入依赖之后，将自动拥有审计、健康检查、Metrics 监控功能。

组件依赖：

```
<dependency>
    <groupId>org.springframework.boot</groupId>
    <artifactId>spring-boot-starter-actuator</artifactId>
</dependency>
```

只需要将其添加到项目的 pom 依赖中即可。此外，还需要在 application.properties 中添加配置：

```
management:
# 禁用安全策略
security:
    enabled: false
    #actuator 开启所有 Web 功能
endpoints:
    web.exposure:.include: "*"
```

运行 Actuator 还需要连接 Redis 服务，因此需要启动 Redis 服务：

```
Administrator@Shawn-Desktop /d/Program Files/Redis
$ ./redis-server.exe
[5176] 17 Feb 21:36:32.246 # oO0OoO0OoO0Oo Redis is starting oO0OoO0OoO0Oo
[5176] 17 Feb 21:36:32.246 # Redis version=5.0.10, bits=64, commit=1c047b68, modified=0, pid=5176, just started
[5176] 17 Feb 21:36:32.246 # Warning: no config file specified, using the default config. In order to specify a config file use d:\program files\redis\redis-server.exe /path/to/redis.conf
```

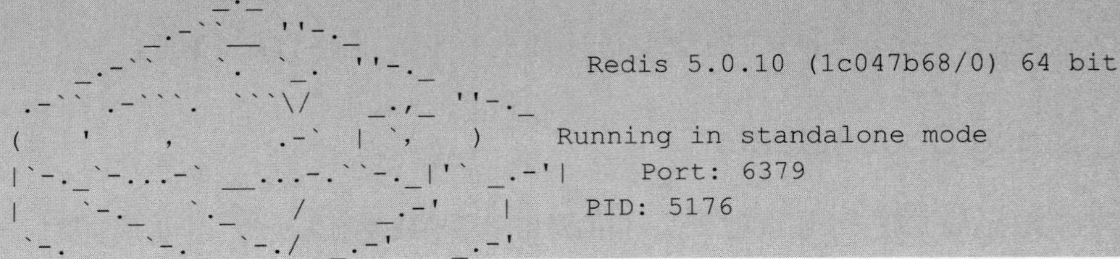

```
 |`-._`-._    `-.__.-'    _.-'_.-'|
 |    `-._`-._        _.-'_.-'    |           http://redis.io
  `-._    `-._`-.__.-'_.-'    _.-'
 |`-._`-._    `-.__.-'    _.-'_.-'|
 |    `-._`-._        _.-'_.-'    |
  `-._    `-._`-.__.-'_.-'    _.-'
      `-._    `-.__.-'    _.-'
          `-._        _.-'
              `-.__.-'

[5176] 17 Feb 21:36:32.251 # Server initialized
[5176] 17 Feb 21:36:32.252 * DB loaded from disk: 0.001 seconds
[5176] 17 Feb 21:36:32.252 * Ready to accept connections
```

Spring Boot Actuator 获取监控的指标数据，可以通过 JMX 或者 HTTP endpoints 来获得。默认 HTTP 端点是启用的。接下来介绍 Actuator 提供的部分端点。

1. 端点（endpoints）

访问 http://localhost:8080/actuator/可以查看所有 Actuator 提供的可以访问的接口：

```
$ curl http://10.168.1.42:8080/actuator | jq "."
{
  "_links": {
    "self": {
      "href": "http://10.168.1.42:8080/actuator",
      "templated": false
    },
    "beans": {
      "href": "http://10.168.1.42:8080/actuator/beans",
      "templated": false
    },
    "caches-cache": {
      "href": "http://10.168.1.42:8080/actuator/caches/{cache}",
      "templated": true
    },
    "caches": {
      "href": "http://10.168.1.42:8080/actuator/caches",
      "templated": false
    },
    "health-path": {
      "href": "http://10.168.1.42:8080/actuator/health/{*path}",
      "templated": true
    },
    "health": {
      "href": "http://10.168.1.42:8080/actuator/health",
      "templated": false
    },
    "info": {
      "href": "http://10.168.1.42:8080/actuator/info",
      "templated": false
    },
```

```
      "conditions": {
        "href": "http://10.168.1.42:8080/actuator/conditions",
        "templated": false
      },
      "configprops": {
        "href": "http://10.168.1.42:8080/actuator/configprops",
        "templated": false
      },
      "env": {
        "href": "http://10.168.1.42:8080/actuator/env",
        "templated": false
      },
      "env-toMatch": {
        "href": "http://10.168.1.42:8080/actuator/env/{toMatch}",
        "templated": true
      },
      "loggers-name": {
        "href": "http://10.168.1.42:8080/actuator/loggers/{name}",
        "templated": true
      },
      "loggers": {
        "href": "http://10.168.1.42:8080/actuator/loggers",
        "templated": false
      },
      "heapdump": {
        "href": "http://10.168.1.42:8080/actuator/heapdump",
        "templated": false
      },
      "threaddump": {
        "href": "http://10.168.1.42:8080/actuator/threaddump",
        "templated": false
      },
      "metrics": {
        "href": "http://10.168.1.42:8080/actuator/metrics",
        "templated": false
      },
      "metrics-requiredMetricName": {
        "href":
"http://10.168.1.42:8080/actuator/metrics/{requiredMetricName}",
        "templated": true
      },
      "scheduledtasks": {
        "href": "http://10.168.1.42:8080/actuator/scheduledtasks",
        "templated": false
      },
      "mappings": {
        "href": "http://10.168.1.42:8080/actuator/mappings",
        "templated": false
      }
    }
  }
```

默认情况下只能访问 /actuator/health 和 /actuator/info 两个端点，由于我们在 application.properties 中配置了 management.endpoints.web.exposure.include: "*"，因此/actuator 会返回所有端点。表 9-1 介绍了一些常用的端点。

表 9-1 常用的端点及其说明

端点	说明
Auditevent	显示应用暴露的审计事件（比如认证进入、订单失败）
info	显示应用的基本信息
health	显示应用的健康状态
metrics	显示应用多样的度量信息
loggers	显示和修改配置的 loggers
logfile	返回 log file 中的内容(如果 logging.file 或者 logging.path 被设置)
httptrace	显示 HTTP 足迹，最近 100 个 HTTP request / response
env	显示当前的环境特性
flyway	显示数据库迁移路径的详细信息
shutdown	优雅地逐步关闭应用
mappings	显示所有的@RequestMapping 路径
scheduledtasks	显示应用中的调度任务
threaddump	执行一个线程 dump
heapdump	返回一个 GZip 压缩的 JVM 堆 dump

更多端点的介绍可以在官网说明查看：https://docs.spring.io/spring-boot/docs/current/reference/html/production-ready-features.html #production-ready-endpoints。

2. 健康状况（health）

访问 http://localhost:8080/actuator/health 查看健康状态，当反馈"status"是"UP"时说明是健康的，"DOWN"则是不健康的。

```
$ curl http://localhost:8080/actuator/health
{"status":"UP"}
```

3. 度量（metrics）

访问/actuator/metrics 端点可以获得所有可以追踪的度量，例如 jvm 内存、CPU 使用、jvm 线程等。

```
$ curl 10.168.1.42:8080/actuator/metrics | jq "."
{
  "names": [
    "hikaricp.connections",
    "hikaricp.connections.acquire",
    "hikaricp.connections.active",
    "hikaricp.connections.creation",
    "hikaricp.connections.idle",
    "hikaricp.connections.max",
    "hikaricp.connections.min",
    "hikaricp.connections.pending",
    "hikaricp.connections.timeout",
```

```
    "hikaricp.connections.usage",
    "jdbc.connections.active",
    "jdbc.connections.idle",
    "jdbc.connections.max",
    "jdbc.connections.min",
    "jvm.buffer.count",
    "jvm.buffer.memory.used",
    "jvm.buffer.total.capacity",
    "jvm.classes.loaded",
    "jvm.classes.unloaded",
    "jvm.gc.live.data.size",
    "jvm.gc.max.data.size",
    "jvm.gc.memory.allocated",
    "jvm.gc.memory.promoted",
    "jvm.gc.pause",
    "jvm.memory.committed",
    "jvm.memory.max",
    "jvm.memory.used",
    "jvm.threads.daemon",
    "jvm.threads.live",
    "jvm.threads.peak",
    "jvm.threads.states",
    "logback.events",
    "process.cpu.usage",
    "process.start.time",
    "process.uptime",
    "system.cpu.count",
    "system.cpu.usage",
    "tomcat.sessions.active.current",
    "tomcat.sessions.active.max",
    "tomcat.sessions.alive.max",
    "tomcat.sessions.created",
    "tomcat.sessions.expired",
    "tomcat.sessions.rejected"
  ]
}
```

要查看某个度量的详细信息时，需要将度量的名称传入/actuator/metrics/:name 接口中，例如查看当前 jvm 使用情况的度量：

```
curl 10.168.1.42:8080/actuator/metrics/jvm.memory.used | jq "."
{
  "name": "jvm.memory.used",
  "description": "The amount of used memory",
  "baseUnit": "bytes",
  "measurements": [
    {
      "statistic": "VALUE",
      "value": 335658720
    }
  ],
  "availableTags": [
```

```
      {
        "tag": "area",
        "values": [
          "heap",
          "nonheap"
        ]
      },
      {
        "tag": "id",
        "values": [
          "Compressed Class Space",
          "PS Survivor Space",
          "PS Old Gen",
          "Metaspace",
          "PS Eden Space",
          "Code Cache"
        ]
      }
    ]
  }
```

4. 日志（loggers）

访问/actuator/loggers 端点可以获取应用中可配置的 loggers 列表和相关日志等级。如果想要获取单个日志的配置信息，可以使用/actuator/loggers/:name 端点获取，把 logger 名称传入即可。

/actuator/loggers 端点同时提供了在应用运行时改变日志级别的能力，比如想要改变某个 logger 的日志等级，可以向/actuator/loggers/:name 端点发送一个 POST 请求，body 内容如下：

```
{
    "configuredLevel": "debug",
    "effectiveLevel": "debug"
}
```

动态修改日志级别的功能对于日常排查问题非常有利。

5. 其他端点

/actuator 返回可以访问的端点列表都是能够通过 HTTP 或者 JMX 来获取数据的，这里不再一一介绍。

6. 自定义路径和端口

如果不喜欢 Actuator 默认的 Actuator 路径，可以通过配置将 Actuator 换成其他路径：

```
management.endpoints.web.base-path=/manage
```

Actuator 管理服务器的端口默认是和应用程序端口一致的，如果有需要，也可以通过配置来更改端口号：

```
management.server.port=8081
```

如果不想公开 HTTP 端点，也可以配置将其禁用，这里只需要将管理服务器端口设为-1：

```
management.server.port=-1
```

关于 Actuator 的更多介绍可以查阅官方文档：https://docs.spring.io/spring-boot/docs/current/reference/html/production-ready-features.html。

9.5.2 APM 监控：链路追踪

1. 概述

链路追踪的概念最早来源于 2010 年谷歌发表的一篇名为 Dapper 的论文。当代的互联网服务通常都是用复杂的、大规模分布式集群来实现的。互联网应用构建在不同的软件模块集上，这些软件模块有可能是由不同的团队开发、使用不同的编程语言来实现、布在了几千台服务器、横跨多个不同的数据中心。因此，需要一些可以帮助理解系统行为、用于分析性能问题的工具。

Google 生产环境下一个被称为 Dapper 的分布式跟踪系统应运而生，随后从 Dapper 发展成为一流的监控系统。Dapper 最初只是一个自给自足的监控工具，最终进化成一个监控平台，这个监控平台促生出多种多样的监控工具，有些甚至已经不是由 Dapper 团队开发的。

Dapper 的出现是为了收集更多的复杂分布式系统的行为信息，然后呈现给 Google 的开发者。分布式系统有一个特殊的好处，就是可以利用大规模低端、廉价的服务器，甚至是个人电脑，成为互联网服务的载体，分布式系统是一个特殊的经济划算的平台。想要在这个上下文中理解分布式系统的行为，就需要监控那些横跨了不同应用、不同服务器之间的关联动作。

下面通过一个跟搜索相关的例子来阐述 Dapper 可以应对哪些挑战。比如一个前段服务可能对上百台查询服务器发起了一个 Web 查询，每一个查询都有自己的 Index。这个查询可能会被发送到多个子系统，这些子系统分别用来处理广告、进行拼写检查或者查找图片、视频或新闻之类的特殊结果。根据每个子系统的查询结果进行筛选，得到最终结果，最后汇总到页面上。我们把这种搜索模型称为“全局搜索”（universal search）。总的来说，这一次全局搜索有可能调用上千台服务器，涉及各种服务，而且用户对搜索的耗时是很敏感的，任何一个子系统的低效都会导致最终的搜索耗时。

如果一个工程师只能知道这个查询耗时不正常，但是他无从知晓这个问题到底是由哪个服务调用造成的，或者为什么这个调用性能差强人意。首先，这个工程师可能无法准确地定位到这次全局搜索是调用了哪些服务，因为新的服务乃至服务上的某个片段都有可能在任何时间上过线或修改过，有可能是面向用户功能，也有可能是一些例如针对性能或安全认证方面的功能改进。其次，不能苛求这个工程师对所有参与这次全局搜索的服务都了如指掌，每一个服务都有可能是由不同的团队开发或维护的。再次，这些暴露出来的服务或服务器有可能同时被其他客户端使用着，所以这次全局搜索的性能问题甚至有可能是由其他应用造成的。举个例子，一个后台服务可能要应付各种各样的请求类型，而一个使用效率很高的存储系统，比如 Bigtable，有可能正被反复读写着，因为上面跑着各种各样的应用。

在上面这个案例中，有两点要求：

- 无所不在的部署和持续的监控。无所不在的重要性不言而喻，因为在使用跟踪系统进行监控时，即便只有一小部分没被监控到，那么人们对这个系统是不是值得信任都会产生巨大的质疑。
- 监控应该是 7×24 小时的。系统异常或是那些重要的系统行为有可能出现过一次就很难甚至不太可能重现。

Dapper 的实现满足了这两个要求，实现了对应用程序透明、低功耗和延展性的设计目标。Dapper 论文的发表引发了 APM 监控系统的变革，受到该论文的启发，各种链路监控系统开始涌现。由于 Dapper 是谷歌内部使用的系统，并不开源，因此这是目前各种链路监控系统纷纷出现和竞争的原因。

目前比较流行的开源链路监控追踪系统有 Twitter 实现的 Zipkin、韩国人开发的 Pinpoint、国人开发的 Skywalking（已贡献给 Apache），还有一些小众产品，比如大众点评的 CAT（开源）、京东的 Hydra、新浪的 Watchman 以及已经被纳入 CNCF 基金会旗下的开源分布式链路追踪标准 Opentracing 的实现 Jeager（Uber 开发）。Opentracing 标准的出现是为了解决不同的分布式追踪系统 API 不兼容的问题，它提供统一的概念和数据标准，提供平台无关、厂商无关的 API，使得开发人员能够方便地添加（或更换）追踪系统的实现。

目前 Zipkin、Apache Skywalking 已经实现了 Opentracing 标准，下面将主要介绍 Zipkin 和 Skywalking。

2. Zipkin

Zipkin 有一些基本概念，具体如下：

（1）brave：用来装备 Java 程序的类库，提供了面向 Standard Servlet、Spring MVC、HTTP Client、JAX RS、Jersey、Resteasy 和 MySQL 等接口的装备能力，可以通过编写简单的配置和代码让基于这些框架构建的应用可以向 Zipkin 报告数据。同时，brave 也提供了非常简单且标准化的接口，在以上封装无法满足要求的时候可以方便扩展与定制。brave 利用 reporter 向 Zipkin 的 Collector 发送 trace 信息。brave 主要是利用拦截器在请求前和请求后分别埋点。例如，Spingmvc 监控使用 Interceptors，MySQL 监控使用 statementInterceptors。同理，Dubbo 的监控是利用 com.alibaba.dubbo.rpc.Filter 来过滤生产者和消费者的请求。

（2）span：基本工作单元。例如，在一个新建的 span 中发送一个 RPC，等同于发送一个回应请求给 RPC，span 通过一个 64 位 ID 唯一标识，trace 以另一个 64 位 ID 标识，span 还有其他数据信息，比如摘要、时间戳事件、关键值注释(tags)、span 的 ID 以及进度 ID（通常是 IP 地址）。span 在不断的启动和停止，同时记录了时间信息。当你创建了一个 span 时，就必须在未来的某个时刻停止它。

（3）trace：一系列 span 组成的一个树状结构。例如，你正在执行一个分布式请求，就可能需要创建一个 trace。

（4）annotation：用来及时记录一个事件的存在，一些核心 annotations 用来定义一个请求的开始和结束。annotation 中包含以下几个概念：

- cs（Client Sent）：客户端发起一个请求，描述 span 的开始。
- sr（Server Received）：服务端获得请求并准备开始处理它，将 sr 减去 cs 时间戳便可得到网络延迟。
- ss（Server Sent）：注解表明请求处理的完成（当请求返回客户端），用 ss 减去 sr 时间戳便可得到服务端需要的处理请求时间。
- cr（Client Received）：表明 span 的结束，客户端成功接收到服务端的回复，用 cr 减去 cs 时间戳便可得到客户端从服务端获取回复的所有需要时间。

当用户发起一次调用时，Zipkin 的客户端会在入口处为整条调用链路生成一个全局唯一的

trace id，并为这条链路中的每一次分布式调用生成一个 span id。span 与 span 之间可以有父子嵌套关系，代表分布式调用中的上下游关系。span 和 span 之间可以是兄弟关系，代表当前调用下的两次子调用。一个 trace 由一组 span 组成，可以看成是由 trace 为根节点、span 为若干个子节点的一棵树。

Zipkin 会将 trace 相关的信息在调用链路上传递，并在每个调用边界结束时异步地把当前调用的耗时信息上报给 Zipkin Server。Zipkin Server 在收到 trace 信息后，将其存储起来。随后 Zipkin 的 Web UI 会通过 API 访问的方式从存储中将 trace 信息提取出来分析并展示。图 9-9 所示是 Zipkin 官方给出的一个追踪调用链图，展示了一个简单的 trace 示例。

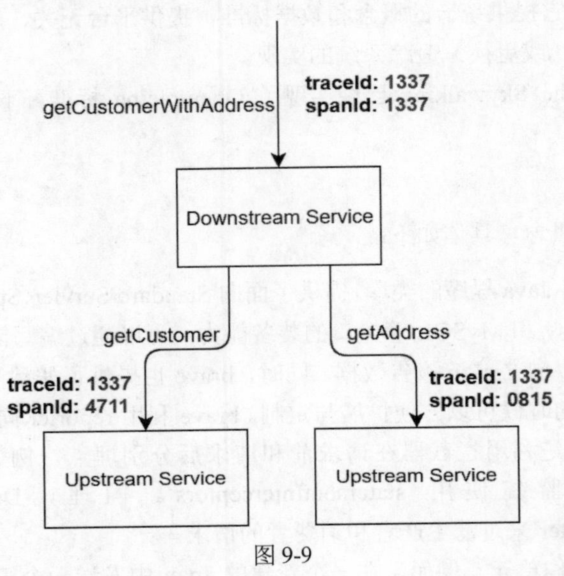

图 9-9

Zipkin Server 下载运行：

```
curl -sSL https://zipkin.io/quickstart.sh | bash -s
java -jar zipkin.jar
```

Zipkin 客户端需要集成到项目中，在项目中添加依赖示例：

```
<!-- 使用 okhttp3 作为 reporter -->
<dependency>
    <groupId>io.zipkin.reporter2</groupId>
    <artifactId>zipkin-sender-okhttp3</artifactId>
    <version>2.8.2</version>
</dependency>
<!--  brave 对 dubbo 的集成 -->
<dependency>
    <groupId>io.zipkin.brave</groupId>
    <artifactId>brave-instrumentation-dubbo-rpc</artifactId>
    <version>5.6.3</version>
</dependency>
<!--  brave 对 mvc 的集成 -->
<dependency>
    <groupId>io.zipkin.brave</groupId>
```

```
    <artifactId>brave-instrumentation-spring-webmvc</artifactId>
    <version>5.6.3</version>
</dependency>
```

Brave 提供了一系列组件接口的支持包，按需引入依赖即可（见图 9-10）。

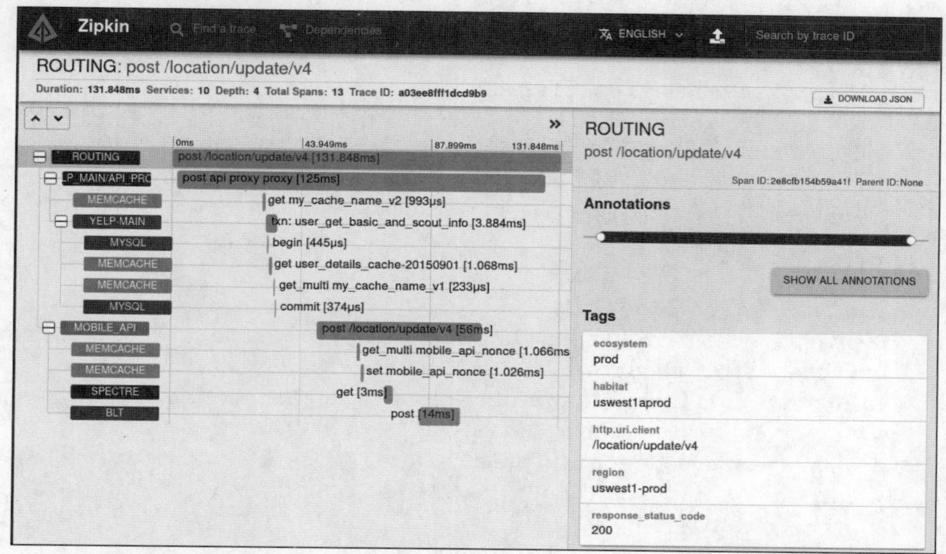

图 9-10

然后在项目中修改 application.yaml 文件，添加 zipkin 的配置：

```
zipkin:
  url: http://127.0.0.1:9411/api/v2/spans
  connectTimeout: 5000
  readTimeout: 10000
  # 取样率，指的是多次请求中有百分之多少传到 zipkin。例如 1.0 是全部取样，0.5 是 50% 取样
  rate: 1.0f
```

创建 ZipkinProperties.java 属性类，内容如下：

示例代码 9-3　ZipkinProperties.java

```
package come.example.demo.config
import …
@Configuration
@ConfigurationProperties("zipkin")
@Data
public class ZipkinProperties {
    @Value("${spring.application.name}")
    private String serviceName;
    private String url;
    private Long connectTimeout;
    private Long readTimeout;
    private Float rate;
}
```

创建 ZipkinProperties 类的目的是读取 application.yaml 中的 Zipkin 配置。

然后创建 ZipkinConfig 配置类，设置相应的拦截器实现 tracing。

示例代码 9-4　ZipkinConfig.java

```
@Configuration
@EnableConfigurationProperties
public class ZipkinConfig {
    @Autowired
    private ZipkinProperties zipkinProperties;
    /**
     * 为了实现 dubbo rpc 调用的拦截
     *
     * @return
     */
    @Bean
    public Tracing tracing() {
        Sender sender = OkHttpSender.create(zipkinProperties.getUrl());
        AsyncReporter reporter = AsyncReporter.builder(sender)
                .closeTimeout(zipkinProperties.getConnectTimeout(),
                        TimeUnit.MILLISECONDS)
                .messageTimeout(zipkinProperties.getReadTimeout(),
                        TimeUnit.MILLISECONDS)
                .build();
        Tracing tracing = Tracing.newBuilder()
                .localServiceName(zipkinProperties.getServiceName())
                .propagationFactory(
                        ExtraFieldPropagation
                            .newFactory(B3Propagation.FACTORY,
"shiliew"))
                .sampler(Sampler.create(zipkinProperties.getRate()))
                .spanReporter(reporter)
                .build();
        return tracing;
    }
    /**
     * MVC Filter，为了实现 SpringMvc 调用的拦截
     * @param tracing
     * @return
     */
    @Bean
    public Filter tracingFilter(Tracing tracing) {
        HttpTracing httpTracing = HttpTracing.create(tracing);
        httpTracing.toBuilder()
            .serverParser(new HttpServerParser() {
                @Override
                public <Req> String spanName(HttpAdapter<Req, ?> adapter, Req req)
{
                        return adapter.path(req);
                }
            })
            .clientParser(new HttpClientParser() {
                @Override
```

```
            public <Req> String spanName(HttpAdapter<Req, ?> adapter, Req req)
{
                    return adapter.path(req);
                }
            }).build();
        return TracingFilter.create(httpTracing);
    }
}
```

这样就做好了集成，能够正常生成所需要的追踪链了。

3. Skywalking

Skywalking 是一个 APM（应用程序性能监视器）系统，专门为微服务、云原生和基于容器（Docker，Kubernetes，Mesos）的体系结构而设计——来自于官方的定义。

Skywalking 是一个开源 APM 系统，包括对 Cloud Native 体系结构中的分布式系统的监视、跟踪、诊断功能。核心功能如下：

- 服务、服务实例、端点指标分析。
- 根本原因分析：在运行时分析代码。
- 服务拓扑图分析。
- 服务、服务实例和端点依赖关系分析。
- 检测到慢速服务和端点。
- 性能优化。
- 分布式跟踪和上下文传播。
- 数据库访问指标。检测慢速数据库访问语句（包括 SQL 语句）。
- 监控告警。
- 浏览器性能监控。
- 基础架构（VM，网络，磁盘等）监控。
- 跨指标，跟踪和日志的协作。

图 9-11 所示展示了 Skywalking 的架构示意图。

SkyWalking 支持从多种来源和多种格式收集遥测（度量、跟踪和日志）数据，包括：

- Java、.NET Core、Node.js、PHP 和 Python 自动仪器代理。
- Go 和 C ++ SDK。
- LUA 代理，尤其是 Nginx、OpenResty 的 LUA 代理。
- 浏览器代理。
- 服务网格的可观察性，控制面板和数据面板。
- 度量系统，包括 Prometheus、OpenTelemetry、Spring Sleuth（Micrometer）、Zabbix。
- 日志。
- Zipkin v1/v2 和 Jaeger gRPC 格式，具有有限的拓扑和指标分析（实验性）。

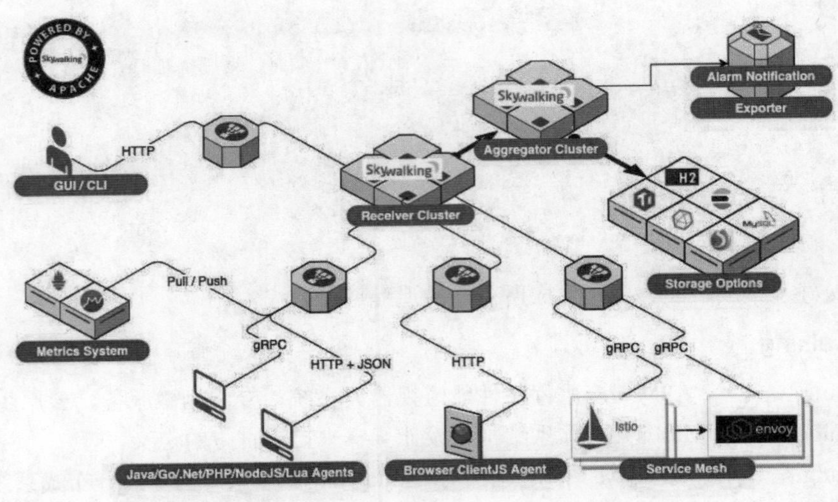

图 9-11

Skywalking 基于 Java 字节码技术实现探针（agent），对应用程序透明无感知，探针用来在应用程序端埋点收集数据并重新格式化以符合 Skywalking 的要求（不同的探针支持不同的来源）。

探针收集的数据将会发送到平台后端，平台后端聚合数据，对数据进行分析、整理，并将数据存储到相应的数据源，比如 ElasticSearch、MySQL 等数据库。然后提供了一个非常简洁美观的 UI 界面来进行数据可视化。图 9-12 展示了探针（Probes/Agent）、平台后端、UI 之间是怎样进行数据交互的。

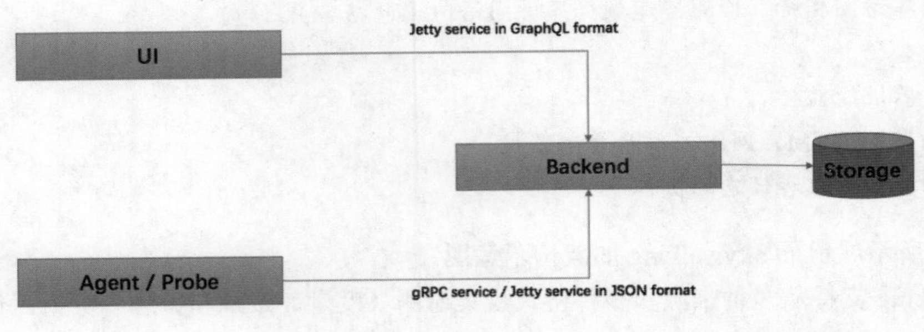

图 9-12

Skywalking APM 的下载地址为 https://mirror.bit.edu.cn/apache/skywalking/8.4.0/apache-skywalking -apm-es7-8.4.0.tar.gz。下载完成后解压，在 bin 目录下运行 startup.bat 或者 startup.sh（根据系统选择），这将会同时启动 oapService（Collector 平台后端服务）、webappService（UI 服务）两个进程。Collector 将会监听端口 11800（gRpc 端口）和 12800（HTTP REST 服务端口），用来接收探针发送过来的数据，UI 服务运行在 8080 端口，默认后端存储采用 H2。

访问 http://localhost:8080/就能看到 Skywalking UI 页面，如图 9-13 所示。现在的页面还是空的，因为没有采集数据。在下一节中将会详细介绍如何在 Spring Boot 应用程序上部署探针采集数据。

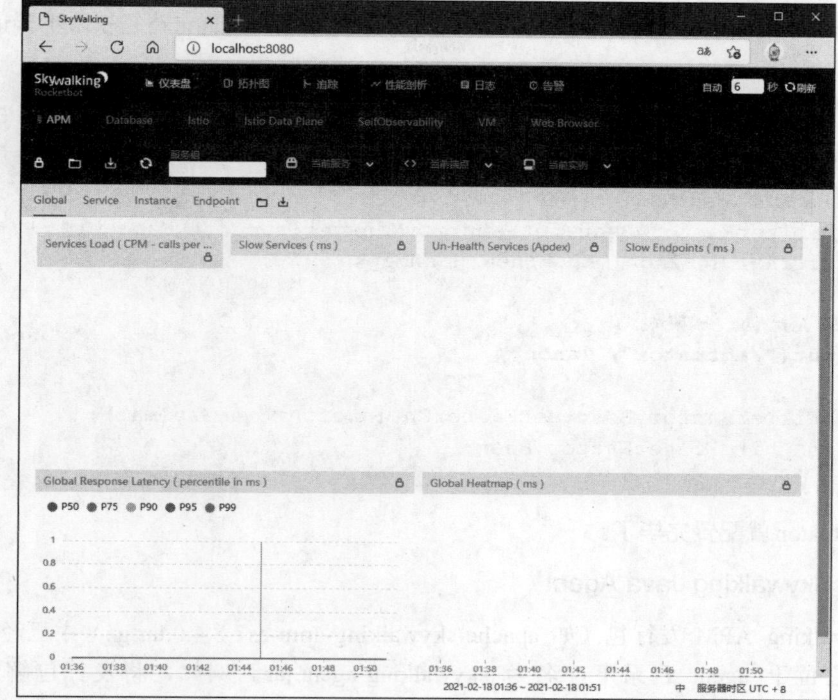

图 9-13

9.5.3 监控 Spring Boot 应用

本节将以 shiro-demo 项目案例为基础，配置实现 Actuator+Skywalking 对其进行监控。

1. 配置 Spring Boot Actuator

首先在 shiro-demo 项目的 pom.xml 中引入依赖：

```
    <!--监控-->
<dependency>
    <groupId>org.springframework.boot</groupId>
    <artifactId>spring-boot-starter-actuator</artifactId>
</dependency>
```

然后在 application.yaml 文件中添加 Actuator 的配置信息：

```
    management:
      # 禁用安全策略
      security:
        enabled: false
      # Actuator 开启所有 Web 功能
      endpoints:
        web.exposure.include: "*"
      # 修改应用监控管理的服务端口为 8081
      server:
  port: 8081
```

修改 ShiroConfig.java，在 filterChainDefinitionMap 中设置/actuator 路径权限不验证（anon）：

```
    @Bean
    public ShiroFilterFactoryBean shiroFilterFactoryBean(SecurityManager
securityManager) {
        ShiroFilterFactoryBean shiroFilterFactoryBean = new
ShiroFilterFactoryBean();
        shiroFilterFactoryBean.setSecurityManager(securityManager);
        Map<String, String> map = new HashMap<>();
        …
        //开放 Actuator 权限
        map.put("/actuator", "anon");
        …
        shiroFilterFactoryBean.setFilterChainDefinitionMap(map);
        return shiroFilterFactoryBean;
    }
```

这样 Actuator 就配置完毕了。

2. 配置 Skywalking Java Agent

在 Skywalking APM 发行包（如 apache-skywalking-apm-es7-8.4.0.tar.gz）中已经包含了 Java Agent 包，它位于 agent 目录下，名为 skywalking-agent.jar。例如，将发行包解压后放置于 D:\code\apache-skywalking-apm-bin-es7 目录下，那么 Java Agent 的路径就是 D:\code\apache-skywalking-apm-bin-es7\agent\skywalking-agent.jar。在 IDEA 中设置 VM options：

```
-javaagent:D:\code\apache-skywalking-apm-bin-es7\agent\skywalking-agent.jar
```

具体设置如图 9-14 所示。

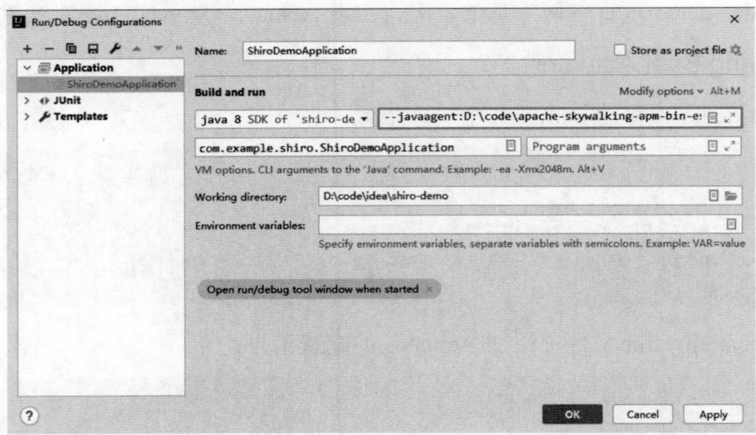

图 9-14

修改 config 目录下的 agent.config 文件，设置 agent.service_name，这一项的值指定 agent 采集的项目服务名称，必须为英文，例如修改 D:\code\apache-skywalking-apm-bin-es7\ config\agent.json：

```
# The service name in UI
agent.service_name=${SW_AGENT_NAME:shiro-demo}
```

将 Skywalking 发行包 agent/optional-plugins 目录下的 apm-spring-annotation-plugin-8.4.0.jar 移

入 agent/plugins 目录中。

至此，Skywalking Java Agent 配置完成。运行 Skywalking 后端平台和 UI 服务（startup.bat/sh）。在运行 shiro-demo 项目之前，先修改端口号：

```
server:
port: 8000
```

因为 8080 端口已经被 Skywalking UI 占用。修改好端口后运行 shiro-demo 项目。

打开浏览器访问 http://localhost:8080/ 发现并没有采集到数据，查看 logs 目录下的 skywalking-api.log 日志文件，发现其中有一行错误提示：

```
    ERROR 2021-02-18 02:43:12:144 http-nio-8000-exec-1 InstMethodsInter :
class[class com.example.shiro.controller.UserRestController] after method[login]
intercept failure
    java.lang.ClassCastException:
org.apache.skywalking.apm.plugin.spring.mvc.commons.JavaxServletRequestHolder
cannot be cast to
org.apache.skywalking.apm.plugin.spring.mvc.commons.RequestHolder
```

出现这个错误的原因是使用热部署插件 spring-boot-devtools 引起了冲突，导致相关类被不同的类加载器加载。解决这个问题的一个办法是移除 spring-boot-devtools，另一个办法是使用命令行运行 jar 包，先执行 mvn clean package 对 shiro-demo 项目进行打包，然后在 target 目录下执行如下命令：

```
    java
-javaagent:D:\code\apache-skywalking-apm-bin-es7\agent\skywalking-agent.jar
-Dskywalking.agent.service_name=shiro-demo -jar shiro-demo-0.0.1-SNAPSHOT.jar
```

成功运行后，使用 curl 请求登录：

```
    curl -H "Content-Type: application/json" -X POST -d '{"username": "admin",
"password": "123"}' http://localhost:8000/user/login
    {"code":1,"msg":"登录成功！
","data":{"id":null,"username":"admin","password":"123","timestamp":null}}
```

然后打开浏览器访问 https://localhost:8080/topology，可以看到生成了如图 9-15 所示的服务拓扑图。

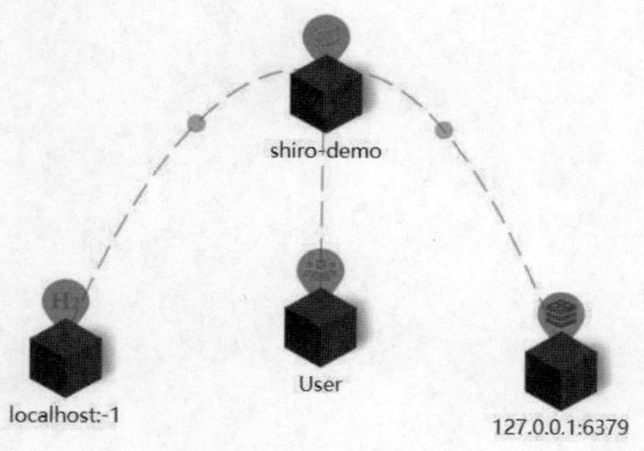

图 9-15

在“追踪”功能中，能看到所有请求的链路，并能发现被请求的/user/login 链路图，如图 9-16 所示。

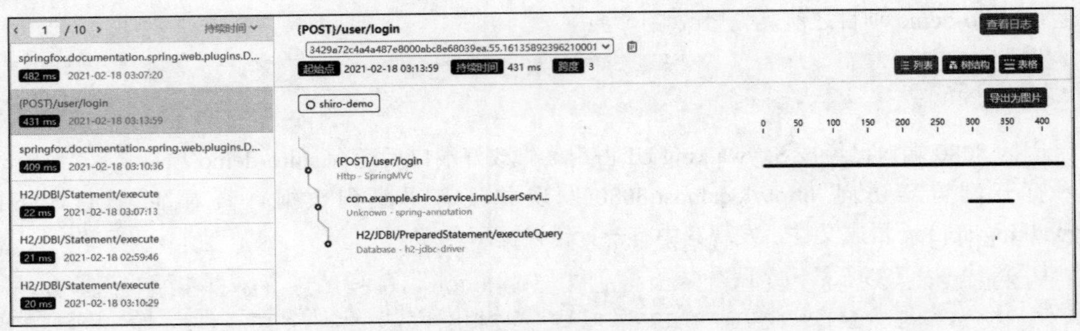

图 9-16

Skywalking 还有许多功能，比如日志分析、监控告警、性能剖析等，这里不再一一展示，有兴趣的话可以尝试部署一下。

第 10 章

综合项目实战

前面章节讲述了 Spring Boot 相关基础，要想学好编程，实战是最佳途径。本章将通过一个简单 Spring Boot 实战案例——图书管理系统，综合应用前面介绍的知识，增强读者对 Spring Boot 的理解、应用能力和成就感。

10.1 项目准备

10.1.1 数据库设计

在企业开发中，应该先根据需求分析设计好数据库，然后才开始进行编码。一个企业级图书管理网站比较复杂，这里就设计如图 10-1 所示的三张表。

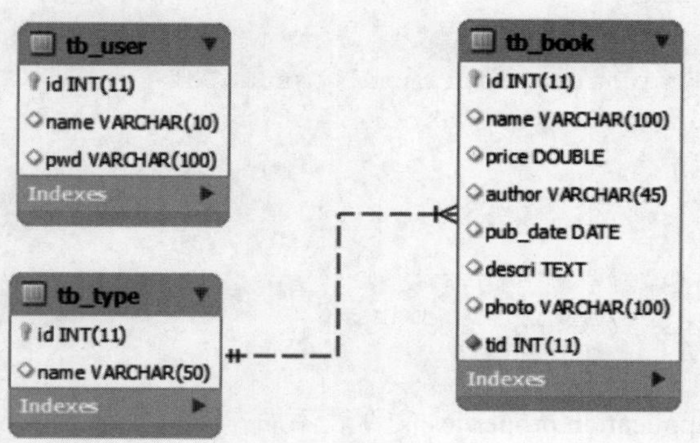

图 10-1

先创建数据库 just_book，然后在该数据库下建立该项目的三张表：

- 用户表：用于存储管理员信息，其中 pwd 使用 Bcrypt 方式加密。
- 图书类型表：实际图书是二级或者三级分类，这里只进行一级分类，该表存储图书的大类。
- 图书表：存储图书基本信息。

10.1.2 项目搭建

下面创建一个基于 Spring Boot+MyBatis+JSP 的项目，项目名称是 bookman，在第 10 章案例源码文件夹下。本小节主要做以下开发工作：

- 添加依赖 Spring Web、MyBatis 和 JSP 依赖。
- 修改 application.properties，添加关于 Web、MyBatis 等的配置。

具体创建步骤如下：

步骤 01 创建 Spring Boot 项目。在 IDEA 中使用 Spring Initializr 向导生成一个 Spring Boot 项目。

步骤 02 添加依赖。选择添加 Spring Web、MyBatis Framework、MySQL Driver、Lombok 依赖。

步骤 03 添加其他依赖。在 pom.xml 中添加 JSP 运行依赖以及验证码，如下代码所示。

示例代码 10-1 pom.xml（部分代码）

```
<!--加密类库-->
<dependency>
    <groupId>org.springframework.security</groupId>
    <artifactId>spring-security-crypto</artifactId>
</dependency>
<!--验证码-->
<dependency>
    <groupId>com.github.penggle</groupId>
    <artifactId>kaptcha</artifactId>
    <version>2.3.2</version>
</dependency>
<!--添加 JSP 编译类库-->
<dependency>
    <groupId>org.apache.tomcat.embed</groupId>
    <artifactId>tomcat-embed-jasper</artifactId>
</dependency>
<!--支持 jstl-->
<dependency>
    <groupId>javax.servlet</groupId>
    <artifactId>jstl</artifactId>
</dependency>
```

步骤 04 修改配置 application.properties 文件。添加连接数据库、Spring MVC 和 MyBatis 相关配置，如下面的代码所示。

示例代码 10-2 application.properties

```
# mybatis
```

```
    spring.datasource.driver-class-name=com.mysql.cj.jdbc.Driver
    spring.datasource.url=jdbc:mysql://localhost:3306/just_book?serverTimezone
=UTC
    spring.datasource.username=root
    spring.datasource.password=root  #注意密码修改
    # mvc
    spring.mvc.view.prefix=/WEB-INF/jsp/
    spring.mvc.view.suffix=.jsp
    server.servlet.context-path=/bookman
    spring.mvc.static-path-pattern=/s/**
    #mybatis 配置：java 中属性命名采用驼峰法，而数据库中一般采用下画线连接多个单词
    # 因而需要完成两者映射
    mybatis.configuration.map-underscore-to-camel-case=true
```

步骤 05 添加 webapp 以及 WEB-INF 目录。默认创建的项目没有这个目录，因而我们需要手动去创建。在 main 目录中先创建目录 webapp，然后在 webapp 目录下新建 WEB-INF 目录。

10.1.3 添加前端依赖

为了快速添加前端依赖 jQuery、bootStrap4、bootstrap-datepicker，这里使用 Node.js 的包管理工具 NPM，具体步骤如下：

步骤 01 安装 Node.Js。根据操作系统选择安装合适版本的 Node.js，如图 10-2 所示，下载完成后安装即可。

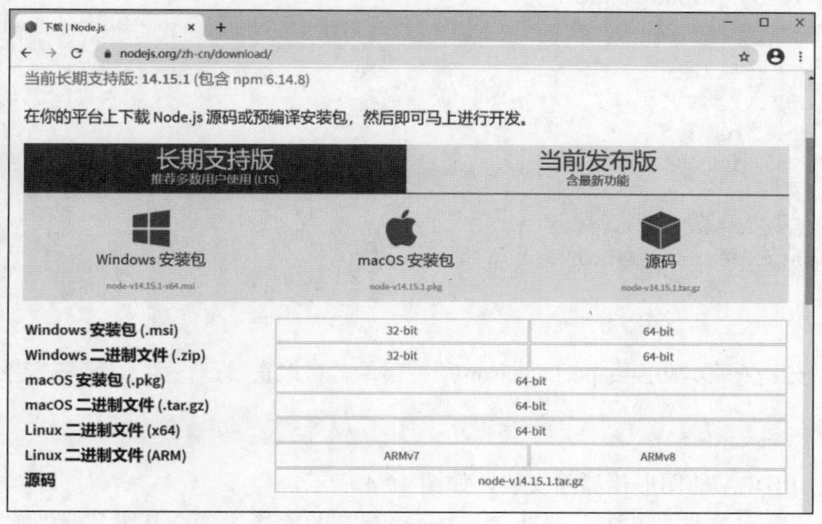

图 10-2

步骤 02 配置 Node.js。安装完后需要在 IDEA 中配置，以便 IDEA 可以找到 Node.js 和 npm。在 IDEA 中选择 File→settings，如图 10-3 所示。

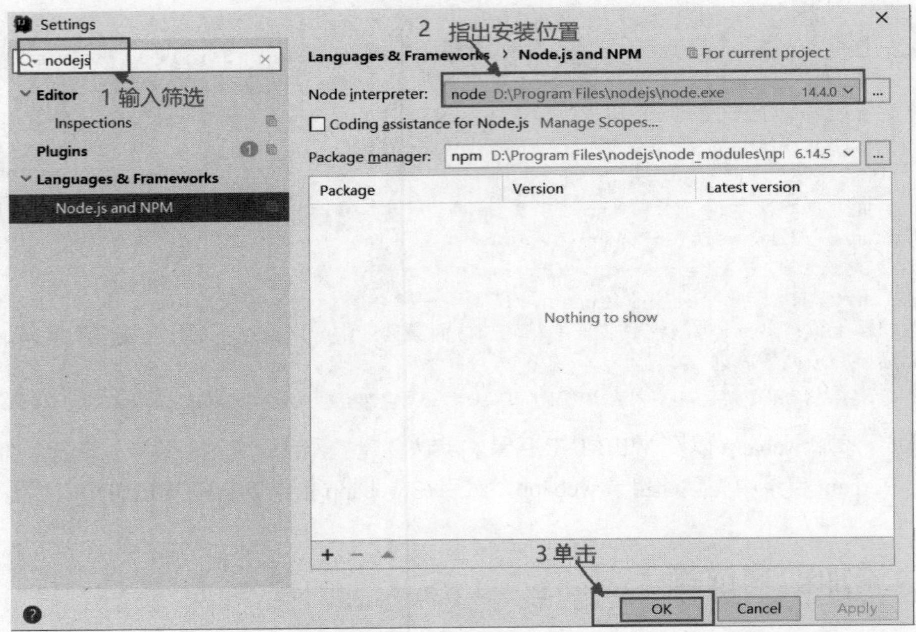

图 10-3

步骤03 创建 package.json 并在 dependencies 部分添加前端依赖。在 resources/static 目录下新建 package.json，内容如示例代码 10-3 所示。

示例代码 10-3 package.json

```
{
  "name": "static",
  "version": "1.0.0",
  "dependencies": {
    "bootstrap": "^4.5.3",
    "jquery": "^3.5.1",
    "bootstrap-datepicker": "^1.9.0",
    "bootstrap4-validator": "^0.0.1"
  }
}
```

步骤04 安装依赖。切换到 package.json 所在目录，可以直接运行如下命令安装依赖：

```
npm install
```

也可以使用 IDEA 的图形化界面安装，如图 10-4 所示。

在 package.json 文件上右击，选择 Run'npm install'命令，就会从服务器把前端依赖下载到 package.json 所在目录下的 node_modules 文件夹中。

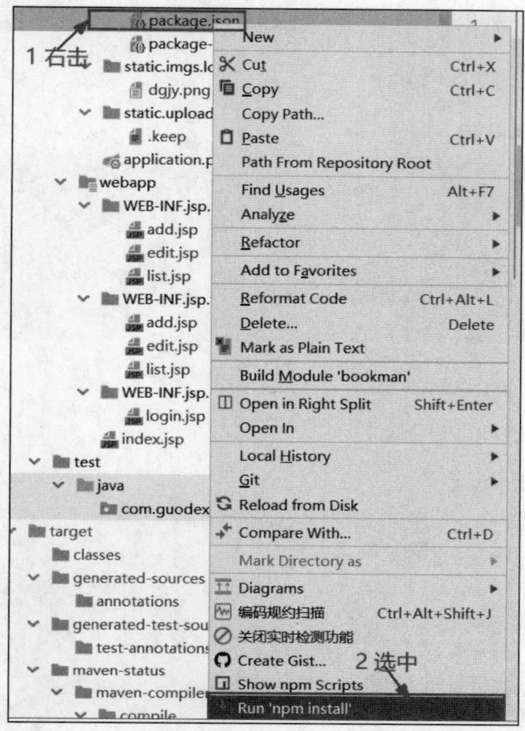

图 10-4

10.1.4　编写实体类

在进行数据库设计时，需要分析出实体关系图（E-R 图）：

- 程序大部分数据要进入关系数据库，实体对应关系数据中的表，关系用外键来表达。
- 实体在 Java 编程中对应 Java 类，称之为实体类，关系用关联属性表达。
- 实体类和数据库表一般一一对应，这里需要如下三个实体类，使用 JavaBean 风格。
- 为了提高开发效率，这里使用 Lombok 自动生成 getter/setter，在 IDEA 中安装 Lombok 插件。
- 为了对用户输入的表单进行验证，这里使用基于注解方式。

1. 实体类 User

这里对应表 tb_user 内容，代码如下。

示例代码 10-4　User.java

```
import lombok.Data;
@Data
public class User {
   private int id;
   private String name;
   private String pwd;
}
```

2. 实体类 Type

这里对应表 tb_type 内容，代码如下。

示例代码 10-5　Type.java

```
import lombok.Data;
@Data
public class Type {
    private int id;
    private String name;
}
```

3. 实体类 Book

这里对应表 tb_book 内容，代码如下。

示例代码 10-6　Book.java

```
import lombok.Data;
import lombok.NoArgsConstructor;
import org.springframework.format.annotation.DateTimeFormat;
import javax.validation.constraints.*;
import java.util.Date;
@Data
@NoArgsConstructor
public class Book {
    private int id;
    @Size(min = 5,max = 30,message = "必填")
    private String name;
    @DecimalMin(value = "0")
    private double price;
    @NotBlank(message = "必填")
    private String author;
    @NotNull(message = "必须")
    @DateTimeFormat(pattern = "yyyy-MM-dd")
    private Date  pubDate;
    @NotBlank(message = "必填")
    private String descri;
    private String photo;
    private int tid;
    private Type type;//关联属性
}
```

10.2　图书添加功能

Spring MVC 开发需要我们提供视图、后端控制器。为了简化视图开发，这里基于 Bootstrap 制作视图。

10.2.1 前端界面制作

在 WEB-INF 目录下新建 jsp 目录，再在其中新建 book 目录，然后在其中新建一个 add.jsp 页面。新建一个基于 Bootstrap 的界面，可以使用在线拖曳工具 layoutit，具体步骤如下。

步骤 01 导入 CSS 和 JavaScript。根据图 10-5 所示的官网 Starter template 模板完成 Bootstrap 使用所需的 JS、CSS 和视口（viewport）配置。

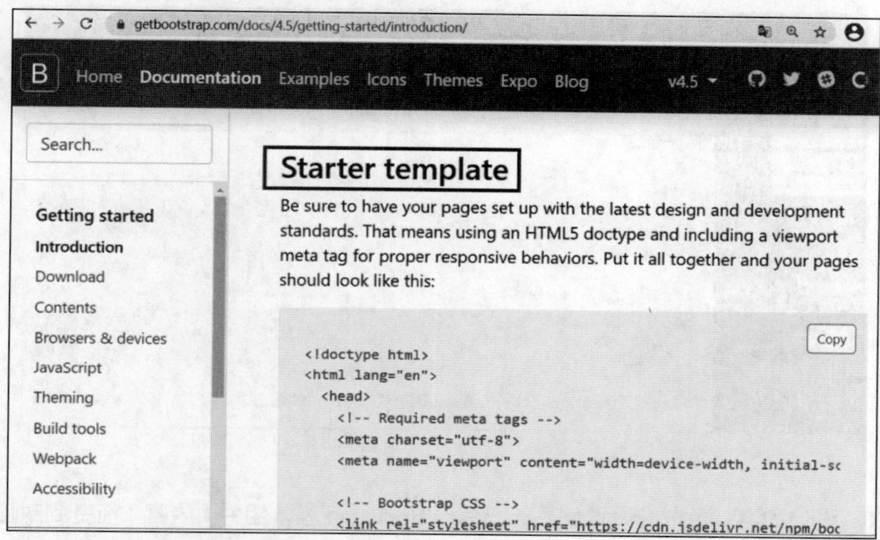

图 10-5

把 CSS 和 JavaScript 地址修改为本地地址，如下代码所示。

示例代码 10-7　add.jsp

```
<%@ page contentType="text/html;charset=UTF-8" %>
<!doctype html>
<html lang="zh">
<head>
    <meta charset="utf-8">
<meta content="width=device-width, initial-scale=1, shrink-to-fit=no"
 name="viewport">
    <!-- Bootstrap CSS -->
    <title>图书添加</title>
<link
 href="${pageContext.request.contextPath}/s/node_modules/bootstrap/dist/cs
 s/bootstrap.min.css"  rel="stylesheet">
</head>
<body>
<!--这里放置基于 bootstrap 的内容-->
<script
 src="${pageContext.request.contextPath}/s/node_modules/jquery/dist/jquery.s
 lim.min.js"></script>
```

```
<script
   src="${pageContext.request.contextPath}/s/node_modules/bootstrap/dist/js/bo
   otstrap.bundle.min.js"></script>
    </body>
    </html>
```

步骤 02 设计布局。正如盖房子首先要对土地进行规划。网页是让浏览器读取的，因而需要布局要放置的内容。Bootstrap 布局使用栅格系统，把每一行划分为 12 等分，每个单元格指定份数即可。这里网页设计为上、中、下三部分，每部分一行一列，如图 10-6 所示。

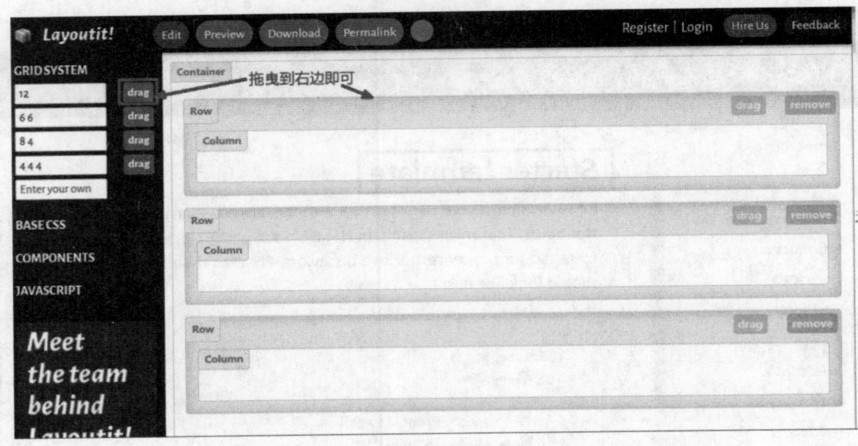

图 10-6

步骤 03 设计内容。当布局设置好后，把基于 Bootstrap 基本组件（内容）拖曳到对应部分（上部是一个导航条，中间是一个表单，底部是一个版权声明）。

- 添加导航条，如图 10-7 所示，找到导航条组件，拖曳到顶部即可。

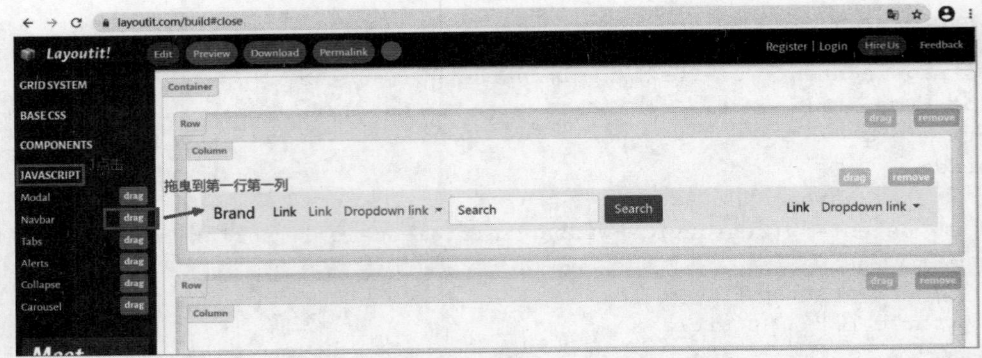

图 10-7

- 添加表单，如图 10-8 所示，找到表单组件，拖曳到中间即可。

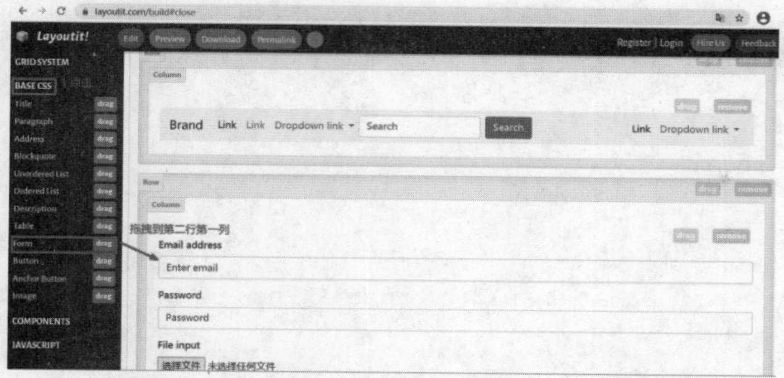

图 10-8

- 添加版权，如图 10-9 所示，找到 title 组件，拖曳到底部即可。

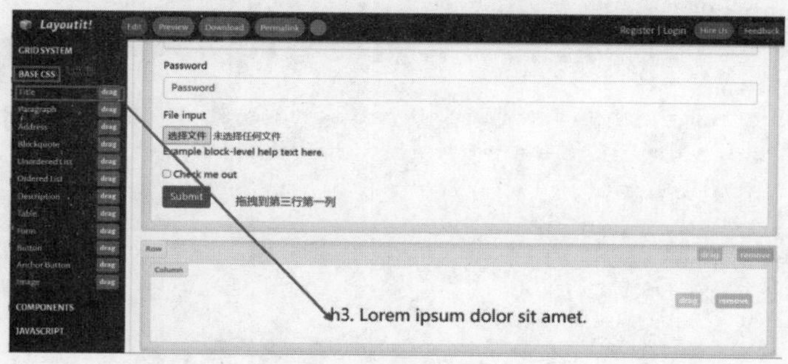

图 10-9

- 下载，设计完毕，可以先单击顶部的 Preview 按钮进行预览，然后单击 Download 按钮下载，如图 10-10 所示。

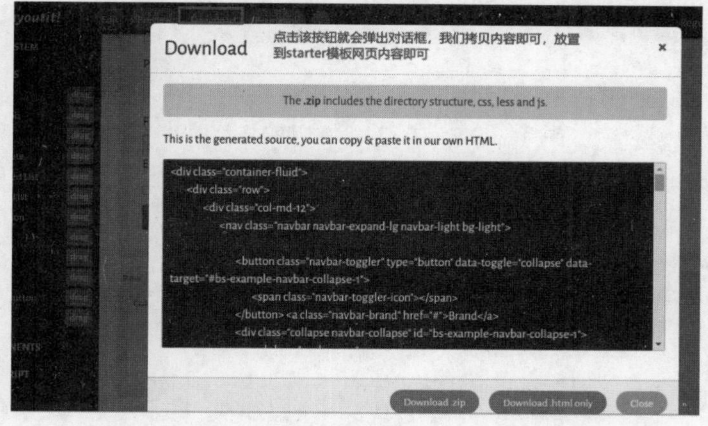

图 10-10

步骤 04　修改下载内容。layoutit 设计后，需要修改表单字段。如果表单验证失败，错误信息回填，我们导入 Spring 的表单标签，具体代码如下。

示例代码 10-7　add.jsp 完整代码

```
<%@ page contentType="text/html;charset=UTF-8" %>
```

```
<%@taglib prefix="c" uri="http://java.sun.com/jsp/jstl/core" %>
<%@taglib prefix="form" uri="http://www.springframework.org/tags/form" %>
<%@taglib prefix="spring" uri="http://www.springframework.org/tags" %>
<!doctype html>
<html lang="zh">
<head>
    <!-- Required meta tags -->
    <meta charset="utf-8">
<meta content="width=device-width, initial-scale=1, shrink-to-fit=no"
      name="viewport">
    <!-- Bootstrap CSS -->
    <title>图书添加</title>
<link
   href="${pageContext.request.contextPath}/s/node_modules/bootstrap/dist/c
   ss/bootstrap.min.css"
         rel="stylesheet">
<link
   href="${pageContext.request.contextPath}/s/node_modules/bootstrap-datepi
   cker/dist/css/bootstrap-datepicker3.standalone.min.css"
         rel="stylesheet">
    <style type="text/css">
        .custom-file-label::after {
            content: "浏览"
        }
    </style>
</head>
<body>
<div class="container-fluid">
    <div class="row">
        <div class="col-md-12">
            <nav class="navbar navbar-expand-lg navbar-light bg-light">

           <button class="navbar-toggler" type="button" data-toggle="collapse"
                   data-target="#bs-example-navbar-collapse-1">
               <span class="navbar-toggler-icon"></span>
           </button>
           <a class="navbar-brand" href="#">
           <img
            src="${pageContext.request.contextPath}/s/imgs/logo/dgjy.png"
            alt=""/></a>
           <div class="collapse navbar-collapse"
                id="bs-example-navbar-collapse-1">
                <ul class="navbar-nav">
                    <li class="nav-item active">
                        <a class="nav-link"
            href="${pageContext.request.contextPath}/book/list">
            图书列表 <span
                              class="sr-only">(current)</span></a>
                    </li>
                    <li class="nav-item">
```

```
                <a class="nav-link"
            href="${pageContext.request.contextPath}/book/add">添加图书
            </a>
                </li>
                <li class="nav-item">
                <a class="nav-link"
            href="${pageContext.request.contextPath}/type/list">类型列表
            </a>
                </li>
                <li class="nav-item">
                <a class="nav-link"
              href="${pageContext.request.contextPath}/type/add">添加类
              型</a>
                </li>
             </ul>
             <ul class="navbar-nav ml-md-auto">
                <li class="nav-item active">
                <a class="nav-link"
            href="${pageContext.request.contextPath}/user/exit">退出
            <span
                          class="sr-only">(current)</span></a>
                </li>
             </ul>
          </div>
       </nav>
    </div>
</div>
<div class="row">
    <div class="col-md-12">
        <div class="card">
            <div class="card-header">图书添加</div>
            <div class="card-body">
               <!--action 配置交给那个小程序处理-->
          <form:form
              action="${pageContext.request.contextPath}/book/add"
              enctype="multipart/form-data"
                     modelAttribute="book">
                 <div class="form-group row">
                <form:label path="name" cssClass="col-sm-2
                    col-form-label">名称:</form:label>
                    <div class="col-sm-10">
                   <form:input path="name"
                    cssClass="form-control"></form:input>
                   <form:errors path="name" element="div"
                    cssClass="invalid-feedback"></form:errors>
                    </div>
                 </div>
                 <div class="form-group row">
                <form:label path="price" cssClass="col-sm-2
                  col-form-label">价格</form:label>
```

```
        <label for="inputPrice" class=""></label>
        <div class="col-sm-10">
       <form:input path="price"
         cssClass="form-control"></form:input>
       <form:errors path="price" element="div"
         cssClass="invalid-feedback"></form:errors>
        </div>
    </div>
    <div class="form-group row">
   <form:label path="author" cssClass="col-sm-2
    col-form-label">作者</form:label>
        <div class="col-sm-10">
       <form:input path="author"
        cssClass="form-control"></form:input>
       <form:errors path="author" element="div"
        cssClass="invalid-feedback"></form:errors>
        </div>
    </div>
    <div class="form-group row">
   <form:label path="pubDate" cssClass="col-sm-2
    col-form-label"
                for="inputPubDate">出版日期</form:label>
        <div class="col-sm-10">
           <div class="input-group">
          <form:input path="pubDate"
           cssClass="form-control" id="inputPubDate"
                       readonly="true"></form:input>
          <form:errors path="pubDate" element="div"
             cssClass="invalid-feedback"></form:errors>
               <div class="input-group-append">
              <span class="input-group-text"
               id="basic-addon2">
               <svg width="1em" height="1em" viewBox="0 0 16
               16" class="bi bi-calendar3"
                        fill="currentColor"
               xmlns="http://www.w3.org/2000/svg">
      <path fill-rule="evenodd"
   d="M14 0H2a2 2 0 0 0-2 2v12a2 2 0 0 0 2 2h12a2 2 0 0 0 2-2V2a2
   2 0 0 0-2-2zM1 3.857C1 3.384 1.448 3 2 3h12c.552 0 1 .384
   1 .857v10.286c0 .473-.448.857-1 .857H2c-.552
   0-1-.384-1-.857V3.857z"/>
      <path fill-rule="evenodd"
   d="M6.5 7a1 1 0 1 0 0-2 1 1 0 0 0 0 2zm3 0a1 1 0 1 0 0-2 1
1 0 0 0 0 2zm3 0a1 1 0 1 0 0-2 1 1 0 0 0 0 2zm-9 3a1 1 0 1 0 0-2
1 1 0 0 0 0 2zm3 0a1 1 0 1 0 0-2 1 1 0 0 0 0 2zm3 0a1 1 0 1 0 0-2
1 1 0 0 0 0 2zm3 0a1 1 0 1 0 0-2 1 1 0 0 0 0 2zm-9 3a1 1 0 1 0
0-2 1 1 0 0 0 0 2zm3 0a1 1 0 1 0 0-2 1 1 0 0 0 0 2zm3 0a1 1 0 1
0 0-2 1 1 0 0 0 0 2z"/>
</svg>
                </span>
```

```
                    </div>
                </div>
            </div>
        </div>
        <div class="form-group row">
        <form:label path="tid" cssClass="col-sm-2
         col-form-label">类型</form:label>
            <div class="col-sm-10">
                <form:select path="tid" cssClass="form-control">
                    <option value="-1">--请选择--</option>
                <form:options items="${types}" itemLabel="name"
                 itemValue="id"></form:options>
                </form:select>
            <form:errors path="tid" element="div"
             cssClass="invalid-feedback"></form:errors>
            </div>
        </div>

        <div class="form-group row">
        <form:label path="descri" cssClass="col-sm-2
         col-form-label">描述</form:label>
            <div class="col-sm-10">
            <form:textarea path="descri"
                cssClass="form-control"></form:textarea>
            <form:errors path="descri" element="div"
             cssClass="invalid-feedback"></form:errors>
            </div>
        </div>
        <div class="form-group row">
        <form:label path="photo" cssClass="col-sm-2
         col-form-label">图片</form:label>
            <div class="col-sm-10">
                <div class="custom-file">
                <input type="file" class="custom-file-input"
                  id="inputPhoto" name="photox">
                <label class="custom-file-label"
                     for="inputGroupFile01">选择文件</label>
                </div>
            </div>
        </div>
        <div class="form-group row">
            <div class="col-sm-2"></div>
            <div class="col-sm-10">
           <button type="submit" class="btn btn-primary">添加
          </button>
            </div>
        </div>
    </form:form>
</div>
<div class="card-footer"></div>
```

```
            </div>
        </div>
    </div>
    <div class="row">
        <div class="col-md-12">
            <div class="text-center">
                Xx 版权所有 Copyright &copy; 2020-2028
            </div>
        </div>
    </div>
</div>
<script src="${pageContext.request.contextPath}/s/node_modules/jquery/
    dist/jquery.slim.min.js"></script>
<script src="${pageContext.request.contextPath}/s/node_modules/bootstrap/
    dist/js/bootstrap.bundle.min.js"></script>
<script src="${pageContext.request.contextPath}/s/node_modules/
    bootstrap-datepicker/dist/js/bootstrap-datepicker.min.js"></script>
<script src="${pageContext.request.contextPath}/s/node_modules/bootstrap-
    datepicker/dist/locales/bootstrap-datepicker.zh-CN.min.js"></script>
<script src="${pageContext.request.contextPath}/s/node_modules/bootstrap4-
        validator/dist/validator.min.js"></script>
<script type="text/javascript">
    $(function () {
        $("#inputPubDate").datepicker({
            format: "yyyy-mm-dd",
            autoclose: true,
            language: 'zh-CN'
        })
    });
</script>
<script type="text/javascript">
    <spring:bind path="book">
    <spring:hasBindErrors name="book">
    <c:forEach  items="${requestScope.errors.fieldErrors }"  var="error">
    //注意不要根据 id，因为标签的 id 可能修改，而 name 属性无法修改
    $("[name='${error.field}']").addClass("is-invalid");
    </c:forEach>
    </spring:hasBindErrors>
    </spring:bind>
</script>
</body>
</html>
```

10.2.2 控制器

编写后端控制器。新建一个 BookController 类添加控制器方法，内容如示例代码 10-8 所示。

示例代码 10-8 BookController.java

```
@Controller
```

```
@RequestMapping("book")
public class BookController {
@Autowired
    private BookService bookService;
    @Autowired
    private TypeService typeService;
    @GetMapping("add")
    public  String add(Model model, @ModelAttribute("book") Book book){
        List<Type> types=typeService.findAllTypes();
        model.addAttribute("types",types);
        return "book/add";
    }
    @PostMapping("/add")
public  String add(@Validated @ModelAttribute("book") Book book, BindingResult
 result, MultipartFile photox, Model model) throws IOException {
        if (result.hasFieldErrors()){
            List<Type> types=typeService.findAllTypes();
            model.addAttribute("types",types);
            return "book/add";
        }
     String newFileName=
     UUID.randomUUID()+"."+FilenameUtils.getExtension(photox.getOriginalFilen
     ame());
     String fullPath=
 ResourceUtils.getFile("classpath:static/upload").getAbsolutePath()+"/"+new
FileName;
        photox.transferTo(new File(fullPath));
        book.setPhoto(newFileName);
        int ret=bookService.saveBook(book);
        if (ret>0){
            return "redirect:/book/list";
        }else {
            return "book/add";
        }
}
```

在编写过程中，IDEA 会提示 BookService 找不到，使用如图 10-11 所示的提示功能自动建类。同时，方法不存在时，也可以自动建立方法。在业务层我们采用同样的策略。

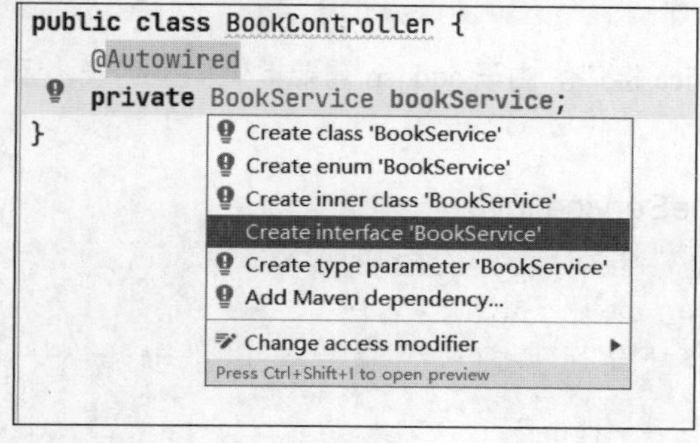

图 10-11

10.2.3 业务层

随着软件越来越复杂，可以采用软件分层策略，各层之间靠接口沟通，因而业务层和 Dao 层都需要提供接口。这也是为什么先开发 Controller 再编写业务层接口的原因。

步骤 01 新建 BookService 接口，对应 Controller 需要发的“通知”，代码如下所示。

示例代码 10-9 BookService.java

```
import com.github.pagehelper.PageInfo;
import com.guodexian.bookman.po.Book;
import org.springframework.transaction.annotation.Transactional;
@Transactional
public interface BookService {
    int saveBook(Book book);
}
```

步骤 02 新建 BookServiceImpl 实现接口 BookService。业务层必须无条件实现上级 Controller 发的“通知”，也就是实现接口，代码如下所示。

示例代码 10-10 BookService.java

```
import com.github.pagehelper.PageHelper;
import com.github.pagehelper.PageInfo;
import com.guodexian.bookman.mapper.BookMapper;
import com.guodexian.bookman.po.Book;
import com.guodexian.bookman.service.BookService;
import org.springframework.beans.factory.annotation.Autowired;
import org.springframework.stereotype.Service;
import java.util.List;
@Service
public class BookServiceImpl implements BookService {
    @Autowired
    private BookMapper bookMapper;
    @Override
    public int saveBook(Book book) {
        return bookMapper.save(book);
    }
}
```

步骤 03 新建 TypeService 接口。由于 add.jsp 需要提供类型列表，因而这里需要定义该接口，代码如下所示。

示例代码 10-11 TypeService.java

```
import com.github.pagehelper.PageInfo;
import com.guodexian.bookman.po.Type;
import org.springframework.transaction.annotation.Transactional;
import java.util.List;
@Transactional//开启注解性事务
public interface TypeService {
```

```
    @Transactional(readOnly = true)
    List<Type> findAllTypes();
}
```

步骤 04 新建 TypeServiceImpl 实现接口 TypeService。

示例代码 10-12　TypeServiceImpl.java

```
import com.github.pagehelper.PageHelper;
import com.github.pagehelper.PageInfo;
import com.guodexian.bookman.mapper.TypeMapper;
import com.guodexian.bookman.po.Type;
import com.guodexian.bookman.service.TypeService;
import org.springframework.beans.factory.annotation.Autowired;
import org.springframework.stereotype.Service;
import java.util.List;
@Service
public class TypeServiceImpl implements TypeService {
    @Autowired
    private TypeMapper typeMapper;
    @Override
    public List<Type> findAllTypes() {
        return typeMapper.findAll();
    }
}
```

10.2.4　Dao 层

Dao 层基于 MyBatis 完成数据存取，只需要编写接口 Mapper 即可，由 MyBatis 自动创建 Mapper 实现类，代码如下所示。

步骤 01 BookMapper。

示例代码 10-13　BookMapper.java

```
import com.guodexian.bookman.po.Book;
import org.apache.ibatis.annotations.*;
import java.util.List;
@Mapper
public interface BookMapper {
@Insert("insert into tb_book
 values(default,#{name},#{price},#{author},#{pubDate},#{descri},#{photo},#
 {tid})")
    int save(Book book);
}
```

步骤 02 TypeMapper。

示例代码 10-14　TypeMapper.java

```
import com.guodexian.bookman.po.Type;
import org.apache.ibatis.annotations.*;
```

```
import java.util.List;
@Mapper
public interface TypeMapper {
    @Select("select * from tb_type")
    List<Type> findAll();
}
```

10.3 图书列表功能

10.3.1 前端界面制作

只需要把图书添加页面内容的第二行第一列换成一个表格和添加分页组件即可。复制 add.jsp，修改其名字为 list.jsp，替换表单为一个表格，并在表格下添加分页组件，代码如下所示。

示例代码 10-15 list.jsp

```
<div class="card">
            <div class="card-header">
                图书列表
            </div>
            <div class="card-body" style="padding: 0px;">
           <table class="table table-bordered table-sm table-hover"
              style="margin-bottom: 0px;">
                  <thead>
                  <tr>
                      <th>
                          #
                      </th>
                      <th>
                          名称
                      </th>
                      <th>
                          价格
                      </th>
                      <th>
                          作者
                      </th>
                      <th>
                          出版日期
                      </th>
                      <th>
                          描述
                      </th>

                      <th>
                          图片
                      </th>
```

```
                <th>
                    类型
                </th>
                <th>
                    操作
                </th>
            </tr>
            </thead>
            <tbody>
            <c:choose>
                <c:when test="${empty pageInfo.list}">
                    <tr>
                    <td colspan="9" class="text-center">数据丢失</td>
                    </tr>
                </c:when>
                <c:otherwise>
                    <c:forEach items="${pageInfo.list}" var="book">
                        <tr>
                            <td>${book.id}</td>
                            <td>${book.name}</td>
                            <td>${book.price}</td>
                            <td>${book.author}</td>
                            <td>
                            <fmt:formatDate value="${book.pubDate}"

                          pattern="yyyy-MM-dd"></fmt:formatDate>
                            </td>
                            <td class="d-inline-block text-truncate"
                           style="max-width:
                        250px;">${book.descri}</td>
                            <td>
                           <img
                          src="${pageContext.request.contextPath}/s/u
                          pload/${book.photo}">
                            </td>
                            <td>${book.type.name}</td>
                            <td>
                               <!--放置删除链接-->
                                  
                                <!--放置编辑链接-->
                            </td>
                        </tr>
                    </c:forEach>
                </c:otherwise>
            </c:choose>
            </tbody>
        </table>
    </div>
    <div class="card-footer" style="padding: 0;">
        <nav aria-label="Page navigation example">
```

```
            <ul class="pagination" style="margin-bottom: 0;">
                <c:choose>
                    <c:when test="${pageInfo.isFirstPage }">
                    <li class="disabled" class="page-item"><a
                    href="javascript:void(0)"

                    class="page-link">&laquo;</a></li>
                    </c:when>
                    <c:otherwise>
                        <li class="page-item">
        <a  href="${pageContext.request.contextPath}/book/list?
pageNum=${pageInfo.prePage }"
                                class="page-link">&laquo;</a></li>
                    </c:otherwise>
                </c:choose>

            <c:forEach items="${pageInfo.navigatepageNums}"
               var="pi">
                    <c:choose>
                        <c:when test="${pageInfo.pageNum==pi}">
                            <li class="page-item active">
                                <a class="page-link"

                  href="${pageContext.request.contextPath}/book/lis
                  t?pageNum=${pi}">${pi}</a>
                            </li>
                        </c:when>
                        <c:otherwise>
                            <li class="page-item">
                                <a class="page-link"

                  href="${pageContext.request.contextPath}/book/lis
                  t?pageNum=${pi}">${pi}</a>
                            </li>
                        </c:otherwise>
                    </c:choose>

                </c:forEach>
                <c:choose>
                    <c:when test="${pageInfo.isLastPage }">
                        <li class="disabled">
                        <a href="javascript:void(0)"
                    class="page-link">&raquo;</a>
                        </li>
                    </c:when>
                    <c:otherwise>
                        <li class="page-item">
                        <a
                  href="${pageContext.request.contextPath}/book/lis
                  t?pageNum=${pageInfo.nextPage}"
```

```
                                    class="page-link">&raquo;</a>
                                </li>
                            </c:otherwise>

                        </c:choose>
                    </ul>
                </div>
</div>
```

10.3.2　控制器

在 BookController 中添加如下代码。

示例代码 10-16　BookController.java

```
@RequestMapping("/list")
public  String list(Model
    model,@RequestParam(required=false,defaultValue="1")int
    pageNum,@RequestParam(required = false,defaultValue = "2") int pageSize){
   PageInfo<Book> pageInfo=bookService.findAllBooks(pageNum,pageSize);
   model.addAttribute("pageInfo",pageInfo);
   return "book/list";
}
```

10.3.3　业务层

步骤 01 在 BookService 接口中添加如下代码。

示例代码 10-17　BookService.java

```
@Transactional(readOnly = true)
PageInfo<Book> findAllBooks(int pageNum,int pageSize);
```

步骤 02 在 BookServiceImpl 中添加如下代码。

示例代码 10-18　BookServiceImpl.java

```
@Override
public PageInfo<Book> findAllBooks(int pageNum,int pageSize) {
   PageHelper.offsetPage((pageNum-1)*pageSize+1-1,pageSize);
   List<Book> ls = bookMapper.findAll();
   PageInfo<Book> pageInfo=new PageInfo<>(ls);
   return pageInfo;
}
```

10.3.4　Dao 层

步骤 01 在BookMapper中添加方法。Book到Type是多对一关系，这里我们配置关联属性type，代码如下所示。

示例代码 10-19　BookMapper.java

```
@Results({
    //配置关联属性 type
    @Result(column="tid",property ="type",one=@One(select=
       "com.guodexian.bookman.mapper.TypeMapper.findById"))
})
@Select("select * from tb_book")
List<Book> findAll();
```

步骤 02 在 TypeMapper 中添加方法。需要根据 BookMapper 传递过来的外键值查找对应的 type，代码如下所示。

示例代码 10-20　TypeMapper.java

```
@Select("select * from tb_type  where id=#{id}")
Type findById(int id);
```

10.4　图书删除功能

10.4.1　前端界面制作

修改 list.jsp 中表单最后操作列的两个 a 标签值，标签体使用 SVG 图标，代码如下所示。

示例代码 10-21　list.jsp（部分代码）

```
<a href="${pageContext.request.contextPath}/book/del?id=${book.id}">
    <svg width="1em" height="1em" viewBox="0 0 16 16" class="bi bi-trash"
        fill="currentColor" xmlns="http://www.w3.org/2000/svg">
      <path d="M5.5 5.5A.5.5 0 0 1 6 6v6a.5.5 0 0 1-1 0V6a.5.5 0 0 1 .5-.5zm2.5
          0a.5.5 0 0 1 .5.5v6a.5.5 0 0 1-1 0V6a.5.5 0 0 1 .5-.5zm3 .5a.5.5 0
          0 0-1 0v6a.5.5 0 0 0 1 0V6z"/>
          <path fill-rule="evenodd"
          d="M14.5 3a1 1 0 0 1-1 1H13v9a2 2 0 0 1-2 2H5a2 2 0 0 1-2-2V4h-.5a1
           1 0 0 1-1-1V2a1 1 0 0 1 1-1H6a1 1 0 0 1 1-1h2a1 1 0 0 1 1 1h3.5a1
           1 0 0 1 1 1v1zM4.118 4L4 4.059V13a1 1 0 0 0 1 1h6a1 1 0 0 0
           1-1V4.059L11.882 4H4.118zM2.5 3V2h11v1h-11z"/>
    </svg>
</a>
```

10.4.2　控制器

示例代码 10-22　BookController.java（部分代码）

```
@RequestMapping("/del")
public  String  del(int id){
    bookService.delBookById(id);
    return "redirect:/book/list";
```

```
}
```

10.4.3　业务层

步骤 01 在 BookService 中添加如下方法。

示例代码 10-23　BookService.java（部分代码）

```
void delBookById(int id);
```

步骤 02 在 BookServiceImpl 中添加如下方法。

示例代码 10-24　BookServiceImpl（部分代码）

```
@Override
public void delBookById(int id) {
bookMapper.del(id);
}
```

10.4.4　Dao 层

在 BookMapper 中添加如下方法：

示例代码 10-25　BookMapper.java（部分代码）

```
@Delete("delete from tb_book where id=#{id}")
void del(int id);
```

10.5　图书编辑功能

10.5.1　前端界面制作

步骤 01 在 list.jsp 页面中添加编辑链接，代码如下。

示例代码 10-26　pom.xml（部分代码）

```
<a href="${pageContext.request.contextPath}/book/edit?id=${book.id}">
    <svg width="1em" height="1em" viewBox="0 0 16 16"
class="bi bi-pencil-square"
fill="currentColor"xmlns="http://www.w3.org/2000/svg">
 <path d="M15.502 1.94a.5.5 0 0 1 0 .706L14.459 3.69l-2-2L13.502.646a.5.5
 0 0 1 .707 0l1.293 1.293zm-1.75 2.456l-2-2L4.939 9.21a.5.5 0 0
.121.196l-.805 2.414a.25.25 0 0 0 .316.316l2.414-.805a.5.5 0 0
0 .196-.121l6.813-6.814z"/>
        <path fill-rule="evenodd"
              d="M1 13.5A1.5 1.5 0 0 0 2.5 15h11a1.5 1.5 0 0 0 1.5-1.5v-6a.5.5
         0 0 0-1 0v6a.5.5 0 0 1-.5.5h-11a.5.5 0 0 1-.5-.5v-11a.5.5 0
         0 1 .5-.5H9a.5.5 0 0 0 0-1H2.5A1.5 1.5 0 0 0 1 2.5v11z"/>
```

```
    </svg>
  </a>
```

步骤 02 新建 edit.jsp。edit.jsp 页面跟 add.jsp 基本相同，只是需要把 id 作为隐藏字段传递给服务器，显示目前该图书已有图片。

示例代码 10-27 edit.jsp（部分代码）

```
<form:form action="${pageContext.request.contextPath}/book/edit"
   enctype="multipart/form-data" modelAttribute="book">
    <form:hidden path="id"></form:hidden>
    .....
      <div class="form-group row">
                            <label for="inputPhoto" class="col-sm-2
col-form-label">图片</label>
                            <div class="col-sm-5">
                                <div class="custom-file">
                               <input type="file" class="custom-file-input"
                                id="inputPhoto" name="photox">
                               <label class="custom-file-label"
                                for="inputPhoto">选择文件</label>
                                </div>
                            <input type="hidden" value="${book.photo}"
                           name="photo">
                             </div>
                             <div class="col-sm-5">
                            <img
                             src="${pageContext.request.contextPath}/s/upload/${
                             book.photo}">
                             </div>
                       </div>
```

10.5.2 控制器

在 BookController 中添加如下两个 edit 方法：

示例代码 10-28 BookController.java（部分代码）

```
    @GetMapping("edit")//转到页面
    public  String edit(int id,Model model){
        List<Type> types=typeService.findAllTypes();
        model.addAttribute("types",types);
        Book book=bookService.findBookById(id);
        model.addAttribute("book",book);
        return "book/edit";
    }
    @PostMapping("/edit")//执行编辑
    public  String edit(@Validated Book book, BindingResult result, MultipartFile
photox,Model model) throws IOException {
        if (result.hasFieldErrors()){
            List<Type> types=typeService.findAllTypes();
```

```
        model.addAttribute("book",book);
        model.addAttribute("types",types);
        return "book/edit";
    }
    //如果用户没有选择图片，就表示不需要修改；选了再去修改
    if (!photox.isEmpty()){
        log.debug(photox);
    String newFileName=
      UUID.randomUUID()+"."+FilenameUtils.getExtension(photox.getOriginalFi
      lename());
    String
    fullPath=ResourceUtils.getFile("classpath:static/upload").getAbsolutePat
    h()+"/"+newFileName;
        photox.transferTo(new File(fullPath));
        book.setPhoto(newFileName);
    }
    int ret=bookService.updateBook(book);
    if (ret>0){
        return "redirect:/book/list";
    }else {
        return "book/edit";
}
```

10.5.3　业务层

步骤 01　在 BookService 中添加方法。

示例代码 10-29　BookService.java（部分代码）

```
//根据 id 查找要编辑的图书
@Transactional(readOnly = true)
Book findBookById(int id);
int updateBook(Book book);
```

步骤 02　在 BookServiceImpl 中添加方法。

示例代码 10-30　BookServiceImpl.java（部分代码）

```
@Override
public Book findBookById(int id) {
    return bookMapper.findById(id);
}
@Override
public int updateBook(Book book) {
    return bookMapper.update(book);
}
```

10.5.4　Dao 层

在 BookMapper 中添加如下代码。

示例代码 10-31 BookMapper.java（部分代码）

```
    @Select("select * from tb_book where id=#{id}")
    Book findById(int id);
@Update("update tb_book set
   name=#{name},price=#{price},author=#{author},pub_date=#{pubDate},descri=#{d
   escri},photo=#{photo},tid=#{tid} where id=#{id}")
    int update(Book book);
```

10.6 登 录

10.6.1 前端界面制作

login.jsp 页面制作在 add.jsp 基础上修改即可，把表单修改为只包含登录信息，代码如下所示。

示例代码 10-32 login.jsp（部分代码）

```
    <!--action 配置交给那个小程序处理-->
    <form action="${pageContext.request.contextPath}/user/login" method="post">
        <div class="form-group row">
          <label for="inputName" class="col-sm-2 col-form-label">名称:</label>
          <div class="col-sm-10">
                         <input type="text" class="form-control"
                     id="inputName" name="name" value="${user.name}">
                          </div>
                       </div>
                       <div class="form-group row">
                      <label for="inputPwd" class="col-sm-2 col-form-label">密
                         码</label>
                          <div class="col-sm-10">
                         <input type="password" class="form-control"
                          id="inputPwd" name="pwd">
                          </div>
                       </div>
                       <div class="form-group row">
                          <label for="inputCode" class="col-sm-2
col-form-label">
                         验证码</label>
                          <div class="col-sm-5">
                         <input type="text" class="form-control"
                          id="inputCode" name="code">
                          </div>
                          <div class="col-sm-5">
                         <img
                          src="${pageContext.request.contextPath}/user/kaptch
                          a" id="codeImg">
                         <a href="javascript:void(0)" class="font-size-12
                        align-bottom" id="freshBtn">刷新验证码</a>
```

```
                    </div>
                </div>
                <div class="form-group row">
                    <div class="col-sm-2"></div>
                    <div class="col-sm-10">
                   <button type="submit" class="btn btn-primary">登录
                    </button>
                    </div>
                </div>
            </form>
    <script type="text/javascript">
        $(function () {
            $("#freshBtn").click(function () {

     $("#codeImg").attr("src","${pageContext.request.contextPath}/user/kapt
     cha?t="+Math.random());
            })
        })
    </script>
```

10.6.2 控制器

这里需要添加两个方法：一个转发到登录界面，另一个执行登录，代码如下所示。

示例代码 10-33 UserController.java（部分代码）

```
    //转发到登录界面
    @GetMapping("/login")
    public String  login(){
        return "user/login";
    }
    @PostMapping("/login")
    public   String login(User user, @RequestParam("code") String  code,
HttpSession session, Model model){
        if (!code.equalsIgnoreCase((String) session.getAttribute("kaptcha"))){
            model.addAttribute("msg","验证码错误");
            model.addAttribute("user",user);
            return "user/login";
         }
         User dbUser=userService.findUserByName(user);
        if (dbUser!=null){
         //登录成功时需要记录已经登录
         session.setAttribute("user",dbUser);
             return "redirect:/book/list";
         }else{
             model.addAttribute("msg","用户名或者密码错误");
             model.addAttribute("user",user);
         return "user/login";
     }
    }
```

10.6.3 业务层

User 的业务层接口代码如下所示。

示例代码 10-34 UserService.java

```
@Transactional
public interface UserService {
    @Transactional(readOnly = true)
    User findUserByName(User user);
}
```

User 业务层接口实现类代码如下所示。

示例代码 10-35 UserServiceImpl.java

```
@Service
public class UserServiceImpl implements UserService {
    @Autowired
    private UserMapper userMapper;
    @Override
    public User findUserByName(User user) {
        BCryptPasswordEncoder encoder=new BCryptPasswordEncoder();
        User dbUser = userMapper.findByName(user.getName());
        //matches 函数内部实现是把输入密码加密和数据库查询的密文比较
   if (encoder.matches(user.getPwd(), dbUser.getPwd())){
            return dbUser;
        }else {
            return null;
        }
    }
}
```

10.6.4 Dao 层

新建 UserMapper 接口，代码如下所示。

示例代码 10-36 UserMapper.java

```
@Mapper
public interface UserMapper {
    /**
     * 查找用户
     * @param user 用户信息
     * @return 是否存在
     */
    @Select("select * from tb_user where name=#{name}")
    User findByName(String name);
}
```

10.6.5 验证码

验证码可以预防机器人程序恶意登录，减轻数据库的压力。下面主要做以下开发工作。

- 添加依赖验证码依赖。
- 编写 Controller 方法。

具体创建步骤如下。

步骤 01 添加依赖。

示例代码 10-37 pom.xml（部分代码）

```
<!--验证码-->
<dependency>
   <groupId>com.github.penggle</groupId>
   <artifactId>kaptcha</artifactId>
   <version>2.3.2</version>
</dependency>
```

步骤 02 编写 Controller 方法。这里添加验证码的生成 Controller 方法，代码如下所示。

示例代码 10-38 UserController.java（部分代码）

```
@GetMapping(path = "/kaptcha")
public void getKaptcha(HttpServletResponse response, HttpSession session) {
 Properties properties = new Properties();
 properties.setProperty("kaptcha.image.width", "100");
      properties.setProperty("kaptcha.image.height", "40");
      properties.setProperty("kaptcha.textproducer.font.size", "32");
      properties.setProperty("kaptcha.textproducer.font.color", "0,0,0");
   properties.setProperty("kaptcha.textproducer.char.string",
   "0123456789ABCDEFGHIJKLMNOPQRSTUVWXYAZ");
      properties.setProperty("kaptcha.textproducer.char.length", "4");
   properties.setProperty("kaptcha.noise.impl",
    "com.google.code.kaptcha.impl.NoNoise");
      DefaultKaptcha kaptchaProducer = new DefaultKaptcha();
      Config config = new Config(properties);
      kaptchaProducer.setConfig(config);
      // 生成验证码
      String text = kaptchaProducer.createText();
      BufferedImage image = kaptchaProducer.createImage(text);
      // 将验证码存入 session
      session.setAttribute("kaptcha", text);
      // 将突图片输出给浏览器
      response.setContentType("image/png");
      try {
      OutputStream os = response.getOutputStream();
      ImageIO.write(image, "png", os);
   } catch (IOException e) {
```

```
        System.out.println("响应验证码失败:" + e.getMessage());
    }
}
```

10.7 权限拦截

一个企业级权限系统包括认证和授权。其中，授权是一个复杂系统，为了简单，本项目仅完成用户授权，读者可以继续使用 Shiro、Spring Security 框架完成。

10.7.1 拦截器

实现接口 HandlerInterceptor 或者继承其默认实现类 HandlerInterceptorAdapter 来完成拦截器定义，代码如下所示。

示例代码 10-39 AuthorInterceptor.java

```
public class AuthorInterceptor extends HandlerInterceptorAdapter {
    @Override
    public boolean preHandle(HttpServletRequest request, HttpServletResponse
         response, Object handler) throws Exception {
        /*
        * 放行:凭卡进入校园，合法用户办卡，不能拦截；否则，办不了卡，不能进入校园。
        * 1 login
        * 2 static 目录下的资源，这些静态资源不用授权就可以查看
        * */
        if (    request.getRequestURI().endsWith("/login")||
                request.getRequestURI().endsWith("/kaptcha")||

            request.getRequestURI().startsWith(request.getContextPath()+"/s/
            ")){
            return true;
        }
        User user= (User) request.getSession().getAttribute("user");
        if (user==null){
            /*还没有登录，请到登录界面*/
            response.sendRedirect(request.getContextPath()+"/user/login");
            return  false;
        }else {
            //继续往下运行
            return true;
        }
    }
}
```

10.7.2　配置拦截器

定义好拦截器后，需要进行配置对哪些资源拦截，代码如下所示。

示例代码 10-40　WebConfig.java

```
    import com.guodexian.bookman.interceptor.AuthorInterceptor;
    import org.springframework.context.annotation.Configuration;
    import org.springframework.web.servlet.config.annotation.
InterceptorRegistry;
    import org.springframework.web.servlet.config.annotation.WebMvcConfigurer;
    @Configuration
    public class WebConfig implements WebMvcConfigurer {
        @Override
        public void addInterceptors(InterceptorRegistry registry) {
            registry.addInterceptor(new  AuthorInterceptor()).
addPathPatterns("/**");//**表示任意层路径
        }
    }
```

10.7.3　添加退出功能

用户完成本次操作后，为防止别人做非法操作需退出本次登录。在 UserController 中添加如下代码。

示例代码 10-41　UserController.java（部分代码）

```
@GetMapping("/exit")
Public  String  exit(HttpSession session){
    session.removeAttribute("user");//删除 session 内容
    return "redirect:/user/login";
}
```

10.8　在 Docker 上部署 Spring Boot 应用

在 Docker 容器中安装 MySQL 数据。

10.8.1　安装 MySQL 镜像

1. 下载镜像

示例代码 10-42　命令行代码

```
C:\Users\HPYH>docker pull mysql
```

2. 创建容器

示例代码 10-43　命令行代码

```
C:\Users\HPYH>docker run -di --name mysql01  -p 33306:3306 -e
MYSQL_ROOT_PASSWORD=root mysq
```

3. 修改允许远程访问

示例代码 10-44　命令行代码

```
docker exec -it mysql01  /bin/bash
root@f7d41e1df938:/# mysql -u root -p root;
mysql> use mysql;
mysql> update user set password_expired = "Y" where user="root";
mysql> ALTER USER'root'@'%' IDENTIFIED WITH mysql_native_password BY 'root';
```

4. 初始化数据库

在上一步的基础上执行如下 MySQL 命令。

示例代码 10-45　MySQL 命令

```
mysql> create database  just_book;
mysql> use  just_book;
```

把下面的 SQL 脚本粘贴到窗口执行。

示例代码 10-46　SQL 语句

```
DROP TABLE IF EXISTS `t_book`;
CREATE TABLE `t_book`  (
  `id` int(11) NOT NULL AUTO_INCREMENT,
  `name` varchar(45) CHARACTER SET utf8 COLLATE utf8_general_ci NULL DEFAULT
NULL,
  `author` varchar(100) CHARACTER SET utf8 COLLATE utf8_general_ci NULL DEFAULT
NULL,
  `tid` int(11) NULL DEFAULT NULL,
  `price` double NULL DEFAULT NULL,
  `descri` text CHARACTER SET utf8 COLLATE utf8_general_ci NULL,
  `photo` varchar(100) CHARACTER SET utf8 COLLATE utf8_general_ci NULL DEFAULT
NULL,
  `pubdate` date NULL DEFAULT NULL,
  PRIMARY KEY (`id`) USING BTREE,
  INDEX `FK_T_BOOK_TID_idx`(`tid`) USING BTREE,
  CONSTRAINT `FK_T_BOOK_TID` FOREIGN KEY (`tid`) REFERENCES `t_type` (`id`) ON
DELETE NO ACTION ON UPDATE NO ACTION
) ENGINE = InnoDB AUTO_INCREMENT = 27 CHARACTER SET = utf8 COLLATE =
utf8_general_ci ROW_FORMAT = Compact;
INSERT INTO `t_book` VALUES (21, '统计学原理111', ' 马立平 ', 3, 36.8, '本书是作
者在多年统计学教学工作经验的基础上编写出来的，用通俗易懂、深入浅出的语言阐述统计学的基本思想与
方法，具有较强的实用性和可操作性。读者通过学习可以掌握统计学中基本的、常用的统计方法的原理与思
想，并能够在此基础上正确地使用统计方法进行统计分析与研究。全书共 16 章，分别为导论、数据收集的方
法、数据的预处理与分组整理、数据特征的统计量描述、数据资料的图形显示、统计指标与多指标综合评价、
概率抽样方法与抽样分布、参数估计、参数的假设检验、方差分析、列联分析与对应分析、相关与回归分析、
```

```
时间数列的描述性分析、时间数列的构成与预测、聚类分析和判别分析、主成分与因子分析。', '',
'2018-11-30');
    INSERT INTO `t_book` VALUES (22, '线性代数习题全解与学习指导', '濮燕敏殷俊锋 ', 3,
36.8, '本书共 5 章，包括线性方程组与矩阵、方阵的行列式、向量空间与线性方程组解的结构、相似矩阵
及二次型、线性空间与线性变换，对配套《线性代数》教材中各章节的习题、测试题进行了详细解答。每章
都配有知识结构、归纳总结、典型例题、习题详解。其中，典型例题中精心挑选了与对应章节相关的全国研
究生入学统一考试试题，并在书中做了标注，便于读者有针对性地练习。',
'74766e33-8991-4707-a782-79c0877e5671.png', '2018-12-03');
    DROP TABLE IF EXISTS `t_role`;
    CREATE TABLE `t_role`  (
      `id` int(11) NOT NULL AUTO_INCREMENT,
      `name` varchar(20) CHARACTER SET utf8 COLLATE utf8_general_ci NULL DEFAULT
NULL,
      `descri` varchar(50) CHARACTER SET utf8 COLLATE utf8_general_ci NULL DEFAULT
NULL,
      PRIMARY KEY (`id`) USING BTREE
    ) ENGINE = InnoDB AUTO_INCREMENT = 3 CHARACTER SET = utf8 COLLATE =
utf8_general_ci ROW_FORMAT = Compact;
    INSERT INTO `t_role` VALUES (1, 'ROLE_USER', '普通用户');
    INSERT INTO `t_role` VALUES (2, 'ROLE_ADMIN', '超级管理员');

    DROP TABLE IF EXISTS `t_type`;
    CREATE TABLE `t_type`  (
      `id` int(11) NOT NULL AUTO_INCREMENT,
      `name` varchar(45) CHARACTER SET utf8 COLLATE utf8_general_ci NULL DEFAULT
NULL,
      PRIMARY KEY (`id`) USING BTREE
    ) ENGINE = InnoDB AUTO_INCREMENT = 4 CHARACTER SET = utf8 COLLATE =
utf8_general_ci ROW_FORMAT = Compact;

    INSERT INTO `t_type` VALUES (1, '电子书');
    INSERT INTO `t_type` VALUES (2, '文学');
    INSERT INTO `t_type` VALUES (3, '数学');

    DROP TABLE IF EXISTS `t_user`;
    CREATE TABLE `t_user`  (
      `id` int(11) NOT NULL AUTO_INCREMENT,
      `name` varchar(20) CHARACTER SET utf8 COLLATE utf8_general_ci NULL DEFAULT
NULL,
      `pwd` varchar(60) CHARACTER SET utf8 COLLATE utf8_general_ci NULL DEFAULT
NULL,
      `birthday` date NULL DEFAULT NULL,
      `age` varchar(255) CHARACTER SET utf8 COLLATE utf8_general_ci NULL DEFAULT
NULL,
      PRIMARY KEY (`id`) USING BTREE
    ) ENGINE = InnoDB AUTO_INCREMENT = 3 CHARACTER SET = utf8 COLLATE =
utf8_general_ci ROW_FORMAT = Compact;

    INSERT INTO `t_user` VALUES (1, 'admin',
'$2a$10$PCClqk3RGNFn7OwKyNIKCecqghneNsm/6.aJgggU24Z2XELKMsiAC', '1999-01-01',
```

```
'23');
    INSERT INTO `t_user` VALUES (2, 'user',
'$2a$10$u9ZRukgj5tpx7hp42DFgjuIntaaiHJZBINJlnXM/.dE2U5dq0gSTi', '3898-08-01',
'23');
    SET FOREIGN_KEY_CHECKS = 1;
```

10.8.2 在 pom.xml 中添加插件

示例代码 10-47 pom.xml（部分代码）

```
    <properties>
        <docker.image.prefix>springio</docker.image.prefix>
    </properties>
    <build>
        <resources>
            <resource>
                <directory>src/main/java</directory>
                <includes>
                    <include>**/*.properties</include>
                    <include>**/*.xml</include>
                </includes>
                <filtering>false</filtering>
            </resource>
            <resource>
                <directory>src/main/resources</directory>
                <includes>
                    <include>**/*</include>
                </includes>
                <filtering>false</filtering>
            </resource>
            <!-- 打包时将 JSP 文件复制到 META-INF 目录下-->
            <resource>
                <!-- 指定 resources 插件处理哪个目录下的资源文件 -->
                <directory>src/main/webapp</directory>
                <!--注意此次必须放在此目录下才能被访问到-->
                <targetPath>META-INF/resources</targetPath>
                <includes>
                    <include>**/**</include>
                </includes>
            </resource>
        </resources>
        <plugins>
    <!--docker 插件-->
            <!-- tag::plugin[] -->
            <plugin>
                <groupId>codockerfile-maven-plum.spotify</groupId>
                <artifactId>gin</artifactId>
                <version>1.4.9</version>
                <configuration>
```

```
    <repository>${docker.image.prefix}/${project.artifactId}</repository>
            </configuration>
        </plugin>
        <!-- end::plugin[] -->
        <!-- tag::unpack[] -->
        <plugin>
            <groupId>org.apache.maven.plugins</groupId>
            <artifactId>maven-dependency-plugin</artifactId>
            <executions>
                <execution>
                    <id>unpack</id>
                    <phase>package</phase>
                    <goals>
                        <goal>unpack</goal>
                    </goals>
                    <configuration>
                        <artifactItems>
                            <artifactItem>
                                <groupId>${project.groupId}</groupId>
                                <artifactId>${project.artifactId}</artifactId>
                                <version>${project.version}</version>
                            </artifactItem>
                        </artifactItems>
                    </configuration>
                </execution>
            </executions>
        </plugin>
        <!-- end::unpack[] -->
        <!-- 资源文件复制插件 -->
        <plugin>
            <groupId>org.apache.maven.plugins</groupId>
            <artifactId>maven-resources-plugin</artifactId>
            <configuration>
                <encoding>UTF-8</encoding>
            </configuration>
        </plugin>
```

10.8.3 新建 Dockerfile

示例代码 10-48　Dockerfile

```
FROM openjdk:8-jdk-alpine
VOLUME /tmp
ARG DEPENDENCY=target/dependency
COPY ${DEPENDENCY}/BOOT-INF/lib /app/lib
COPY ${DEPENDENCY}/META-INF /app/META-INF
COPY ${DEPENDENCY}/BOOT-INF/classes /app
RUN apk add --update font-adobe-100dpi ttf-dejavu fontconfig
ENTRYPOINT
["java","-cp","app:app/lib/*","com.guodexian.bookman.BookManApplication"]
```

10.8.4 修改数据库 URL

示例代码 10-49 application.properties（部分代码）

```
spring.datasource.url=jdbc:mysql://mysql01:3306/just_book?serverTimezone=UTC
```

10.8.5 配置允许 Maven 直接上传镜像

为了构建和启动 docker 镜像，docker 插件默认访问端口 2375 发送命令给 Docker Daemon，但是该端口默认是关闭的。在 Windows 下开启 Docker Daemon 端口比较容易，按照如图 10-12 所示，选中红框部分，然后点击“Apply & restart”即可。

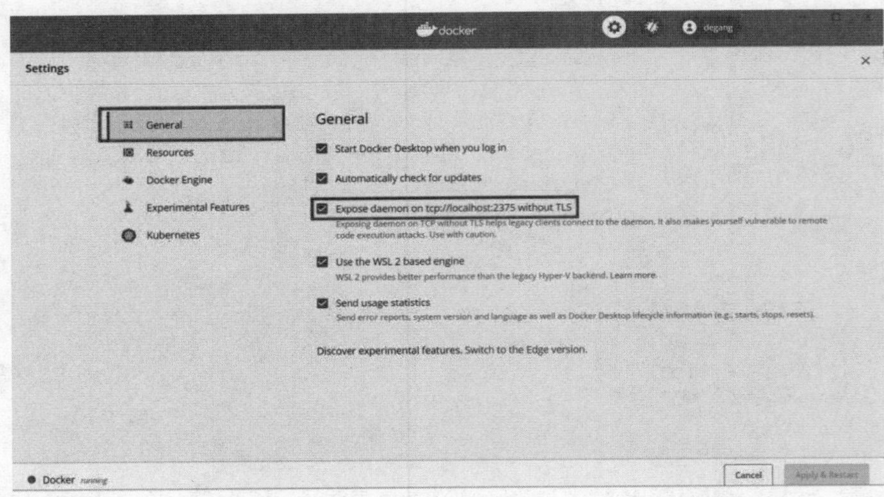

图 10-12

10.8.6 执行 Maven 命令

新建 Maven 运行配置步骤如下：

步骤 01 如图 10-13 所示单击顶部添加配置“Add Configuration...”，弹出对话框。

图 10-13

步骤 02 如图 10-14 所示，在弹出的对话框中选择 Maven。

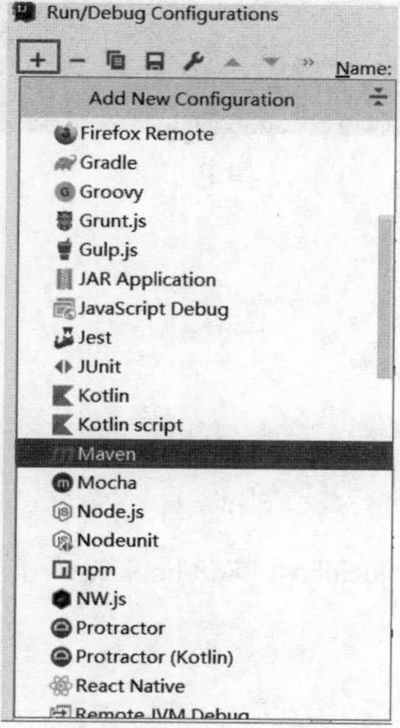

图 10-14

步骤 03　如图 10-15 所示，在 Command Line 中输入 clean package dockerfile:build 命令。

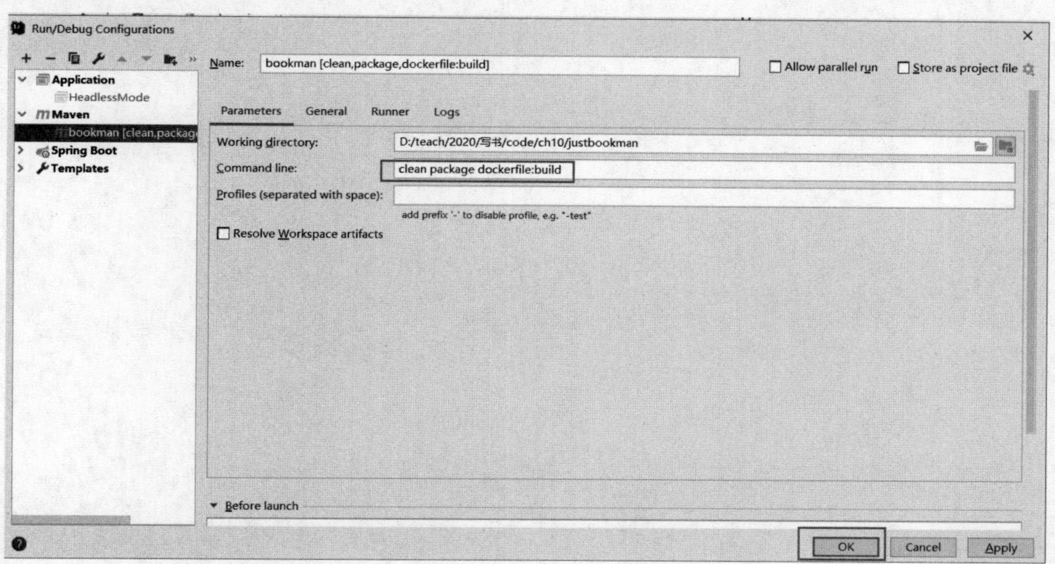

图 10-15

步骤 04　如图 10-16 所示，单击按钮运行即可。

图 10-16

步骤 05　如果成功，就会看到如图 10-17 所示的提示。

```
[INFO]
[INFO] Detected build of image with id 387594b8fcdc
[INFO] Building jar: D:\study\2020\[illegible]a\justbookman\target\bookman-0.0.1-SNAPSHOT-docker-info.jar
[INFO] Successfully built degang/bookman:latest
```

图 10-17

10.8.7 运行镜像

示例代码 10-50 命令行代码

```
C:\Users\hp>docker run --name spring-boot-docker -d -p 8888:8888 -link
      mysql01:mysql01 springio/bookman
7b0ddfbf10788d20e4670bba82df6c57ced9e53d9720c4cd1a6a4ef98b051856
```

在浏览器地址栏中输入“http://localhost:8888/book/list”，会出现一个登录页面，允许你登录系统。